Remote Sensing Observations to Improve Knowledge of Lithosphere–Atmosphere–Ionosphere Coupling during the Preparatory Phase of Earthquakes

Remote Sensing Observations to Improve Knowledge of Lithosphere–Atmosphere–Ionosphere Coupling during the Preparatory Phase of Earthquakes

Editors

Dedalo Marchetti

Kaiguang Zhu

Yunbin Yuan

MDPI

Basel • Beijing • Wuhan • Barcelona • Belgrade • Novi Sad • Cluj • Manchester

Editors
Dedalo Marchetti
INGV–Italian National Institute of Geophysics and Volcanology
Roma
Italy

Kaiguang Zhu
Jilin University
Changchun
China

Yunbin Yuan
Chinese Academy of Sciences
Wuhan
China

Editorial Office
MDPI AG
Grosspeteranlage 5
4052 Basel, Switzerland

This is a reprint of articles from the Special Issue published online in the open access journal *Remote Sensing* (ISSN 2072-4292) (available at: https://www.mdpi.com/journal/remotesensing/special_issues/LAIC_EarthquakePrecursors).

For citation purposes, cite each article independently as indicated on the article page online and as indicated below:

Lastname, A.A.; Lastname, B.B. Article Title. *Journal Name* **Year**, *Volume Number*, Page Range.

ISBN 978-3-7258-2103-7 (Hbk)
ISBN 978-3-7258-2104-4 (PDF)
doi.org/10.3390/books978-3-7258-2104-4

Cover image courtesy of Dedalo Marchetti

Contents

About the Editors

Dedalo Marchetti

Dedalo Marchetti received a bachelor's and master's degree in Physics and Astrophysics from La Sapienza, Rome, Italy, in 2009 and 2012, respectively. He obtained a Ph.D. in Physics from Roma Tre, Rome, Italy, in 2016, with a thesis focused on the realization of a prototype of an instrument to measure the polarization of exoplanets.

Since 2016, he has focused on the research of possible seismo-induced effects in the ionosphere, working in the Italian National Institute of Geophysics and Vulcanology in the SwArm For Earthquake study (SAFE) project and its extension e-SAFE funded by ESA. His focus is primarily on geophysical interactions in the Earth's system and, in particular, studying the preparatory phase of medium and large earthquakes by satellite and ground measurements to detect possible lithosphere–atmosphere–ionosphere coupling effects.

From September 2018, he worked in China for about five years. For the first 2 years, he was enrolled at Nanjing University of Information Science and Technology (NUIST) and the following three years at Jilin University. He developed several investigations dealing with the analysis of the China Seismo Electromagnetic Satellite launched on 2 February 2018 and is still in orbit. The principal focus of his current research is to analyze variations in the Earth's magnetic field to search for possible precursors of earthquakes by satellite and ground measurements. He is also developing statistical and case study analyses for research on other earthquake effects in the atmosphere and ionosphere.

Presently, he is enrolled at the Italian National Institute of Geophysics and Volcanology (INGV) within the European Plate Observing System (EPOS) group to implement multiparametric research, with the EPOS Platform also using machine learning and artificial intelligence.

Kaiguang Zhu

Kaiguang Zhu is a distinguished professor at the College of Instrumentation and Electrical Engineering at Jilin University, located in Changchun, China. With a robust academic and professional background, prof. Zhu is recognized for her expertise in electromagnetic signal processing and the extraction of pre-earthquake-induced signals.

Kaiguang Zhu earned her PhD in Earth Exploration and Information Technology from Jilin University, China, in 2002. Following her doctoral studies, she furthered her research as a postdoctoral fellow at the University of Windsor in Canada from 2005 to 2006.

Throughout her career, prof. Zhu has played a pivotal role in numerous high-impact research projects. From 2020 to 2023, she was the Principal Investigator (PI) for the project "Research on Seismo-Electromagnetic Satellite Anomalies Extraction Based on Blind Source Separation" and since 2024, she has been the PI of a new project titled "Research on spatio-temporal and time-frequency correlation methods of multi-layer earthquake electromagnetic precursor anomalies", both funded by the National Natural Science Foundation of China. These projects are dedicated to a better understanding of electromagnetic phenomena related to seismic activities.

Yunbin Yuan

Yunbin Yuan is a Professor and the Director of the Global Navigation Satellite System (GNSS) Application and Research Group, Innovation Academy for Precision Measurement Science and Technology, Chinese Academy of Sciences, Wuhan, China. His current research interests include GNSS-based spatial environmental monitoring and analysis, high-precision GNSS satellite navigation and positioning, GNSSs in orbit-determination applications, system imitation tests, and integrated navigation.

Preface

This content is a reprint of the Special Issue "Remote Sensing Observations to Improve Knowledge of Lithosphere–Atmosphere–Ionosphere Coupling during the Preparatory Phase of Earthquakes" which contains 16 contributions, in the form of 15 papers and an editorial. The Special Issue derives from the exigence of covering a relatively "young" research field, which is the study of the possible coupling between the lithosphere, atmosphere, and ionosphere before earthquakes. This field has grown in recent years due to the benefit of new instrumentation, especially from space (remote sensing satellites), and we launched this Special Issue in order to collect new results and future perspectives, as well as a review of the story in the field. We suggest reading this Special Issue Reprint to researchers who want to start studying the preparatory phase of earthquakes from a wide geophysical perspective and experienced scientists for new ideas and results. You will find contributions that deal specifically with each geo-layer (lithosphere, atmosphere, and ionosphere) and other ones that focus more on interactions and newly proposed methods.

We acknowledge the journal Remote Sensing and its staff who provided uncountable help in all the phases of realization of this Special Issue and reprint and all the colleagues who supplied any suggestions, contributions, and ideas. Without all of these provisions, this work would not be possible!

We hope you enjoy reading this Special Issue Reprint and your feedback is truly appreciated!

Dedalo Marchetti, Kaiguang Zhu, and Yunbin Yuan
Editors

remote sensing

MDPI

Editorial

Editorial of Special Issue "Remote Sensing Observations to Improve Knowledge of Lithosphere–Atmosphere–Ionosphere Coupling during the Preparatory Phase of Earthquakes"

Dedalo Marchetti [1,*], Yunbin Yuan [2] and Kaiguang Zhu [3]

1 Independent Researcher, 00145 Rome, Italy
2 State Key Laboratory of Geodesy and Earth's Dynamics, Innovation Academy for Precision Measurement Science and Technology (APM), Chinese Academy of Sciences, Wuhan 430077, China
3 College of Instrumentation and Electrical Engineering, Jilin University, Changchun 130061, China
* Correspondence: dedalo.marchetti@ingv.it or dedalo.marchetti.work@gmail.com; Tel.: +86-18936866083

1. Introduction

We launched this Special Issue with the aim of collecting papers that use satellite data and new methodologies to understand the preparatory phase of medium–large earthquakes in the world. In recent decades, several satellite observations have been used for the precise estimation of co-seismic effects (such as ground displacement estimated using the InSAR technique) and to search for possible precursor signals [1]. Some satellites have been launched for this purpose, such as CSES-01 (China Seismo Electromagnetic Satellite), successfully in orbit since 2 February 2018 [2]. Other satellites especially dedicated to Earth observation (e.g., meteorological observatories and Earth geomagnetic field monitoring) can be used to investigate eventual pre-earthquake pieces of evidence [3]. Several models have been proposed to explain these phenomena with different physical or chemical mechanisms [4,5]. In addition, statistical studies support the existence of seismo-induced phenomena [6,7] as well as single earthquake investigations such as the case of the widely investigated M7.8 Nepal 2015 earthquake [8–10]. This Special Issue addresses these points in a modern geophysical approach, searching for possible signatures of Lithosphere–Atmosphere–Ionosphere Coupling (LAIC) before the earthquake's occurrence, including new methodologies and points of view.

Citation: Marchetti, D.; Yuan, Y.; Zhu, K. Editorial of Special Issue "Remote Sensing Observations to Improve Knowledge of Lithosphere–Atmosphere–Ionosphere Coupling during the Preparatory Phase of Earthquakes". *Remote Sens.* **2024**, *16*, 1064. https://doi.org/10.3390/rs16061064

Received: 3 March 2024
Accepted: 12 March 2024
Published: 18 March 2024

2. Content and Coverage of the Published Papers in This Special Issue

This Special Issue comprises 15 papers and this Editorial, a list of which is provided after Conclusion in section "List of contributions", and a graphical representation is shown in Figure 1. In total, 71 authors from 19 institutions and 8 countries (China, Italy, Spain, Japan, Iran, United Kingdom, United States of America, and Kazakhstan) ensured the success of this Special Issue, and we, the Guest Editors, are grateful to all of them.

Due to the wide coverage of this topic, the papers in this Special Issue can be read and downloaded as an open-access book and are even available as a printed book. So, we suggest the "List of contributions" an order of reading, but each contribution is independent, and the reader can choose a different order to read the contents of this Special Issue.

A graphical representation of the contributions in this Special Issue is provided in Figure 1. For each paper, a figure has been reproduced in a dimension of a stamp. These are the graphical abstracts where available or a significant figure from the content of the specific paper. The authors, the year of publication, and a few words about the content of the paper are reported. The papers have been organised in Figure 1 according to their main layer of investigation. Still, four of them have been placed in a separate box on the left, representing the papers that focused more on multi-layer investigations. Despite this, the papers on the right side also investigated mutual interactions, especially with the lithosphere. Hence, the distinction is not strict, but we think it will help readers to search for specific content and also illustrates that all the layers and their possible investigations have been discussed in the papers in this Special Issue.

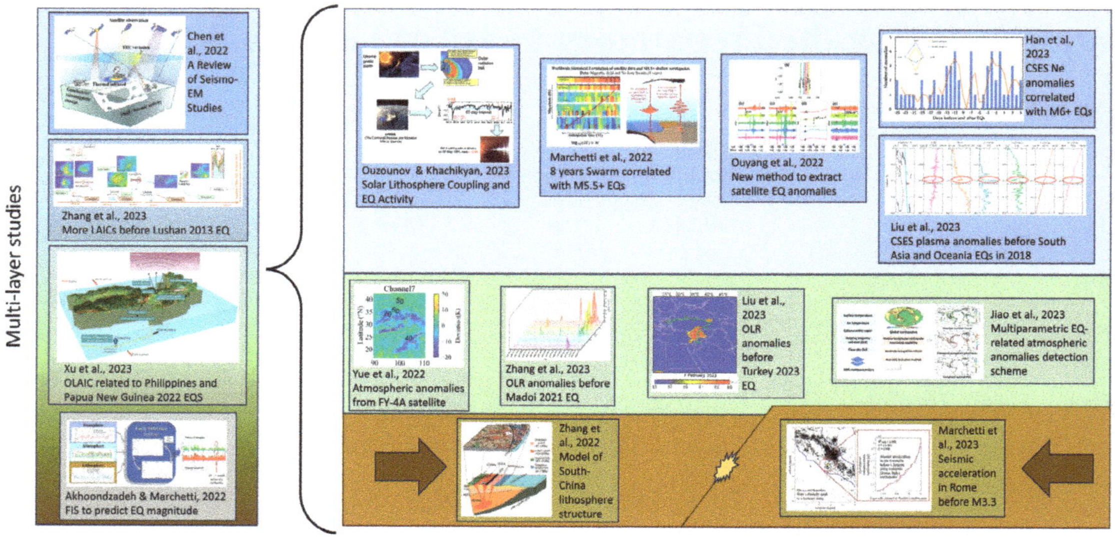

Figure 1. A visual summary of all the contributions in this Special Issue. The studies have been divided into those that focus more on multi-layer investigation (on the **left**) and those that focus more on a specific layer among the lithosphere, atmosphere, and ionosphere. This division is not as strict as dividing all the papers according to whether they are related to the observation of the earthquake or the lithosphere, but we think it may help the reader in providing a quick reference of what to expect to find in each specific paper.

3. Summary of the Content of This Special Issue

A wide and comprehensive review of seismo-electromagnetic pre-earthquake processes has been provided in Paper #1, and it constitutes a very good introduction to the topic of this Special Issue, also providing some future perspectives on how machine learning and artificial intelligence could contribute to the topic. Paper #2 presented a seismic tomography method to reconstruct the lithospheric structure of Southern China. This paper, even if focused on the lithosphere, reminds the LAIC community how important a precise and detailed knowledge of the lithosphere is in the area that the source of events under investigation is located. Paper #3 investigated the lithosphere, atmosphere, and ionosphere six months before a small earthquake of magnitude 3.3 occurred in the surrounding area of Rome (Italy). The only result likely connected to the earthquake in this "Communication" is a seismic acceleration, and for this reason, it is considered a paper that contributed to the lithosphere analysis.

Paper #4 proposes an interesting investigation of infrared data acquired by the Chinese satellite FengYun-4 (FY-4A). The authors searched for possible thermal infrared anomalies eventually related to the M7.4 Madoi (also known as Madou) and M6.4 Yangbi earthquakes that occurred on the same day (21 May 2021) in China. In the last part of the paper, the authors provided a spatio-temporal statistical analysis of thermal infrared anomalies on eleven seismic events in China with a magnitude between M5.0 and M7.4 from March to September 2021, showing a good correlation for nine out of eleven events with an estimated gain of 1.9 with respect to random expectation. A complementary work on longwave infrared anomalies before the Madoi earthquake is provided in Paper #5, showing that the intensity and number of anomalies increased before the mainshock. Paper #6 analysed Outgoing Longwave Radiation (OLR) with an added component: a comparison with tidal stress showing cross-interactions of the geolayers and tidal forces before and after the occurrence of the recent catastrophic earthquake that occurred on 6 February 2023 in Turkey. Finally, Paper #7 provided a method based on machine learning to systematically search for atmospheric multiparametric (surface and air temperature, column water vapour OLR and clear sky OLR) anomalies from remote sensing data of the AQUA/AIRS satellite possibly related to shallow (depth $\leq$ 0 km) global M6+ earthquakes that occurred from 2006 to 2020. The authors provided statistical coefficients of prediction capability divided into three earthquake location types: inland, oceanic, and coastal.

Regarding the ionosphere, Paper #8 proposed a new vision of the earthquake phenomenon from a different point of view, i.e., including a possible "trigger" from geomagnetic activity. In fact, it showed the formation of a new radiation belt with a compatible L-shell before large seismic events. Paper #9 confirmed previous studies reporting significant ionospheric electromagnetic disturbances detected by Swarm satellites before global shallow (depth $\leq$ 50 km) M5.5+ earthquakes and that their anticipation time increases with magnitude. In addition, it was found that the frequency of magnetic anomalies seems higher for anomalies recorded before sea earthquakes compared with land seismic events with slower-frequency magnetic signals. Analogously to the previous study, Paper #10 statistically investigated the CSES-01 Ne anomalies before the M6.8+ global and M6.0+ China earthquakes in the first five years of the mission. Also, the work considers the classification of anomalies and earthquakes as a function of marine/land, magnitude, and hemisphere of occurrence, suggesting that anomalies seem to point from the epicentre toward the equator (southward in the Northern Hemisphere and northward in the Southern Hemisphere). Paper #11 proposed a new method to extract Swarm magnetic anomalies likely related to earthquakes, finding linear polarisation to be a possible seismo-induced feature in good candidates after excluding anomalies resulting from geomagnetic activity and other known sources. Paper #12 investigated the composition of ionospheric plasma before four strong earthquakes in Southeast Asia and North Oceania, identifying promising pre-earthquake ionospheric disturbances.

Finally, we have three contributions in this Special Issue that explicitly investigated possible interactions between the lithosphere, atmosphere, and ionosphere. In particular, Paper #13 investigated six months of data from the three geolayers before the Lushan (China) 2013 earthquake. It identified three possible LAICs 130 days, 48~40 days, and 20~6 days before the mainshock, which are explainable by different mechanisms due to the involved parameters, suggesting that different LAIC models are not in contrast to each other but describe different ways of coupling. Paper #14 proposed the use of a continuous logic system known as Fuzzy (instead of Boolean) to predict the incoming earthquake magnitude, through analysing the anomalies recorded in the lithosphere, atmosphere, and ionosphere, finding promising indications around one month before the five investigated strong earthquakes. Finally, Paper #15 investigated two large earthquakes on the sea coast, proposing a new framework of LAIC: Ocean–Lithosphere–Atmosphere–Ionosphere Coupling (OLAIC). The paper identified an important role of the oceanic mass in influencing the anomalies and how they propagate to the atmosphere up to the ionosphere.

4. Conclusions

An important point provided by different contributions in this Special Issue is the effect of geomagnetic activity. We would clarify that the exclusion of the anomalies likely induced by geomagnetic disturbances as carried out, for example, in Papers #9–11 is not in conflict with the new vision proposed by Ouzounov and Khachikyan in Paper #8 as this second phenomenon would be a trigger for the earthquake, and the source of the ionospheric anomalies produced by "classic" LAIC would be the nucleation source of the earthquakes or their preparation area in the lithosphere. However, excluding the geomagnetic active time could obscure critical phenomena induced by the lithosphere in the ionosphere, even though, due to more complex ionospheric conditions, it is undoubtedly more difficult to distinguish between external and possible internal disturbance induced by an earthquake. With this Special Issue, several contributions statistically confirmed the existence of pre-earthquake anomalies, and specific papers investigated possible patterns of anomalies (see Papers #13, #15), suggesting several LAIC models. It is still unclear why several earthquakes show different patterns, and this needs to be a major research direction in the near future. Some suggestions from this Special Issue also suggested that tectonic settings and water (ocean) may play a major role (e.g., Paper #9, #15) but are insufficient to explain the different results completely.

Funding: The Guest Editors of this Special Issue received funding from the China Postdoctoral Science Foundation (grant number 2021M691190) and the National Natural Science Foundation of China (grant number 41974084). D.M. acknowledges ISSI (Bern, Switzerland) and ISSI-BJ (Beijing, China) for the support to the International Team 553 "CSES and Swarm Investigation of the Generation Mechanisms of Low Latitude Pi2 Waves" led by Essam Ghamry and Zeren Zhima and International Team 23-583/57 "Investigation of the Lithosphere Atmosphere Ionosphere Coupling (LAIC) Mechanism before the Natural Hazards" led by Dedalo Marchetti and Essam Ghamry.

Acknowledgments: A special acknowledgement to the Academic Editors Salvatore Stramondo, Stephan Havemann, José Fernández, Michael E. Gorbunov, and Roberto Orosei, who helped to evaluate several papers published in this Special Issue.

Conflicts of Interest: The authors declare no conflicts of interest. The funders had no role in the design of the study; in the collection, analyses, or interpretation of data; in the writing of the manuscript; or in the decision to publish the results.

List of Contributions:

Paper number #:

1. Chen, H.; Han, P.; Hattori, K. Recent Advances and Challenges in the Seismo-Electromagnetic Study: A Brief Review. *Remote Sens.* **2022**, *14*, 5893. https://doi.org/10.3390/rs14225893.

Studies about lithosphere

2. Zhang, X.; Qian, Y.; Shen, X.; Huang, H.; Chai, H. Shallow Crustal Structure of S-Wave Velocities in the Coastal Area of South China Constrained by Receiver Function Amplitudes. *Remote Sens.* **2022**, *14*, 2760. https://doi.org/10.3390/rs14122760.
3. Marchetti, D.; Zhu, K.; Marchetti, L.; Zhang, Y.; Chen, W.; Cheng, Y.; Fan, M.; Wang, S.; Wang, T.; Wen, J.; et al. Quick Report on the ML = 3.3 on 1 January 2023 Guidonia (Rome, Italy) Earthquake: Evidence of a Seismic Acceleration. *Remote Sens.* **2023**, *15*, 942. https://doi.org/10.3390/rs15040942.

Studies about atmosphere

4. Yue, Y.; Chen, F.; Chen, G. Pre-seismic anomalies detection from multichannel infrared images of FY-4A satellite. *Remote Sens.* **2023**, *15*, 259. https://doi.org/10.3390/rs15010259.
5. Zhang, J.; Sun, K.; Zhu, J.; Mao, N.; Ouzounov, D. Application of Model-Based Time Series Prediction of Infrared Longwave Radiation Data for Exploring the Precursory Patterns Associated with the 2021 Madoi Earthquake. *Remote Sens.* **2023**, *15*, 4748. https://doi.org/10.3390/rs15194748.
6. Liu, J.; Cui, J.; Zhang, Y.; Zhu, J.; Huang, Y.; Wang, L.; Shen, X. Study of the OLR Anomalies before the 2023 Turkey M7.8 Earthquake. *Remote Sens.* **2023**, *15*, 5078. https://doi.org/10.3390/rs15215078.
7. Jiao, Z.; Hao, Y.; Shan, X. A Spatially Self-Adaptive Multiparametric Anomaly Identification Scheme Based on Global Strong Earthquake. *Remote Sens.* **2023**, *15*, 3803. https://doi.org/10.3390/rs15153803.

Studies about ionosphere

8. Ouzounov, D.; Khachikyan, G. Studying the Impact of the Geospace Environment on Solar Lithosphere Coupling and Earthquake Activity. *Remote Sens.* **2024**, *16*, 24. https://doi.org/10.3390/rs16010024.
9. Marchetti, D.; De Santis, A.; Campuzano, S.A.; Zhu, K.; Soldani, M.; D'Arcangelo, S.; Orlando, M.; Wang, T.; Cianchini, G.; Di Mauro, D.; et al. Worldwide Statistical Correlation of Eight Years of Swarm Satellite Data with M5.5+ Earthquakes: New Hints about the Preseismic Phenomena from Space. *Remote Sens.* **2022**, *14*, 2649. https://doi.org/10.3390/rs14112649.

10. Han, C.; Yan, R.; Marchetti, D.; Pu, W.; Zhima, Z.; Liu, D.; Xu, S.; Lu, H.; Zhou, N. Study on Electron Density Anomalies Possibly Related to Earthquakes Based on CSES Observations. *Remote Sens.* **2023**, *15*, 3354. https://doi.org/10.3390/rs15133354.
11. Ouyang, X.-Y.; Wang, Y.-F.; Zhang, X.-M.; Wang, Y.-L.; Wu, Y.-Y.A. A New Analysis Method for Magnetic Disturbances Possibly Related to Earthquakes Observed by Satellites. *Remote Sens.* **2022**, *14*, 2709. https://doi.org/10.3390/rs14112709.
12. Liu, D.; Zeren, Z.; Huang, H.; Yang, D.; Yan, R.; Wang, Q.; Shen, X.; Liu, C.; Guan, Y. The Ionospheric Plasma Perturbations before a Sequence of Strong Earthquakes in Southeast Asia and Northern Oceania in 2018. *Remote Sens.* **2023**, *15*, 5735. https://doi.org/10.3390/rs15245735.

Multiparametric studies on lithosphere atmosphere and ionosphere

13. Zhang, Y.; Wang, T.; Chen, W.; Zhu, K.; Marchetti, D.; Cheng, Y.; Fan, M.; Wang, S.; Wen, J.; Zhang, D.; et al. Are There One or More Geophysical Coupling Mechanisms before Earthquakes? The Case Study of Lushan (China). *Remote Sens.* **2023**, *15*, 1521. https://doi.org/10.3390/rs15061521.
14. Akhoondzadeh, M.; Marchetti, D. Developing a Fuzzy Inference System Based on Multi-Sensor Data to Predict Powerful Earthquake Parameters. *Remote Sens.* **2022**, *14*, 3203. https://doi.org/10.3390/rs14133203.
15. Xu, X.; Wang, L.; Chen, S. Analysis of Ocean–Lithosphere–Atmosphere–Ionosphere Coupling Related to Two Strong Earthquakes Occurring in June–September 2022 on the Sea Coast of Philippines and Papua New Guinea. *Remote Sens.* **2023**, *15*, 4392. https://doi.org/10.3390/rs15184392.

References

1. Astafyeva, E. Ionospheric Detection of Natural Hazards. *Rev. Geophys.* **2019**, *57*, 1265–1288. [CrossRef]
2. Shen, X.; Zong, Q.-G.; Zhang, X. Introduction to Special Section on the China Seismo-Electromagnetic Satellite and Initial Results. *Earth Planet. Phys.* **2018**, *2*, 439–443. [CrossRef]
3. Tramutoli, V.; Cuomo, V.; Filizzola, C.; Pergola, N.; Pietrapertosa, C. Assessing the Potential of Thermal Infrared Satellite Surveys for Monitoring Seismically Active Areas: The Case of Kocaeli (İzmit) Earthquake, August 17, 1999. *Remote Sens. Environ.* **2005**, *96*, 409–426. [CrossRef]
4. Liperovsky, V.A.; Pokhotelov, O.A.; Meister, C.-V.; Liperovskaya, E.V. Physical Models of Coupling in the Lithosphere-Atmosphere-Ionosphere System before Earthquakes. *Geomagn. Aeron.* **2008**, *48*, 795–806. [CrossRef]
5. Pulinets, S.; Ouzounov, D. Lithosphere–Atmosphere–Ionosphere Coupling (LAIC) Model—An Unified Concept for Earthquake Precursors Validation. *J. Asian Earth Sci.* **2011**, *41*, 371–382. [CrossRef]
6. Yan, R.; Parrot, M.; Pinçon, J.-L. Statistical Study on Variations of the Ionospheric Ion Density Observed by DEMETER and Related to Seismic Activities: Ionospheric Density and Seismic Activity. *J. Geophys. Res. Space Phys.* **2017**, *122*, 12421–12429. [CrossRef]
7. De Santis, A.; Marchetti, D.; Pavón-Carrasco, F.J.; Cianchini, G.; Perrone, L.; Abbattista, C.; Alfonsi, L.; Amoruso, L.; Campuzano, S.A.; Carbone, M.; et al. Precursory Worldwide Signatures of Earthquake Occurrences on Swarm Satellite Data. *Sci. Rep.* **2019**, *9*, 20287. [CrossRef] [PubMed]
8. Ouzounov, D.; Pulinets, S.; Davidenko, D.; Rozhnoi, A.; Solovieva, M.; Fedun, V.; Dwivedi, B.N.; Rybin, A.; Kafatos, M.; Taylor, P. Transient Effects in Atmosphere and Ionosphere Preceding the 2015 M7.8 and M7.3 Gorkha–Nepal Earthquakes. *Front. Earth Sci.* **2021**, *9*, 757358. [CrossRef]
9. De Santis, A.; Balasis, G.; Pavón-Carrasco, F.J.; Cianchini, G.; Mandea, M. Potential Earthquake Precursory Pattern from Space: The 2015 Nepal Event as Seen by Magnetic Swarm Satellites. *Earth Planet. Sci. Lett.* **2017**, *461*, 119–126. [CrossRef]
10. Wu, L.; Qi, Y.; Mao, W.; Lu, J.; Ding, Y.; Peng, B.; Xie, B. Scrutinizing and Rooting the Multiple Anomalies of Nepal Earthquake Sequence in 2015 with the Deviation–Time–Space Criterion and Homologous Lithosphere–Coversphere–Atmosphere–Ionosphere Coupling Physics. *Nat. Hazards Earth Syst. Sci.* **2023**, *23*, 231–249. [CrossRef]

remote sensing

MDPI

Review

Recent Advances and Challenges in the Seismo-Electromagnetic Study: A Brief Review

Hongyan Chen [1], Peng Han [1,2,*] and Katsumi Hattori [3,4]

1 Department of Earth and Space Sciences, Southern University of Science and Technology, Shenzhen 518055, China
2 Guangdong Provincial Key Laboratory of Geophysical High-Resolution Imaging Technology, Southern University of Science and Technology, Shenzhen 518055, China
3 Graduate School of Science, Chiba University, Chiba 263-8522, Japan
4 Center for Environmental Remote Sensing, Chiba University, Chiba 263-8522, Japan
* Correspondence: hanp@sustech.edu.cn; Tel.: +86-755-8801-5515

Abstract: Due to their potential application in earthquake forecasting, seismo-electromagnetic phenomena were intensively studied for several decades all over the world. At present, measurements from ground to space have accumulated a large amount of observation data, proving an excellent opportunity for seismo-electromagnetic study. Using a variety of analytical methods to examine past earthquake events, many electromagnetic changes associated with earthquakes have been independently reported, supporting the existence of pre-earthquake anomalies. This study aimed to give a brief review of the seismo-electromagnetic studies preceding earthquakes and to discuss possible ways for the application of seismo-electromagnetic signals at the current stage. In general, seismo-electromagnetic signals can be classified into electric and magnetic changes in the lithosphere and perturbations in the atmosphere. We start with seismo-electromagnetic research in the lithosphere, and then we review the studies in the lower atmosphere and upper atmosphere, including some latest topics that aroused intense scholarly interest. The potential mechanisms of seismo-electromagnetic phenomena are also discussed. It was found that although a number of statistical tests show that electromagnetic anomalies may contain predictive information for major earthquakes, with probability gains of approximately 2–6, it is still difficult to make use of seismo-electromagnetic signals efficiently in practice. To address this, finally, we put forward some preliminary ideas about how to apply the seismo-electromagnetic information in earthquake forecasting.

Keywords: seismo-electromagnetic phenomena; precursors; earthquake forecast

Citation: Chen, H.; Han, P.; Hattori, K. Recent Advances and Challenges in the Seismo-Electromagnetic Study: A Brief Review. *Remote Sens.* **2022**, *14*, 5893. https://doi.org/10.3390/rs14225893

Academic Editor: Stephan Havemann

Received: 26 October 2022
Accepted: 18 November 2022
Published: 21 November 2022

1. Introduction

Earthquake forecast is one of the most challenging scientific issues. The key to earthquake forecasting is to find the precursors that help predict impending earthquakes [1]. Science presented the great unsolved scientific mysteries of our time in the Special Issue marking the journal's 125th anniversary. Whether earthquake precursors that can help in prediction exist is one of them. To date, there have been many studies on earthquake forecasts, and most of them adopt a catalog-based probabilistic approach [2–5]. Although a certain amount of precursory information is available in earthquake catalogs [6], earthquake forecasting, especially short-term prediction, is still far from adequate [7,8]. Nevertheless, non-seismological approaches may provide additional useful information and broaden the knowledge for prediction [9,10]. The content of earthquake precursors includes many physical parameters such as mechanical deformation, gas emissions, groundwater level variations, ground temperature variations, fluctuations of the electromagnetic field, and so on. The detection of electromagnetic perturbations before fault ruptures was proposed as a useful way to monitor crustal activities and it has been proven to be a promising phenomenon preceding earthquakes. To date, there have been many reports on electromagnetic field changes associated with earthquakes in a very wide frequency range from megahertz to quasi-DC [11–13]. These phenomena, known as seismo-electromagnetic phenomena, have been intensively studied for some decades [12,14–17] all over the world, from measurements on the ground to space detection. Based on the observation instruments in different spheres of the earth, seismo-electromagnetic phenomena can generally be classified into electric and magnetic changes in the lithosphere (measurement on the surface and/or underground), perturbations in the lower atmosphere (measurement above the surface and below the ionosphere), and the upper atmosphere (partially ionized or fully ionized regions).

2. Seismo-Electromagnetic Phenomena in the Lithosphere

The seismo-electromagnetic phenomena in the lithosphere are commonly recorded by passive ground-based observations of electric fields and/or magnetic fields [18]. Early seismo-electromagnetic studies measured perturbations of the electric potential difference between two electrodes buried in the ground. The case study of an M5.6 earthquake (depth = 17 km) was reported by Miyakoshi [19]. The observation carbon rod electrodes were set at an epicentral distance of 3 km. They found that potential differences changed clearly prior to the earthquake. This kind of signal was known as the seismic electric signal (SES). The most famous SES may be the one reported from Greece by Varotsos, Alexoupoulos, and Nicomos (VAN) [20–22]. They observed anomalous electric signals on both short (10–200 m) and long (1–3 km) dipoles with electric field amplitudes of up to 20 μV/m and durations of a few minutes prior to several moderate–large earthquakes [21]. Varotsos et al. [22] issued some predictions of earthquakes since 1990 using the SES. These predictions soon drew lots of attention and aroused active debates (see details in Vol. 23 Issue 11, *Geophysical Research Letters*, 1996). In addition to seismic electric signals in Greece, there are some examples of SES observed in various earthquake-prone areas worldwide [23]. Recently, the natural time analysis of geoelectric field and seismicity changes were combined to study the precursory SES activity [24,25].

Magnetic field changes associated with earthquakes were also observed. One of the most promising candidates is the measurement of ultralow frequency (ULF) electromagnetic phenomena (less than 1 Hz), because of their deeper skin depths [12,14,15]. To date, now, a large number of ULF electromagnetic phenomena associated with earthquakes have been reported [10,11,26–41]. One of the most well-known ULF seismo-magnetic phenomena was the magnetic field changes related to the Ms7.1 Loma Prieta earthquake that happened on 18 October 1989, reported by Fraser-Smith et al. [40]. They found that the amplitude of ULF magnetic fields, especially at the frequency of 0.01 Hz, began to increase significantly about two weeks before the main shock. Due to the limitations of observation equipment, only the averaged amplitude of the magnetic field in the horizontal component was registered, which made it very difficult to analyze detailed waveforms for further study. Other case studies of the 1988 Spitak earthquake (M6.9) and the 1993 Guam earthquake (M 8.0) reported by the Russian group [41] and the Japanese group [42], respectively, also showed potential changes in the magnetic field before earthquakes. Hattori et al. [43] analyzed the geomagnetic data associated with the 1997 Kagoshima earthquakes (M6.5 and M6.3). By employing several reference stations, they eliminated the influence of global ionospheric disturbances and detected anomalous changes in the magnetic field, possibly related to the two large earthquakes.

The 2000 Izu earthquake swarm was another important case for seismo-electromagnetic study, as there were both electric and magnetic measurements in the vicinity of the epicenters. By analyzing these data simultaneously, Uyeda et al. [44] reported very robust results for seismo-electromagnetic phenomena. The ratio of electric potential between the two different observation dipoles at the frequency of 0.01 Hz has shown clear anomalous changes about two months before the earthquake swarm. These anomalies were unique against a 3-year-long background. The principal component analysis (PCA) of geomagnetic data also showed anomalies before the swarm. The electromagnetic phenomena associated with the 2000 Izu earthquake swarm may be one of the most plausible results, as many reports have been published using different kinds of methodologies, such as PCA [44,45], singular spectral analysis (SSA) [46], fractal analysis [31,47,48], spectral density analysis [49], and direction-finding analysis [26]. In general, in addition to the above methods used for seismo-electromagnetic anomaly recognition, there are also other methods such as polarization analysis [49,50], wavelet analysis [51], diurnal variation analysis [35], and so on.

Whether these magnetic anomalies contain precursory information and how they can improve the forecasting of sizable earthquakes were not fully discussed. A sensitive geomagnetic network was installed in Kanto, Japan, to verify and clarify the ULF electromagnetic phenomena preceding large earthquakes. Plenty of data associated with moderate-large earthquakes have been accumulated. Researchers have analyzed the geomagnetic data observed during the past decade in the Izu and Boso Peninsulas. Hattori et al. [52] conducted some statistical studies of ULF seismo-magnetic phenomena in these areas and verified the correlation between ULF magnetic anomalies and sizeable local earthquakes. Previous studies showed that the earthquake-related magnetic signals may decrease with the epicentral distance and increase with the magnitude of the earthquake event [12,13,30,53], suggesting that augmenting earthquake forecasting with magnetic anomalies might involve their distance and magnitude dependences. The performance of trial forecasts based on ULF geomagnetic anomalies from the Kakioka Geomagnetic Observatory was evaluated by Han et al. [13]. They used the deep-night 0.01 Hz band data to avoid man-made noise, and the results showed that approximately 20% of the target earthquakes were preceded by the hypothesized type of magnetic anomaly related to the earthquake's occurrence. The authors rejected the concern that the successful forecasts were aftershocks and post-seismic disturbances of earthquakes, by demonstrating that trial forecasting based on recent seismicity is no better than random. Recently, Han et al. evaluated the optimal parameter for earthquake forecasting in the Kanto region, Japan [10], by utilizing long-term magnetic data observed in the region during the period 2000–2010 and applying Molchan's error diagram [54,55].

In the United States, a research project named QuakeFinder (QF) is carried out by the aerospace company Stellar Solutions. The goal was to develop a methodology to detect and analyze pre-earthquake electromagnetic (EM) signals in order to produce earthquake forecasts. As of April 2018, the array comprised 168 stations, most located in California, along the San Andreas Fault, and some located in Peru, Chile, Greece, and Sumatra [56]. The researchers discovered increases in ultra-low frequency magnetic pulse activity starting two weeks before the nearby seismic events in some cases and disappearing after the event [57]. The discovery is consistent with the statistical results of ULF seismo-magnetic phenomena in Japan [13]. In addition to the single earthquake's electromagnetic signals, some studies used 15+ years of data to statistically test whether electromagnetic signals do exist before significant earthquakes (close to the magnetometer stations) [58,59]. The algorithmic framework for investigating the temporal relationship of magnetic field pulses and earthquakes has been proposed, and some studies of searching for pre-earthquake signals such as long-term air ion monitoring have been investigated [58]. Recently, the case–control study on a decade of ground-based magnetometers in California conducted by QF and Google revealed signals in the frequency domain of 0.016–25 Hz before earthquakes [59], and the artificial intelligence method used can improve the efficiency and accuracy of anomaly detection. When compared to the statistical results of papers by Han et al. [13], the above results added to the evidence that there is a legitimate electromagnetic signal prior to medium-to-large earthquakes. However, more work needs to be conducted to refine the algorithms and to characterize and remove the known noise sources.

In China, a mobile geomagnetic monitoring network has been established in the last 20 years, covering most of the country's potential earthquake zones, and the geomagnetic anomaly provides very important support for the prediction of earthquake location [60]. Some research about the pre-seismic anomaly characteristics of geomagnetic changes was reported. For example, the amplitude ratio of geomagnetic harmonic waves was significantly enhanced before the Ms5.5 Eryuan earthquake in 2013 [61], and the geomagnetic and geoelectric data show abnormal variations before the Wenchuan earthquake [33]. There were also some reports about the apparent resistivity changes before and after earthquakes [62–65]. The apparent resistivity change revealed by magnetotelluric (MT) measurements is a well-accepted technique, and some tests on the relation between earthquakes and apparent resistivity changes were revealed by repeated MT methods [66,67]. Further rigorous analysis based on continuous MT observation is required to ensure the reliability of the results.

On the other hand, some researchers claim that there were no electromagnetic changes prior to earthquakes. For example, Kappler et al. [68] analyzed the geomagnetic data associated with the 2004 Parkfield Earthquake (September 28, Mw 6.0, depth 8 km) by several methods, and they found no clear change of magnetic fields before the main shock. At present, there are still active debates in the geophysical community on seismo-electromagnetic phenomena [69–71].

3. Seismo-Electromagnetic Phenomena in the Lower Atmosphere

The seismo-electromagnetic phenomena in the lower atmosphere include atmospheric electric field perturbations, earthquake lights [72,73], thermal anomalies [74–76], and so on. The anomaly of atmospheric electric fields, possibly associated with earthquakes, is an interesting seismo-electromagnetic phenomenon in the lower atmosphere. There are some independent observations of earthquake-related anomalies of the atmospheric electric field after the Ms 7.8 Tangshan earthquake in China. To verify the reliability of the atmospheric electric precursors, one research group of the Institute of Geophysics has made continuous observations and carried out some studies on this topic since 1976. They obtained empirical relationships between the anomalies of the atmospheric electric field and earthquakes, and applied them to some earthquake prediction tests [77]. Recently, anomalous negative signals observed by ground-based atmospheric electric field instruments under fair weather conditions were considered a novel earthquake prediction approach [78]. One of the best-documented observations of earthquake lights is from Japanese earthquakes, where bright lights at the ground level were photographed from 20 to 200 m in diameter with a duration from 10 s to 2 min, restricted to mountain summits in a quartz-diorite faulted rock [73]. Luminous phenomena were reported, starting at approximately nine months before the 6 April 2009 earthquake and continuing until approximately five months after the shock [72]. A statistical study of lightning activity and $M \geq 5.0$ earthquakes in the Taiwan region showed lightning activities tending to appear around the forthcoming epicenter and being enhanced a few days before shallow land earthquakes [79]. Possible explanations about earthquake lights include ultrashort-period air oscillations and generation of potential differences by the piezoelectric effect [80].

The thermal anomalies include atmospheric temperature, brightness temperature, surface latent heat flux (SLHF), outgoing longwave radiation, thermal infrared spectral range (TIR), etc. For example, the preseismic anomalies of SLHF were observed by Qin et al. [81], and they reported surface temperature anomalies before two major earthquakes [82]. Ouzounov et al. [76,83] reported thermal radiation anomalies associated with some major earthquakes, showing that infrared signals were observed near the epicenter areas and stayed from several hours to a few days. Genzano et al. [84] adopted the broadly used robust satellite technique data analysis methodology to identify the significant sequence of thermal anomalies acquired by multi-functional transport satellite (MTSAT) over Japan, confirming a noncausal correlation between anomalies and earthquakes. Tramutoli et al. [74] proposed a statistically well-founded definition of TIR anomalies for satellite TIR surveys. A probability gain of up to 3.7 was achieved as far as only significant sequences of TIR anomalies preceding earthquakes were considered. In practice, the remote sensing of thermal infrared anomalies can be affected by cloud coverage [85], and sometimes it is hard to discriminate thermal contributions that are possibly connected to seismic activity in cloudy conditions [86].

4. Seismo-Electromagnetic Phenomena in the Upper Atmosphere

The seismo-electromagnetic phenomena in the upper atmosphere are usually detected by using satellite observations and/or ground-based transmitter signals [17,87–91]. Since the 1990s, Hayakawa et al. continuously published some research papers that confirm the correlation between very low frequency (VLF) (3–30 kHz)/low frequency (LF) (30–300 kHz) radio signals and earthquakes [42,92–94]. They found significant shifts in the terminator times and phase before the main shock for the signal's propagation characteristics. Their statistical study suggested that the VLF/LF ionospheric perturbation in terms of amplitude (trend) show a statistically significant precursory behavior (3–5 days before the earthquake) and the enhancement of dispersion (fluctuation) is clearly visible for $M \geq 6.0$ earthquakes. The statistical results also confirm and support previous results by Rozhnoi et al. [95]. The VLF from the direction of the source of earthquakes was reported by Asada et al. [96] though without establishing statistical significance as 'earthquake-preceding phenomena'. Five megahertz-sampling rates of VLF recorders were developed in Japan, and Nagao et al. determined the source location using the arrival time difference [97]. They examined the data for 5 days before and 1 day after an earthquake on 25 June 2017, and found seven similar pulses located within 20 km of the hypocenter. There was no light in that time window. There are also two sites observing VLF pulses, ultra-low frequency, and atmospheric electric fields at Kuroshio-town in Kochi, Japan [98]. In the past ten years, the application of ground-based and satellite signals in seismic monitoring was developed in China, and VLF/LF artificial source signal monitoring has become one of the main technical means of seismic disturbance analysis [99,100].

The French satellite DEMETER was the first seismo-electromagnetic satellite devoted to investigating seismo-electromagnetic phenomena. The pre-earthquake disturbances in the ionosphere were observed by the DEMETER satellite. Parrot et al. [101] and Sarkar et al. [17] demonstrated several examples of variations in the plasma parameters recorded by DEMETER satellites over the epicenters of earthquakes before their occurrence. Zeng et al. [102] analyzed the DEMETER data and their results indicated that the electron concentration, temperature, and oxygen ion changed rapidly 5 days before the Wenchuan earthquake. Akhoondzadeh et al. [103] investigated the electron and ion density variations before strong earthquakes ($M > 6.0$) using DEMETER and GPS data. They found that the anomalous deviations prior to earthquakes had different signs from case to case, and their amplitude depended on the magnitude of the earthquake. There were positive and negative anomalies in both DEMETER and GPS-TEC data for 1–5 days before all the studied earthquakes during quiet geomagnetic conditions. These results were verified by subsequent studies [104–109], suggesting that the ionosphere might have some response prior to large earthquakes. Additionally, the VLF field intensity observed by the DEMETER satellite was found to be statistically significant [110]. The field intensity was found to take a minimum at the time when the satellite was closest to the epicenter of the upcoming earthquake. The pre-seismic changes were observed by satellite only after stacking many cases due to the noises.

After the DEMETER satellite ended in 2010, the European Space Agency launched the Swarm satellite in 2013. Marchetti et al. [111–113] reported the seismo-ionospheric anomalies from Swarm satellite data around the time of the strong earthquake and the worldwide statistical correlation of eight years of satellite data with M5.5+ earthquakes. The precursory worldwide signatures study found some clear concentrations of electron density and magnetic anomalies from more than two months to some days before the earthquake occurrences, and the epicenter location was related to the detected anomalies [112,114]. There were also some reports about the detection of ionospheric signals in the region magnetically conjugate to the earthquake zone [115].

The China Seismo-Electromagnetic Satellite (CSES) mission, launched in 2018, is the first satellite in China to measure seismo-associated phenomena in the near-Earth electromagnetic environment, having a lot of application prospects in the study of seismology, geophysics, space sciences, and so on [91]. This was planned to operate for 5 years. During its five-year operation time, CSES records many kinds of data, such as multi-band waveform and the spectra of electric and magnetic fields; in situ plasma parameters (including the densities and temperatures of electrons and ions); electron density profiles and tomography; energetic particle flux and energy spectra. Many reports showed that the performance of the payload can satisfy the requirement of the scientific objectives of the mission [116–122]. Next, we discuss some of the last published results. Yan et al. [123] reported the observations of unusual ionospheric irregularities by CSES before four earthquakes. Zhang et al. [65] investigated ionospheric perturbations in very low-frequency transmitters recorded by CSES, and they claimed that the perturbations to the satellite were more likely related to the overlapped electric field in the earthquake preparation area. An investigation into the characteristics of pre-earthquake ionospheric effects related to four shallow-focus earthquakes in Indonesia was conducted by Song et al. [124]. Significant positive perturbations 1–7 days before earthquakes were revealed. Recently, Zhu et al. [125] published the CSES data analysis of electron density (Ne) and temperature (Te), to support a temporal-spatial correlation between the earthquakes and the observation of positive Ne variations/negative Te variations before the seismic events. Xiong et al. [126,127] and Zhong et al. [128] analyzed the pre-earthquake electromagnetic, ionospheric, or space electric field perturbations by machine learning. However, similarly to other reports on earthquake precursors, the anomalous variation of parameters has not been revealed in all studied earthquakes, and the perturbations seem to occur at various temporal and spatial distances to the earthquake. Other possible sources of the observed anomalies in the ionospheric and geomagnetic dynamics driven by solar activity may not be fully ruled out. On the other hand, the satellite orbits the earth at a high speed, and the observed anomalous variations cover a large zone even up to a 2000 km range. Overall, due to the relatively low resolutive power, the applications based on the satellite may face a challenge of the better constraint size and location of earthquake preparation areas. Even though the reported analyses from CSES are not conclusive now, the second mission of the CSES series, CSES-02, will be launched in 2023. It will have the same multi-instrumental payload and similar orbital parameters as the first satellite, but will allow for reducing the revisit time of the same geographic region. The large number of observations will help in studying the frequencies, amplitudes, and characteristics of ionospheric perturbations related to earthquakes and generation mechanisms.

Among the seismo-electromagnetic phenomena in the ionosphere, anomalous changes in the ionospheric total electron content (TEC) are one of the most frequently reported phenomena preceding large earthquakes [129–140]. TEC is a measure of integrated density along a straight line between the satellite and ground GPS stations. One TEC unit (TECU) is equal to 10^{16} electrons/m^2. Various GPS stations can be utilized for ground-based TEC observations, including the International GNSS Station (IGS), the Crustal Movement Observation Network of China (CMONC), the GPS Earth Observation Network (GEONET), and so on. Two-dimensional TEC maps derived from ground-based observations are widely used, such as the data from the Center for Obit Determination of Europe (CODE), and they are regarded as one of the most precise available Global Ionosphere Maps (GIM) and treated as a high-accuracy reference [141]. In contrast, space-borne TEC data are rare due to the lack of processing TEC data in the satellite platform. However, satellite systems can overcome the problem that ground stations are not available in the ocean [142], and act as a supplement to GIM-TEC.

The method to detect anomalies in the ionosphere includes the mean/median statistical envelope method [143,144], cubic fit analysis [140], pattern recognition analyses [139,145,146], etc. The mean/median statistical envelope methods use the N-day mean or median techniques to detect anomalies, becoming a common and popular method. However, there is no clear justification for the selection of N-day used in studies, and the method faces the challenge of accounting for TEC changes owing to different sources such as space weather and seismic effects. The confidence level, which defines the normal ionospheric variations, also needs to be ascertained in the mean statistical envelope methods. The cubic fit analysis was adopted by Heki to obtain the anomalies about five TECU several hours/minutes before the 2011 Tohoku earthquake. The study indicated that similar preseismic TEC anomalies were seen in the 2010 Chile earthquake (Mw8.8), the 2004 Sumatra (Mw9.2), and the 1994 Hokkaido (Mw8.3) earthquake, but not in smaller earthquakes [140]. The anomalies even increased to 10 TECU before the Sumatra earthquake (Mw9.2). Some scholars point out that it might yield erroneous interpretations about anomalies [147,148]. Iwata and Umeno introduced the correlation analysis method and detected the preseismic TEC anomalies before the 2011 Tohoku earthquake and the 2016 Kumamoto earthquake, whose characteristic patterns do not depend on the choice of extrapolation functions for nonlinear regression [149,150].

In the research on seismo-electromagnetic phenomena in the ionosphere, the effects of other sources should be considered because the TEC perturbations are the integrated product of space weather and other kinds of sources such as earthquakes, tsunamis, volcanic eruptions [151], etc. The choice of space weather indices could be justified based on the location, depending on the data availability and station coverage [152]. When defining the space effect levels, some consensus values should be used based on the authoritative parameters from space physics, and all relevant space weather indices should be verified for quiet or low activity in order to ensure the space-related influences on ionospheric perturbations [142]. Some technological methods were developed to eliminate the effect of space weather activity without the use of arbitrary space indices to detect pre-earthquake TEC anomalies, including wavelet analyses [153], singular spectrum analysis [151,154], etc. In addition, demonstrating the three-dimensional structure of ionospheric disturbances can better reshape the morphological characteristics [155,156]. In general, it is very important to establish a reliable normal background field, exclude the interference effects, and identify reliable anomalies related to seismic activity [157].

In the review report of Pulinets et al. [16], the total electron content (TEC) over the epicenter usually increased or decreased for approximately 4–6 h on 1–5 days before earthquakes, and the TEC anomalous amplitude was related to the earthquake magnitude and local time, ranging from ±30% to ±100%. Meanwhile, some statistical investigations into the relationship between earthquakes and TEC anomalies were performed. Liu et al. [158] statistically examined the ionospheric plasma frequency (or electron density) in ionograms recorded by a local ionosonde and found that the critical frequency of the F2-peak, foF2, significantly decreased 1–5 days prior to 13 $M \geq 6.0$ earthquakes in Taiwan during the period 1994–1999. Kon et al. [159] analyzed the TEC anomalies before the earthquakes in Japan and found that the positive TEC anomalies were usually observed 1–5 days before $Mw > 6.0$ earthquakes with a focal depth of less than 40 km. Hirooka et al. [135,136] found that the electron density in the F-layer decreased by approximately 6 TECU for 3 days before the earthquakes in the low geomagnetic latitudinal regions. Chen et al. [160] gave the statistical evidence that the electron density decreased 1–5 days before $M \geq 6.0$ earthquakes in mainland China. Zhu et al. [161] found that the occurrence rates of TEC disturbances within 5 days before the inland earthquakes were higher than those after earthquakes, and the biggest difference of spatial distribution lies in the epicenter area. He and Heki [162] found possible ionospheric electron density enhancement detected before the large earthquakes by analyzing the vertical total electron content data observed by GNSS stations near epicenters, and these preseismic signals were mostly within 5 TECU. It is quite interesting to find that the TEC anomalies prior to large earthquakes can be either positive (Japan region) or negative (Taiwan region). In addition, TEC variability in seismically active regions is greater than in non-seismically active regions [142].

There are also some studies that focused on the changes associated with earthquakes in the magnetosphere, for example, a study of charged particle precipitation from van Allen radiation belts during a land earthquake and an ocean earthquake was reported [163]. Particle bursts (PBs) were found to be accumulated only around the earthquake day, reaching their maximum on the day of the earthquake and with a complete absence of such events on days away from the event day. Some statistical studies showed significant bursts of high-energy particles observed by SOHO satellite in the magnetosphere before earthquakes [164]. In general, since the magnetosphere is mostly affected by solar activity, it might be more complicated to study seismo-electromagnetic phenomena there.

Regarding the preparation zone of the earthquake, the most widely used equation is $R = 10^{0.43M_w}$, where R is the radius (km) and M_w is the magnitude [165]. The equation gained only used 24 earthquake events and did not account for ionospheric anomalies, although the authors concluded that it could be used to estimate the effective precursor manifestation zone [165]. Based on data from Benioff (i.e., crustal) strains, Bowman et al. found that the exponent in the equation should be increased for $M_w \geq 6.0$ earthquakes as they rupture the entire crust [166], so using the equation straightforwardly might result in errors. In addition, most scholars agreed that pre-earthquake TEC anomalies are most significant close to the epicenter [167], while other studies reported that they are greatest at the edge of the preparation zone [168]. Liu and Xu [169] revealed that the most perturbed ionospheric regions were located south of the epicenter and the zone was greater than 1000 km. Consequently, arguments have been put forward that earthquake preparation on the ionosphere/magnetosphere could have a larger radius [170]. The most complicated part of earthquake forewarning to use the ionosphere/magnetosphere perturbance is the large preparation zone based on the anomalies [142], so the surface observation should be combined to constrain potential earthquake locations.

5. Possible Mechanisms of Seismo-Electromagnetic Phenomena

To understand the mechanisms of seismo-electromagnetic phenomena in the lithosphere, various experiments and simulations have been conducted [171–181]. The mechanisms proposed for the generation of electromagnetic signals include: electrokinetic effects [182,183], piezo-electric effects [184], piezomagnetic effects [185], triboelectric effects [186], induction effects [187,188], microfracture effects [89,189], and P-holes [190–193]. For most of these models, electromagnetic signals are more or less related to stress or strain changes. The numerical simulation of coupled seismic and electromagnetic waves in multilayered porous media were developed to explain the coseismic electromagnetic signals [194,195]. However, field observations can hardly record strong, clear co-rupture signals [11]. The selectivity phenomenon of SES might explain why some electromagnetic anomalies are observed before earthquakes and others are not [175]. One of the most studied issues in the physics of earthquake precursors is the role of seismic-induced radon exhalation in the generation of electromagnetic disturbances. It has been proposed that the enhancement of the total rock surface due to failure would increase the emissions of radon and other gases from grains and facilitate migration. This would be in agreement with the enhancement of radon concentration observed in aftershocks and some laboratory experiments [196] which showed the growth of radon emissions of granites under compressional stress.

Although a number of studies have discussed possible mechanisms of pre-earthquake disturbances in the atmosphere [129,192], the generation process is still not well understood. Most hypotheses were based on the propagation of EM waves, which originated from the ground due to certain mechanisms such as piezo-electric effects [80], the tribo-electric effect [186], and positive holes (P-holes) [190]. Pulinets and Boyarchuk [129] discussed a model based on the emissions of radioactive gases or metallic ions. Pulinets and Ouzounov [197] proposed a unified concept for the pre-earthquake atmospheric disturbances, named the "Lithosphere–Atmosphere–Ionosphere Coupling (LAIC)" model. According to the LAIC model, radon can generate local ionization in the lower layers of the atmosphere that can facilitate water vapor condensation, leading to the release of latent heat exhalation (that could explain thermal fluctuations) and local variation of the conductivity, thereby impacting the electric circuit over the earthquake preparation zone and generating the observed seismo-associated ionospheric anomalies. Harrison et al. [198] argued that enhanced ionization in the lower atmosphere would increase the vertical current flow in fair weather, which might help understand earthquake–cloud coupling [199] through the global atmospheric electric circuit [200]. Kuo et al. [201,202] performed some numerical simulations and showed ionosphere plasma bubbles and density variations induced by pre-earthquake rock currents and associated surface charges. Other mechanisms including the atmospheric gravity waves (AGWs) triggered by gas releases or thermal anomalies [203] and electric current of seismic origin due to the convective transfer of charged soil aerosols [204,205] can also explain pre-earthquake disturbances in the atmosphere to some extent [206]. The coseismal lower atmospheric and ionospheric perturbance triggered by seismic waves through acoustic and/or gravity waves was verified for the M9.0 Tohoku earthquake [207].

Some researchers present a potential physics-based research approach that takes into account the interaction among observation, data processing methodology, and physical model [208]. The coupling between the lithosphere and atmosphere requires multiple instruments installed on and near the Earth's surface and space satellites. Recently, a novel system for monitoring vibrations and perturbations in the lithosphere, lower atmosphere and upper atmosphere was established in the countryside of Leshan City, Sichuan Province, China, in 2021 [209]. The system combines all kinds of observation instruments collecting data from the lithosphere and atmosphere, to gain abundant measurements across different layers preceding earthquakes. As a novel systematic observing platform, it can provide researchers with more information about the preparation process for forthcoming earthquakes in the future.

6. Applications of Seismo-Electromagnetic Signals in Earthquake Forecasting

There are some methods to evaluate whether the seismo-electromagnetic phenomena contain precursory information, among which the receiver operating characteristic (ROC) curve and Molchan's error diagram are the two most widely used [160,210]. Statistical tests have shown that the electromagnetic anomalies may contain predictive information for major earthquakes, with probability gains of approximately 2–6 [74,160,210]. In addition to seismo-electromagnetic phenomena, other precursors were also be observed. A representative event of which various geophysical precursors were documented is the 2011 Tohoku earthquake (Mw 9.0) [38,140,211–220]. Locations of seismo-geomagnetic anomalies in spatial distribution showed a good correlation with the epicenter of the earthquake [38,220]. These spatiotemporal results were consistent with those obtained from other independent observations such as the groundwater level and GPS displacements [219]. The related precursory characteristics of the earthquake are listed in the Table 1 below. The coupling of multiple pre-earthquake phenomena provides new insights on earthquake preparation process, suggesting that incorporating multi-geophysical information might be a possible means of improving earthquake forecast.

Table 1. Representative precursory characteristics before the 2011 Tohoku earthquake.

Observation	Leading Time	Characteristic	Reference
Seismic activity	23 years	Seismic quiescence in rupture region	(Katsumata, 2011 [211])
	6 years	Decade-scale decrease in b value	(Nanjo et al., 2012 [212])
	A few months	Seismicity exhibited distinct minima	(Sarlis et al., 2013 [213])
	1 month	Foreshock sequence and slow slip	(Kato et al., 2012 [214])
Deformation	2 months	Crustal movement changes	(Chen et al., 2014 [215])
	1 month	Slow slip event	(Ito et al., 2013 [216])
	1 month	GPS surface motion	(Hattori et al., 2014 [217])
Fluid	3 months	Groundwater level decrease	(Orihara et al.,2016 [218])
Geomagnetism	2 months	Geomagnetic diurnal variation changes	(Han et al., 2016 [219])
Ionosphere	40 min	TEC enhancement	(Heki, 2011 [140])

As the earthquake process is rather complex, the earthquake preparation phase cannot be easily monitored using a single approach. To improve the reliability of earthquake forecasting, the two following models that could integrate precursory information from different approaches were proposed. One is the earthquake model that combines self-exciting and mutually exciting elements from the Hawkes process [62,221,222]. The core idea of this combined model is that the status of the earthquake event at present is controlled by the event itself (self-exciting) and all the external factors (mutually exciting) in the past. The conditional intensity function is a time-varying point process with rate $\lambda(t)$ written as

$$\lambda(t) = \mu + \lambda_s(t) + \lambda_E(t) \tag{1}$$

where μ represents the constant background rate, $\lambda_s(t)$ is the self-exciting term that models the information from past seismic events, and $\lambda_E(t)$ is the external excitation term that models the information from non-seismic observations such as the seismo-electromagnetic anomalies. The other is a network model based on machine learning, where multi-geophysical precursor information is taken as the input layer, and the factors of earthquakes such as time, location, and magnitude are taken as the output layers, as shown in Figure 1. Some important features in discriminating pre-earthquake perturbations may be established by the machine learning network. Overall, the forecast model combining the integrated observations (shown in Figure 2) is a worthwhile study in the future.

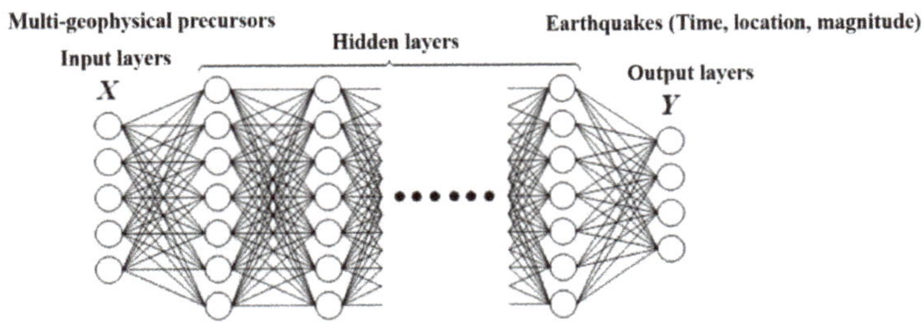

Figure 1. Earthquake forecast network based on machine learning.

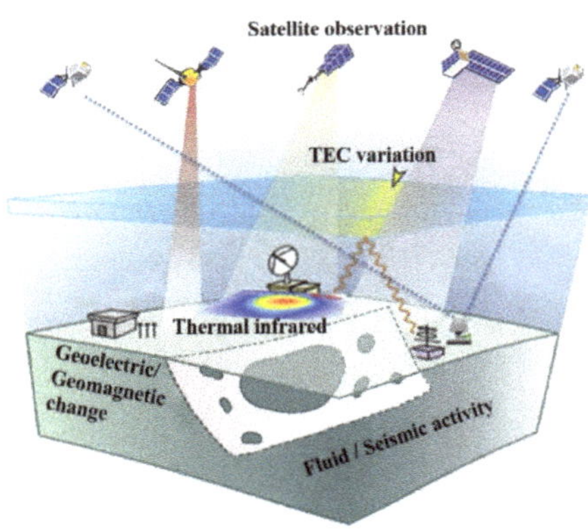

Figure 2. Multi-geophysical observation system for earthquake modeling and forecasting (modified from [83]).

7. Conclusions

As a kind of potential precursor of the earthquake, the seismo-electromagnetic phenomena were paid considerable attention and play a crucial role in the study of seismic hazards, although there are some controversial problems deserving further study. After reviewing the seismo-electromagnetic study in the lithosphere, the lower and upper atmosphere, and discussing the possible mechanisms, we present the related achievements and the potential applications in implementing seismo-electromagnetic phenomena for earthquake forecasting. The challenges include the variability of the detected intensity, frequencies, spatial, and temporal distribution around the epicenter, etc. Since the electromagnetic signal contains certain pieces of precursor information, achieving a better understanding of the physics of seismo-electromagnetic phenomena and developing multi-geophysical integrated forecast models is worth the efforts being made by the scientific community worldwide.

Author Contributions: Conceptualization, P.H., K.H. and H.C.; methodology, H.C.; software, H.C.; validation, P.H. and K.H.; formal analysis, H.C.; investigation, H.C.; resources, P.H.; data curation, P.H.; writing—original draft preparation, P.H. and H.C.; writing—review and editing, P.H. and K.H.; visualization, H.C.; supervision, P.H. and K.H.; project administration, P.H.; funding acquisition, P.H. All authors have read and agreed to the published version of the manuscript.

Funding: This research is partly supported by National Natural Science Foundation of China (41974083), China Seismic Experimental Site (DQJB22Z01-09), Shenzhen Science and Technology Program (Grant No. KQTD20170810111725321), and Guangdong Provincial Key Laboratory of Geophysical High-Resolution Imaging Technology (2022B1212010002).

Institutional Review Board Statement: Not applicable.

Informed Consent Statement: Not applicable.

Data Availability Statement: Not applicable.

Acknowledgments: The authors would like to thank Jiancang Zhuang for helpful discussions.

Conflicts of Interest: The authors declare no conflict of interest.

References

1. Ma, J. Whether earthquake precursors help for prediction do exist. *Sci. China Press Chin. Ed.* **2016**, *61*, 409–414.
2. Jordan, T.H. Earthquake predictability, brick by brick. *Seismol. Res. Lett.* **2006**, *77*, 3–6. [CrossRef]
3. Davis, C.; Keilis-Borok, V.; Kossobokov, V.; Soloviev, A. Advance prediction of the 11 March 2011 Great East Japan Earthquake: A missed opportunity for disaster preparedness. *Int. J. Disaster Risk Reduct.* **2012**, *1*, 17–32. [CrossRef]
4. Iacoletti, S.; Cremen, G.; Galasso, C. Validation of the epidemic-type aftershock sequence (ETAS) models for simulation-based seismic hazard assessments. *Seismol. Res. Lett.* **2022**, *93*, 1601–1618. [CrossRef]
5. Ogata, Y. Prediction and validation of short-to-long-term earthquake probabilities in inland Japan using the hierarchical space-time ETAS and space-time Poisson process models. *Earth Planets Space* **2022**, *74*, 110. [CrossRef]
6. Zhuang, J. Next-day earthquake forecasts for the Japan region generated by the ETAS model. *Earth Planets Space* **2011**, *63*, 207–216. [CrossRef]
7. Zechar, J.D.; Jordan, T.H. Testing alarm-based earthquake predictions. *Geophys. J. Int.* **2008**, *172*, 715–724. [CrossRef]
8. Shebalin, P.; Narteau, C.; Holschneider, M. From alarm-based to rate-based earthquake forecast models. *Bull. Seimolog. Soc. Am.* **2012**, *102*, 64–72. [CrossRef]
9. Wang, T.; Zhuang, J.; Kato, T.; Bebbington, M. Assessing the potential improvement in short-term earthquake forecasts from incorporation of GPS data. *Geophys. Res. Lett.* **2013**, *40*, 2631–2635. [CrossRef]
10. Han, P.; Zhuang, J.; Hattori, K.; Chen, C.-H.; Febriani, F.; Chen, H.; Yoshino, C.; Yoshida, S. Assessing the potential earthquake precursory information in ULF magnetic data recorded in Kanto, Japan during 2000–2010: Distance and magnitude dependences. *Entropy* **2020**, *22*, 859. [CrossRef]
11. Uyeda, S.; Nagao, T.; Kamogawa, M. Short-term earthquake prediction: Current status of seismo-electromagnetics. *Tectonophysics* **2009**, *470*, 205–213. [CrossRef]
12. Hattori, K. ULF geomagnetic changes associated with large earthquakes. *Terr. Atmos. Ocean. Sci.* **2004**, *15*, 329–360. [CrossRef]
13. Han, P.; Hattori, K.; Hirokawa, M.; Zhuang, J.; Chen, C.; Febriani, F.; Yamaguchi, H.; Yoshino, C.; Liu, J.; Yoshida, S. Statistical analysis of ULF seismomagnetic phenomena at Kakioka, Japan, during 2001–2010. *J. Geophys. Res.* **2014**, *119*, 4998–5011. [CrossRef]
14. Park, S.K. Precursors to earthquakes: Seismoelectromagnetic signals. *Surv. Geophys.* **1996**, *17*, 493–516. [CrossRef]
15. Johnston, M.J.S. Review of electric and magnetic fields accompanying seismic and volcanic activity. *Surv. Geophys.* **1997**, *18*, 441–476. [CrossRef]
16. Pulinets, S.A.; Legen'ka, A.D.; Gaivoronskaya, T.V.; Depuev, V.K. Main phenomenological features of ionospheric precursors of strong earthquakes. *J. Atmos. Sol. Terr. Phys.* **2003**, *65*, 1337–1347. [CrossRef]
17. Sarkar, S.; Gwal, A.K.; Parrot, M. Ionospheric variations observed by the DEMETER satellite in the mid-latitude region during strong earthquakes. *J. Atmos. Sol. Terr. Phys.* **2007**, *69*, 1524–1540. [CrossRef]

18. Hayakawa, M.; Molchanov, O.A.; Team, N.; Team, U.E.C. Summary report of NASDA's earthquake remote sensing frontier project. *Phys. Chem. Earth* **2004**, *29*, 617–625. [CrossRef]
19. Miyakoshi, J. Anomalous time-variation of the self-potential in the fractured zone of an active fault preceding the earthquake occurrence. *J. Geomagn. Geoelectr.* **1986**, *38*, 1015–1030. [CrossRef]
20. Varotsos, P.; Alexopoulos, K.; Nomicos, K.; Lazaridou, M. Earthquake prediction and electric signals. *Nature* **1986**, *322*, 120. [CrossRef]
21. Varotsos, P.; Alexopoulos, K.; Lazaridou, M. Latest aspects of earthquake prediction in Greece based on seismic electric signals, II. *Tectonophysics* **1993**, *224*, 1–37. [CrossRef]
22. Varotsos, P.; Alexopoulos, K.; Lazaridouvarotsou, M.; Nagao, T. Earthquake predictions issued in Greece by seismic electric signals since 6 February 1990. *Tectonophysics* **1993**, *224*, 269–288. [CrossRef]
23. Sarlis, N.V.; Varotsos, P.A.; Skordas, E.S.; Uyeda, S.; Zlotnicki, J.; Nagao, T.; Rybin, A.; Lazaridou-Varotsos, M.S.; Papadopoulou, K.A. Seismic electric signals in seismic prone areas. *Earthq. Sci.* **2018**, *31*, 44–51. [CrossRef]
24. Sarlis, N.V.; Skordas, E.S. Study in natural time of geoelectric field and seismicity changes preceding the Mw6.8 earthquake on 25 October 2018 in Greece. *Entropy* **2018**, *20*, 882. [CrossRef] [PubMed]
25. Sarlis, N.V.; Skordas, E.S.; Christopoulos, S.-R.G.; Varotsos, P.A. Natural time analysis: The area under the receiver operating characteristic curve of the order parameter fluctuations minima preceding major earthquakes. *Entropy* **2020**, *22*, 583. [CrossRef] [PubMed]
26. Ismaguilov, V.S.; Kopytenko, Y.A.; Hattori, K.; Hayakawa, M. Variations of phase velocity and gradient values of ULF geomagnetic disturbances connected with the Izu strong earthquakes. *Nat. Hazards Earth Syst. Sci.* **2003**, *3*, 211–215. [CrossRef]
27. Nagao, T.; Enomoto, Y.; Fujinawa, Y.; Hata, M.; Hayakawa, M.; Huang, Q.; Izutsu, J.; Kushida, Y.; Maeda, K.; Oike, K. Electromagnetic anomalies associated with 1995 Kobe earthquake. *J. Geodyn.* **2002**, *33*, 401–411. [CrossRef]
28. Hattori, K.; Takahashi, I.; Yoshino, C.; Isezaki, N.; Iwasaki, H.; Harada, M.; Kawabata, K.; Kopytenko, E.; Kopytenko, Y.; Maltsev, P.; et al. ULF geomagnetic field measurements in Japan and some recent results associated with Iwateken Nairiku Hokubu earthquake in 1998. *Phys. Chem. Earth* **2004**, *29*, 481–494. [CrossRef]
29. Serita, A.; Hattori, K.; Yoshino, C.; Hayakawa, M.; Isezaki, N. Principal component analysis and singular spectrum analysis of ULF geomagnetic data associated with earthquakes. *Nat. Hazards Earth Syst. Sci.* **2005**, *5*, 685–689. [CrossRef]
30. Zhuang, J.; Vere-Jones, D.; Guan, H.; Ogata, Y.; Ma, L. Preliminary analysis of observations on the ultra-low frequency electric field in the Beijing region. *Pure Appl. Geophys.* **2005**, *162*, 1367–1396. [CrossRef]
31. Telesca, L.; Hattori, K. Non-uniform scaling behavior in ultra-low-frequency (ULF) earthquake-related geomagnetic signals. *Phys. A Stat. Mech. Its Appl.* **2007**, *384*, 522–528. [CrossRef]
32. Bleier, T.E.; Dunson, C.; Maniscalco, M.; Bryant, N.; Bambery, R.; Freund, F. Investigation of ULF magnetic pulsations, air conductivity changes, and infra red signatures associated with the 30 October Alum Rock M5.4 earthquake. *Nat. Hazards Earth Syst. Sci.* **2009**, *9*, 585–603. [CrossRef]
33. Xie, T.; Liu, J.; Lu, J.; Li, M.; Yao, L.; Wang, Y.; Yu, C. Retrospective analysis on electromagnetic anomalies observed by ground fixed station before the 2008 Wenchuan Ms8.0 earthquake. *Chin. J. Geophys. Chin. Ed.* **2018**, *61*, 1922–1937.
34. Chen, C.; Hsu, H.L.; Wen, S.; Yeh, T.; Chang, F.Y.; Wang, C.; Liu, J.; Sun, Y.; Hattori, K.; Yen, H. Evaluation of seismo-electric anomalies using magnetic data in Taiwan. *Nat. Hazards Earth Syst. Sci.* **2013**, *13*, 597–604. [CrossRef]
35. Huang, Q. Retrospective investigation of geophysical data possibly associated with the Ms8.0 Wenchuan earthquake in Sichuan, China. *J. Asian Earth Sci.* **2011**, *41*, 421–427. [CrossRef]
36. Huang, Q. Forecasting the epicenter of a future major earthquake. *Proc. Natl. Acad. Sci. USA* **2015**, *112*, 944–945. [CrossRef]
37. Wen, S.; Chen, C.; Yen, H.; Yeh, T.; Liu, J.; Hattori, K.; Peng, H.; Wang, C.; Shin, T.C. Magnetic storm free ULF analysis in relation with earthquakes in Taiwan. *Nat. Hazards Earth Syst. Sci.* **2012**, *12*, 1747–1754. [CrossRef]
38. Xu, G.; Han, P.; Huang, Q.; Hattori, K.; Febriani, F.; Yamaguchi, H. Anomalous behaviors of geomagnetic diurnal variations prior to the 2011 off the Pacific coast of Tohoku earthquake (Mw9.0). *J. Asian Earth Sci.* **2013**, *77*, 59–65. [CrossRef]
39. Ouyang, X.Y.; Parrot, M.; Bortnik, J. ULF Wave Activity Observed in the Nighttime Ionosphere Above and Some Hours Before Strong Earthquakes. *J. Geophys. Res. Space Phys.* **2020**, *125*, e2020JA028396. [CrossRef]
40. Fraser-Smith, A.C.; Bernardi, A.; McGill, P.R.; Ladd, M.E.; Helliwell, R.A.; Villard, O.G., Jr. Low-frequency magnetic field measurements near the epicenter of the Ms 7.1 Loma Prieta Earthquake. *Geophys. Res. Lett.* **1990**, *17*, 1465–1468. [CrossRef]
41. Kopytenko, Y.A.; Matiashvili, T.G.; Voronov, P.M.; Kopytenko, E.A.; Molchanov, O.A. Detection of ultra-low-frequency emissions connected with the Spitak earthquake and its aftershock activity, based on geomagnetic pulsations data at Dusheti and Vardzia observatories. *Phys. Earth Planet. Inter.* **1993**, *77*, 85–95. [CrossRef]
42. Hayakawa, M.; Kawate, R.; Molchanov, O.A.; Yumoto, K. Results of ultra-low-frequency magnetic field measurements during the Guam Earthquake of 8 August 1993. *Geophys. Res. Lett.* **1996**, *23*, 241–244. [CrossRef]
43. Hattori, K.; Akinaga, Y.; Hayakawa, M.; Yumoto, K.; Nagao, T.; Uyeda, S. ULF magnetic anomaly preceding the 1997 Kagoshima earthquake. *Seism. Electromagn.* **2002**, *353*, 19–28.
44. Uyeda, S.; Hayakawa, M.; Nagao, T.; Molchanov, O.; Hattori, K.; Orihara, Y.; Gotoh, K.; Akinaga, Y.; Tanaka, H. Electric and magnetic phenomena observed before the volcano-seismic activity in 2000 in the Izu Island Region, Japan. *Proc. Natl. Acad. Sci. USA* **2002**, *99*, 7352–7355. [CrossRef] [PubMed]

45. Hattori, K.; Serita, A.; Gotoh, K.; Yoshino, C.; Harada, M.; Isezaki, N.; Hayakawa, M. ULF geomagnetic anomaly associated with 2000 Izu Islands earthquake swarm, Japan. *Phys. Chem. Earth* **2004**, *29*, 425–435. [CrossRef]
46. Hattori, K.; Serita, A.; Yoshino, C.; Hayakawa, M.; Isezaki, N. Singular spectral analysis and principal component analysis for signal discrimination of ULF geomagnetic data associated with 2000 Izu Island Earthquake Swarm. *Phys. Chem. Earth* **2006**, *31*, 281–291. [CrossRef]
47. Gotoh, K.; Hayakawa, M.; Smirnova, N. Fractal analysis of the ULF geomagnetic data obtained at Izu Peninsula, Japan in relation to the nearby earthquake swarm of June–August 2000. *Nat. Hazards Earth Syst. Sci.* **2003**, *3*, 229–236. [CrossRef]
48. Telesca, L.; Lapenna, V.; Macchiato, M.; Hattori, K. Investigating non-uniform scaling behavior in Ultra Low Frequency (ULF) earthquake-related geomagnetic signals. *Earth Planet. Sci. Lett.* **2008**, *268*, 219–224. [CrossRef]
49. Ismaguilov, V.S.; Kopytenko, Y.A.; Hattori, K.; Voronov, P.M.; Molchanov, O.A.; Hayakawa, M. ULF magnetic emissions connected with under sea bottom earthquakes. *Nat. Hazards Earth Syst. Sci.* **2001**, *1*, 23–31. [CrossRef]
50. Kotsarenko, A.; Molchanov, O.; Hayakawa, M.; Koshevaya, S.; Grimalsky, V.; Enriquez, R.W.; Cruz-Abeyro, J.A.L. Investigation of ULF magnetic anomaly during Izu earthquake swarm and Miyakejima volcano eruption at summer 2000, Japan. *Nat. Hazards Earth Syst. Sci.* **2005**, *5*, 63–69. [CrossRef]
51. Han, P.; Hattori, K.; Huang, Q.; Hirano, T.; Ishiguro, Y.; Yoshino, C.; Febriani, F. Evaluation of ULF electromagnetic phenomena associated with the 2000 Izu Islands earthquake swarm by wavelet transform analysis. *Nat. Hazards Earth Syst. Sci.* **2011**, *11*, 965–970. [CrossRef]
52. Hattori, K.; Han, P.; Yoshino, C.; Febriani, F.; Yamaguchi, H.; Chen, C. Investigation of ULF Seismo-Magnetic Phenomena in Kanto, Japan During 2000–2010: Case Studies and Statistical Studies. *Surv. Geophys.* **2013**, *34*, 293–316. [CrossRef]
53. Schekotov, A.; Molchanov, O.A.; Hayakawa, M.; Fedorov, E.; Chebrov, V.N.; Sinitsin, V.; Gordeev, E.; Belyaev, G.G.; Yagova, N. ULF/ELF magnetic field variations from atmosphere induced by seismicity. *Radio Sci.* **2007**, *42*, 1–13. [CrossRef]
54. Molchan, G. Structure of optimal strategies in earthquake prediction. *Tectonophysics* **1991**, *193*, 267–276. [CrossRef]
55. Molchan, G.M.; Kagan, Y.Y. Earthquake prediction and its optimization. *J. Geophys. Res.* **1992**, *97*, 4823–4838. [CrossRef]
56. Warden, S.; Bleier, T.; Kappler, K. Long term air ion monitoring in search of pre-earthquake signals. *J. Atmos. Sol. Terr. Phys.* **2019**, *186*, 47–60. [CrossRef]
57. Dunson, J.C.; Bleier, T.E.; Roth, S.; Heraud, J.; Alvarez, C.H.; Lira, A. The Pulse Azimuth effect as seen in induction coil magnetometers located in California and Peru 2007–2010, and its possible association with earthquakes. *Nat. Hazards Earth Syst. Sci.* **2011**, *11*, 2085–2105. [CrossRef]
58. Kappler, K.N.; Schneider, D.D.; MacLean, L.S.; Bleier, T.E.; Lemon, J.J. An algorithmic framework for investigating the temporal relationship of magnetic field pulses and earthquakes applied to California. *Comput. Geosci.* **2019**, *133*, 104317. [CrossRef]
59. Heavlin, W.D.; Kappler, K.; Yang, L.; Fan, M.; Hickey, J.; Lemon, J.; MacLean, L.; Bleier, T.; Riley, P.; Schneider, D. Case-control study on a decade of ground-based magnetometers in California reveals modest signal 24–72 h prior to earthquakes. *J. Geophys. Res. Solid Earth* **2022**, *127*, e2022JB024109. [CrossRef]
60. Yuan, H.H.; Chen, B.; Gu, Z.W.; Ni, Z.; Su, S.P.; Xing, C.J.; Song, C.K. Mobile seismic and geomagnetic monitoring network in China. In Proceedings of the 2018 Annual Meeting of Chinese Geoscience Union, Beijing, China, 21–24 October 2018.
61. Li, Q.; Yuan, Y.; Yang, X.; Cai, S.; Sun, W. Variation of the geomagnetic harmonic wave amplitude ratio before the MS5.5 Eryuan earthquake in 2013. *Acta Seismol. Sin. Chin. Ed.* **2016**, *38*, 122–129.
62. Ma, Q. Multi-dipole observation system and study on the abnormal variation of the geoelectric field observed at Capital network before the 2006 Wen'an, Hebei of China, Ms 5.1 earthquake. *Acta Seismol. Sin. Chin. Ed.* **2008**, *30*, 615–625.
63. Qian, J.; Ma, Q.; Li, S. Further study on the anomalies in apparent resistivity in the NE configuration at Chengdu station associated with Whenchuan Ms 8.0 earthquake. *Acta Seismol. Sin. Chin. Ed.* **2013**, *35*, 4–17.
64. Ma, Q.; Fang, G.; Li, W.; Zhou, J. Electromagnetic anomalies before the 2018 Lushan Ms 7.0 earthquake. *Acta Seismol. Sin. Chin. Ed.* **2021**, *35*, 717–730.
65. Du, X.; Ye, Q.; Ma, Z.; Li, N.; Chen, J.; Tan, D. The detection depth of symmetric four-electrode resistivity observation in/near the epicentral region of strong earthquakes. *Chin. J. Geophys. Chin. Ed.* **2008**, *51*, 1943–1949. [CrossRef]
66. Tang, J.; Zhao, G.; Jijun, W.; Wenjun, L.; Yan, Z. Variation and analysis of resistivity before and after the Zhangbei-Shangyi earthquake. *Seismol. Geol. Chin. Ed.* **1998**, *20*, 164–171.
67. Chen, J.; Du, X.; Tan, D.; An, Z. Magnetotelluric monitoring of earthquakes in the laohushan fault zone, gansu province. *Earthq. Chin. Ed.* **2009**, *29*, 79–85.
68. Kappler, K.N.; Morrison, H.F.; Egbert, G.D. Long-term monitoring of ULF electromagnetic fields at Parkfield, California. *J. Geophys. Res. Solid Earth* **2010**, *115*, 406. [CrossRef]
69. Campbell, W.H. Natural magnetic disturbance fields, not precursors, preceding the Loma Prieta earthquake. *J. Geophys. Res. Space Phys.* **2009**, *114*, 307. [CrossRef]
70. Thomas, J.N.; Love, J.J.; Johnston, M.J.S. On the reported magnetic precursor of the 1989 Loma Prieta earthquake. *Phys. Earth Planet. Inter.* **2009**, *173*, 207–215. [CrossRef]
71. Masci, F.; Thomas, J.N. Are there new findings in the search for ULF magnetic precursors to earthquakes? *J. Geophys. Res. Space Phys.* **2015**, *120*, 10289–10304. [CrossRef]
72. Fidani, C. The earthquake lights (EQL) of the 6 April 2009 Aquila earthquake, in Central Italy. *Nat. Hazards Earth Syst. Sci.* **2010**, *10*, 967–978. [CrossRef]

73. Derr, J. Earthquake lights: A review of observations and present theories. *Bull. Seismol. Soc. Am.* **1973**, *63*, 2177–2187.
74. Tramutoli, V.; Cuomo, V.; Filizzola, C.; Pergola, N.; Pietrapertosa, C. Assessing the potential of thermal infrared satellite surveys for monitoring seismically active areas: The case of Kocaeli (İzmit) earthquake, 17 August 1999. *Remote Sens. Environ.* **2005**, *96*, 409–426. [CrossRef]
75. Ouzounov, D.; Bryant, N.; Logan, T.; Pulinets, S.; Taylor, P. Satellite thermal IR phenomena associated with some of the major earthquakes in 1999–2003. *Phys. Chem. Earth Parts A/B/C* **2006**, *31*, 154–163. [CrossRef]
76. Ouzounov, D.; Liu, D.; Chunli, K.; Cervone, G.; Kafatos, M.; Taylor, P. Outgoing long wave radiation variability from IR satellite data prior to major earthquakes. *Tectonophysics* **2007**, *431*, 211–220. [CrossRef]
77. Jianguo, H. Near earth surface anomalies of the atmospheric electric field and earthquakes. *Acta Seismol. Sin. Engl. Ed.* **1989**, *2*, 289–298.
78. Chen, T.; Li, L.; Zhang, X.-X.; Ma, Q.-M.; Li, W.; Ti, S.; Wu, H.; Li, R.-K.; Luo, J.; Su, J.-F. Near-epicenter weather conditions several hours before strong earthquakes (Ms $\geq$ 6). *Nat. Hazards* **2022**, *110*, 57–68. [CrossRef]
79. Liu, J.Y.; Chen, Y.I.; Huang, C.H.; Ho, Y.Y.; Chen, C.H. A statistical study of lightning activities and M $\geq$ 5.0 earthquakes in Taiwan during 1993–2004. *Surv. Geophys.* **2015**, *36*, 851–859. [CrossRef]
80. Finkelstein, D.; Hill, R.D.; Powell, J.R. The Piezoelectric Theory of earthquake lightning. *J. Geophys. Res.* **1973**, *78*, 992–993. [CrossRef]
81. Qin, K.; Wu, L.; De Santis, A.; Wang, H. Surface latent heat flux anomalies before the MS 7.1 New Zealand earthquake 2010. *Chin. Sci. Bull.* **2011**, *56*, 3273. [CrossRef]
82. Kai, Q. Preliminary analysis of surface temperature anomalies that preceded the two major Emilia 2012 earthquakes (Italy). *Ann. Geophys.* **2012**, *55*, 823–828.
83. Ouzounov, D.; Pulinets, S.; Hattori, K.; Taylor, P. *Pre-Earthquake Processes: A Multidisciplinary Approach to Earthquake Prediction Studies*; John Wiley & Sons: Hoboken, NJ, USA, 2018; Volume 234.
84. Genzano, N.; Filizzola, C.; Hattori, K.; Pergola, N.; Tramutoli, V. Statistical correlation analysis between thermal infrared anomalies observed from MTSATs and large earthquakes occurred in Japan (2005–2015). *J. Geophys. Res. Solid Earth* **2021**, *126*, e2020JB020108. [CrossRef]
85. Blackett, M.; Wooster, M.J.; Malamud, B.D. Exploring land surface temperature earthquake precursors: A focus on the Gujarat (India) earthquake of 2001. *Geophys. Res. Lett.* **2011**, *38*, 303. [CrossRef]
86. Zhang, Y.; Meng, Q. A statistical analysis of TIR anomalies extracted by RSTs in relation to an earthquake in the Sichuan area using MODIS LST data. *Nat. Hazards Earth Syst. Sci.* **2019**, *19*, 535–549. [CrossRef]
87. Galperin, Y.I.; Hayakawa, M. On the magnetospheric effects of experimental ground explosions observed from AUREOL-3. *J. Geomagn. Geoelectr.* **1996**, *48*, 1241–1263. [CrossRef]
88. Hayakawa, M.; Molchanov, O.A.; Ondoh, T.; Kawai, E. Anomalies in the sub-ionospheric VLF signals for the 1995 Hyogo-ken Nanbu earthquake. *J. Phys. Earth* **1996**, *44*, 413–418. [CrossRef]
89. Molchanov, O.A.; Hayakawa, M. Subionospheric VLF signal perturbations possibly related to earthquakes. *J. Geophys. Res. Space Phys.* **1998**, *103*, 17489–17504. [CrossRef]
90. Akhoondzadeh, M.; Parrot, M.; Saradjian, M.R. Investigation of VLF and HF waves showing seismo-ionospheric anomalies induced by the 29 September 2009 Samoa earthquake (M-w = 8.1). *Nat. Hazards Earth Syst. Sci.* **2010**, *10*, 1061–1067. [CrossRef]
91. Shen, X.H.; Zhang, X.M.; Yuan, S.G.; Wang, L.W.; Cao, J.B.; Huang, J.P.; Zhu, X.H.; Piergiorgio, P.; Dai, J.P. The state-of-the-art of the China Seismo-Electromagnetic Satellite mission. *Sci. China Technol. Sci.* **2018**, *61*, 634–642. [CrossRef]
92. Hayakawa, M.; Molchanov, O.A.; Ondoh, T.; Kawai, E. On the precursory signature of kobe earthquake on VLF subionospheric signals. In Proceedings of the 1997 International Symposium on Electromagnetic Compatibility, Beijing, China, 21–23 May 1997; pp. 72–75.
93. Hayakawa, M.; Ohta, K.; Maekawa, S.; Yamauchi, T.; Ida, Y.; Gotoh, T.; Yonaiguchi, N.; Sasaki, H.; Nakamura, T. Electromagnetic precursors to the 2004 Mid Niigata Prefecture earthquake. *Phys. Chem. Earth Parts A/B/C* **2006**, *31*, 356–364. [CrossRef]
94. Hayakawa, M. VLF/LF radio sounding of ionospheric perturbations associated with earthquakes. *Sensors* **2007**, *7*, 1141–1158. [CrossRef]
95. Rozhnoi, A.; Solovieva, M.S.; Molchanov, O.A.; Hayakawa, M. Middle latitude LF (40 kHz) phase variations associated with earthquakes for quiet and disturbed geomagnetic conditions. *Phys. Chem. Earth Parts A/B/C* **2004**, *29*, 589–598. [CrossRef]
96. Asada, T.; Baba, H.; Kawazoe, M.; Sugiura, M. An attempt to delineate very low frequency electromagnetic signals associated with earthquakes. *Earth Planets Space* **2001**, *53*, 55–62. [CrossRef]
97. Nagao, T.; Kamogawa, M.; Izutsu, J.; Baba, H.; Narushima, T.; Takamura, N.; Sakurada, T.; Uehara, H. First report of the electromagnetic wave detection system in VLF range, Tokai University—Proven for the existence of preseismic phenomena. *Bull. Inst. Ocean. Rresearch Dev.* **2016**, *37*, 29–36.
98. Moriya, T.; Mogi, T.; Takada, M. Anomalous pre-seismic transmission of VHF-band radio waves resulting from large earthquakes, and its statistical relationship to magnitude of impending earthquakes. *Geophys. J. Int.* **2010**, *180*, 858–870. [CrossRef]
99. Zhang, X. The development in seismic application research of VLF/LF radio waves. *Acta Seismol. Sin. Chin. Ed.* **2021**, *43*, 656.
100. Zhao, G.Z.; Wang, L.F.; Tang, J.; Chen, X.B.; Zhan, Y.; Xa, Q.B.; Wang, J.J.; Cai, J.T.; Xu, G.J.; Wan, Z.S.; et al. New experiments of CSELF electromagnetic method for earthquake monitoring. *Chin. J. Geophys. Chin. Ed.* **2010**, *53*, 479–486.

101. Parrot, M.; Berthelier, J.J.; Lebreton, J.P.; Sauvaud, J.A.; Santolik, O.; Blecki, J. Examples of unusual ionospheric observations made by the DEMETER satellite over seismic regions. *Phys. Chem. Earth Parts A/B/C* **2006**, *31*, 486–495. [CrossRef]
102. Zeng, Z.C.; Zhang, B.; Fang, G.Y.; Wang, D.F.; Yin, H.J. The analysis of ionospheric variations before Wenchuan earthquake with DEMETER data. *Chin. J. Geophys. Chin. Ed.* **2009**, *52*, 11–19. [CrossRef]
103. Akhoondzadeh, M.; Parrot, M.; Saradjian, M.R. Electron and ion density variations before strong earthquakes (M>6.0) using DEMETER and GPS data. *Nat. Hazards Earth Syst. Sci.* **2010**, *10*, 7–18. [CrossRef]
104. He, Y.; Yang, D.; Qian, J.; Parrot, M. Response of the ionospheric electron density to different types of seismic events. *Nat. Hazards Earth Syst. Sci. Discuss.* **2011**, *11*, 2173–2180.
105. Sarkar, S.; Tiwari, S.; Gwal, A.K. Electron density anomalies associated with M $\geq$ 5.9 earthquakes in Indonesia during 2005 observed by DEMETER. *J. Atmos. Sol. -Terr. Phys.* **2011**, *73*, 2289–2299. [CrossRef]
106. Parrot, M. Statistical analysis of automatically detected ion density variations recorded by DEMETER and their relation to seismic activity. *Ann. Geophys.* **2012**, *55*, 149–155.
107. Li, M.; Parrot, M. Statistical analysis of an ionospheric parameter as a base for earthquake prediction. *J. Geophys. Res. Space Phys.* **2013**, *118*, 3731–3739. [CrossRef]
108. Pisa, D.; Němec, F.; SantolíK, O.; Parrot, M.; Rycroft, M. Additional attenuation of natural VLF electromagnetic waves observed by the DEMETER spacecraft resulting from preseismic activity. *J. Geophys. Res. Space Phys.* **2013**, *118*, 5286–5295. [CrossRef]
109. Rozhnoi, A.; Solovieva, M.; Parrot, M.; Hayakawa, M.; Biagi, P.-F. VLF/LF signal studies of the ionospheric response to strong seismic activity in the Far Eastern region combining the DEMETER and ground-based observations. *Phys. Chem. Earth Parts A/B/C* **2015**, *85–86*, 141–149. [CrossRef]
110. Němec, F.; Santolík, O.; Parrot, M.; Berthelier, J.J. Spacecraft observations of electromagnetic perturbations connected with seismic activity. *Geophys. Res. Lett.* **2008**, *35*, 109. [CrossRef]
111. Marchetti, D.; Akhoondzadeh, M. Analysis of Swarm satellites data showing seismo-ionospheric anomalies around the time of the strong Mexico (M-w = 8.2) earthquake of 8 September 2017. *Adv. Space Res.* **2018**, *62*, 614–623. [CrossRef]
112. Marchetti, D.; De Santis, A.; Campuzano, S.A.; Zhu, K.; Soldani, M.; D'Arcangelo, S.; Orlando, M.; Wang, T.; Cianchini, G.; Di Mauro, D.; et al. Worldwide statistical correlation of eight years of swarm satellite data with M5.5+ earthquakes: New hints about the preseismic phenomena from space. *Remote Sens.* **2022**, *14*, 2649. [CrossRef]
113. Marchetti, D.; De Santis, A.; D'Arcangelo, S.; Poggio, F.; Piscini, A.; Campuzano, S.A.; De Carvalho, W.V.J.O. Pre-earthquake chain processes detected from ground to satellite altitude in preparation of the 2016–2017 seismic sequence in Central Italy. *Remote Sens. Environ.* **2019**, *229*, 93–99. [CrossRef]
114. De Santis, A.; Marchetti, D.; Pavón-Carrasco, F.J.; Cianchini, G.; Perrone, L.; Abbattista, C.; Alfonsi, L.; Amoruso, L.; Campuzano, S.A.; Carbone, M.; et al. Precursory worldwide signatures of earthquake occurrences on Swarm satellite data. *Sci. Rep.* **2019**, *9*, 20287. [CrossRef] [PubMed]
115. Li, M.; Parrot, M. Statistical analysis of the ionospheric ion density recorded by DEMETER in the epicenter areas of earthquakes as well as in their magnetically conjugate point areas. *Adv. Space Res.* **2018**, *61*, 974–984. [CrossRef]
116. Cao, J.B.; Zeng, L.; Zhan, F.; Wang, Z.G.; Wang, Y.; Chen, Y.; Meng, Q.C.; Ji, Z.Q.; Wang, P.F.; Liu, Z.W.; et al. The electromagnetic wave experiment for CSES mission: Search coil magnetometer. *Sci. China Technol. Sci.* **2018**, *61*, 653–658. [CrossRef]
117. Li, M.; Shen, X.H.; Parrot, M.; Zhang, X.M.; Zhang, Y.; Yu, C.; Yan, R.; Liu, D.P.; Lu, H.X.; Guo, F.; et al. Primary joint statistical seismic influence on ionospheric parameters recorded by the CSES and DEMETER satellites. *J. Geophys. Res. Space Phys.* **2020**, *125*, e2020JA028116. [CrossRef]
118. Liu, J.; Guan, Y.B.; Zhang, X.M.; Shen, X.H. The data comparison of electron density between CSES and DEMETER satellite, swarm constellation and IRI model. *Earth Space Sci.* **2021**, *8*, e2020EA001475. [CrossRef]
119. Yang, Y.Y.; Hulot, G.; Vigneron, P.; Shen, X.H.; Zhima, Z.; Zhou, B.; Magnes, W.; Olsen, N.; Toffner-Clausen, L.; Huang, J.P.; et al. The CSES global geomagnetic field model (CGGM): An IGRF-type global geomagnetic field model based on data from the China seismo-electromagnetic satellite. *Earth Planets Space* **2021**, *73*, 45. [CrossRef]
120. Yan, R.; Zhima, Z.R.; Xiong, C.; Shen, X.H.; Huang, J.P.; Guan, Y.B.; Zhu, X.H.; Liu, C. Comparison of electron density and temperature from the CSES satellite with other space-borne and ground-based observations. *J. Geophys. Res. Space Phys.* **2020**, *125*, e2019JA027747. [CrossRef]
121. Shen, X.H.; Huang, J.P.; Lin, J.; Luo, Z.C.; Yue, H.J.; Wu, L.X.; Zhang, X.M.; Cui, J. Project plan and research on data analysis and processing technology of geophysical exploration satellite and application research of earthquake prediction. *Prog. Earthq. Sci. Chin. Ed.* **2022**, *52*, 1–25.
122. Zhima, Z.; Zhou, B.; Zhao, S.; Wang, Q.; Huang, J.; Zeng, L.; Lei, J.; Chen, Y.; Li, C.; Yang, D.; et al. Cross-calibration on the electromagnetic field detection payloads of the China seismo-electromagnetic satellite. *Sci. China Technol. Sci.* **2022**, *65*, 1415–1426. [CrossRef]
123. Rui, Y.; XuHui, S.; JianPing, H.; Qiao, W.; Wei, C.; DaPeng, L.; YanYan, Y.; HengXin, L.; Song, X. Examples of unusual ionospheric observations by the CSES prior to earthquakes. *Earth Planet. Phys.* **2018**, *2*, 515–526.
124. Song, R.; Hattori, K.; Zhang, X.M.; Sanaka, S. Seismic-ionospheric effects prior to four earthquakes in Indonesia detected by the China seismo-electromagnetic satellite. *J. Atmos. Sol. Terr. Phys.* **2020**, *205*, 105291. [CrossRef]
125. Zhu, K.; Zheng, L.; Yan, R.; Shen, X.; Zeren, Z.; Xu, S.; Chu, W.; Liu, D.; Zhou, N.; Guo, F. The variations of electron density and temperature related to seismic activities observed by CSES. *Nat. Hazards Res.* **2021**, *1*, 88–94. [CrossRef]

126. Xiong, P.; Long, C.; Zhou, H.Y.; Battiston, R.; Zhang, X.M.; Shen, X.H. Identification of Electromagnetic Pre-Earthquake Perturbations from the DEMETER Data by Machine Learning. *Remote Sens.* **2020**, *12*, 3643. [CrossRef]
127. Xiong, P.; Long, C.; Zhou, H.; Battiston, R.; De Santis, A.; Ouzounov, D.; Zhang, X.; Shen, X. Pre-earthquake ionospheric perturbation identification using CSES data via transfer learning. *Front. Environ. Sci.* **2021**, *9*, 514. [CrossRef]
128. Li, Z.; Yang, B.; Huang, J.; Yin, H.; Yang, X.; Liu, H.; Zhang, F.; Lu, H. Analysis of pre-earthquake space electric field disturbance observed by CSES. *Atmosphere* **2022**, *13*, 934. [CrossRef]
129. Pulinets, S.; Boyarchuk, K. *Ionospheric Precursors of Earthquakes*; Springer: Cham, The Netherlands, 2004.
130. Liu, J.Y.; Chen, Y.I.; Jhuang, H.K.; Lin, Y.H. Ionospheric foF2 and TEC anomalous days associated with $M \geq 5.0$ earthquakes in Taiwan during 1997–1999. *Terr. Atmos. Ocean. Sci.* **2004**, *15*, 371–383. [CrossRef]
131. Liu, J.Y.; Chen, Y.I.; Chuo, Y.J.; Chen, C.S. A statistical investigation of preearthquake ionospheric anomaly. *J. Geophys. Res. Space Phys.* **2006**, *111*, 304. [CrossRef]
132. Liu, J.Y.; Chen, Y.I.; Chen, C.H.; Liu, C.Y.; Chen, C.Y.; Nishihashi, M.; Li, J.Z.; Xia, Y.Q.; Oyama, K.I.; Hattori, K.; et al. Seismoionospheric GPS total electron content anomalies observed before the 12 May 2008 M(w)7.9 Wenchuan earthquake. *J. Geophys. Res. Space Phys.* **2009**, *114*, 320. [CrossRef]
133. Liu, J.Y.; Chen, Y.I.; Chen, C.H.; Hattori, K. Temporal and spatial precursors in the ionospheric global positioning system (GPS) total electron content observed before the 26 December 2004 M9.3 Sumatra-Andaman Earthquake. *J. Geophys. Res. Space Phys.* **2010**, *115*, 312. [CrossRef]
134. Liu, J.Y.; Le, H.; Chen, Y.I.; Chen, C.H.; Liu, L.; Wan, W.; Su, Y.Z.; Sun, Y.Y.; Lin, C.H.; Chen, M.Q. Observations and simulations of seismoionospheric GPS total electron content anomalies before the 12 January 2010 M7 Haiti earthquake. *J. Geophys. Res. Space Phys.* **2011**, *116*, 302.
135. Hirooka, S.; Hattori, K.; Nishihashi, M.; Takeda, T. Neural network based tomographic approach to detect earthquake-related ionospheric anomalies. *Nat. Hazards Earth Syst. Sci.* **2011**, *11*, 2341–2353. [CrossRef]
136. Hirooka, S.; Hattori, K.; Nishihashi, M.; Kon, S.; Takeda, T. Development of ionospheric tomography using neural network and its application to the 2007 Southern Sumatra earthquake. *Electr. Eng. Jpn.* **2012**, *181*, 9–18. [CrossRef]
137. Astafyeva, E.; Shalimov, S.; Olshanskaya, E.; Lognonne, P. Ionospheric response to earthquakes of different magnitudes: Larger quakes perturb the ionosphere stronger and longer. *Geophys. Res. Lett.* **2013**, *40*, 1675–1681. [CrossRef]
138. Guo, J.; Li, W.; Yu, H.; Liu, Z.; Kong, Q. Impending ionospheric anomaly preceding the Iquique Mw8.2 earthquake in Chile on 1 April 2014. *Geophys. J. Int.* **2015**, *203*, 1461–1470. [CrossRef]
139. Akyol, A.A.; Arikan, O.; Arikan, F. A machine learning-based detection of earthquake precursors using ionospheric data. *Radio Sci.* **2020**, *55*, 1–21. [CrossRef]
140. Heki, K. Ionospheric electron enhancement preceding the 2011 Tohoku-Oki earthquake. *Geophys. Res. Lett.* **2011**, *38*, 312. [CrossRef]
141. Mukhtarov, P.; Pancheva, D.; Andonov, B.; Pashova, L. Global TEC maps based on GNSS data: 1. Empirical background TEC model. *J. Geophys. Res. Space Phys.* **2013**, *118*, 4594–4608. [CrossRef]
142. Lim, B.J.M.; Leong, E.C. Challenges in the detection of ionospheric pre-earthquake total electron content anomalies (PETA) for earthquake forewarning. *Pure Appl. Geophys.* **2019**, *176*, 2425–2449. [CrossRef]
143. Liu, J.Y.; Chen, C.H.; Chen, Y.I.; Yang, W.H.; Oyama, K.I.; Kuo, K.W. A statistical study of ionospheric earthquake precursors monitored by using equatorial ionization anomaly of GPS TEC in Taiwan during 2001–2007. *J. Asian Earth Sci.* **2010**, *39*, 76–80. [CrossRef]
144. Astafyeva, E.; Heki, K. Vertical TEC over seismically active region during low solar activity. *J. Atmos. Sol. Terr. Phys.* **2011**, *73*, 1643–1652. [CrossRef]
145. Akhoondzadeh, M. Application of Artificial Bee Colony algorithm in TEC seismo-ionospheric anomalies detection. *Adv. Space Res.* **2015**, *56*, 1200–1211. [CrossRef]
146. Sompotan, A.; Puspito, N.; Joelianto, E.; Hattori, K. Analysis of ionospheric precursor of earthquake using GIM-TEC, kriging and neural network. *Asian J. Earth Sci.* **2015**, *8*, 32–44. [CrossRef]
147. Kamogawa, M.; Kakinami, Y. Is an ionospheric electron enhancement preceding the 2011 Tohoku-Oki earthquake a precursor? *J. Geophys. Res. Space Phys.* **2013**, *118*, 1751–1754. [CrossRef]
148. Masci, F.; Thomas, J.N.; Villani, F.; Secan, J.A.; Rivera, N. On the onset of ionospheric precursors 40 min before strong earthquakes. *J. Geophys. Res. Space Phys.* **2015**, *120*, 1383–1393. [CrossRef]
149. Iwata, T.; Umeno, K. Correlation analysis for preseismic total electron content anomalies around the 2011 Tohoku-Oki earthquake. *J. Geophys. Res. Space Phys.* **2016**, *121*, 8969–8984. [CrossRef]
150. Iwata, T.; Umeno, K. Preseismic ionospheric anomalies detected before the 2016 Kumamoto earthquake. *J. Geophys. Res. Space Phys.* **2017**, *122*, 3602–3616. [CrossRef]
151. Chen, H.; Miao, M.; Chang, Y.; Wang, Q.; Shen, X.; Hattori, K.; Han, P. Singular spectrum analysis of the total electron content changes prior to $M \geq 6.0$ earthquakes in the Chinese mainland during 1998–2013. *Front. Earth Sci.* **2021**, *9*, 1–12. [CrossRef]
152. Jhuang, H.-K.; Ho, Y.-Y.; Kakinami, Y.; Liu, J.; Oyama, K.-I.; Parrot, M.; Hattori, K.; Nishihashi, M.; Zhang, D. Seismo-ionospheric anomalies of the GPS-TEC appear before the 12 May 2008 1 M7.9 Wenchuan Earthquake. *Int. J. Remote Sens.* **2010**, *31*, 3579. [CrossRef]

153. He, L.M.; Wu, L.X.; Pulinets, S.; Liu, S.J.; Yang, F. A nonlinear background removal method for seismo-ionospheric anomaly analysis under a complex solar activity scenario: A case study of the M9.0 Tohoku earthquake. *Adv. Space Res.* **2012**, *50*, 211–220. [CrossRef]
154. Guo, J.; Shi, K.; Liu, X.; Sun, Y.; Li, W.; Kong, Q. Singular spectrum analysis of ionospheric anomalies preceding great earthquakes: Case studies of Kaikoura and Fukushima earthquakes. *J. Geodyn.* **2019**, *124*, 1–13. [CrossRef]
155. He, L.M.; Heki, K. Three-dimensional distribution of ionospheric anomalies prior to three large earthquakes in Chile. *Geophys. Res. Lett.* **2016**, *43*, 7287–7293. [CrossRef]
156. Song, R.; Hattori, K.; Zhang, X.; Liu, J.-Y.; Yoshino, C. Detecting the ionospheric disturbances in Japan using the three-dimensional computerized tomography. *J. Geophys. Res. Space Phys.* **2021**, *126*, e2020JA028561. [CrossRef]
157. Zhao, G.Z.; Zhang, X.M.; Cai, J.T.; Zhan, Y.; Ma, Q.Z.; Tang, J.; Du, X.B.; Han, B.; Wang, L.F.; Chen, X.B.; et al. A review of seismo-electromagnetic research in China. *Sci. China Earth Sci.* **2022**, *52*, 1229–1246. [CrossRef]
158. Liu, J.Y.; Chen, Y.I.; Pulinets, S.A.; Tsai, Y.B.; Chuo, Y.J. Seismo-ionospheric signatures prior to M $\geq$ 6.0 Taiwan earthquakes. *Geophys. Res. Lett.* **2000**, *27*, 3113–3116. [CrossRef]
159. Kon, S.; Nishihashi, M.; Hattori, K. Ionospheric anomalies possibly associated with M $\geq$ 6.0 earthquakes in the Japan area during 1998–2010: Case studies and statistical study. *J. Asian Earth Sci.* **2011**, *41*, 410–420. [CrossRef]
160. Chen, Y.-I.; Huang, C.-S.; Liu, J.-Y. Statistical evidences of seismo-ionospheric precursors applying receiver operating characteristic (ROC) curve on the GPS total electron content in China. *J. Asian Earth Sci.* **2015**, *114*, 393–402. [CrossRef]
161. Zhu, F.; Jiang, Y. Investigation of GIM-TEC disturbances before M $\geq$ 6.0 inland earthquakes during 2003–2017. *Sci. Rep.* **2020**, *10*, 18038. [CrossRef]
162. He, L.M.; Heki, K. Ionospheric anomalies immediately before M(w)7.0–8.0 earthquakes. *J. Geophys. Res. Space Phys.* **2017**, *122*, 8659–8678. [CrossRef]
163. Chakraborty, S.; Sasmal, S.; Basak, T.; Chakrabarti, S.K. Comparative study of charged particle precipitation from Van Allen radiation belts as observed by NOAA satellites during a land earthquake and an ocean earthquake. *Adv. Space Res.* **2019**, *64*, 719–732. [CrossRef]
164. Marchitelli, V.; Harabaglia, P.; Troise, C.; De Natale, G. On the correlation between solar activity and large earthquakes worldwide. *Sci. Rep.* **2020**, *10*, 1–10.
165. Dobrovolsky, I.P.; Zubkov, S.I.; Miachkin, V.I. Estimation of the size of earthquake preparation zones. *Pure Appl. Geophys.* **1979**, *117*, 1025–1044. [CrossRef]
166. Bowman, D.D.; Ouillon, G.; Sammis, C.G.; Sornette, A.; Sornette, D. An observational test of the critical earthquake concept. *J. Geophys. Res. Solid Earth* **1998**, *103*, 24359–24372. [CrossRef]
167. Jiang, W.; Ma, Y.; Zhou, X.; Li, Z.; An, X.; Wang, K. Analysis of ionospheric vertical total electron content before the 1 April 2014 Mw 8.2 Chile earthquake. *J. Seismol.* **2017**, *21*, 1599–1612. [CrossRef]
168. Liu, J.Y.; Chen, Y.I.; Chuo, Y.J.; Tsai, H.F. Variations of ionospheric total electron content during the Chi-Chi Earthquake. *Geophys. Res. Lett.* **2001**, *28*, 1383–1386. [CrossRef]
169. Liu, W.J.; Xu, L. Statistical analysis of ionospheric TEC anomalies before global M(w) $\geq$ 7.0 earthquakes using data of CODE GIM. *J. Seismol.* **2017**, *21*, 759–775. [CrossRef]
170. Karia, S.P.; Pathak, K.N. Change in refractivity of the atmosphere and large variation in TEC associated with some earthquakes, observed from GPS receiver. *Adv. Space Res.* **2011**, *47*, 867–876. [CrossRef]
171. Fitterman, D. Calculations of self-potential anomalies near vertical contacts. *Geophysics* **1979**, *44*, 195–205. [CrossRef]
172. Yoshida, S. Convection current generated prior to rupture in saturated rocks. *J. Geophys. Res.* **2001**, *106*, 2103–2120. [CrossRef]
173. Yoshida, S.; Ogawa, T. Electromagnetic emissions from dry and wet granite associated with acoustic emissions. *J. Geophys. Res. Solid Earth* **2004**, *109*, 204. [CrossRef]
174. Freund, F.; Heraud, J.; Centa, V.; Scoville, J. Mechanism of unipolar electromagnetic pulses emitted from the hypocenters of impending earthquakes. *Eur. Phys. J. Spec. Top* **2021**, *230*, 47–65. [CrossRef]
175. Huang, Q.; Lin, Y. Selectivity of seismic electric signal (SES) of the 2000 Izu earthquake swarm: A 3D FEM numerical simulation model. *Proc. Jpn. Acad. Ser. B* **2010**, *86*, 257–264. [CrossRef]
176. Enomoto, Y. Coupled interaction of earthquake nucleation with deep Earth gases: A possible mechanism for seismo-electromagnetic phenomena. *Geophys. J. Int.* **2012**, *191*, 1210–1214. [CrossRef]
177. Shinbrot, T.; Kim, N.H.; Thyagu, N.N. Electrostatic precursors to granular slip events. *Proc. Natl. Acad. Sci. USA* **2012**, *109*, 10806–10810. [CrossRef]
178. Ren, H.; Chen, X.; Huang, Q. Numerical simulation of coseismic electromagnetic fields associated with seismic waves due to finite faulting in porous media. *Geophys. J. Int.* **2012**, *188*, 925–944. [CrossRef]
179. Ren, H.; Wen, J.; Huang, Q.; Chen, X. Electrokinetic effect combined with surface-charge assumption: A possible generation mechanism of coseismic EM signals. *Geophys. J. Int.* **2015**, *200*, 837–850. [CrossRef]
180. Ren, H.; Huang, Q.; Chen, X. Numerical simulation of seismo-electromagnetic fields associated with a fault in a porous medium. *Geophys. J. Int.* **2016**, *206*, 205–220. [CrossRef]
181. Leeman, J.; Scuderi, M.M.; Marone, C.; Saffer, D.M.; Shinbrot, T. On the origin and evolution of electrical signals during frictional stick slip in sheared granular material. *J. Geophys. Res.* **2014**, *119*, 4253–4268. [CrossRef]

182. Mizutani, H.; Ishido, T.; Yokokura, T.; Ohnishi, S. Electrokinetic phenomena associated with earthquakes. *Geophys. Res. Lett.* **1976**, *3*, 365–368. [CrossRef]
183. Fenoglio, M.A.; Johnston, M.J.S.; Byerlee, J.D. Magnetic and electric fields associated with changes in high pore pressure in fault zones: Application to the Loma Prieta ULF emissions. *J. Geophys. Res. Solid Earth* **1995**, *100*, 12951–12958. [CrossRef]
184. Huang, Q. One possible generation mechanism of co-seismic electric signals. *Proc. Jpn. Acad. Ser. B* **2002**, *78*, 173–178. [CrossRef]
185. Sasai, Y. Piezomagnetic fields produced by dislocation sources. *Surv. Geophys.* **1994**, *15*, 363–382. [CrossRef]
186. Gershenzon, N.I.; Gokhberg, M.B.; Yunga, S.L. On the electromagnetic field of an earthquake focus. *Phys. Earth Planet. Inter.* **1993**, *77*, 13–19. [CrossRef]
187. Draganov, A.B.; Inan, U.S.; Taranenko, Y.N. ULF magnetic signatures at the Earth surface due to ground water flow: A possible precursor to earthquakes. *Geophys. Res. Lett.* **1991**, *18*, 1127–1130. [CrossRef]
188. Surkov, V. ULF electromagnetic perturbations resulting from the fracture and dilatancy in the earthquake preparation zone. *Atmos. Ionos. Electromagn. Phenom. Assoc. Earthq.* **1999**, *1*, 371–382.
189. Molchanov, O.A.; Hayakawa, M. Generation of ULF electromagnetic emissions by microfracturing. *Geophys. Res. Lett.* **1995**, *22*, 3091–3094. [CrossRef]
190. Freund, F. Time-resolved study of charge generation and propagation in igneous rocks. *J. Geophys. Res. Solid Earth* **2000**, *105*, 11001–11019. [CrossRef]
191. Freund, F. Charge generation and propagation in igneous rocks. *J. Geodyn.* **2002**, *33*, 543–570. [CrossRef]
192. Freund, F. Pre-earthquake signals: Underlying physical processes. *J. Asian Earth Sci.* **2011**, *41*, 383–400. [CrossRef]
193. Freund, F.; Takeuchi, A.; Lau, B.W.S. Electric currents streaming out of stressed igneous rocks—A step towards understanding pre-earthquake low frequency EM emissions. *Phys. Chem. Earth* **2006**, *31*, 389–396. [CrossRef]
194. Ren, H.; Huang, Q.; Chen, X. A new numerical technique for simulating the coupled seismic and electromagnetic waves in layered porous media. *Earthq. Sci.* **2010**, *23*, 167–176. [CrossRef]
195. Gao, Y.X.; Harris, J.M.; Wen, J.; Huang, Y.H.; Twardzik, C.; Chen, X.F.; Hu, H.S. Modeling of the coseismic electromagnetic fields observed during the 2004 M-w 6.0 Parkfield earthquake. *Geophys. Res. Lett.* **2016**, *43*, 620–627. [CrossRef]
196. Koike, K.; Yoshinaga, T.; Suetsugu, K.; Kashiwaya, K.; Asaue, H. Controls on radon emission from granite as evidenced by compression testing to failure. *Geophys. J. Int.* **2015**, *203*, 428–436. [CrossRef]
197. Pulinets, S.; Ouzounov, D. Lithosphere-Atmosphere-Ionosphere Coupling (LAIC) model—An unified concept for earthquake precursors validation. *J. Asian Earth Sci.* **2011**, *41*, 371–382. [CrossRef]
198. Harrison, R.G.; Aplin, K.L.; Rycroft, M.J. Atmospheric electricity coupling between earthquake regions and the ionosphere. *J. Atmos. Sol. Terr. Phys.* **2010**, *72*, 376–381. [CrossRef]
199. Harrison, R.; Aplin, K.; Rycroft, M. Brief Communication: Earthquake-cloud coupling through the global atmospheric electric circuit. *Nat. Hazards Earth Syst. Sci. Discuss.* **2013**, *1*, 7271–7283. [CrossRef]
200. Rycroft, M.J.; Nicoll, K.A.; Aplin, K.L.; Giles Harrison, R. Recent advances in global electric circuit coupling between the space environment and the troposphere. *J. Atmos. Sol. Terr. Phys.* **2012**, *90–91*, 198–211. [CrossRef]
201. Kuo, C.L.; Huba, J.D.; Joyce, G.; Lee, L.C. Ionosphere plasma bubbles and density variations induced by pre-earthquake rock currents and associated surface charges. *J. Geophys. Res. Space Phys.* **2011**, *116*, 317. [CrossRef]
202. Kuo, C.L.; Lee, L.C.; Huba, J.D. An improved coupling model for the lithosphere-atmosphere-ionosphere system. *J. Geophys. Res. Space Phys.* **2014**, *119*, 3189–3205. [CrossRef]
203. Shvets, A.V.; Hayakawa, M.; Molchanov, O.A.; Ando, Y. A study of ionospheric response to regional seismic activity by VLF radio sounding. *Phys. Chem. Earth Parts A/B/C* **2004**, *29*, 627–637. [CrossRef]
204. Liperovsky, V.A.; Meister, C.V.; Liperovskaya, E.V.; Vasil'eva, N.E.; Alimov, O. On spread-E-s effects in the ionosphere before earthquakes. *Nat. Hazards Earth Syst. Sci.* **2005**, *5*, 59–62. [CrossRef]
205. Sorokin, V.M.; Chmyrev, V.M.; Yaschenko, A.K. Theoretical model of DC electric field formation in the ionosphere stimulated by seismic activity. *J. Atmos. Sol. Terr. Phys.* **2005**, *67*, 1259–1268. [CrossRef]
206. Zhang, X.M.; Shen, X.H. The development in seismo-ionospheric coupling mechanism. *Prog. Earthq. Sci. Chin. Ed.* **2022**, *52*, 193–202.
207. Liu, J.Y.; Chen, C.H.; Sun, Y.Y.; Chen, C.H.; Tsai, H.F.; Yen, H.Y.; Chum, J.; Lastovicka, J.; Yang, Q.S.; Chen, W.S.; et al. The vertical propagation of disturbances triggered by seismic waves of the 11 March 2011 M9.0 Tohoku earthquake over Taiwan. *Geophys. Res. Lett.* **2016**, *43*, 1759–1765. [CrossRef]
208. Huang, Q.; Han, P.; Hattori, K.; Ren, H. Electromagnetic Signals Associated With Earthquakes. In *Seismoelectric Exploration*; John Wiley & Sons: Hoboken, NJ, USA, 2020; pp. 415–436.
209. Chen, C.-H.; Sun, Y.-Y.; Lin, K.; Zhou, C.; Xu, R.; Qing, H.; Gao, Y.; Chen, T.; Wang, F.; Yu, H.; et al. A new instrumental array in Sichuan, China, to monitor vibrations and perturbations of the lithosphere, atmosphere, and ionosphere. *Surv. Geophys.* **2021**, *42*, 1425–1442. [CrossRef]
210. Han, P.; Hattori, K.; Zhuang, J.; Chen, C.; Liu, J.; Yoshida, S. Evaluation of ULF seismo-magnetic phenomena in Kakioka, Japan by using Molchan's error diagram. *Geophys. J. Int.* **2017**, *208*, 482–490. [CrossRef]
211. Katsumata, K. A long-term seismic quiescence started 23 years before the 2011 off the Pacific coast of Tohoku Earthquake (M = 9.0). *Earth Planets Space* **2011**, *63*, 36. [CrossRef]

212. Nanjo, K.Z.; Hirata, N.; Obara, K.; Kasahara, K. Decade-scale decrease inb value prior to the M9-class 2011 Tohoku and 2004 Sumatra quakes. *Geophys. Res. Lett.* **2012**, *39*, 304. [CrossRef]
213. Sarlis, N.V.; Skordas, E.S.; Varotsos, P.A.; Nagao, T.; Kamogawa, M.; Tanaka, H.; Uyeda, S. Minimum of the order parameter fluctuations of seismicity before major earthquakes in Japan. *Proc. Natl. Acad. Sci. USA* **2013**, *110*, 13734–13738. [CrossRef]
214. Kato, A.; Obara, K.; Igarashi, T.; Tsuruoka, H.; Nakagawa, S.; Hirata, N. Propagation of slow slip leading up to the 2011 Mw 9.0 Tohoku-Oki earthquake. *Science* **2012**, *335*, 705–708. [CrossRef]
215. Chen, C.-H.; Wen, S.; Liu, J.-Y.; Hattori, K.; Han, P.; Hobara, Y.; Wang, C.-H.; Yeh, T.-K.; Yen, H.-Y. Surface displacements in Japan before the 11 March 2011 M9.0 Tohoku-Oki earthquake. *J. Asian Earth Sci.* **2014**, *80*, 165–171. [CrossRef]
216. Ito, Y.; Hino, R.; Kido, M.; Fujimoto, H.; Osada, Y.; Inazu, D.; Ohta, Y.; Iinuma, T.; Ohzono, M.; Miura, S.; et al. Episodic slow slip events in the Japan subduction zone before the 2011 Tohoku-Oki earthquake. *Tectonophysics* **2013**, *600*, 14–26. [CrossRef]
217. Hattori, K.; Han, P. Investigation on preparation process of the 2011 off the Pacific Coast of Tohoku Earthquake (Mw 9.0) by GPS data. In Proceedings of the 2014 American Geophysics Union Fall Meeting, San Francisco, CA, USA, 15–19 December 2014.
218. Orihara, Y.; Kamogawa, M.; Nagao, T. Preseismic Changes of the level and temperature of confined groundwater related to the 2011 Tohoku earthquake. *Sci. Rep.* **2014**, *4*, 6907. [CrossRef] [PubMed]
219. Han, P.; Hattori, K.; Huang, Q.H.; Hirooka, S.; Yoshino, C. Spatiotemporal characteristics of the geomagnetic diurnal variation anomalies prior to the 2011 Tohoku earthquake (Mw 9.0) and the possible coupling of multiple pre-earthquake phenomena. *J. Asian Earth Sci.* **2016**, *129*, 13–21. [CrossRef]
220. Han, P.; Hattori, K.; Xu, G.J.; Ashida, R.; Chen, C.H.; Febriani, F.; Yamaguchi, H. Further investigations of geomagnetic diurnal variations associated with the 2011 off the Pacific coast of Tohoku earthquake (Mw 9.0). *J. Asian Earth Sci.* **2015**, *114*, 321–326. [CrossRef]
221. Ogata, Y.; Akaike, H.; Katsura, K. The application of linear intensity models to the investigation of causal relations between a point process and another stochastic process. *Ann. Inst. Stat. Math.* **1982**, *34*, 373–387. [CrossRef]
222. Zhuang, J.; Matsu'ura, M.; Han, P. Critical zone of the branching crack model for earthquakes: Inherent randomness, earthquake predictability, and precursor modelling. *Eur. Phys. J. Spec. Top.* **2021**, *230*, 409–424. [CrossRef]

remote sensing

MDPI

Article

Shallow Crustal Structure of S-Wave Velocities in the Coastal Area of South China Constrained by Receiver Function Amplitudes

Xin Zhang [1], Yinping Qian [1], Xuzhang Shen [2,*], He Huang [2] and Haibin Chai [3]

1 Guangdong Earthquake Agency, Guangzhou 510070, China; zxdqwl@163.com (X.Z.); yinpingqian@163.com (Y.Q.)
2 School of Earth Science and Engineering, Sun Yat-Sen University, Zhuhai 519000, China; huangh365@mail.sysu.edu.cn
3 Fourth Habour Engineering Investigation and Design Institute of the Ministry of Transport, China Communications Construction Company Limited, Guangzhou 510000, China; chaihb@fhdigz.com
* Correspondence: shenxzh5@mail.sysu.edu.cn

Abstract: As a traditional method, passive seismic exploration is used to construct the body-wave velocity structure of the upper crust, but it is cost-ineffective and depth-limited when applied to large areas. In this study, we use another more economical method to determine the S-wave velocity (SWV) of the upper crust based on the principle that the amplitude of the direct P-wave on the teleseismic receiver function is sensitive to the upper crust. Using the amplitudes of the massive receiver functions from permanent broadband seismic stations, the SWV structure of the upper crust is obtained in the coastal area of South China (CASC). A pattern of high to low SWVs is exhibited across the study area, with SWVs varying about 2.5–3.7 km/s from west to east. In the profile parallel to the coastline, lateral variations in the SWV correspond to the fault zone, indicating that the cutting depth of most coastal faults is approximately 10 km. Referring to previous studies, we deduce that the low SWV in most sub-areas can be interpreted as the joint effect of the sedimentary layer of the alluvial plain and the accumulation of underground heat flows, in addition to multistage fracturing tectonism. Moreover, the gradual change in the SWV in each profile from the surface to approximately 10 km is correlated with multiple invasions and the coverage of volcanic rocks, to a certain extent.

Keywords: S-wave velocity (SWV); multistage intrusive rocks; coastal area of South China; amplitude information of the receiver function

Citation: Zhang, X.; Qian, Y.; Shen, X.; Huang, H.; Chai, H. Shallow Crustal Structure of S-Wave Velocities in the Coastal Area of South China Constrained by Receiver Function Amplitudes. *Remote Sens.* **2022**, *14*, 2760. https://doi.org/10.3390/rs14122760

Academic Editors: Kaiguang Zhu, Yunbin Yuan and Dedalo Marchetti

Received: 27 March 2022
Accepted: 5 June 2022
Published: 8 June 2022

1. Introduction

The coastal area of South China (CASC) is near the northern margin of the South China Sea and typically has small earthquakes. Due to complex tectonic activity between the Cathaysia block and the Pacific plate, this area is important for the eastern margin structural framework [1,2]. Faults in the CASC can be divided into three groups, namely, NNE-NE-trending, EW-ENE-trending and NW-WNW-trending, in which the NNE-NE-trending faults have the most obvious tectonic features [3,4]. Currently, some faults exhibit strong activity and remain important earthquake locations. Locations with seismicity depend on geological tectonic activity processes, crustal structures and material properties. The bottom of the upper crust in the CASC is remarkably undulating, lies at depths of approximately 10 km under the Yangtze block and 6–7 km under the Cathaysia block [5]. Zhao et al. [6] and Ye et al. [7] suggested that the average lithology of the upper crust in both blocks is granite-granodiorite and biotite gneiss. The granite-granodiorite layer beneath the Cathaysia block is much thicker. Abundant geophysical data [8] have revealed that the Yangtze and Cathaysia blocks have crustal structures that are characterized by an average composition of granite gneiss for the upper crust, with the presence of mica quartz schist, felsic granulite, paragranulite, and granite-granodiorite beneath the Yangtze block. To explore the upper crust, ambient noise imaging and artificial source seismic methods are the main strategies for inspecting large-scale material compositions and Cathaysia block-to-sea crust transitions [9–13]. However, the S-wave velocity (SWV) of the upper crust in the CASC, including the constraint of the receiver function method, still needs to be explored in detail.

As one of the most effective seismologic methods, the receiver function is used to study the crustal thicknesses and velocity structures under seismic stations. In the traditional process, the arrival time of the seismic phase is easy to measure in a relatively stable manner. Various studies have obtained the velocity structure based on the arrival time [14–17], while the H-K superposition is a classic strategy for utilizing the temporal information of the receiver function. Moreover, the amplitude of the receiver function can also be adapted to obtain the SWV of the upper crust, as well as the velocity and density contrasts of the Moho discontinuity. Referring to previous studies on receiver function amplitudes [18–20], Qian et al. [21] first proposed this idea of data processing, and developed a method to inverse the SWV structure of the upper crust with the amplitude of the receiver function [22,23].

The study area exhibits episodic tectonism between the Pacific and Eurasian plates [24,25], where the outcrops of late Mesozoic igneous rocks extend beyond the boundary of the CASC block. The CASC was the transitional zone between the Cathaysian block and oceanic crust, and was influenced by the subduction of the Pacific plate. The back-arc faulting and seafloor expansion of the CASC were associated with multiple tectonic movements since the Cretaceous [26,27]. A large number of Mesozoic intrusive rocks developed along these NE-trending faults. The NW-trending faults formed later than the Cretaceous intrusive events, resulting in strong neotectonics [26]. However, the magmatic suite that intruded in the Mesozoic had a mantle source composed of late Archean and Mesoproterozoic crustal components of xenoliths. Moreover, as a result of collisions and regional metamorphism, the volcanic suite and sediments were altered, forming abundant metamorphic rocks in the Lianhuashan fault (LHSF) zone. Moreover, heat flows and strong metamorphic activity are manifested by dozens of faults. Many strong earthquakes, such as the largest reservoir-induced earthquake, recorded as Ms 6.1 in the Xinfengjiang reservoir (XFJR), and the largest oceanic earthquake, recorded as Ms 7.2 near Nan'ao Island, have caused great damage to the CASC. In order to elucidate the heat flows and seismogenic structure, we use the amplitude of the receiver function to obtain the SWV structure of the upper crust. By comparing our findings with previous studies, we attempt to establish a model for the evolution of the structural environment.

2. Data and Methods

2.1. Geological Setting

The distribution of seismic stations covers the whole CASC, as shown in Figure 1a. The faults in Figure 1c mainly have NE-SW strikes, of which the typical faults are the LHSF, coastal faults, the Heyuan fault, and the Xijiang fault (XIJF). The XFJR has the characteristics of frequent occurrences of small earthquakes and dense stations, and its seismogenic structures are evaluated later.

2.2. Method of the Direct Wave Amplitude of the Receiver Function

In this study, seismic waveform data from 68 broadband stations in the CASC during 2007–2011 are used to calculate the receiver function. The seismic data recorded on three-component digital broadband stations. These stations were equipped with 24-digit broadband/very broadband seismometers (60 s or 120 s) with a sampling rate of 100 per second. Teleseismic events from the waveform data are intercepted at Ms > 5.5, with epicentral distances ranging from 30° to 90°. The distributions of seismic events used in the receiver function calculation are shown in Figure 1b. Each teleseismic event in the

30°–90° range can be assumed to be an incident wave with a steep angle, and the wave likely propagated to the station as a plane wave.

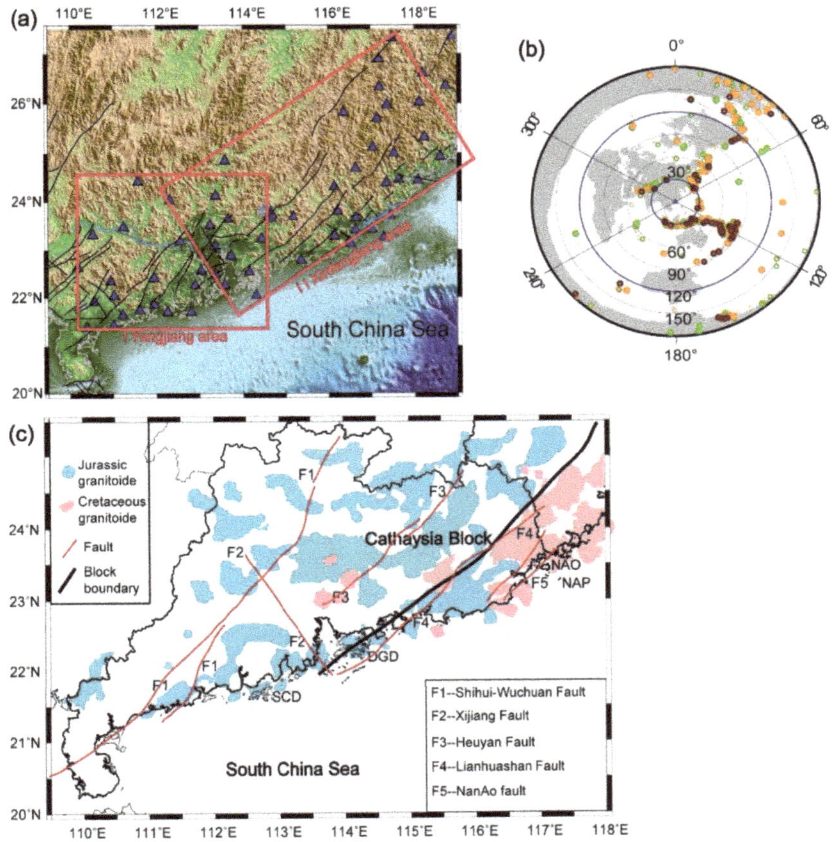

Figure 1. Spatial distributions of seismic stations (**a**) and seismic events (**b**) in the study area. The seismic events that are used in calculating the receiver function are shown in (**b**). The distribution of Mesozoic granites in (**c**) was retrieved from Li [28], while the late Mesozoic volcanic arcs were retrieved from Zhou et al. [3] and Li et al. [29]. LHSF represents the Lianghuashan fault; SWCF represents the Sihui-Wuchuan fault. XIJF and HEYF represent the Xijiang fault and Heyuan fault, respectively.

As the Pp arrives at the seismic station, it forms a direct P-wave (Pp) or is converted into an S-wave (Ps) at the interface. It also forms multiple waves when reflected at the discontinuities below the surface. In such multiple reflections, not only is conversion between the P-wave and S-wave adopted, but PpPs and PpSs + PsPs combinations are also generated.

When the P-wave spreads through the 1-D velocity model with horizontally layered structures, it produces a ground displacement response [14]. The displacement response can be expressed in the time domain as:

$$\mathrm{Z(t)} = \sum_{k=0}^{n} z_k s(t - t_k) \tag{1}$$

$$R(t) = \sum_{k=0}^{n} r_k s(t - t_k) \quad (2)$$

where Z(t) and R(t) represent displacements of the vertical and radial components of motion, respectively; $s(t)$ represents the time function of the effective source; t_k represents the arrival time of the seismic phase, where $k = 0$ indicates the direct P-wave; and z_k and r_k represent the amplitudes of the vertical and radial components of the seismic phase, respectively, and represent the values of the reflection coefficients obtained by multiplying the transmission coefficients. Through deconvolution of the horizontal component by the vertical component of the surface displacement, the medium structural response in the frequency domain is expressed as the following:

$$H(\omega) = \frac{S(\omega)R(\omega)}{S(\omega)Z(\omega)} = \frac{R(\omega)}{Z(\omega)} \quad (3)$$

where ω represents the ray frequency, $S(\omega)$ represents the source spectrum, and $H(\omega)$ represents the medium structural response and radial receiver function. Through Fourier transformation of the receiver function, the following are obtained:

$$R(\omega) = r_0 \sum_{k=0}^{n} \hat{r}_k e^{-iwt_k} \quad (4)$$

$$Z(\omega) = z_0 \sum_{k=0}^{n} \hat{z}_k e^{-iwt_k} \quad (5)$$

where $\hat{z}_k$ represents the k-phase amplitude after processing the direct P-wave pulse amplitude in relation to the vertical component; that is, $\hat{z}_k = z_k/z_0$ and $\hat{r}_k = r_k/r_0$.

The receiver function indicates seismic properties; therefore, the discontinuity structure can be deduced by calculating the radial receiver function. Given the simple model of the horizontal single layer, there are only direct P-waves, P-S converted waves and resonant multiple waves; thus, Equations (4) and (5) hold and can be simplified as follows:

$$R(\omega) = r_0 \left[1 + \hat{r}_p e^{-iwt_p} + \hat{r}_s e^{-iwt_s}\right] \quad (6)$$

$$Z(\omega) = z_0 \left[1 + \hat{z}_p e^{-iwt_p} + \hat{z}_s e^{-iwt_s}\right] \quad (7)$$

Substituting (6) and (7) into (3) and ignoring the higher-order terms, the following is obtained:

$$H(\omega) = \frac{r_0}{z_0} \frac{1 + \hat{r}_s e^{-iwt_s} + 2\hat{z}_p coswt_p}{1 + 2\hat{z}_p coswt_p} = \frac{r_0}{z_0}\left[1 + \hat{r}_s e^{-iwt_s}\right] \quad (8)$$

Equation (8) shows the absolute amplitude of the direct P-wave on the free surface ($t = 0$). The amplitude ratio, r_0/z_0, denotes the velocity structure near the surface. It is very important to add the amplitude ratio information into the receiver function. Assume a constant velocity near the surface. We can combine the equations for the free-surface displacement due to an incident P wave [30]:

$$\frac{r_0}{z_0} = \frac{2p\eta_{V_{s0}}}{Vs_0^{-2} - 2p^2} \quad (9)$$

where $\eta_{V_{s0}} = \left(1/(Vs_0)^2 - p^2\right)^{1/2}$ represents the vertical slowness of the S-wave and p represents the horizontal slowness (i.e., ray parameter), while Vs_0 represents the SWV near the surface. Equation (9) shows the relationship between the amplitude and average velocity at low frequencies. For higher-frequency bands, amplitudes of the seismic phase must be calculated using synthetic seismograms; however, the direct P-wave amplitude ratio is dependent on the near-surface velocity [31].

The direct P-wave amplitude is correlated only with the SWV (Vs_0) and ray parameter (p) near the surface. The ray parameter can be calculated from the epicentral distance and focal depth [32]. When the epicentral distance is provided, the direct P-wave amplitude is mainly affected by the SWV near the surface. Based on the amplitude value of the P-wave receiver function, the SWV of the upper crust can be calculated. Since the receiver function is associated only with SWVs under seismic stations, the receiver functions of different epicentral distances are similar for stations with uniform transverse structures of the crust. Therefore, the amplitudes of the direct P-waves at different epicentral distances at the same station can be applied in order to constrain the SWV of the upper crust.

2.3. Method of the SWV

In order to verify the feasibility of the actual data processing process, we compare the theoretical seismogram [33] with the actual data processing results. First, we establish a single-layer crustal velocity model (Figure 2a). The SWV of the crust is 3.75 km/s, while the P-S velocity ratio is 1.732. The velocity structure under the crust refers to the IASP91 model [34]. Second, we select an epicentral distance of 70° and calculate the theoretical seismogram by the reflection transmissivity method [33]. Then, we select the Gaussian filter factor (alpha = 1.5) and calculate the theoretical receiver function using the time domain iterative deconvolution method [35]. Finally, we measure the Pp phase amplitude in the theoretical receiver function. The initial movement of the P-wave is taken as the starting time. The maximum value of the wave peak near the starting time (±2 s) is the P-wave amplitude.

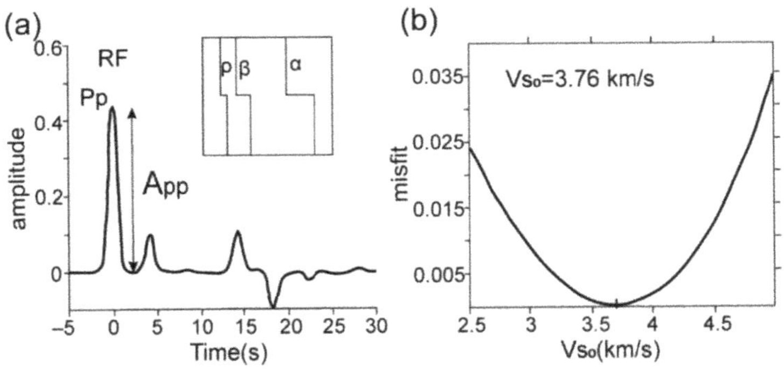

Figure 2. Receiver function of a single epicentral distance (**a**) and the subsurface SWV (**b**).

We use the grid search method and the Pp-wave amplitude to determine the SWV. The values are 2.5–3.5 km/s, while the search step is 0.01 km/s. For each search, the theoretical value of the absolute Pp-wave amplitude can be calculated by using Equation (9). In the search process, the theoretical value of the Pp-wave amplitude with different epicentral distances can be calculated by the Vs_0 in each step. The difference between the observed and theoretical values can be expressed as the objective function, $\text{misfit}(Vs_0)$:

$$\text{misfit}(Vs_0) = \sqrt{\sum_i \left(A_i^{obs} - A_i^{syn}\right)^2} \tag{10}$$

where A^{obs} represents the observed value of the Pp amplitude, A^{syn} represents the theoretical value of the Pp amplitude, and Σ is the grid scanning superposition of different ray parameters. The optimal value can be obtained by calculating the misfit (Vs_0) according to Equation (10) and taking the horizontal coordinate corresponding to the minimum value. The Pp-wave amplitude in the R-direction receiver function can be measured in order to obtain the Vs_0 (Figure 2a).

A_{Pp} represents the direct wave (Pp) amplitude of the receiver function, while the "+" symbol represents the optimal Vs_0 to be searched in the abscissa.

2.4. Depth and Alpha

We use a Gaussian filter ($\exp(-\frac{\omega^2}{4\alpha^2})$) to suppress the high-frequency noise when extracting the receiver function. The effective cutoff frequency is defined as the frequency while the filter amplitude is reduced to its $\frac{1}{e}$.

Without depth information while calculating Vs_0 based on Equation (9), we cannot directly determine its specific depth. In order to test the depth distribution range of SWV, we tested four models in Figure 3. In Figure 3a,c, the velocity above the 35 km discontinuity increases in gradient, in which the difference between them is that the increasing trend is different; the growth is faster and the value on the surface is greater in Figure 3a. In Figure 3b,d, the depths of the two discontinuities are 20 km and 35 km, respectively, and the velocity remains unchanged below 35 km and increases linearly above 20 km; the difference between them is that the increasing trend is different; the growth is faster and the value on the surface is greater in Figure 3b. Figure 3e shows the SWV corresponding to multiple Gaussian filter factors obtained by our method. When the Gaussian filter factor becomes larger (meaning shorter wavelength), the corresponding S-wave velocity becomes smaller, and the depth corresponding to the S-wave velocity gradually becomes shallower. When comparing the calculated SWV with the average SWV in these models, the calculated value is close to the average value of a certain depth of the model, reflecting the depth that corresponds to a filter factor (as shown in Figure 3e). With this processing approach, we select different filter factors in order to calculate the receiving function, and then obtain the conversion of SWV in depth.

Gaussian filtering is the classical smoothing method in the deconvolution process of receiver functions. This study's method essentially uses the dispersion of receiver function amplitude. The relationship between alpha and the depth of SWV has no quantitative result in mathematics, so we finally give the most sensitive depth of SWV based on reasonable numerical experiments (Figure 3). When alpha is smaller, it indicates the average value of SWV from the surface to the deeper crust; when alpha is larger, it indicates the average value of SWV from the surface to the shallower crust (Figure 3). Wang et al. [22] showed the model in which different alpha factors correspond to the SWV at a certain depth in the upper crust by calculating the amplitude of the receiver function. This method involves the following two steps: first, the high-frequency approximate amplitude formula of direct P-waves in receiver functions of individual stations is used to fit the observed amplitude distribution against the ray parameters at different frequencies; and second, the S-wave velocity depth profile beneath each station is constrained according to an empirical correlation between frequency and depth. In this study, we referred to the results reported by Wang et al. [22,23].

There are divergences in receiver function amplitudes corresponding to different alpha factors in Figure 3. In order to test the influence of different Gaussian factors (alpha), we choose the values of 1.0, 1.5, 2.5, 3 and 5. Based on the model shown in Figure 3a–d, we use two groups of receiver functions to detect the influence of different filter frequency windows on the receiver function amplitude. Each receiver function for each epicentral distance can yield a Vs_0; thus the optimal Vs_0, which is determined by all receiver functions, is used. The least squares fit error (ε) is calculated by the following:

$$\varepsilon = S_y \sqrt{\frac{1}{N} \frac{\left(V_{s0}(i) - \overline{V}_{s0}\right)^2}{\sum_{i=1}^{i=N} \left(V_{s0}(i) - \overline{V}_{s0}\right)^2}} \tag{11}$$

where N represents the number of waveforms with different epicentral distances at each station; $\overline{V}_{s0}$ represents the average value of all $V_{s0}(i)$ values; $S_y = \sqrt{U/(N-2)}$ represents the residual mean square deviation; U represents the sum of the regression squares of different V_{s0} values; and the error percentage is $\delta = \epsilon/\overline{V}_{s0} \cdot 100\%$.

In general, a larger alpha indicates a shallower SWV (Figure 3e). For example, alpha = 1.5 shows that the SWV is at a depth of approximately 5 km; alpha = 6 indicates that at a depth of 1 km, the SWV is near the surface; and alpha = 1 shows that the SWV is at a depth of 8 km in the upper crust. Although four velocity-density models are used in the test process, the trend of alpha to the Vs_0 depth is similar. When alpha < 2, the results tend to be consistent, which means that the velocity structure of the S-wave is more reliable below 4 km. According to the numerical test results, different *alphas* indicate S-wave velocities at different depths; therefore, the comprehensive results can show the SWV structure of the upper crust.

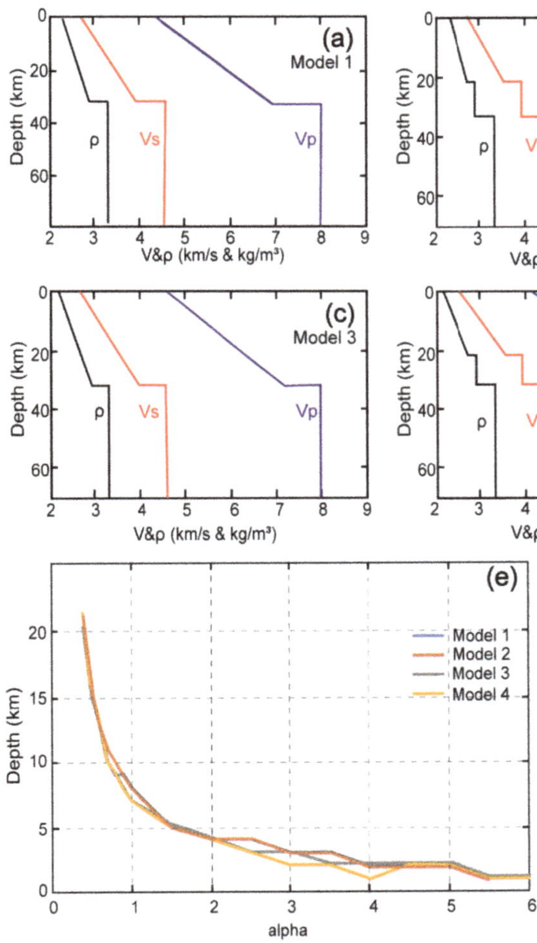

Figure 3. Gaussian filtering factors (alpha) corresponding to SWVs at different depths. (**a**–**d**) Crustal structure model with different velocity-density values; (**e**) test results of the models.

2.5. Data Processing Procedures

Our data processing procedures consist of the following 5 steps:

(i) Events with epicentral distances of 30° to 90° and M $\geq$ 5.5 are selected from the continuous waveform data. Waveforms with high signal-to-noise ratios and clear initial motions are extracted and preprocessed for filtering, deburring, mean value removal, and linear trend removal, among others.

(ii) Based on the earthquake azimuth, the waveform from the N-E-Z component is rotated to the R-T-Z component. Then, the data from 20 s before to 150 s after the initial P-wave movement are intercepted. Gaussian filter factors are selected (such as alpha = 1.0, 1.5, 2.5, 3, 5, and 7) in order to filter the signal of all stations. The receiver function is calculated by the time domain deconvolution method and filtering process.

(iii) Using the vertical component of the receiver function to deconvolute itself, its maximum amplitude value is obtained to unify the radial receiver function, in order to obtain the absolute amplitude [15].

(iv) Direct wave amplitudes of receiver functions at different stations in different directions are measured as observation values (A^{obs}). For each Vs_0, the theoretical value (A^{syn}) of the direct wave amplitude is calculated using Equation (1). The smallest value of the misfit (Vs_0) corresponds to the optimal solution of the Vs_0. Finally, the SWV near the surface of the area is obtained.

(v) In order to improve the signal-to-noise ratio, receiver functions of different epicentral distances are evenly distributed in 60° epicentral distance units, while receiver functions in each unit are superposed into every 1° grid size. Figure 4a shows the superposition results of multiple receiver functions at different epicentral distances obtained by the DNB station, with a high signal-to-noise ratio and clear seismic phases. Figure 4c shows the SWV $Vs_0 = 3.21 \pm 0.22$ km/s of the DNB station, with a percentage error of $\delta = 6.8\%$. Figure 4b,d show the superposition results of multiple receiver functions at different epicentral distances from the YGX station, with $Vs_0 = 2.78 \pm 0.21$ km/s and $\delta = 7.5\%$.

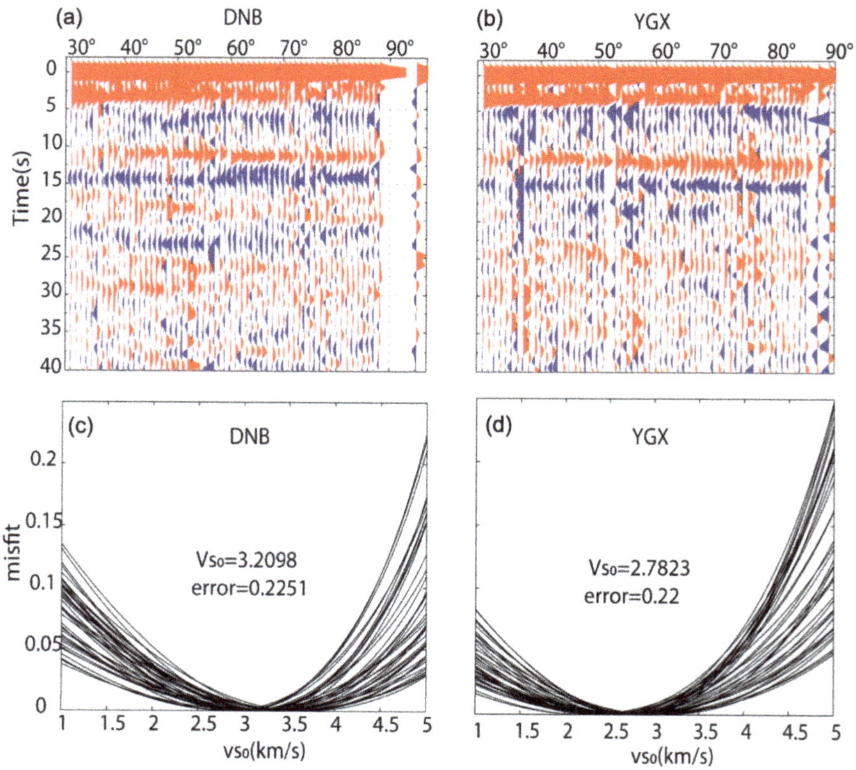

Figure 4. Receiver functions of the DNB and YGX stations (**a**,**b**) as well as their subsurface SWVs (**c**,**d**). Receiver functions of different epicentral distances to the DNB and YGX stations (alpha = 1.5). The thin line indicates the Vs_0 search process, while the "+" symbol represents the optimal Vs_0 to be searched in the abscissa.

3. Results

3.1. Spatial Distribution of the SWV

According to the epicentral distance, the receiver functions of each station are stacked in each 1° range; therefore, the SWV is the mean value of the area adjacent to each station. In Figure 5, the SWVs of the upper crust in the CASC are 2.6–3.6 km/s. From west to east, the upper crust of the CASC can be divided into four regions from high- to low-SWV trends, corresponding to the distribution of intrusive rocks in the four periods. In the northeastern part of the study area, the highest value is 3.6 km/s, which is much higher than those in the central areas (2.5–2.8 km/s). The northern part (ZHJ-CTTZ station) is also part of the low-SWV region, which is less than 2.8 km/s. The SWVs of the western area are generally 2.8–3.5 km/s, and are lower in the south and higher in the north, which may be correlated with widespread underground thermal activities and intrusive rocks.

The SWV of the area containing faults extending in the NW-SE direction should be higher than that of other areas, with a value greater than 3.2 km/s. These faults thicken in Figure 6, and are located in the adjacent areas of the DNB station, YND station, and YSCG station. Speculating that the fault zone in this direction may have formed high SWVs and was not damaged by intrusive rocks is reasonable. This phenomenon is caused by the NW-trending faults that formed later than the Cretaceous intrusive events, and induced strong neotectonics [26].

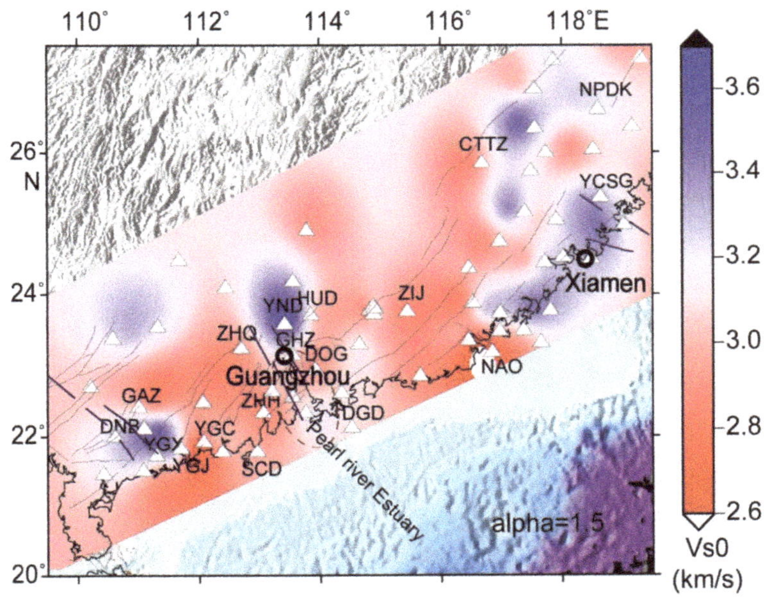

Figure 5. SWVs after the interpolation of each station (alpha = 1.5, H ≅ 5 km) around the study area. The white triangle represents the seismic station. The tight line represents the fault, and the thick line represents the fault extending in the NW-SE direction.

3.2. SWV Distribution with Depth

The SWV exhibits an increasing trend with depth in most areas from 1 km to 8 km (in Figures 6 and 7), corresponding to the loose geological sedimentary layer, loose locally weathered rock layer, and dense bedrock layer. In order to analyze the longitudinal SWV structure, Figure 6 shows longitudinal slices of the SWV structure in the XFJR and its adjacent areas. The western XFJR is a high-velocity area, whose range extends to 8 km underground, whereas the eastern XFJR is a low-velocity area. The phenomenon that the SWV in the center (~5 km) is lower than that in the shallow zone (>1 km) and in the deeper zone (~11 km) indicates that heat flows and strong rheological effects exist in the deep zone. However, a high- to low-velocity transition area appears in the eastern XFJR, in which the surface velocity (~1 km) is significantly lower than that in the deep zone (~8 km). It may be affected by the impoundment of the reservoir, resulting in the shallow layer being more broken and filled by the fluid.

Figure 7 shows the longitudinal slice of the SWV structure in Yangjiang and its adjacent areas. In the horizontal direction, the western DNB station is an obvious high-velocity area, whereas the southern area is a relatively low-velocity area. The SWV in the coastal area is higher than that in the inland area and is relatively stable in the longitudinal direction. The lateral variation in the SWV in the deep layer is smaller, while the lateral variation in the shallow layer is more complex than that in the deep zone.

As shown in Figure 8, profile AA′ is the ZHJ-YCSG station, and profile BB′ is the YLS-NPDK station, which represents the penetration depth of the fault zone. In profile AA′ (Figure 8b), the SWV under the HEY station has an obvious lateral difference, while the relatively high-velocity body with an overall eastward dip extends from the lower crust to the surface. The same trend appears in profile BB′ (Figure 8a), except that the lower SWV layer becomes shallower (from 10 to 7 km).

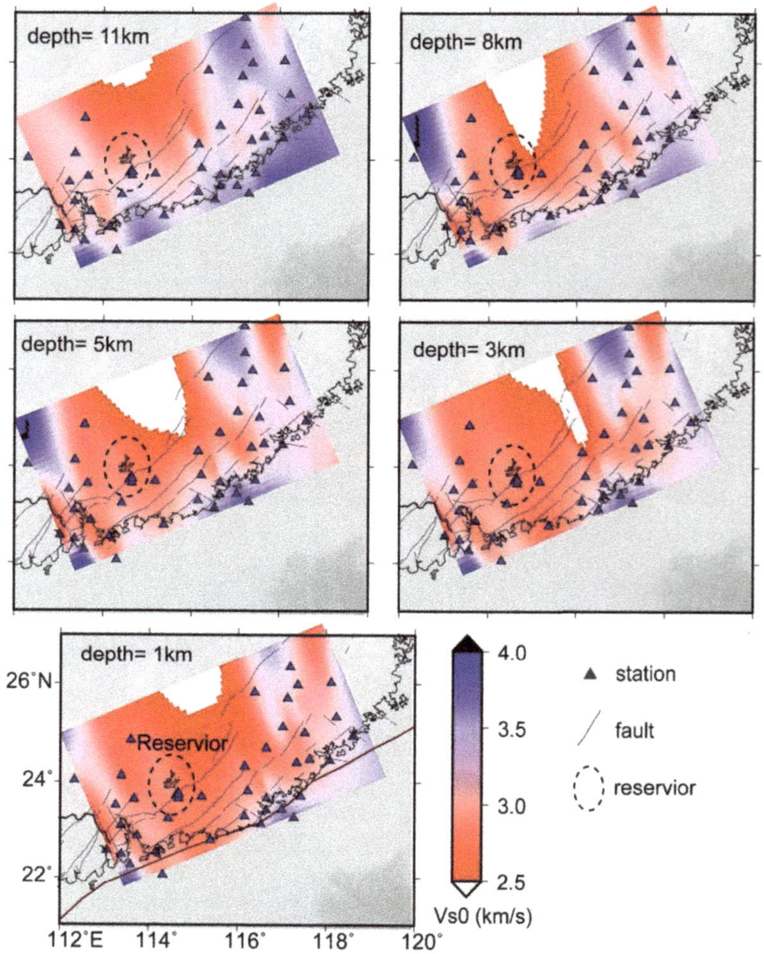

Figure 6. SWV tomographic structure in the XFJR and its adjacent area. The depth range of 11 to 1 km corresponds to alpha values of 0.5 to 5.

Generally, the fault zone contains a transition zone between high and low SWVs, and the penetration depths of most faults reach the lower crust, such as that of the SHWF (Figure 8b). However, the penetration depth of the LHSF is controversial. Geological studies [28,29] show that the Upper Jurassic group, which is developed in the LHSF zone, is a continental intermediate–acidic volcanic rock modified by dynamic metamorphism, with thicknesses of 1284–6419 m, which is equivalent to the depth of the shallow low- to high-velocity body transformation zone in Profile AA′. In addition, the low-velocity area (over 5 km) on the east side of the LHSF may be the area in which heat flows are concentrated, which softens the formation as a result of concentrated thermal subsidence. In addition, the locations of XIJF and HEYF are clear transformation zones of low to high velocities in profile BB′. The depths of the two faults are less than 5 km, and the high-velocity layer above 3.4 km/s does not appear below 9 km.

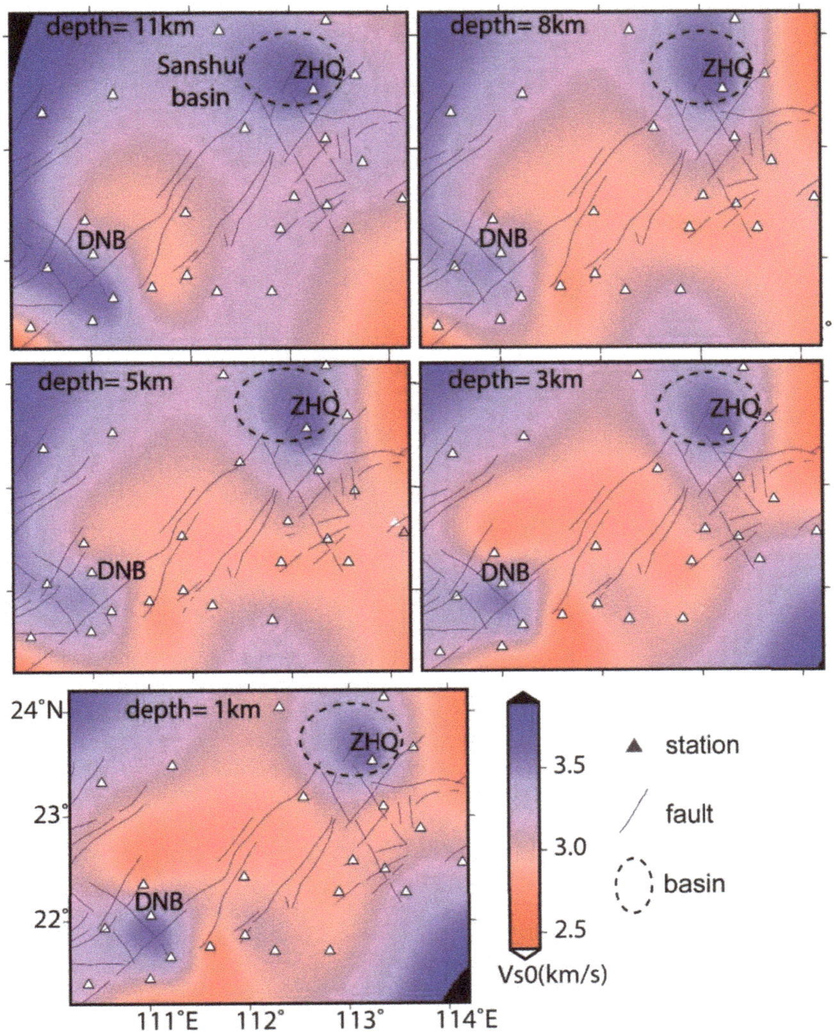

Figure 7. SWV tomographic structure in Yangjiang and its adjacent area.

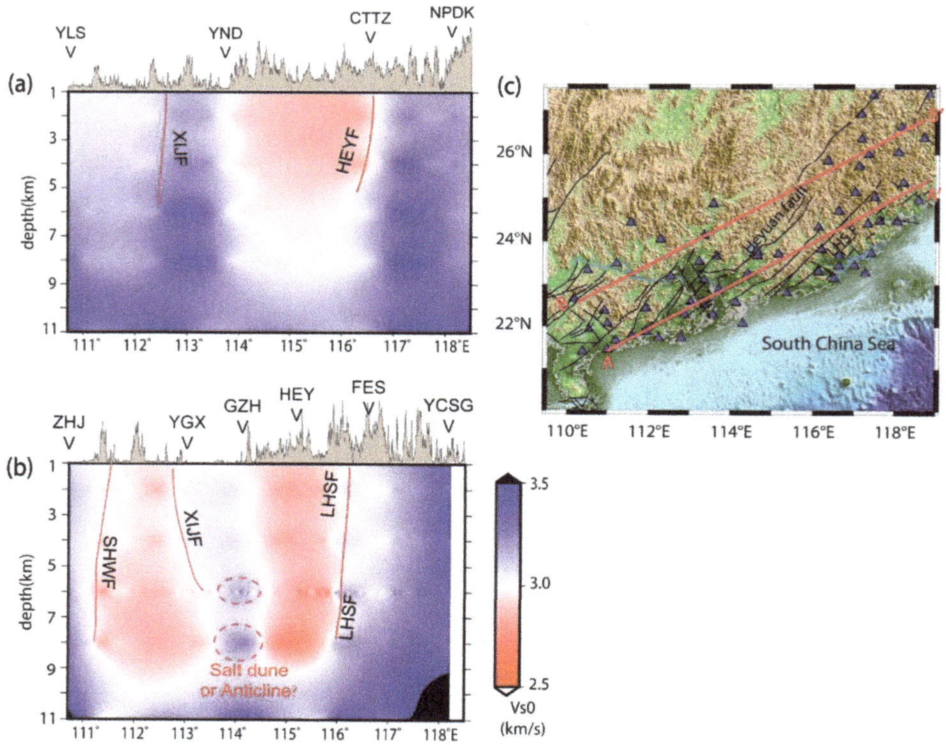

Figure 8. SWV structure profile parallel to the coastline. (**a**) SWV structure of the YLS-NPDK profile (BB'); (**b**) SWV structure of the ZHJ-YCSG profile (AA'); (**c**) locations of the two profiles. LHSF represents the Lianhuashan fault; SWCF represents the Sihui-Wuchuan fault. XIJF and HEYF represent the Xijiang fault and Heyuan fault, respectively.

4. Discussion

Compared with previous findings, some overlaps were present between our study area and those of Zhang et al. [36], Shen et al. [37] and Huang et al. [38]. According to the Rayleigh wave sensitivity depth curve [36], the velocity group within the 2–8 s period reflects the SWV structure at 10 km or deeper. In particular, the velocity group is near the 5 s period, which is most sensitive to a depth of 5 km. Figure 9a shows the mean SWV value with alpha = 1.5 in the central area, which is close to that of the Rayleigh wave velocity group with a period of 5 s. The similarity of the two results is attributed to the switch between high and low velocities, indicating that the SWV structure in the upper crust is relatively complex in the transverse direction. In the eastern part of the study area, Zhang et al. [36] reported that the velocity group in the 5 s period was approximately 2.9 km/s, and Huang et al. [38] reported that the velocity group in the 5 s period was approximately 2.8–3.2 km/s (Figure 9a). In this study, the SWVs are approximately 2.9–3.1 km/s, which are comparable with previous results and closer to the results of Huang et al. [38].

A large cluster of exposed intrusive rocks is located in the eastern part of the study area, where the sedimentary layer is very thin in the entire area. However, it is difficult to establish the existence of a thinner sedimentary layer and a low SWV via ambient noise imaging or receiver functions. The most typical example is the NAP station, which is located on a volcanic island over magmatic intrusive rocks; however, its SWV above 8 km is less than 3.0 km/s, which does not correspond with the sedimentary layer thickness. The most reasonable explanation for the low SWV is that an area with seismicity where earthquakes have high activity levels in the middle and lower crust may cause the lithosphere to fracture.

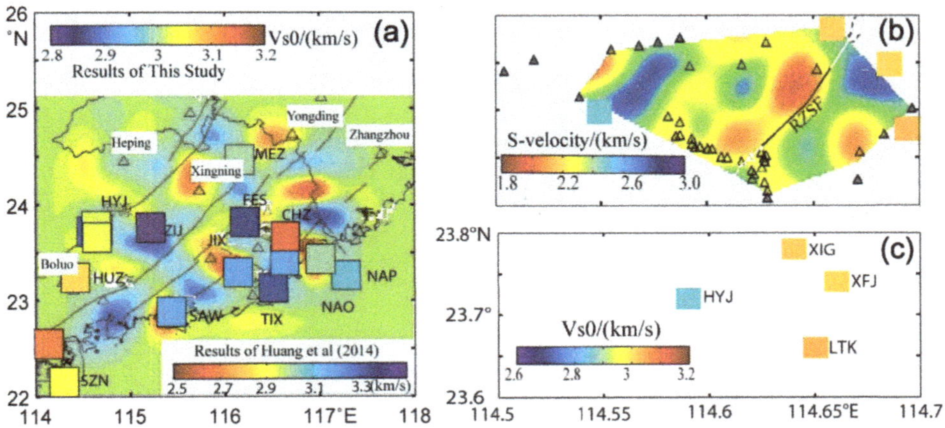

Figure 9. Comparisons of SWVs and Rayleigh wave velocity groups of ambient noise tomography in a similar area. The two colored bars in (**a**) are consistent. (**a**) The bottom map is the result of ambient noise tomography by Huang et al. [38] in eastern Guangdong Province; (**b**) the bottom map is a result of ambient noise tomography by He et al. [13] in XFJR; (**c**) finding of Vs_0 in this study in the XFJR.

The SWV is a useful parameter for buildings design [39,40], especially in the XFJR area which may experience large earthquakes. Wang et al. [41] reported that SWVs greater than 1 km are less than 3.0 km/s based on ambient noise imaging in a super dense array. In this paper, alpha = 7 shows an SWV near 1 km. In Figure 8b, the SWV (alpha = 7) of the XFJ station is close to 3.0 km/s, which is greater than the noise imaging result; that of the HYJ station is 2.85 km/s, which is close to the noise imaging result; and that of the LTK station is 2.8 km/s, which is close to the noise imaging result on the right side of the HEYF. These results indicate that the SWV near the XFJR surface is uniform in a small zone. However, the SWV in this study is larger than that of the noise imaging of a super dense array. As shown in Figure 8b, the results of noise imaging are 1.6–3.0 km/s, while our findings are approximately 2.9–3.1 km/s (Figure 9c). In addition, since the single station receiver function in this study is the average value of the adjacent area, differences in the S-wave velocities of several stations in the XFJR are very small, but the value is larger than the result obtained by noise imaging. Moreover, the accuracy of the SWV near the shallow surface (~1 km) given by the amplitude information of the receiver function is relatively low and thus is not as good as the noise imaging of the super dense array.

Two main explanations can be given for the differences in S-wave velocities near the surface of the South China Sea. The first is the inverse correlation with the sedimentary layer thickness, resulting in a thicker sedimentary layer indicating a smaller SWV [21,40]. The second is the distribution of the heat flows in the upper crust. High heat flows soften the regional strata, resulting in a low SWV [9]. Therefore, the SWV is higher in the region without heat flows. These two factors may account for the most obvious differences in the SWV structure. However, at a few stations, the results are not consistent with the two explanations. Therefore, other possible reasons, such as the impact of earthquakes in some fault zones, as well as the effects of multiple intrusive rocks in coastal areas, should be considered.

In the Yangjiang area and the Sanshui Basin near the ZHQ station, the thickness of the sedimentary layer indicates a low-SWV area. Under the Sanshui Basin near the ZHQ station, the sedimentary layer is relatively thick, while the SWV is less than 3.0 km/s, and the low-SWV body extends to less than 5 km. In other areas, the low-SWV area is deeper, which may not be due to the thick sedimentary layer and should be analyzed further. On the other hand, the low-SWV area is not always a thick sedimentary layer in the study area. For example, hot springs are exposed areas near FES station, where the SWV is less than

2.8 km/s. Another explanation is that heat flow activity in the upper crust may have led to a low SWV, especially at the FES and PUN stations along the LHSF. However, similar to the western region of the study area, the region with a high SWV contains more hot springs. Although the spatial distributions exhibited obvious regional and zonal characteristics, it is difficult to explain the low SWV bodies in order. In the following study, we will focus on whether the basins come from the invasion and transformation of geological bodies, by establishing more detailed array observation to build 3D crustal structures, as shown in the lithosphere of Italian Central Apennine [42].

The discontinuity of the SWV in the upper crust can indicate the geological structure caused by multistage magma intrusion. Throughout the geological history of the CASC, there have been many stages of intrusive rocks. The high- to low-SWV areas are correlated with the distribution of multistage intrusive rocks, sedimentary layers, structural metamorphic belts, and earthquake fractures caused by intrusive forces (Figure 10).

Since 250 Ma, the CASC has been subjected to several magmatic intrusions [3,4,26] of Triassic granite (250–200 Ma), Jurassic granite (200–150 Ma), Cretaceous granite and Cretaceous volcanic rocks (139–80 Ma), which correspond to multiple subductions and roll-backs of the Paleo-Pacific plate [43]. The previous two intrusions and the latter Cretaceous volcanic rocks are divided by the LHSF (Figure 10). The magmatic intrusion model contains three steps. First, the magma chamber in the mantle upwelled due to the subduction of the Pacific plate. Second, the magma began to break through the Moho surface and the crust, and then it upwelled by utilizing some structural zones as a channel to intrude into the sedimentary rock area. Third, with the strengthened subduction of the Pacific plate, multiple stages of magma pushed and covered each other, forming the present pattern of volcanic rocks, sedimentary rocks, and the contact metamorphic zone (near the fault). In addition, some channels in which magma rose acted as ways for rising mantle heat flows, leading to the current geothermal developments in some areas.

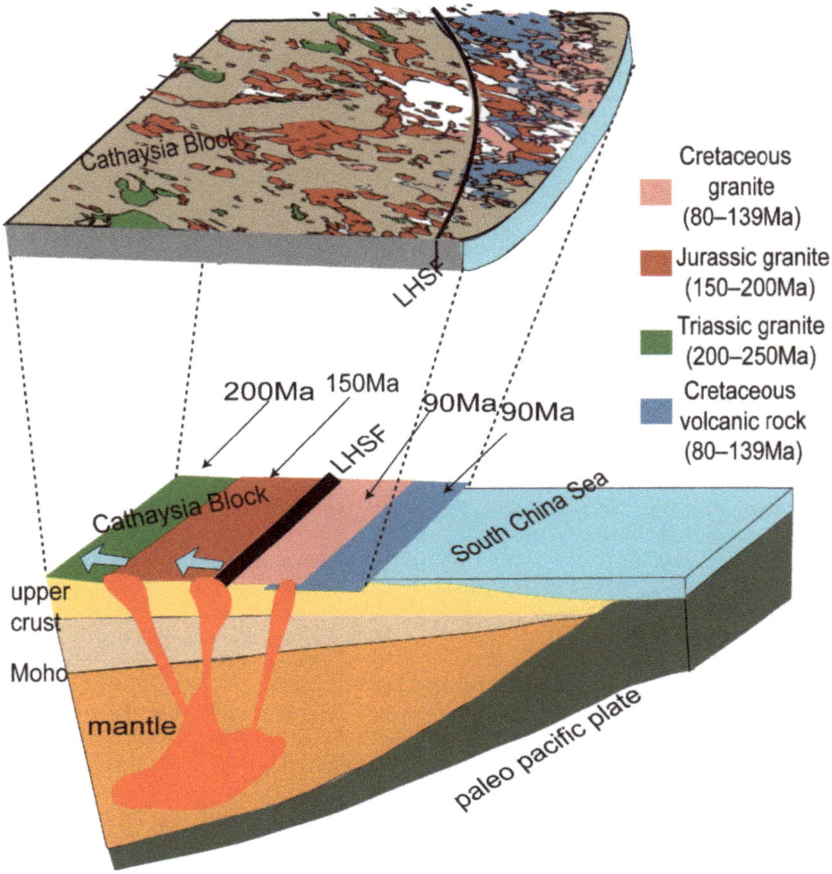

Figure 10. Modeling of several intrusive rocks. Since 250 Ma, the CASC has been subjected to three magmatic intrusions: Triassic granite (250–200 Ma), Jurassic granite (200–150 Ma), and Cretaceous granite and Cretaceous volcanic rocks (139–80 Ma).

5. Conclusions

Based on amplitude information of receiver functions, the SWV structure of the upper crust over 8 km in the CASC was obtained. In conjunction with the existing geological and petrological research results, SWV structure reveals the intrusion mechanism of multistage subvolcanic rocks near the surface in CASC. We obtained the following conclusions:

(i) We propose a method of using the amplitude of the receiver function to obtain the SWV that corresponds to the Gaussian filter factor. This method is stable, economical, environmentally friendly and efficient, while comparing with the active source surveying in the upper crust. Meanwhile, the accuracy and separation rate of this method need more numerical experiments and data processing to verify, especially in the area with complex velocity structure.

(ii) In the study area, we deduce that the low SWV in most sub-areas can be interpreted as the joint effect of the sedimentary layer of the alluvial plain and the accumulation of underground heat flows, in addition to multistage fracturing tectonism. The gradual change in the SWV in each profile from the surface to approximately 10 km is correlated with multiple invasions and the coverage of volcanic rocks to a certain extent, indicating weathering layer coverage and structural transformations between multistage intrusive rocks in the CASC.

Author Contributions: Conceptualization, X.S. and Y.Q.; data collecting and processing, H.C., X.Z. and Y.Q.; methodology, X.S. and Y.Q.; writing and original draft, X.Z. and H.H. All authors have read and agreed to the published version of the manuscript.

Funding: This study was financially supported by the National Natural Science Foundation of China (Grant 41874052 and 41730212), the Second Tibetan Plateau Scientific Expedition and Research Program (STEP) (2019QZKK0701); project supported by Innovation Group Project of Southern Marine Science and Engineering Guangdong Laboratory (Zhuhai) (No.311021002), Guangdong Province Introduced Innovative R&D Team (2016ZT06N331, 2017ZT07Z066), Guangdong Collaborative Innovation Center for Earthquake Prevention and Mitigation (2018B020207011), and Natural Science Foundation of Guangdong Province (2022A1515011105).

Data Availability Statement: Data sharing is not applicable to this article.

Acknowledgments: The raw tele-seismic waveforms in this study were provided by the Guangdong Seismic Network Center of the Guangdong Earthquake Agency. Some figures were plotted using Generic Mapping Tools (Wessel and Smith, 1995).

Conflicts of Interest: The authors declare no conflict of interest.

References

1. Gilder, S.A.; Gill, J.; Coe, R.S.; Zhao, X.X.; Liu, Z.W.; Wang, G.X. Isotopic and paleomagnetic constraints on the Mesozoic tectonic evolution of south China. *J. Geophys. Res. Solid Earth* **1996**, *101*, 16137–16154. [CrossRef]
2. Xia, S.; Zhao, M.; Qiu, X.; Xu, H.; Shi, X. Crustal structure in an onshore-offshore transitional zone near Hong Kong, northern south china sea. *J. Asian Earth Sci.* **2010**, *37*, 460–472. [CrossRef]
3. Zhou, Z.M.; Ma, C.Q.; Xie, C.F.; Wang, L.X.; Liu, Y.Y.; Liu, W. Genesis of highly fractionated I-type granites from Fengshun complex: Implications to tectonic evolutions of South China. *J. Asian Earth Sci.* **2016**, *27*, 444–460. [CrossRef]
4. Li, Z.X.; Li, X.H. Formation of the 1300-km-wide intracontinental Orogen and Postorogenic magmatic province in Mesozoic South China: A flat-slab subduction model. *Geology* **2007**, *35*, 179–182. [CrossRef]
5. Huang, L.T.; Shen, X.Z.; Zheng, W.J. Moho properties of western Ordos block and surrounding regions constrained by teleseismic receiver functions and its tectonic implication. *Chin. J. Geophys.* **2020**, *63*, 871–885.
6. Zhao, X.L.; Yu, M.G.; Liu, K. The Magmatic and Genetic Evolution of Early Cretaceous Granitoids in Eastern Guangdong Province. *Geol. Rev.* **2012**, *58*, 965–977.
7. Ye, H.M.; Mao, J.R.; Zhao, X.L.; Liu, K.; Chen, D.D. Revisiting early-middle Jurassic igneous activity in the Nanling mountains, south China: Geochemistry and implications for regional geodynamics. *J. Asian Earth Sci.* **2013**, *72*, 108–117. [CrossRef]
8. Zhang, Z.; Zhang, X.; Badal, J. Composition of the crust beneath southeastern China derived from an integrated geophysical data set. *J. Geophys. Res. Solid Earth* **2008**, *113*, 1–25. [CrossRef]
9. Huang, H.B.; Guo, X.W.; Xia, S.H.; Qiu, X.L. Study of Crustal Thickness and Poisson's Ratio in the Coastal Area of South China. *Chin. J. Geophys.* **2014**, *57*, 860–871.
10. Xia, S.H.; Cao, J.H.; Sun, J.L.; Lv, J.S.; Xu, H.L.; Zhang, X.; Wan, K.Y.; Fan, C.Y.; Zhou, P.X. Seismogenic structures of the 2006 M_L4.0 Dangan Island earthquake offshore Hong Kong. *J. Ocean Univ. China* **2018**, *17*, 169–176. [CrossRef]
11. Xia, S.; Zhou, P.; Zhao, D.; Cao, J. Seismogenic structure in the source zone of the 1918 m7.5 nanao earthquake in the northern South China sea. *Phys. Earth Planet. Inter.* **2020**, *302*, 106472. [CrossRef]
12. Zhang, X.; Ye, X.W.; Lv, J.S.; Sun, J.; Wang, X. Crustal structure revealed by a deep seismic sounding profile of Baijing-Gaoming-Jinwan in the Pearl river delta. *J. Ocean Univ. China* **2018**, *17*, 186–194. [CrossRef]
13. He, L.P.; Sun, X.; Yang, H.F.; Qin, J.L.; Shen, Y.S.; Ye, X.W. Upper crustal structure and earthquake mechanism in the Xinfengjiang water reservoir, Guangdong, China. *J. Geophys. Res. Solid Earth* **2018**, *123*, 3799–3813. [CrossRef]
14. Langston, C.A. Structure under Mount Rainier, Washington, inferred form tele-seismic body waves. *J. Geophys. Res. -Atmos.* **1979**, *84*, 4749–4762. [CrossRef]
15. Ammon, C.J.; Randall, G.E.; Zandt, G. On the non-uniqueness of receiver function inversions. *J. Geophys. Res.* **1990**, *95*, 15303–15318. [CrossRef]
16. Zhu, L.P. Crustal structure across the San Andreas Fault, southern California from tele-seismic converted waves. *Earth Planet. Sci. Lett.* **2000**, *179*, 183–190. [CrossRef]
17. Ma, Y.L.; Zhou, H.L. Crustal thicknesses and poisson's ratios in china by joint analysis of tele-seismic RFS and Rayleigh wave dispersion. *Geophys. Res. Lett.* **2007**, *34*. [CrossRef]
18. Julià, J. Constraining velocity and density contrasts across the crust–mantle boundary with receiver function amplitudes. *Geophys. J. Int.* **2007**, *141*, 286–301. [CrossRef]
19. Park, S.; Tsai, V.C.; Ishii, M. Frequency-dependent P wave polarization and its subwavelength near-surface depth sensitivity. *Geophys. Res. Lett.* **2019**, *46*, 14377–14384. [CrossRef]
20. Qian, Y.P.; Shen, X.Z. The approach and application of constraining velocity and density contrasts across the Moho using receiver functions. *Chin. J. Geophys.* **2017**, *60*, 2980–2992.

21. Qian, Y.P.; Shen, X.Z.; Li, C.Q. Constraining the sub-surface SWV of the northeastern margin of Tibetan Plateau with receiver functions. *Chin. J. Geophys.* **2018**, *61*, 3951–3963.
22. Wang, X.; Chen, L.; Ling, Y.; Gao, Y.; Zhang, J.; Yao, H. A new method to constrain shallow crustal s-wave velocities based on direct P-wave amplitudes in receiver functions and its application in northeastern Tibet. *Sci. China Earth Sci.* **2018**, *62*, 1819–1831. [CrossRef]
23. Wang, X.; Chen, L.; Yao, H. A new body-wave amplitude ratio-based method for imaging shallow crustal structure and its application in the Sichuan Basin, Southwestern China. *Geophys. Res. Lett.* **2021**, *48*, e2021GL095186. [CrossRef]
24. Zhou, X.M.; Sun, T.; Shen, W.Z.; Shu, L.S.; Niu, Y.L. Petrogenesis of Mesozoic granitoids and volcanic rocks in South China: A response to tectonic evolution. *Episodes* **2006**, *29*, 26–33. [CrossRef]
25. Huang, H.B.; Xiong, H.; Qiu, X.L.; Li, Y.H. Crustal structure and magmatic evolution in the Pearl River Delta of the Cathaysia Block: New constraints from receiver function modeling. *Tectonophysics* **2018**, *778*, 228365. [CrossRef]
26. Yao, Y.; Zhan, W.; Liu, Z.; Zhang, Z.; Zhan, M.; Sun, J. Neotectonics and its relations to the evolution of the Pearl river delta, Guangdong, China. *J. Coast. Res.* **2013**, *66*, 1–11. [CrossRef]
27. Zheng, W.; Mao, J.W.; Zhao, C.S.; Yu, X.F.; Zhao, H.; Ouyang, Z.X.; Wu, X. Early Cretaceous magmatism and associated polymetallic mineralization in South China: The Tiantang example. *Int. Geol. Rev.* **2017**, *60*, 1560–1580. [CrossRef]
28. Li, X.H. Cretaceous magmatism and lithospheric extension in Southeast China. *J. Asian Earth Sci.* **2000**, *18*, 293–305. [CrossRef]
29. Li, F.C.; Sun, Z.; Yang, H.F. Possible spatial distribution of the Mesozoic volcanic arc in the present-day South China Sea continental margin and its tectonic implications. *J. Geophys. Res. Solid Earth* **2018**, *123*, 6215–6235. [CrossRef]
30. Aki, K.; Richards, P.G. Quantitative seismology-theory and methods. *Freeman* **1980**, *1*, 116–119.
31. Ammon, C.J. The isolation of receiver effects from teleseismic P waveforms. *Bull. Seismol. Soc. Am.* **1991**, *81*, 2504–2510. [CrossRef]
32. Crotwell, H.P.; Owens, T.J.; Ritsema, J. The TauP Toolkit: Flexible seismic travel-time and ray-path utilities. *Seismol. Res. Lett.* **1997**, *70*, 154–160. [CrossRef]
33. Ma, Y.L. YASEIS: Yet Another computer program to calculate synthetic SEISmograms for a spherically multi-layered Earth model. In Proceedings of the 2013 EGU General Assembly Conference, Vienna, Austria, 7–12 April 2013; Volume 15.
34. Kennett, B.L.N.; Engdahl, E.R.; Buland, R. Constraints on seismic velocities in the Earth from traveltimes. *Geophys. J. Int.* **1995**, *122*, 108–124. [CrossRef]
35. Ligorria, J.P.; Ammon, C.J. Iterative deconvolution and receiver-function estimation. *Bull. Seismol. Soc. Am.* **1999**, *89*, 1395–1400. [CrossRef]
36. Zhang, L.N.; Luo, Y.; Chen, Z.Y.; Xie, Z.Z.; Zhou, L.J.; Zeng, X.F. Surface wave group velocity tomography imaging for Fujian area from ambient noise. *Earthquake* **2018**, *38*, 136–145.
37. Shen, Y.S.; Kang, Y. Surface wave tomography of Guangdong and its adjacent areas from ambient seismic noise. *Acta Seismol. Sin.* **2014**, *36*, 826–836.
38. Huang, Y.M.; Shen, Y.S.; Yang, M.L. Surface wave group velocity tomography from ambient seismic noise in east region of Guangdong Province. *Earthq. Res. China* **2012**, *4*, 360–369.
39. Gičev, V.; Trifunac, M.D.; Todorovska, M.I.; Kocaleva, M.M.; Stojanova, A.; Kokalanov, V. Ambient vibration measurements in an irregular building. *Soil Dyn. Earthq. Eng.* **2021**, *141*, 106484. [CrossRef]
40. Kokalanov, V.; Trifunac, M.D.; Gičev, V.; Kocaleva, M.; Stojanova, A. High frequency calibration of a finite element model of an irregular building via ambient vibration measurements. *Soil Dyn. Earthq. Eng.* **2022**, *153*, 107005. [CrossRef]
41. Wang, S.; Sun, X.L.; Qin, J.L.; He, L.P.; Deng, Y.F. Fine fault structure of Xinfengjiang water reservoir area from high-frequency ambient noise tomography. *Chin. J. Geophys.* **2018**, *61*, 593–603.
42. Bruno, P.P.G.; Villani, F.; Improta, L. High-resolution seismic imaging of fault-controlled basins: A case study from the 2009 Mw 6.1 central Italy earthquake. *Tectonics* **2022**, *41*, e2022TC007207. [CrossRef]
43. Liu, J.X.; Shuo, W.; Wang, X.L.; Du, D.H.; Xing, G.F.; Fu, J.M.; Chen, X.; Sun, Z.M. Refining the spatio-temporal distributions of Mesozoic granitoids and volcanic rocks in SE China. *J. Asian Earth Sci.* **2020**, *201*, 1367–9120. [CrossRef]

remote sensing

MDPI

Communication

Quick Report on the ML = 3.3 on 1 January 2023 Guidonia (Rome, Italy) Earthquake: Evidence of a Seismic Acceleration

Dedalo Marchetti [1], Kaiguang Zhu [1,*], Laura Marchetti [2], Yiqun Zhang [1], Wenqi Chen [1], Yuqi Cheng [1], Mengxuan Fan [1], Siyu Wang [1], Ting Wang [1], Jiami Wen [1], Donghua Zhang [1] and Hanshuo Zhang [1]

1 College of Instrumentation and Electrical Engineering, Jilin University, Changchun 130061, China
2 Noviconsult s.a.s., 15060 Basaluzzo, AL, Italy
* Correspondence: zhukaiguang@jlu.edu.cn

Abstract: This paper investigates possible anomalies on the occasion of the ML = 3.3 earthquake which occurred on 1st January 2023 close to Guidonia Montecelio (Rome, Italy). This earthquake followed another very close event recorded on 23 December 2022 with a magnitude ML = 3.1 (epicentral distance of less than 1 km). Seismological investigations clearly show an acceleration of seismicity in the preceding six months in a circular area of about a 60 km radius. Two conclusions coming from the time-to-failure power law fitting to the cumulative Benioff strain curve are the most probable: the ML3.3 of 1 January 2023 is the mainshock of the seismic sequence, or an incoming earthquake of a magnitude of about 4.1 provides a slightly better fit (higher determination coefficient) of the seismic data. Further investigations are necessary to assess whether the accumulated stress has been totally released. No atmospheric anomalies related to this seismic activity have been identified, even if some SO_2 emissions seem to be induced by the tectonic and volcanic sources in the south Tyrrhenian Sea. Swarm satellite magnetic data show about 20 anomalous tracks six months before the Guidonia earthquake. In particular, on 16 December 2022, anomalous oscillations of the east component of the geomagnetic field are temporally compatible with the seismic acceleration, but other sources are also possible. Other anomalous magnetic signals are more likely to be associated with the ongoing seismic activity offshore of the Marche region in Italy (the strongest event up to now is Mw = 5.5 on 9 November 2022 close to Pesaro Urbino).

Keywords: seismic acceleration; Guidonia; earthquake; Benioff cumulative stress; swarm; pre-earthquake processes

Citation: Marchetti, D.; Zhu, K.; Marchetti, L.; Zhang, Y.; Chen, W.; Cheng, Y.; Fan, M.; Wang, S.; Wang, T.; Wen, J.; et al. Quick Report on the ML = 3.3 on 1 January 2023 Guidonia (Rome, Italy) Earthquake: Evidence of a Seismic Acceleration. *Remote Sens.* **2023**, *15*, 942. https://doi.org/10.3390/rs15040942

Academic Editor: Salvatore Stramondo

Received: 3 January 2023
Revised: 4 February 2023
Accepted: 7 February 2023
Published: 8 February 2023

1. Introduction

On 1 January 2023 at 13:07:46 Universal Time—UT (14:07 Central European Time—CET)—an earthquake of magnitude ML = 3.3 was localised northeast of the capital of Italy, Rome, by the Italian National Institute of Geophysics and Volcanology (INGV). In particular, the epicenter was 3 km southeast of Guidonia Montecelio town, at the coordinates of 41.982°N and 12.750°E. The earthquake's depth was shallow, estimated at about 9 km (http://cnt.rm.ingv.it/event/33781261, accessed on 2 January 2023). The event was relatively low in magnitude, but it obtained particular interest because it followed another seismic event with a slightly lower magnitude ML = 3.1 that occurred on 23 December 2022 at 16:35:13 UT (17:35 CET) with almost the same epicentre (41.984°N and 12.756°E, distance from the abovementioned event of about 0.5 km). This event's hypocentral depth was estimated at 10 km (http://terremoti.ingv.it/event/33714971, accessed on 2 January 2023). Based on the data on the INGV website, at least 681 people (last update 1 January 2023 at 19:31 CET) felt the event on 23 December 2022, filling out the form on https://www.hsit.it/ (accessed on 2 January 2023) and 765 people after the event on 1 January 2023 (last update 2 January 2023 at 3:55 CET). From 1977, around this event, within ~50 km, previous earthquakes were characterised by different focal mechanisms. Some thrust earthquakes have been historically recorded towards the northeast, while strike-slip and normal fault events have been localised on the southeast and west sides. In particular, the Guidonia earthquakes that occurred on 23 December 2022 and 1 January 2023 could have been produced by a strike-slip fault proposed by Faccenna et al. [1] starting on the Cornicolani Mounts and ending at the Colli Albani volcano district, crossing the city of Guidonia city (dashed line in Figure 1 of their paper, depicted here as the green dashed-dot line in Figure 1b). In the box delimited by 41.9°N ≤ latitude ≤ 42.1°N and 12.6°E ≤ longitude ≤ 12.8°E, the highest magnitude event between 1985 and 2023 was Md = 3.3 on 16 February 1999, 2 km north of Tivoli, which seems located on a northeast parallel fault. Other magnitude 2.9 earthquakes were recorded on the same fault close to Tivoli town on 22 September 2005 (doublet) and on 20 August 2013. After 1 January 2023, an event of magnitude Mw = 3.2 was recorded 8 km northwest of Colonna town on 22 January 2023 at 1:46:11 UT (http://cnt.rm.ingv.it/event/33960061, accessed on 29 January 2023). This event was localised on the southern side of the same strike-slip fault (dashed-dot green line in Figure 1b). In conclusion, the ML3.3 Guidonia earthquake of 1 January 2023 is presently the strongest event ever reported with the instrumental catalogue (from 1985) on this fault.

One basic question is whether the events are parts of a seismic sequence or a swarm. A seismic sequence is characterised by a higher magnitude event (the mainshock) followed by aftershocks, and could be preceded by foreshocks of lower magnitude. In contrast, a seismic swarm is a chain of comparable and low-magnitude earthquakes without a clear mainshock [2]. If this is a seismic sequence, it is crucial to know if the event of magnitude ML = 3.3 of 1 January 2023 is the mainshock or if a future event of higher magnitude is likely to occur. Furthermore, for seismic sequences, the decay of the aftershocks generally follows the Omori-Utsu law [3]; in contrast, a seismic swarm has an almost constant seismic rate. The peculiarities of the seismic earthquakes of 22 December 2022 and 1 January 2023 localised close to Guidonia (Rome, Italy) motivates a further investigation that we report in this short paper. The research is based on our previous experience in the field of lithosphere, atmosphere, and ionosphere analysis prior to seismic events.

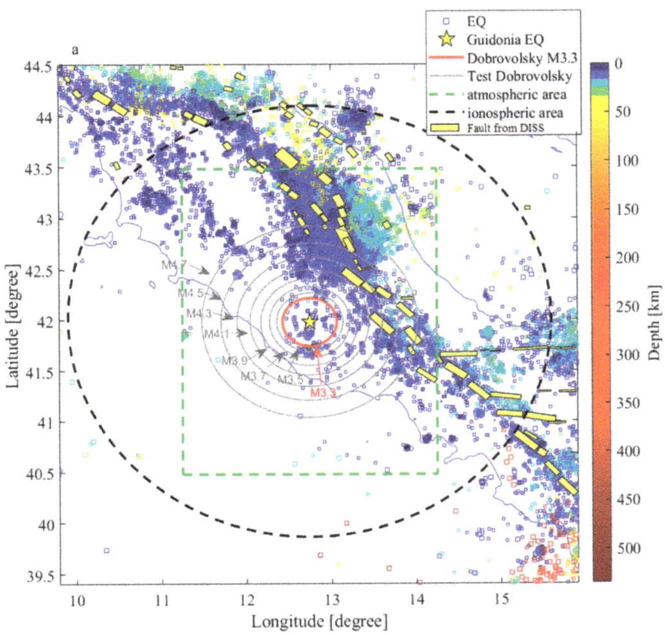

Figure 1. *Cont.*

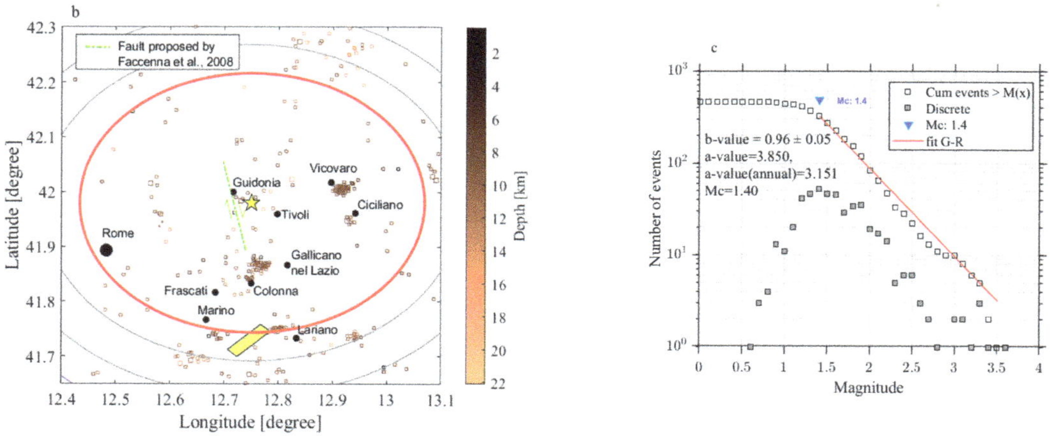

Figure 1. Earthquakes that occurred in Central Italy from 1st January 2018 until 1st January 2023. (**a**) Geographical map of the earthquake (EQ) events with active seismic fault from DISS: the yellow star represents the event of Guidonia 1 January 2023, ML = 3.3; the red circle is its Dobrovolsky area; grey circles are the tested Dobrovolsky areas; the green dashed rectangle is the area used to calculate atmospheric time series, and the black dashed circle is the research area for ionospheric investigations. The adapted colour bar scale shows the hypocentral depth. (**b**) Enlargement of the map around the Guidonia (Rome) earthquake. In addition to previous details, the principal local towns are shown in relation to recorded seismicity. The transcurrent fault proposed by Faccenna [1] is reported with a dashed-dot green line. (**c**) The Gutenberg–Richter (G–R) distribution of the magnitude of the earthquakes, excluding the seismicity in the northeast corner of the map, i.e., within a radius of 50 km from Guidonia epicentres. The magnitude of completeness and G–R parameters are reported.

2. Materials and Methods

In this paper, we investigated the earthquake catalogue, the atmospheric data from a climatological archive, and the magnetic field and electron density data from the Swarm satellite constellation.

2.1. Earthquake Catalogue

The earthquake catalogue was acquired from the INGV ISIDE [4] earthquake web portal (http://cnt.rm.ingv.it/, accessed on 2 January 2023) by the selection of the events inside the square box delimited by 34.4°N $\leq$ latitude $\leq$ 44.5°N and 9.8°E $\leq$ longitude $\leq$ 15.9°E from 1st January 2018 until 1st January 2023. No constraints were selected for magnitude and depth. The geographical positions of all the events are shown in Figure 1a. The seismicity induced by the very long Italian Seismic Sequence of Amatrice-Norcia-Capitigliano (Montereale) 2016–2017 [5] is visible in the diagonal from the northwest to the southeast sides of the map of Figure 1a.

The completeness magnitude "Mc" was analysed by the ZMap software [6], excluding the events associated with the Amatrice-Norcia-Capitigliano Italian Seismic sequence (i.e., the event more than 50 km from the Guidonia 2023 earthquake) to avoid a bias in the analysis. In fact, the Apennines central region of Italy, affected by this sequence, was more thoroughly covered with permanent seismic stations and temporary stations after the start of the same sequence. The completeness magnitude was estimated to be Mc = 1.4, as visible in the graph in Figure 1c. In addition, this graph represents the Gutenberg–Richter law [7] with "a" and "b" values, which represent the amount of recorded seismicity and the ratio between smaller and higher magnitude events in the region, respectively. The "a" value is the decimal logarithm of the amount of M0+ earthquakes in the investigated area in the analysed time, or during one year. The b-value is generally found to be equal to 1 at a global or regional scale. Before large earthquakes, its value can decrease, indicating a lack of larger-magnitude events in the statistic [8,9]. Only earthquakes with magnitude $M \geq Mc$ will be considered in the following analyses. The cumulative Benioff stress was computed for the events inside the Dobrovolsky radius. Benioff stress is an indication of the stress accumulating on a fault [10]. The Dobrovolsky radius is a distance typically used to estimate the preparation area of an earthquake [11].

In particular, Dobrovolsky's radius in kilometres is

$$r_{Dob} = 10^{(0.43 \cdot M)}$$

The Benioff cumulative stress *S(t)* could be empirically estimated by the following formula:

$$S(t) = \sum_{i=1}^{N(t)} \sqrt{10^{(1.5 \cdot M_i + 4.8)}},$$

where "*i*" is the *i*-th earthquake with magnitude M_i, and *N(t)* is the total number of earthquakes that occurred until the time *t*. According to Mignan et al. [12] and Cianchini et al. [13], the following analytical model, in the form of a time-to-failure power law, has been fitted to the Benioff cumulative stress:

$$S(t) = A + B \cdot \left(t_f - t\right)^m,$$

where parameters *A* and *B* are proportional to the final and initial accumulated stress, respectively, and m is an exponential factor in the range from 0 to 1, typically equal to 0.3 [13]. We underline that the time-to-failure power law analytical model assumes that the mainshock is incoming at the time t_f, and it is not expected to work in the case of the aftershock sequence, a swarm of earthquakes, or a seismic quiescence.

Fitting such a curve is equivalent to resolving the following equation:

$$A + B \cdot \left(t_f - t\right)^m = \sum_{i=1}^{N(t)} \sqrt{10^{(1.5 \cdot M_i + 4.8)}}.$$

In this work, *m* has been fixed to $m = 3$ to guarantee the stability of the power law fit. The graphs report "t_f", which is the time to failure, i.e., the likelihood estimation of the mainshock time.

2.2. Atmospheric Data Processing

The data from the climatological archive MERRA-2 by NASA [14] have been investigated to monitor the atmosphere. With the experience of the earthquakes previously studied, the MEANS software was used, with the removal of a linear trend to take into account the possible effect of global warming, as done in De Santis et al. [14] for the Ridgecrest, US, 2019 earthquake analysis. The MEANS algorithm was used for the first time by Piscini et al. [15] to investigate volcano eruptions, and later applied to several other volcano eruptions and earthquakes [14,16–20]. The algorithm essentially estimates the typical value of the atmospheric parameter for the day and region, constructing a historical time series, particularly the mean and standard deviation of the parameter for the area under study. If, in the year with the natural hazard event, the parameter exceeds two standard deviations of the historical time series, it is considered anomalous. This report investigated surface air temperature, aerosol, surface total energy flux, sulphur dioxide, surface specific humidity, and total precipitation.

2.3. Ionospheric Data Processing

The ionosphere is monitored by the European Space Agency (ESA) Swarm constellation. The Swarm mission is composed of three identical satellites in low-Earth quasi-polar orbits, called Alpha, Bravo and Charlie, aiming to measure the geomagnetic field with state-of-the-art precision [21]. The use of Swarm data to study earthquake preparation has been explored by several researchers in the last few years by using a single case study approach [20,22–33], or by statistically correlating the magnetic and electron density anomalies of Swarm data with M5.5+ earthquakes that occurred in the first 4.7 years of Swarm mission (De Santis et al. [34]), or the first eight years (Marchetti et al. [35]), or by using machine learning (Xiong et al. [36]).

In this report, we applied the MASS algorithm to Swarm magnetic field data defined with all the details in [29,33–35], and the NeLOG algorithm defined with all the details in [29], which was used recently in [37] to study the 2020 Samos (Greece) earthquake. MASS analyses the data track-by-track removing the background, by estimating the derivative and subtracting a cubic spline. Such data processing also intrinsically removes the long-term trend of the geomagnetic field from the residuals. The residuals are monitored within a moving window comparing the root mean square of the window with the root mean square of the entire track along −50°S and +50°N geomagnetic latitude. Higher latitude values are excluded to avoid the influence of polar activity.

NeLOG investigates the electron density latitudinal profiles, removing a polynomial of ten degrees from the investigated track and searching for tracks with at least ten anomalous residual samples. The sample is marked as anomalous if its value exceeds "kt" times the standard deviation of the whole residuals.

3. Results

Here, we discuss the results of the seismological, atmospheric, and ionospheric investigations.

3.1. Seismological Investigation

The cumulative Benioff stress was calculated inside the Dobrovolsky area, testing several mainshocks from magnitude 3.3 to 5.5, as listed in Table 1. The lower limit corresponds to the assumption that the 1 January 2023 event is the mainshock. The higher-magnitude events correspond to the hypothesis that a stronger event is incoming. In addition, Table 1 provides the b-value of the Gutenberg–Richter distribution, calculated in the real or tested Dobrovoslky area by ZMap software [6], using the maximum curvature method for more stable solutions. The detailed graphs are reported in Supplementary Materials Figure S1. As previously mentioned, a low b-value could indicate a lack of higher-magnitude events. This low b-value was found in Dobrovolsky's area of tested events with a magnitude between 3.7 and 4.1.

Table 1. List of the real (in bold) and tested earthquake magnitudes with the corresponding Dobrovolsky's radius. All the events are centred at the epicentre of the ML3.3 Guidonia 1 January 2023 earthquake (41.982°N, 12.750°E). The b-value of the Gutenberg–Richter distribution is computed during the six months before the Guidonia earthquake by the maximum curvature method, and details and graphs are provided in Supplementary Materials Figure S1.

	Estimated Magnitude	Dobrovolsky Radius [km]	b-Value (In Previous 6 Months)	C [1] (In the Previous 6 Months)
Real earthquake	**3.3**	**26.2**	**1.0 ± 0.3**	**0.318**
Tested earthquakes	3.5	32.0	1.0 ± 0.3	0.359
	3.7	39.0	0.7 ± 0.1	0.342
	3.9	47.5	0.7 ± 0.1	0.326
	4.1	57.9	0.7 ± 0.1	0.356
	4.3	70.6	1.0 ± 0.2	0.714
	4.5	86.1	1.04 ± 0.05	0.943
	4.7	105.0	1.15 ± 0.03	1.075

[1] C is the coefficient of acceleration defined in the text, and it indicates an acceleration when C < 1. In particular, as it approaches 0, the stress curve indicates a more accelerating behaviour.

For each fit, the adjusted statistic coefficient of determination "adj-R^2" and the estimated time of failure are reported in Figures S2 and S3 (in Supplementary Materials), and in Figure 2 for 5 years, 1 year, and 6 months before the ML = 3.3 Guidonia 1 January 2023 earthquake, respectively.

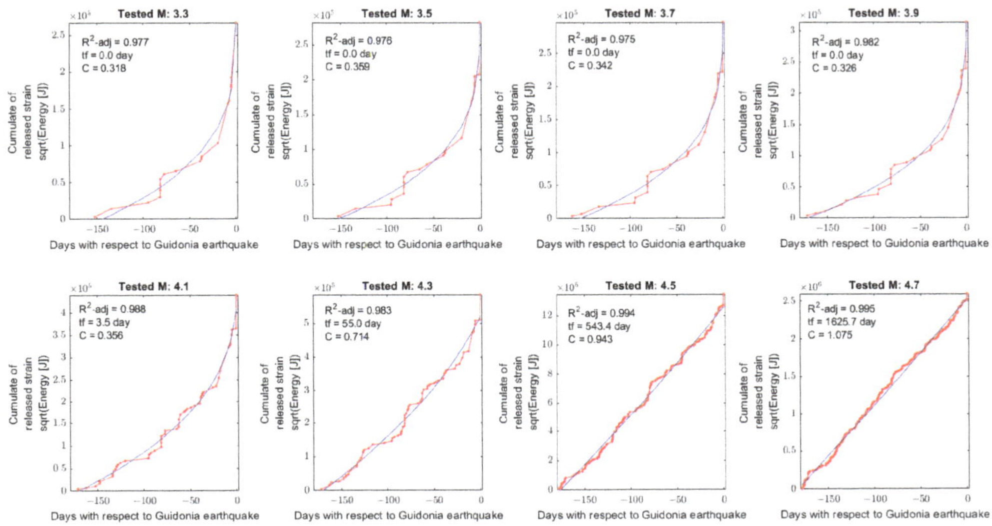

Figure 2. Benioff cumulative stress curves (in red) calculated within six months of the ML = 3.3 1 January 2023 Guidonia earthquake. Each graph is based on a circular area centred on the 1 January 2023 ML3.3 Guidonia earthquake with a Dobrovolsky's radius of earthquakes with a magnitude from 3.3 to 4.7. For each plot, the blue curve represents the time-to-failure power law fit, whose adjusted coefficient of determination R^2-adj and the time to failure (tf) are reported. The acceleration coefficient C is also reported.

In addition, the acceleration coefficient C defined by Cianchini et al. [13] has been reported in Figures S2 and S3 and Figure 2, and Table 1, calculated as:

$$C = \frac{rms_{power\ law}}{rms_{linear\ fit}},$$

where "$rms_{power\ law}$" and "$rms_{linear\ fit}$" are the root mean square errors of samples "S_i" ($i = 1 \ldots N$) with respect to their fits "f"; in particular, rms is estimated as:

$$rms = \sqrt{\frac{\sum_i (S_i - f_i)^2}{N}}.$$

For lower values of C, the fit represents more acceleration. On the other hand, C = 1 indicates that the trend is linear, and C > 1 indicates that a linear fit is better than the power law. Such a situation could be due to a trend influenced by the aftershocks of another seismic sequence, such as the one of Amatrice-Norcia-Capitigliano (Montereale) Central Italy 2016–2017, in particular for a long trend and a larger radius of analysis.

The investigations for 5 years and 1 year before ML = 3.3 Guidonia 1st January 2023 earthquake do not provide interesting results (Supplementary Materials Figures S2 and S3). In fact, the trends are dominated by another important seismicity which occurred in the area (the Mw = 3.3 11 April 2022, Ciciliano, Rome event, http://cnt.rm.ingv.it/event/30543871, accessed on 2 January 2023). The Apennine residual sequence instead dominates the tests, with a larger magnitude cited above. This is the reason why a magnitudes higher than 4.7

are not reliable to be tested, as the crossing with seismicity from the Central Apennines region becomes predominant.

On the other hand, the investigation in the 6 months before the ML = 3.3 Guidonia 1st January 2023 earthquake, reported in Figure 2, provided interesting results. From magnitude 3.3 to 3.9, the estimated time-to-failure is 0.0 days, and the maximum acceleration reported for M = 3.3 is C = 0.318. Considering that no higher events were recorded in this area on 1 January 2023, among these four analyses, the M = 3.3 test, i.e., the hypothesis that the past earthquake was the mainshock, must be considered the most reliable. Despite this, the analysis for M = 4.1 and M = 4.3 provided good fits, with a time to failure of a few days or weeks (for M4.1, tf = 3.5 days and for M4.3, tf = 55 days). The investigations of M4.5 and M4.7 provided very good fits (R^2-adj very close to 1), although they are almost linear (as underlined by their C being close to 1). So, selecting the analysis with the highest adjusted R^2 and $C \ll 1$, the most reliable result is the analysis of M4.1 with R^2-adj = 0.988 and C = 0.356. Even though this could appear as a prediction, that interpretation is not totally correct. Firstly, it is necessary to estimate the maximum magnitude that the fault could generate. Secondly, it is crucial to determine if these events have totally or partially released the accumulated stress. Furthermore, the time-to-failure power law could not account for several geological and tectonic factors, so it cannot be used alone. Finally, the proper interpretation for the authors is that, in the past six months, in the circular area of a radius of 58 km centred on the Guidonia 1 January 2023 earthquake, a clear seismic acceleration was recorded, but further studies are necessary to propose eventual future scenarios.

3.2. Atmospheric Investigations

In this section, we present the atmospheric investigation six months the ML3.3 Guidonia 1st January 2023 earthquake. The results in terms of time series and maps of surface air temperature, aerosol, surface total energy flux (also known as surface latent heat flux), SO_2, surface specific humidity, and total precipitation are provided in Supplementary Materials in Figures S4–S7, Figures 3–6. All the time series were calculated in a square box centred on the Guidonia 1st January 2023 earthquake with a 3° side. The surface air temperature time series has some anomalous values. Still, none have multi-day persistence, a criterion found effective for Central Italy by Piscini et al. [38].

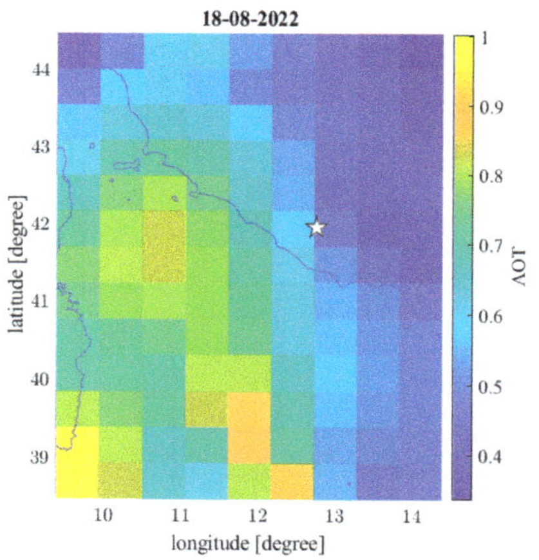

Figure 3. The aerosol map on 18 August 2022. The epicentre of Guidonia 1 January 2023 earthquake is shown with a white star.

The aerosol, humidity, total precipitation, and especially SO_2 investigations revealed several anomalous days, and we provided some geographical maps to better investigate if the increase in these quantities was local or global. The aerosol on 18 August 2022 was particularly anomalous, and its map is shown in Figure 3. From the aerosol distribution, it seems that all of the Tyrrhenian Sea recorded a high value of aerosol, probably due to particular weather conditions. In any case, it does not seem related to the earthquake.

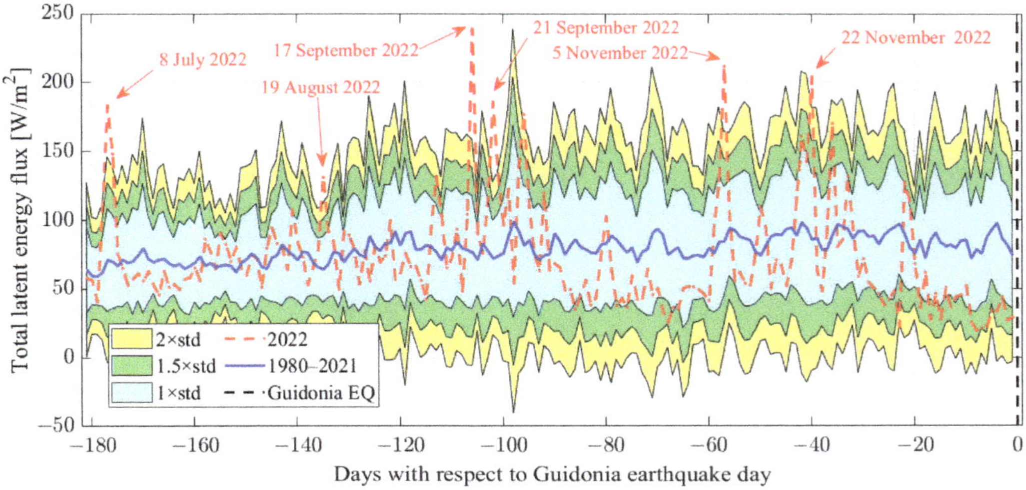

Figure 4. Total latent energy flux time series from 4 July 2022 to 31 December 2022 (i.e., from six months before until the day before the ML = 3.3 Guidonia 1 January 2023 earthquake, red dashed line). The blue line represents the historical mean calculated from 1980 to 2021 in the same period and region. Cyan, green, and yellow bands represent 1.0, 1.5, and 2.0 standard deviations of the historical values from 1980 to 2021. The horizontal axis represents the day before the Guidonia earthquake for the red dashed line (i.e., the 2022 time series), and the mean and standard deviation of the historical values are computed on the same day of the previous years. The vertical dashed black line represents the Guidonia earthquake (EQ) occurrence time. A red arrow with the specific date is reported for the anomalous days, i.e., the values that exceeded two standard deviations.

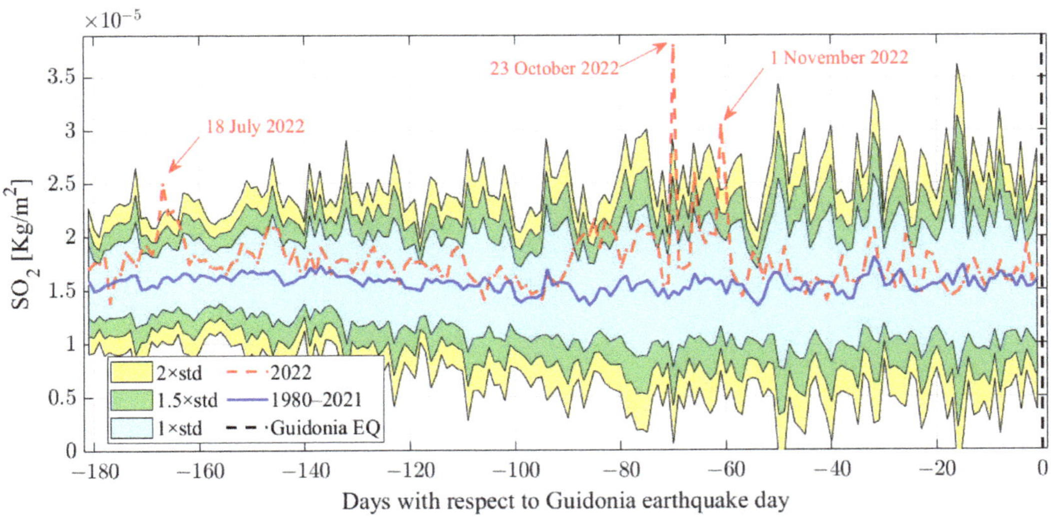

Figure 5. SO_2 time series from 4 July 2022 until 31 December 2022 (i.e., from six months before until the day before the ML = 3.3 1 January 2023 Guidonia earthquake). The years 1985, 1989, and 1993 have been excluded automatically as particularly anomalous. The representation graphic is the same as Figure 4.

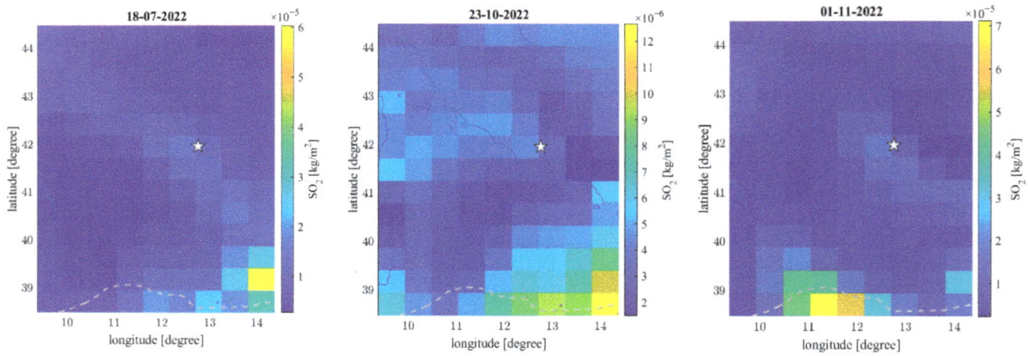

Figure 6. SO_2 maps on 17 July, 23 October and 1 November 2022. The epicentre of Guidonia 1 January 2023 earthquake is shown with a white star.

Figure 4 reports the time series of the surface total energy flux, known as surface latent heat flux (SLHF), in the six months before the ML3.3 Guidonia 1 January 2023 earthquake. Several days exceeded the two-standard-deviation threshold, and they can be considered anomalous. Their occurrence dates (8 July, 19 August, 17 and 21 September, and 5 and 22 November 2022) are reported in red in Figure 4.

Figure 5 shows the time series of sulphur dioxide in the six months before the ML3.3 Guidonia 1 January 2023 earthquake. Figure 6 reports three anomalous days of SO_2 in the investigated area, i.e., 18 July, 23 October, and 1 November 2022. The 18 July and 1 November dates show a slightly higher value of SO_2 along almost all the coastline, excluding a local phenomenon. Despite this, in all three maps, the highest values are recorded in the sea, corresponding to an important subduction slab of the Tyrrhenian Sea (dashed grey line). The emission of this SO_2 could be due to the stress on the subduction slab, and the important volcanic activity of the several volcanos close to this area (the Stromboli in the Aeolian Islands, submersed Marsili, or Etna, in Sicily). These volcanos are well known to emit SO_2, and contribute significantly to the global budget of SO_2 emissions [39].

A summary of all the identified anomalies in the atmosphere is reported in Table 2. An "X" in the table reports any day for which the surface air temperature, aerosol, SO_2 surface total energy latent heat flux, or total precipitation parameter exceeded the positive threshold of 2 standard deviations of the 40-year background (1980–2021). For surface specific humidity, we extracted the anomalous days when the parameter was below the negative threshold of 2 standard deviations. This choice relied on the fact that the lithosphere, atmosphere, and ionosphere coupling mechanism proposed by Pulinets and Ouzonouv [40] predicts an anomalous decrease in humidity due to the chemical bonds of water vapour molecules in the air with the ionised charges produced by the release of radon or the positive holes emitted from the fault according to Freund [41].

Table 2. List of the atmospheric anomalies identified inside the square area of 3° sides, centred on the ML3.3 Guidonia 1 January 2023 earthquake. The parameter which is anomalous on a specific day is marked with a "X" in the table.

Date	Atmospheric Parameter					
	Surface Air Temperature	Surface Specific Humidity	Aerosol	SO_2	Surface Total Energy Latent Heat Flux	Total Precipitation
4 July 2022	X					
7 July 2022						X
8 July 2022					X	

Table 2. *Cont.*

Date	Atmospheric Parameter					
	Surface Air Temperature	Surface Specific Humidity	Aerosol	SO_2	Surface Total Energy Latent Heat Flux	Total Precipitation
18 July 2022				X		
20 July 2022				X		
17 August 2022	X					
18 August 2022			X			X
19 August 2022					X	
31 August 2022						X
8 September 2022	X					X
15 September 2022			X			
17 September 2022		X			X	
21 September 2022		X			X	
22 September 2022		X				
27 September 2022					X	
8 October 2022				X		
10 October 2022			X			
23 October 2022				X		
1 November 2022				X		
5 November 2022					X	
13 November 2022						X
22 November 2022					X	
25 November 2022						X
2 December 2022						X
3 December 2022			X			X

Finally, we compared the volcanic activity in nearby regions in the same period to determine if the atmospheric anomalies could be caused by wind transport of volcanic gases and ash above the research area. Figure 7 compares the SO_2 time series we obtained in the research area with the SO_2 measured by the Italian National Institute of Geophysics and Volcanology (INGV) for three volcanoes (Etna, Stromboli, and Vulcano) and provided in the official weekly or monthly active volcano reports. Considering that the data from the Etna volcano are already normalised, we decided to normalise the other data also, including our time series. Notably, the SO_2 anomalies identified in July 2022 followed some Vulcano degassing. Consequently, they could be due to volcanic activity. On the other hand, the SO_2 anomalies of October and November 2022 could be excluded as not being induced by volcanic activity, as some emissions from the calderas of these volcanos did not precede, but followed these anomalies. We propose that the emissions recorded in October and November 2022 could be from a tectonic source in the Tyrrhenian Sea subduction slab.

3.3. *Ionospheric Investigations*

Although the ionosphere is typically investigated inside Dobrovolsky's area, we take into account that, in this particular case, such an area would be too small (see Figure 1) compared with the altitude of the Swarm satellites (about 470 km above sea level for Alpha and Charlie). So, we decided to consider an area with a diameter equal to the altitude of the two lower-orbit satellites, i.e., a radius equal to 235 km. The MASS algorithm was run for the data from 3 July 2022 to 1 January 2023. In order to avoid external disturbances, only quiet geomagnetic time was considered, using geomagnetic indexes shown in Figure 8. It is possible to note that there are several periods with geomagnetic activity, which is reasonable, as the time under investigation is during the solar activity increase of the 25th solar cycle.

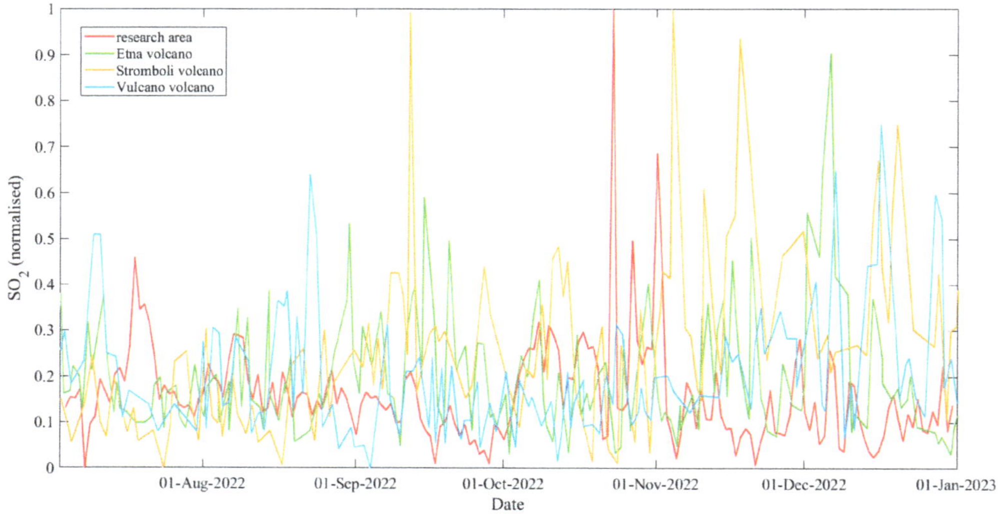

Figure 7. Comparison of SO_2 time series in the research area with SO_2 measured in the craters of three active close volcanos: Etna, Stromboli, and Vulcano. All the data were normalised in the range from 0 to 1.

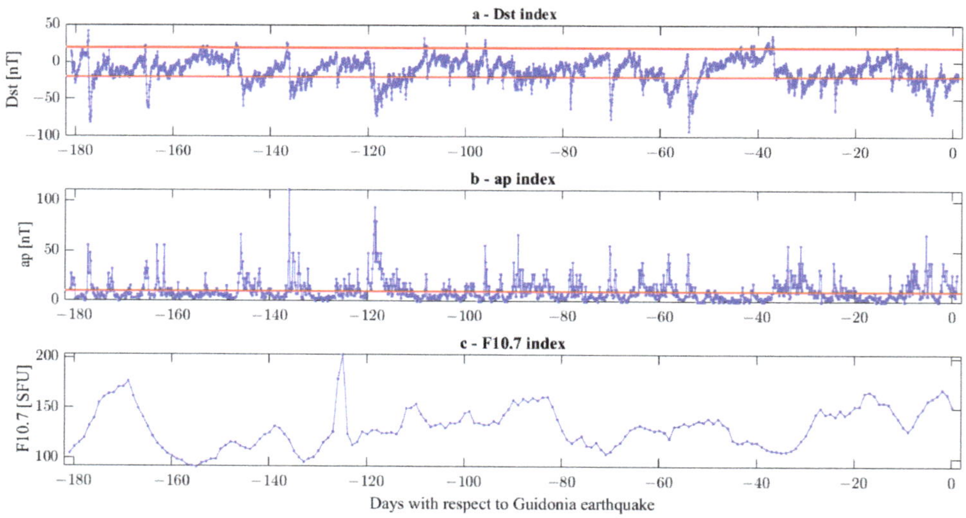

Figure 8. Geomagnetic indexes Dst (**a**), ap (**b**) and solar activity F10.7 index (**c**) in the six months before the ML = 3.3 Guidonia 1 January 2023 earthquake. Red lines show the thresholds used in the automatic codes to distinguish quiet or disturbed conditions ($|Dst| > 20$ nT or ap > 10 nT).

After processing all the tracks of Swarm Alpha, Bravo and Charlie, about 20 tracks present some anomalies inside the research area in quiet geomagnetic conditions ($|Dst| \leq 20$ nT and ap ≤ 10 nT), using a window of 1° latitude (found optimal in the Central Italy Swarm investigation [42]), and a threshold of rms inside the window of more than 2.5 times the threshold of rms of the whole track. Table 3 reports the list of anomalous tracks, and their cumulate number versus time is shown in Figure S8 in the supplementary materials. No anomalies were identified in the absolute scalar intensity of the geomagnetic field with the selected parameters.

Table 3. List of the Swarm anomalous tracks in the circular research area of 235 km of radius centred on the ML3.3 Guidonia 1 January 2023 earthquake.

Swarm	Date	Time UT	Local Time	Anomalous Component	Dst [nT]	ap [nT]	F10.7 [SFU]	Figure
Bravo	9 July 2022	19:25:45	20:19:32	X-North, Y-East, Z-Center	−15	6	139.9	Figure S9
Bravo	10 July 2022	07:38:42	08:33:08	X-North, Y-East, Z-Center	−6	9	155.3	Figure S10
Bravo	16 July 2022	18:51:07	19:43:40	X-North, Y-East, Z-Center	−7	3	176.9	Figure S11
Bravo	17 July 2022	07:04:03	07:57:16	X-North	−8	3	163.7	Figure S12
Bravo	31 July 2022	05:54:17	06:45:33	Y-East	14	5	95.1	Figure S13
Bravo	6 August 2022	17:06:22	17:56:06	Y-East	20	3	116.2	Figure 9
Bravo	14 August 2022	04:44:07	05:33:49	X-North	−20	7	125.7	Figure S14
Charlie	26 August 2022	14:18:41	15:05:31	Y-East	1	4	120.4	Figure S15
Alpha	30 August 2022	13:53:05	14:46:46	X-North	−4	7	124.4	Figure S16
Charlie	26 September 2022	11:19:43	12:19:28	X-North, Y-East, Z-Center	2	4	135.7	Figure S17
Alpha	3 October 2022	10:56:32	11:44:51	Y-East	−11	5	158.3	Figure S18
Charlie	26 October 2022	08:42:02	09:38:43	X-North	10	0	122.8	Figure S19
Alpha, Charlie	13 November 2022	18:44:39	19:43:53	X-North, Y-East, Z-Center	−8	6	136.1	Figures 10 and S20
Alpha	16 November 2022	18:46:36	19:27:45	Y-East	5	5	131.5	Figure S21
Bravo	8 December 2022	18:50:27	19:34:44	X-North, Y-East	−14	7	142.3	Figure S22
Alpha	16 December 2022	15:54:40	16:47:10	Y-East	−1	2	164.0	Figure 11

We note that the track of Swarm Bravo on 9 July 2022 followed one day after the surface latent heat flux anomaly reported in Table 2 and Figure 4, suggesting a bottom-up transmission of the anomalous phenomenon.

The three most interesting tracks in the Y-East component of the geomagnetic field are reported in Figures 9–11, and the other anomalous tracks are presented in Supplementary Materials from Figure S9 to Figure S22. In particular, Figure 9 shows an anomalous signal of about 3 nT/s peak-peak intensity, but the geomagnetic index Dst is at the upper boundary limit of 20 nT. In fact, looking at the trend of Dst (e.g., on https://wdc.kugi.kyoto-u.ac.jp/dst_realtime/202208/index.html, accessed on 2 January 2023), it continues to increase in the following hours, and reaches −58 nT on 8 August 2022, so this track was acquired at the sudden commencement of a moderate geomagnetic storm. Although the signal was perfectly centred on the incoming earthquake, the external source was the most reliable one to explain this signal.

Figure 10 reports an interesting track acquired on 13 November 2022 by the Swarm Alpha satellite, with an anomaly superposed on the seismic activity recorded in the North Adriatic Sea close to the Marche Italian region, especially in November 2022 but still ongoing (in January 2023). In particular, on 9 November 2022, offshore Pesaro Urbino town, a shallow (depth estimated equal to 5 km) Mw = 5.5 earthquake was recorded with epicentral coordinates 43.983°N, 13.324°E (http://cnt.rm.ingv.it/event/33301831, accessed on 28 January 2023). The anomaly depicted by Swarm Alpha in the geomagnetic field could have been induced by the aftershock activity recorded at the same coordinates, due to the compression of the two regional tectonic plates. Noticing that anomalous precipitation had been recorded on the same day (see Table 2 and Figure S7 in Supplementary Materials), we checked if there was evidence of lightning activity. We looked at the data from the Lightning Image Sensor (LIS) onboard the International Space Station (ISS) [43], acquired just one hour after the Swarm track above Italy, and presented this in Figure S32 in Supplementary Materials. There is no evidence of lightning activity in Italy, as it was detected only in Brazil and the Papua New Guinea region. We tend to exclude that a thunderstorm could have induced this Swarm anomaly, and that a seismo-induced origin is still plausible.

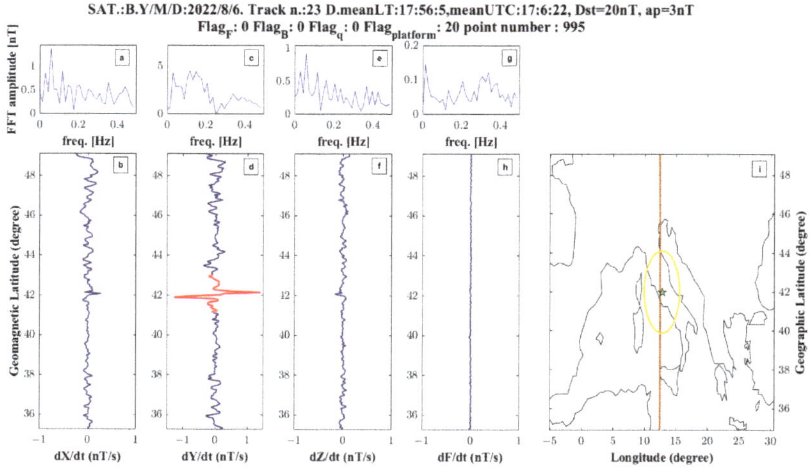

Figure 9. Swarm Bravo magnetic field analysis on 6 August 2022, track 23. The title reports the first satellite letter, date of acquisition, track number, mean local time and UT time, geomagnetic indexes, and the total number of original samples acquired with anomalous ESA Swarm data quality flags. Map (**i**) shows the projection of the satellite track in brown, and eventual anomalous flags are marked with red. The green star represents the epicentre of the M3.3 Guidonia 1 January 2023 earthquake, and the yellow circle is the Dobrovolsky of an earthquake of magnitude 4.1. The residuals of geomagnetic signals for X-North (**b**), Y-East (**d**), Z-Center (**f**) and F absolute scalar intensity (**h**) are represented with blue lines on the left, in correspondence with the map, and eventual anomalies inside the research area are marked with red lines. A spectrum of the signal made by Fast Fourier Transform (FFT) inside the minimum and maximum latitudes of the area is reported in (**a**), (**c**), (**e**) and (**g**) subfigures for X, Y, Z and F, respectively.

Figure 11 shows a track of Swarm Bravo acquired during very quiet geomagnetic conditions (Dst = −1 nT, ap = 2 nT), with an anomalous geomagnetic field Y-East component signal recorded above Central Italy. Though the track preceded by about one week the first shock of ML3.1 on 23 December 2022, and was acquired during the seismic acceleration phase, the diffuse nature of the anomaly makes it difficult to confirm that it could be seismo-induced, though that cannot be excluded.

The analysis of electron density measured by Swarm Alpha, Bravo, and Charlie in the six months before the ML3.3 Guidonia 1 January 2023 earthquake is provided in supplementary materials. In particular, Figure S23 shows the cumulate of all the anomalous tracks extracted by applying the NeLOG algorithm with a threshold kt = 2.0. The algorithm selected five anomalous tracks, shown in detail in Figures S24–S28. Among them, the only track which could have an interesting signal is the one acquired by Swarm Bravo on 16 July 2022 (Figure S24), which show oscillations with long period of electron density from a geographical latitude of 42.5°N to 44.5°N. The oscillation is between the Central Apennines and Marche regions. Furthermore, the same track reported anomalies in the geomagnetic field in all three components (Figure S11). Contemporary anomalies in the magnetic field and electron density could be produced by plasma bubbles, which are a possible pre-earthquake phenomenon, as simulated by Kuo et al. [44]. Finally, the anomalous electromagnetic signals recorded by Swarm Bravo on 16 July 2022 could be considered good candidates as pre-seismo-induced phenomena of the seismic activity of November 2022–January 2023 in the Adriatic Sea and Marche region. The other anomalous tracks presented in Figures S25–S28 show an anomalous signal typical of some period in Swarm electron density from the Langmuir probe already found in [29] for the M6.9

Aegean Sea (Greece) 2015 earthquake. Such a signal is surely not due to earthquakes, as it affects all the orbits of the Swarm satellites globally in certain periods. Further specific investigations are necessary to identify its nature, particularly if it is some instrumental noise or ionospheric signal.

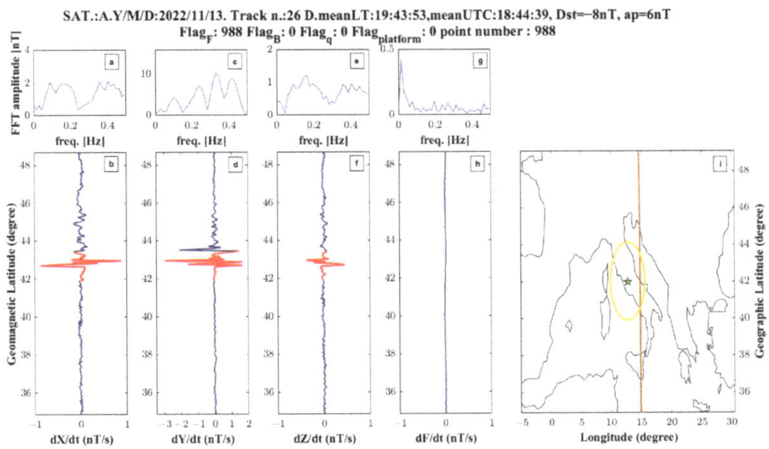

Figure 10. Swarm Alpha magnetic field analysis on 13 November 2022, track 26. The title reports the first satellite letter, date of acquisition, track number, mean local time and UT time, geomagnetic indexes, and the total number of original samples acquired with anomalous ESA Swarm data quality flags. Map (**i**) shows the projection of the satellite track in brown, and eventual anomalous flags are marked with red. The green star represents the epicentre of the M3.3 Guidonia 1 January 2023 earthquake, and the yellow circle is the Dobrovolsky of an earthquake of magnitude 4.1. The residuals of geomagnetic signals for X-North (**b**), Y-East (**d**), Z-Center (**f**) and F absolute scalar intensity (**h**) are represented with blue lines on the left, in correspondence with the map, and eventual anomalies inside the research area are marked with red lines. A spectrum of the signal made by Fast Fourier Transform (FFT) inside the minimum and maximum latitudes of the area is reported in (**a**), (**c**), (**e**) and (**g**) subfigures for X, Y, Z and F, respectively.

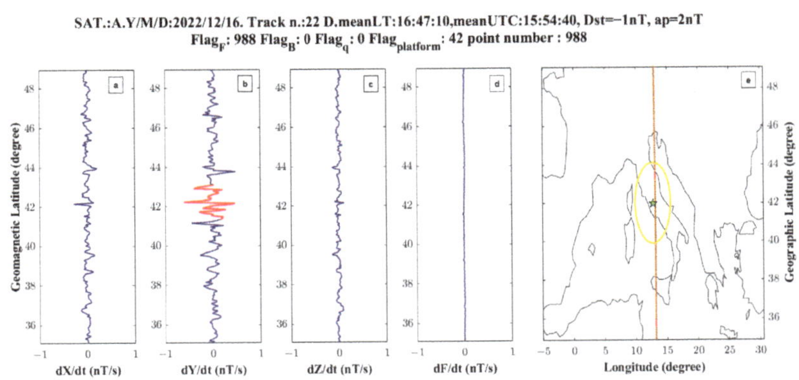

Figure 11. Swarm Bravo magnetic field analysis on 16 December 2022, track 22. The reported information is the same as in Figure 9, except for the absence of FFT spectra due to some badly flagged samples. The X, Y, and Z components are shown in (**a**–**c**), while the absolute scalar intensity is shown in (**d**) and the map with satellite track projection is shown in (**e**). The part of the track with bad samples is marked with red colour instead of brown.

4. Discussion and Conclusions

In this paper, we investigated the ML = 3.3 Guidonia (Rome, Italy) earthquake that occurred on 1 January 2023 by the analysis of seismic, atmospheric and ionospheric data. From a preliminary investigation, seismological analysis identified an acceleration of the seismicity around this epicentre, especially in the previous six months. On the other hand, the atmospheric and ionospheric investigations did not provide any results that seemed to be linked to this seismic event, but some anomalies could be seismo-induced by the other seismic activity in the same period in the Adriatic Sea and Marche Italian region. The absence of possible coupling of the lithosphere with the atmosphere and ionosphere is in agreement with the low magnitude of the Guidonia seismic event (ML = 3.3). Such a low magnitude is, in fact, not expected to produce any ionospheric effect, per Marchetti et al. [35], which proposed at least M = 5.5 or even M = 6.0, and per Perevalova [45], which suggested M6.5.

Finally, the seismological investigation not only clearly identified a seismic acceleration, but suggested that these events were likely part of a seismic sequence, and the analyses were compatible with two main hypotheses:

1. The ML3.3 of 1 January 2023 is the mainshock of the seismic quiescence (R^2-adj = 0.977 and acceleration coefficient C = 0.32);
2. The mainshock, of magnitude M4.1 (R^2-adj = 0.988 and acceleration coefficient C = 0.36), could be an incoming event in the following weeks/months.

The second hypothesis seems mathematically slightly more probable, but both hypotheses describe well the seismicity recorded in the previous months around Guidonia (Rome, Italy). In any case, the acceleration nature ($C \ll 1$) of all the reliable solutions tends to exclude that the events are part of a seismic swarm.

Even though some speculation on the evolution of the seismic activity can be done, it cannot be considered as a prediction because no tests have been ever made, so, without a validation of the method by an independent body (e.g., CSEP, https://cseptesting.org/, accessed on 2 January 2023), the analysis has no predictive value. Furthermore, the correct interpretation of the time-to-failure power law fit is that a seismic acceleration during the last six months, in a 58 km radius circle centred on Guidonia, has been detected. Further investigations are absolutely necessary to propose the future evolution of the seismic sequence. Due to the peculiarity observed in the past months, the installation of geochemical and seismic multiparametric stations could be evaluated to detect eventual gases or fluids released from the ground and improve the completeness magnitude of the area.

Finally, this example shows how these techniques could be applied in real time for monitoring (but not yet prediction) purposes. In particular, Figure 12 shows a conceptual idea of a multiparametric platform to investigate the lithosphere, atmosphere, and ionosphere. This platform could automatically download new data when available; specific algorithms need to explore the particular dataset. Finally, more complex modules would search for possible LAIC, considering the mechanism's physics (analogous to Wu et al. [46]). An external clock would trigger the several stages of the multiparametric platform. For example, in the first submission of this paper, atmospheric data for December 2022 were unavailable, and we produced the limited time series to 30 November (see Figures S29–S31 and the preprint [47] of the first submission). In the revision stage, the atmospheric data of December 2022 became available, permitting us to provide the full time series. In the future, all of these stages will be automatically executed in order to concentrate on the upper level of the geophysical interpretation and investigation of the results. Beyond this, a real-time geophysical remote sensing multiparametric platform could permit us to provide operative products, not only to better understand and monitor our vulnerable planet, but to mitigate possible negative consequences of natural hazards.

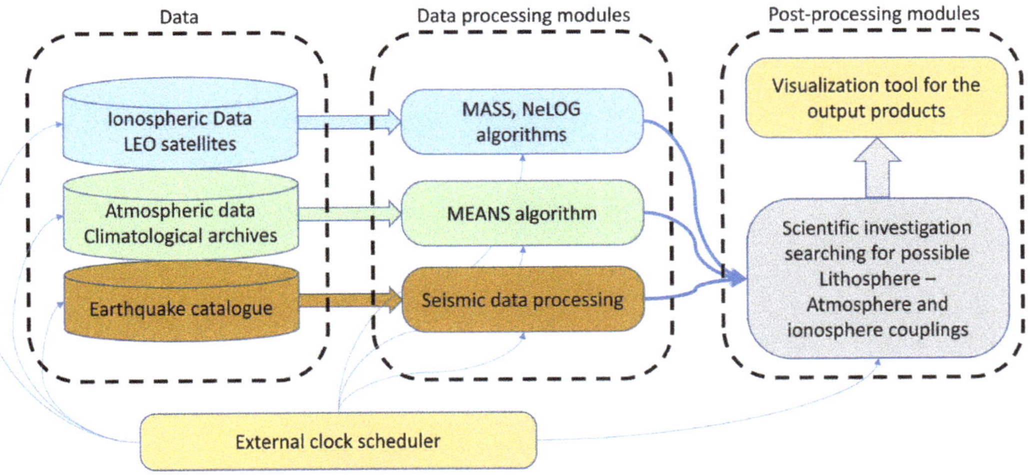

Figure 12. Concept of a real-time geophysical multiparametric platform to investigate and systematically search for lithosphere, atmosphere, and ionosphere coupling phenomena.

Supplementary Materials: The following supporting information can be downloaded at: https://www.mdpi.com/article/10.3390/rs15040942/s1. Figure S1: Gutenberg-Richeter distributions of the earthquake in the areas listed in Table 1; Figures S2 and S3: Benioff cumulative stress curves in the 5 years and 1 year before the ML=3.3 1 January 2023 Guidonia earthquake; Figure S4: Surface air temperature time series; Figure S5: Aerosol time series; Figure S6: Surface specific humidity time series; Figure S7: Total precipitation time series; Figure S8: cumulate number of Swarm Alpha, Bravo and Charlie geomagnetic field anomalies; Figures S9–S22: Swarm anomalous geomagnetic field tracks; Figure S23: cumulate number of Swarm Alpha, Bravo and Charlie electron density (Ne) anomalies; Figures S24–S28: Swarm anomalous Ne tracks; Figures S29–S31: Preliminary atmospheric time series processed immediately after the EQ occurrence (on 2 January 2023); Figure S32: Lightning activity recorded by Lightning Image Sensors (LIS) mounted on International Space Station (ISS) on 13 November 2023 between 19:29:59 UT and 21:02:59 UT. Reference [6] is cited in the supplementary materials.

Author Contributions: Conceptualisation, methodology, software, writing—original draft preparation, formal analysis, investigation, data curation, visualisation, D.M.; supervision, K.Z.; validation, L.M.; resources, project administration, funding acquisition, K.Z. and D.M; writing—review and editing, L.M., D.M., Y.Z., K.Z., W.C., Y.C., M.F., T.W., S.W., J.W., D.Z. and H.Z. All authors have read and agreed to the published version of the manuscript.

Funding: This research was funded by the National Natural Science Foundation of China, grant number 41974084; the China Postdoctoral Science Foundation, grant number 2021M691190; the International Cooperation Project of the Department of Science and Technology of Jilin Province, grant number 20200801036GH. The APC has been funded by Dedalo Marchetti.

Data Availability Statement: We acknowledge the INGV for the seismic catalogue used in this paper, freely available at http://cnt.rm.ingv.it/, accessed on 3 January 2023. MERRA-2 data can be downloaded from https://disc.gsfc.nasa.gov/datasets?project=MERRA-2 (accessed on 2 January 2023) with Earth Observation NASA free credential. Swarm data are freely available via ftp and http at swarm-diss.eo.esa.int server (accessed on 2 January 2023). We acknowledge INGV for free provision of Database of Individual Seismogenic Sources (DISS) and available at https://diss.ingv.it/ (accessed on 22 January 2023). We acknowledge Lang, Timothy 2022. Non-Quality Controlled Lightning Imaging Sensor (LIS) on International Space Station (ISS) Science Data. Dataset available online from the NASA Global Hydrometeorology Resource Center DAAC, Huntsville, Alabama, U.S.A. doi: http://dx.doi.org/10.5067/LIS/ISSLIS/DATA110.

Acknowledgments: We would acknowledge Alessandro Piscini, Rita Di Giovambattista, Gianfranco Cianchini, Francisco Javier Pavón-Carrasco, Saioa Arquero Campuzano, Dario Sabbagh, Loredana Perrone, Maurizio Soldani, and Angelo De Santis for their contributions in the preparation, writing and optimisation of some codes re-used in this work, originally developed for some of the cited papers and constructive scientific discussion on previous work whose concepts are here re-applied. We also acknowledge the International Space Science Institute (ISSI at Bern, Switzerland) for supporting International Team 553 led by Essam Ghamry and Zeren Zhima.

Conflicts of Interest: The authors declare no conflict of interest. The funders had no role in the design of the study; in the collection, analyses, or interpretation of data; in the writing of the manuscript; or in the decision to publish the results.

References

1. Faccenna, C.; Soligo, M.; Billi, A.; De Filippis, L.; Funiciello, R.; Rossetti, C.; Tuccimei, P. Late Pleistocene Depositional Cycles of the Lapis Tiburtinus Travertine (Tivoli, Central Italy): Possible Influence of Climate and Fault Activity. *Glob. Planet. Chang.* **2008**, *63*, 299–308. [CrossRef]
2. Frepoli, A.; Cimini, G.B.; De Gori, P.; De Luca, G.; Marchetti, A.; Monna, S.; Montuori, C.; Pagliuca, N.M. Seismic Sequences and Swarms in the Latium-Abruzzo-Molise Apennines (Central Italy): New Observations and Analysis from a Dense Monitoring of the Recent Activity. *Tectonophysics* **2017**, *712–713*, 312–329. [CrossRef]
3. Utsu, T.; Ogata, Y.; Matsu'ura, S.R. The Centenary of the Omori Formula for a Decay Law of Aftershock Activity. *J. Phys. Earth* **1995**, *43*, 1–33. [CrossRef]
4. ISIDe Working Group. *Italian Seismological Instrumental and Parametric Database (ISIDe)*; Istituto Nazionale di Geofisica e Vulcanologia (INGV): Rome, Italy, 2007. [CrossRef]
5. Chiaraluce, L.; Di Stefano, R.; Tinti, E.; Scognamiglio, L.; Michele, M.; Casarotti, E.; Cattaneo, M.; De Gori, P.; Chiarabba, C. Monachesi, G.; et al. The 2016 Central Italy Seismic Sequence: A First Look at the Mainshocks, Aftershocks, and Source Models *Seismol. Res. Lett.* **2017**, *88*, 757–771. [CrossRef]
6. Wiemer, S. A Software Package to Analyse Seismicity: ZMAP. *Seismol. Res. Lett.* **2001**, *72*, 373–382. [CrossRef]
7. Gutenberg, B. *Seismicity of the Earth and Associated Phenomena*; Read Books Ltd.: Worcestershire, UK, 2013.
8. Montuori, C.; Murru, M.; Falcone, G. Spatial Variation of the B-Value Observed for the Periods Preceding and Following the 24 August 2016, Amatrice Earthquake (ML 6.0) (Central Italy). *Ann. Geophys.* **2016**, *59*, 12. [CrossRef]
9. De Santis, A.; Cianchini, G.; Favali, P.; Beranzoli, L.; Boschi, E. The Gutenberg-Richter Law and Entropy of Earthquakes: Two Case Studies in Central Italy. *Bull. Seismol. Soc. Am.* **2011**, *101*, 1386–1395. [CrossRef]
10. Mignan, A. The Stress Accumulation Model: Accelerating Moment Release and Seismic Hazard. In *Advances in Geophysics*; Elsevier: Amsterdam, The Netherlands, 2008; Volume 49, pp. 67–201. ISBN 978-0-12-374231-5.
11. Dobrovolsky, I.P.; Zubkov, S.I.; Miachkin, V.I. Estimation of the Size of Earthquake Preparation Zones. *Pure Appl. Geophys.* **1979**, *117*, 1025–1044. [CrossRef]
12. Mignan, A.; King, G.C.P.; Bowman, D. A Mathematical Formulation of Accelerating Moment Release Based on the Stress Accumulation Model. *J. Geophys. Res. Solid Earth* **2007**, *112*, B07308. [CrossRef]
13. Cianchini, G.; De Santis, A.; Di Giovambattista, R.; Abbattista, C.; Amoruso, L.; Campuzano, S.A.; Carbone, M.; Cesaroni, C.; De Santis, A.; Marchetti, D.; et al. Revised Accelerated Moment Release Under Test: Fourteen Worldwide Real Case Studies in 2014–2018 and Simulations. *Pure Appl. Geophys.* **2020**, *177*, 4057–4087. [CrossRef]
14. De Santis, A.; Cianchini, G.; Marchetti, D.; Piscini, A.; Sabbagh, D.; Perrone, L.; Campuzano, S.A.; Inan, S. A Multiparametric Approach to Study the Preparation Phase of the 2019 M7.1 Ridgecrest (California, United States) Earthquake. *Front. Earth Sci.* **2020**, *8*, 540398. [CrossRef]
15. Piscini, A.; Marchetti, D.; De Santis, A. Multiparametric Climatological Analysis Associated with Global Significant Volcanic Eruptions During 2002–2017. *Pure Appl. Geophys.* **2019**, *176*, 3629–3647. [CrossRef]
16. Marchetti, D.; de Santis, A.; D'Arcangelo, S.; Poggio, F.; Piscini, A.; Campuzano, S.A.; de Carvalho, W.V.J.O. Pre-Earthquake Chain Processes Detected from Ground to Satellite Altitude in Preparation of the 2016–2017 Seismic Sequence in Central Italy. *Remote Sens. Environ.* **2019**, *229*, 93–99. [CrossRef]
17. Marchetti, D.; Zhu, K.; Zhang, H.; Zhima, Z.; Yan, R.; Shen, X.; Chen, W.; Cheng, Y.; He, X.; Wang, T.; et al. Clues of Lithosphere, Atmosphere and Ionosphere Variations Possibly Related to the Preparation of La Palma 19 September 2021 Volcano Eruption. *Remote Sens.* **2022**, *14*, 5001. [CrossRef]
18. Akhoondzadeh, M.; De Santis, A.; Marchetti, D.; Wang, T. Developing a Deep Learning-Based Detector of Magnetic, Ne, Te and TEC Anomalies from Swarm Satellites: The Case of Mw 7.1 2021 Japan Earthquake. *Remote Sens.* **2022**, *14*, 1582. [CrossRef]
19. Akhoondzadeh, M.; Marchetti, D. Developing a Fuzzy Inference System Based on Multi-Sensor Data to Predict Powerful Earthquake Parameters. *Remote Sens.* **2022**, *14*, 3203. [CrossRef]
20. Marchetti, D.; de Santis, A.; Shen, X.; Campuzano, S.A.; Perrone, L.; Piscini, A.; Di Giovambattista, R.; Jin, S.; Ippolito, A.; Cianchini, G.; et al. Possible Lithosphere-Atmosphere-Ionosphere Coupling Effects Prior to the 2018 Mw = 7.5 Indonesia Earthquake from Seismic, Atmospheric and Ionospheric Data. *J. Asian Earth Sci.* **2020**, *188*, 104097. [CrossRef]

21. Friis-Christensen, E.; Lühr, H.; Hulot, G. Swarm: A Constellation to Study the Earth's Magnetic Field. *Earth Planets Space* **2006**, *58*, 351–358. [CrossRef]
22. Akhoondzadeh, M.; De Santis, A.; Marchetti, D.; Piscini, A.; Cianchini, G. Multi Precursors Analysis Associated with the Powerful Ecuador (MW = 7.8) Earthquake of 16 April 2016 Using Swarm Satellites Data in Conjunction with Other Multi-Platform Satellite and Ground Data. *Adv. Space Res.* **2018**, *61*, 248–263. [CrossRef]
23. Zhu, K.; Li, K.; Fan, M.; Chi, C.; Yu, Z. Precursor Analysis Associated With the Ecuador Earthquake Using Swarm A and C Satellite Magnetic Data Based on PCA. *IEEE Access* **2019**, *7*, 93927–93936. [CrossRef]
24. Zhu, K.; Fan, M.; He, X.; Marchetti, D.; Li, K.; Yu, Z.; Chi, C.; Sun, H.; Cheng, Y. Analysis of Swarm Satellite Magnetic Field Data Before the 2016 Ecuador (Mw = 7.8) Earthquake Based on Non-Negative Matrix Factorization. *Front. Earth Sci.* **2021**, *9*, 621976. [CrossRef]
25. Fan, M.; Zhu, K.; De Santis, A.; Marchetti, D.; Cianchini, G.; Piscini, A.; He, X.; Wen, J.; Wang, T.; Zhang, Y.; et al. Analysis of Swarm Satellite Magnetic Field Data for the 2015 Mw 7.8 Nepal Earthquake Based on Nonnegative Tensor Decomposition. *IEEE Trans. Geosci. Remote Sens.* **2022**, *60*, 2006119. [CrossRef]
26. Ghamry, E.; Mohamed, E.K.; Sekertekin, A.; Fathy, A. Integration of Multiple Earthquakes Precursors before Large Earthquakes: A Case Study of 25 April 2015 in Nepal. *J. Atmos. Sol. Terr. Phys.* **2023**, *242*, 105982. [CrossRef]
27. Marchetti, D.; De Santis, A.; Campuzano, S.A.; Soldani, M.; Piscini, A.; Sabbagh, D.; Cianchini, G.; Perrone, L.; Orlando, M. Swarm Satellite Magnetic Field Data Analysis Prior to 2019 Mw = 7.1 Ridgecrest (California, USA) Earthquake. *Geosciences* **2020**, *10*, 502. [CrossRef]
28. De Santis, A.; Perrone, L.; Calcara, M.; Campuzano, S.A.; Cianchini, G.; D'Arcangelo, S.; Di Mauro, D.; Marchetti, D.; Nardi, A.; Orlando, M.; et al. A Comprehensive Multiparametric and Multilayer Approach to Study the Preparation Phase of Large Earthquakes from Ground to Space: The Case Study of the June 15 2019, M7.2 Kermadec Islands (New Zealand) Earthquake. *Remote Sens. Environ.* **2022**, *283*, 113325. [CrossRef]
29. De Santis, A.; Marchetti, D.; Spogli, L.; Cianchini, G.; Pavón-Carrasco, F.J.; de Franceschi, G.; Di Giovambattista, R.; Perrone, L.; Qamili, E.; Cesaroni, C.; et al. Magnetic Field and Electron Density Data Analysis from Swarm Satellites Searching for Ionospheric Effects by Great Earthquakes: 12 Case Studies from 2014 to 2016. *Atmosphere* **2019**, *10*, 371. [CrossRef]
30. Marchetti, D.; Akhoondzadeh, M. Analysis of Swarm Satellites Data Showing Seismo-Ionospheric Anomalies around the Time of the Strong Mexico (Mw = 8.2) Earthquake of 08 September 2017. *Adv. Space Res.* **2018**, *62*, 614–623. [CrossRef]
31. Akhoondzadeh, M.; de Santis, A.; Marchetti, D.; Shen, X. Swarm-TEC Satellite Measurements as a Potential Earthquake Precursor Together With Other Swarm and CSES Data: The Case of Mw7.6 2019 Papua New Guinea Seismic Event. *Front. Earth Sci.* **2022**, *10*, 820189. [CrossRef]
32. Christodoulou, V.; Bi, Y.; Wilkie, G. A Tool for Swarm Satellite Data Analysis and Anomaly Detection. *PLoS ONE* **2019**, *14*, e0212098. [CrossRef]
33. de Santis, A.; Balasis, G.; Pavón-Carrasco, F.J.; Cianchini, G.; Mandea, M. Potential Earthquake Precursory Pattern from Space: The 2015 Nepal Event as Seen by Magnetic Swarm Satellites. *Earth Planet. Sci. Lett.* **2017**, *461*, 119–126. [CrossRef]
34. de Santis, A.; Marchetti, D.; Pavón-Carrasco, F.J.; Cianchini, G.; Perrone, L.; Abbattista, C.; Alfonsi, L.; Amoruso, L.; Campuzano, S.A.; Carbone, M.; et al. Precursory Worldwide Signatures of Earthquake Occurrences on Swarm Satellite Data. *Sci. Rep.* **2019**, *9*, 20287. [CrossRef]
35. Marchetti, D.; De Santis, A.; Campuzano, S.A.; Zhu, K.; Soldani, M.; D'Arcangelo, S.; Orlando, M.; Wang, T.; Cianchini, G.; Di Mauro, D.; et al. Worldwide Statistical Correlation of Eight Years of Swarm Satellite Data with M5.5+ Earthquakes: New Hints about the Preseismic Phenomena from Space. *Remote Sens.* **2022**, *14*, 2649. [CrossRef]
36. Xiong, P.; Marchetti, D.; De Santis, A.; Zhang, X.; Shen, X. SafeNet: SwArm for Earthquake Perturbations Identification Using Deep Learning Networks. *Remote Sens.* **2021**, *13*, 5033. [CrossRef]
37. Sasmal, S.; Chowdhury, S.; Kundu, S.; Politis, D.Z.; Potirakis, S.M.; Balasis, G.; Hayakawa, M.; Chakrabarti, S.K. Pre-Seismic Irregularities during the 2020 Samos (Greece) Earthquake (M = 6.9) as Investigated from Multi-Parameter Approach by Ground and Space-Based Techniques. *Atmosphere* **2021**, *12*, 1059. [CrossRef]
38. Piscini, A.; de Santis, A.; Marchetti, D.; Cianchini, G. A Multiparametric Climatological Approach to Study the 2016 Amatrice–Norcia (Central Italy) Earthquake Preparatory Phase. *Pure Appl. Geophys.* **2017**, *174*, 3673–3688. [CrossRef]
39. Allard, P.; Carbonnelle, J.; Métrich, N.; Loyer, H.; Zettwoog, P. Sulphur Output and Magma Degassing Budget of Stromboli Volcano. *Nature* **1994**, *368*, 326–330. [CrossRef]
40. Pulinets, S.; Ouzounov, D. Lithosphere–Atmosphere–Ionosphere Coupling (LAIC) Model—An Unified Concept for Earthquake Precursors Validation. *J. Asian Earth Sci.* **2011**, *41*, 371–382. [CrossRef]
41. Freund, F. Pre-Earthquake Signals: Underlying Physical Processes. *J. Asian Earth Sci.* **2011**, *41*, 383–400. [CrossRef]
42. Marchetti, D.; De Santis, A.; D'Arcangelo, S.; Poggio, F.; Jin, S.; Piscini, A.; Campuzano, S.A. Magnetic Field and Electron Density Anomalies from Swarm Satellites Preceding the Major Earthquakes of the 2016–2017 Amatrice-Norcia (Central Italy) Seismic Sequence. *Pure Appl. Geophys.* **2020**, *177*, 305–319. [CrossRef]
43. Lang, T. *Non-Quality Controlled Lightning Imaging Sensor (LIS) on International Space Station (ISS) Science Data*; NASA Global Hydrology Resource Center DAAC: Huntsville, AL, USA, 2022. [CrossRef]
44. Kuo, C.L.; Lee, L.C.; Huba, J.D. An Improved Coupling Model for the Lithosphere-Atmosphere-Ionosphere System. *J. Geophys. Res. Space Phys.* **2014**, *119*, 3189–3205. [CrossRef]

45. Perevalova, N.P.; Sankov, V.A.; Astafyeva, E.I.; Zhupityaeva, A.S. Threshold Magnitude for Ionospheric TEC Response to Earthquakes. *J. Atmos. Sol. Terr. Phys.* **2014**, *108*, 77–90. [CrossRef]
46. Wu, L.; Qi, Y.; Mao, W.; Lu, J.; Ding, Y.; Peng, B.; Xie, B. Scrutinizing and Rooting the Multiple Anomalies of Nepal Earthquake Sequence in 2015 with the Deviation–Time–Space Criterion and Homologous Lithosphere–Coversphere–Atmosphere–Ionosphere Coupling Physics. *Nat. Hazards Earth Syst. Sci.* **2023**, *23*, 231–249. [CrossRef]
47. Marchetti, D.; Zhu, K.; Marchetti, L.; Zhang, Y.; Chen, W.; Cheng, Y.; Fan, M.; Wang, S.; Wang, T.; Wen, J.; et al. Quick Report of the ML = 3.3 on 1 January 2023 Guidonia (Rome, Italy) Earthquake: Evidence of a Seismic Acceleration. *Preprints* **2023**, 2023010067. [CrossRef]

remote sensing MDPI

Article

Pre-Seismic Anomaly Detection from Multichannel Infrared Images of FY-4A Satellite

Yingbo Yue [1,2], Fuchun Chen [1,*] and Guilin Chen [1]

1 Key Laboratory of Infrared System Detection and Imaging Technology, Shanghai Institute of Technical Physics, Chinese Academy of Sciences, Shanghai 200083, China
2 University of Chinese Academy of Sciences, Beijing 100049, China
* Correspondence: fuchun.chen@mail.sitp.ac.cn

Abstract: Research of seismic infrared remote sensing has been undertaken for several decades, but there is no stable and effective earthquake prediction method. A new algorithm combining the long short-term memory and the density-based spatial clustering of applications with noise models is proposed to extract the anomalies from the multichannel infrared remote sensing images of the Fengyun-4 satellites. A statistical analysis is used to validate the correlation between the anomalies and earthquakes. The results show that the correlation rate is 64.29%, the hit rate is 68.75%, and the probability gain is about 1.91. In the Madoi and YangBi earthquake cases, the infrared anomaly detected in this paper is correlated with the TEC anomaly found in the previous research. This indicates that it is feasible to combine multi-source data to improve the accuracy of earthquake prediction in future studies.

Keywords: pre-seismic anomaly; anomaly detection multichannel infrared images; remote sensing

Citation: Yue, Y.; Chen, F.; Chen, G. Pre-Seismic Anomaly Detection from Multichannel Infrared Images of FY-4A Satellite. *Remote Sens.* **2023**, *15*, 259. https://doi.org/10.3390/rs15010259

Academic Editor: Salvatore Stramondo

Received: 10 November 2022
Revised: 26 December 2022
Accepted: 28 December 2022
Published: 1 January 2023

1. Introduction

Earthquake prediction is a complicated issue [1]. The application of multi-source data may benefit to collect more detailed surface and atmospheric information and interpret pre-seismic anomalies [2]. Researchers have found multiple types of pre-seismic anomalies, including crustal deformation [3], infrared radiation [4], temperature [5,6], humidity [7], electromagnetic field [8,9], atmospheric composition [10], and so on. Huang [11] found changes in the seismicity pattern before a strong earthquake.

The previous research on seismic infrared remote sensing includes three aspects: the mechanism of anomaly [12–14], anomaly detection methods [15], and correlation analysis [16]. According to experiments and observations, changes in the earth's crust, its atmosphere, and the earth's electric field may cause anomalies of infrared radiation [14,17]. The data from the NOAA [18], Terra [19], Aqua [20], and Fengyun satellites [4,21,22] could capture infrared images continuously over a large area for a long time and have been applied to analyze earthquake precursors. The earth's system is complex and infrared radiation is affected by several factors [23]. The changes in infrared radiation caused by earthquakes are difficult to observe directly. It is necessary to eliminate the background field and high-frequency noise before anomaly detection. Many signal processing methods have also been used to detect this pre-seismic anomaly, such as robust satellite techniques (RST) [24,25], wavelet transforms [26], power spectrum [4], and other classical signal processing methods. Machine learning has also been used to detect anomalies related to earthquakes. Akhoondzadeh et al. [27] compared the classical methods and artificial intelligence in the pre-seismic anomaly detection from infrared data. Researchers have used the Kalman filter [28], auto-regressive integrated moving average [29], support vector machines [30], neural networks, and genetic algorithm [31] to predict the time series of infrared or total electron content (TEC) data for pre-seismic anomaly detection. Zhai et al. [32] used the ARIMA model to detect the infrared anomalies before two earthquakes. Jing et al. [33]

found anomalies in microwave brightness temperature and outgoing longwave radiation before the Madoi earthquake. Yang et al. [34] found the thermal infrared brightness temperature anomalies before this event.

Most of the infrared data used for earthquake prediction are long-wave infrared images, and a few are medium-wave infrared images [35]. In this paper, the multichannel infrared remote sensing images from the FY-4A satellite were applied, which included long-wave and medium-wave infrared images, and a new algorithm based on the long short-term memory (LSTM) and the density-based spatial clustering of applications with noise (DBSAN) models was proposed to detect the anomalies. The anomalies and earthquakes were analyzed statistically to validate the performance of this method in earthquake prediction.

2. Materials and Methods

2.1. Data and Study Area

The Fengyun-4 satellites are China's latest generation of geostationary-orbit meteorological satellites. The first one (FY-4A) among the Fengyun-4 satellites was launched in December 2016 and has been providing data since 2018. The FY-4A satellite has two main optical detectors. One is the advanced geostationary radiation imager (AGRI), and the other is the geostationary interferometric infrared sounder (GIIRS). The AGRI of FY-4A could obtain images from 14 channels, including visible and infrared light [36]. The wavelengths of these channels are shown in Table 1. In this paper, the data of the FY-4A satellite from 2019 to 2021 were used to analyze the multichannel infrared anomalies before earthquakes. The former two-year (2019–2020) data were used to train the model, and the last one-year (2021) data were used to detect anomalies. The radiations emitted by the earth's surface are mainly medium-wave infrared and long-wave infrared light, so only nighttime data from the seventh to the fourteenth channel were applied to avoid the influence of sunshine.

Table 1. The wavelengths of 14 channels of the advanced geostationary radiation imager (AGRI).

Channel	Wavelength (μm)
1	0.45~0.49
2	0.55~0.75
3	0.75~0.90
4	1.36~1.39
5	1.58~1.64
6	2.10~2.35
7	3.50~4.00
8	3.50~4.00
9	5.80~6.70
10	6.90~7.30
11	8.00~9.00
12	10.30~11.30
13	11.50~12.50
14	13.20~13.80

China is one of the countries with frequent earthquakes. The China earthquake networks center (www.ceic.ac.cn (accessed on 11 August 2022)) provided the earthquake list. There were 106 earthquakes with a magnitude over 5 in China and its surrounding area (from 70°E to 140°E, from 0°N to 60°N) in 2021. The seismic information is collected in Supplementary Materials. The geographic locations of the epicenters are shown in Figure 1. The strongest earthquake in the study area is the M 7.4 Madoi earthquake on 22 May 2021 (UTC+8). The Yangbi earthquake with a magnitude of 6.4 occurred about 4 h before the Madoi earthquake. There may be a series of earthquakes in the same region during a short term. The strongest earthquake in the earthquake sequence is the mainshock, and multiple foreshocks or aftershocks may occur before and after it. Taking the Madoi M7.4 and Yangbi M6.4 earthquakes in May 2021 as examples, these two earthquakes and their foreshock and aftershocks are shown in Table 2. The Yangbi earthquake with a magnitude of 6.4

occurred at 21:48:34 on 21 May 2021 (UTC+8). The epicenter was at 99.87°E, 25.67°N. The depth of the hypocenter was 8 km. There was one foreshock and two aftershocks with a magnitude of over five around it. The Madoi earthquake with a magnitude of 7.4 occurred at 02:04:11 on 22 May 2021, (UTC+8). The epicenter was at 98.34°E, 34.59°N. The depth of the hypocenter was 17 km. There was one aftershock with a magnitude of over five around it. The study area in this paper is the region in the red box in Figure 1 (from 90°E to 110°E, from 17°N to 42°N), and there were 20 earthquakes with a magnitude of over 5, including 16 mainshocks.

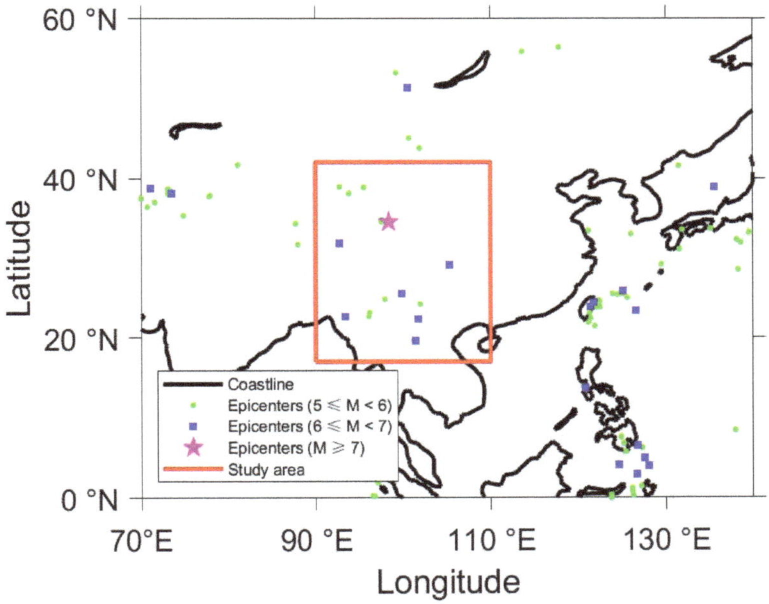

Figure 1. The epicenters of the earthquakes in 2021.

Table 2. Seismic list around Madoi M7.4 and Yangbi M6.4 earthquakes.

Time (UTC+8)	Magnitude	Latitude (°N)	Longitude (°E)	Depth (km)	Type
22 May 2021, 10:29:34	5.1	34.85	97.5	10	Aftershock
22 May 2021, 02:04:11	7.4	34.59	98.34	17	Mainshock
21 May 2021, 22:31:10	5.2	25.59	99.97	8	Aftershock
21 May 2021, 21:55:28	5.0	25.67	99.89	8	Aftershock
21 May 2021, 21:48:34	6.4	25.67	99.87	8	Mainshock
21 May 2021, 21:21:25	5.6	25.63	99.92	10	Foreshock

2.2. *Anomaly Detection*

2.2.1. Long Short-Term Memory (LSTM)

The data with a spatial resolution of 4 km are used in this paper. The changes in infrared radiation in the nearby region are similar, and the seismic infrared anomalies commonly covered a large region [37]. The original data were down-sampled with an interval of 5 pixels to improve the efficiency of data processing. The brightness temperatures were obtained by looking up the calibration table, and the everyday average was calculated with the brightness temperature during nighttime (from 23:00 to 04:00 UTC+8). The preprocessed data are four-dimension data (two-dimensional geographic coordinates, time, and channel).

In this paper, the model based on the LSTM is used to predict the multichannel infrared data. The model comprised two layers (an LSTM layer and a fully connected layer). The structure of the LSTM layer is shown in Figure 2 [38]. $X_{t,m,n}$ is the input variable comprised of three-day and eight-channel data, as Equations (1) and (2) show [39]:

$$X_{t,m,n} = \begin{bmatrix} S_{t-3,m,n} \\ S_{t-2,m,n} \\ S_{t-1,m,n} \end{bmatrix}, \tag{1}$$

$$S_{t-k,m,n} = \left[BT_{t-k,m,n,7}, BT_{t-k,m,n,8}, \ldots, BT_{t-k,m,n,14}\right], \tag{2}$$

where $BT_{t-k,m,n,p}$ denotes the brightness temperature of the data at the mth line and the nth column at the $t-k$ moment in the pth channel. In Figure 2, $H_{t,m,n}$ is the output of the LSTM layer at the t moment. It was transferred to the state at the $t+1$ moment, and it was also inputted into the next layer to calculate the predicted value $\hat{Y}_{t,m,n}$, which has the same dimension as the output variables $Y_{t,m,n}$, as Equations (3) and (4) show:

$$\hat{Y}_{t,m,n} = \left[\widehat{BT}_{t,m,n,7}, \widehat{BT}_{t,m,n,8}, \ldots, \widehat{BT}_{t,m,n,14}\right], \tag{3}$$

$$Y_{t,m,n} = \left[BT_{t,m,n,7}, BT_{t,m,n,8}, \ldots, BT_{t,m,n,14}\right], \tag{4}$$

where $BT_{t,m,n,p}$ denotes the brightness temperature of the data at the mth line and the nth column at the t moment in the pth channel; $\widehat{BT}_{t,m,n,p}$ denotes the predicted value at the mth line and the nth column at the t moment in the pth channel.

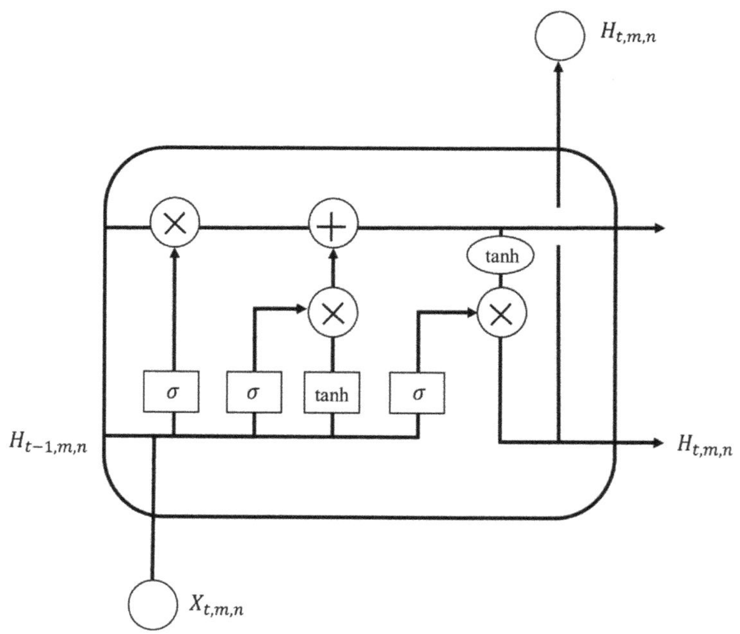

Figure 2. The structure of the LSTM layer.

The root-mean-square error of the predicted and observed values is shown in Equation (5) [32]:

$$RMSE_P = \sqrt{\frac{\sum_{m=1}^{M}\sum_{n=1}^{N}\sum_{t=1}^{T}\left(\widehat{BT}_{t,m,n,p} - BT_{t,m,n,p}\right)^2}{M \times N \times T}}, \tag{5}$$

where $RMSE_P$ is the root-mean-square error in the pth channel; M and N are the numbers of the lines and columns of the image; and T is the time length of the data.

2.2.2. Density-Based Spatial Clustering of Application with Noise (DBSCAN)

The difference between the observed and predicted values is shown in Equation (6):

$$\begin{aligned}\Delta Y_{t,m,n} &= Y_{t,m,n} - \widehat{Y}_{t,m,n} \\ &= \left[BT_{t,m,n,7} - \widehat{BT}_{t,m,n,7}, BT_{t,m,n,8} - \widehat{BT}_{t,m,n,8}, \ldots, BT_{t,m,n,14} - \widehat{BT}_{t,m,n,14}\right] \\ &= \left[\Delta BT_{t,m,n,7}, \Delta BT_{t,m,n,8}, \ldots, \Delta BT_{t,m,n,14}\right]\end{aligned} \tag{6}$$

where $\Delta BT_{t,m,n,p}$ is the difference between the observed and predicted values at the mth line and the nth column at the t moment in the pth channel. The data in different channels were normalized, as Equations (7)–(10) show:

$$R_{t,m,n} = \left[R_{t,m,n,7}, R_{t,m,n,8}, \ldots, R_{t,m,n,14}\right], \tag{7}$$

$$R_{t,m,n,p} = \frac{\Delta BT_{t,m,n,8} - \mu_p}{\sigma_p}, \tag{8}$$

$$\mu_p = \frac{\sum_{m=1}^{M}\sum_{n=1}^{N}\sum_{t=1}^{T}\Delta BT_{t,m,n,p}}{M \times N \times T}, \tag{9}$$

$$\sigma_p = \sqrt{\frac{\sum_{m=1}^{M}\sum_{n=1}^{N}\sum_{t=1}^{T}\left(\Delta BT_{t,m,n,p} - \mu_p\right)^2}{M \times N \times T}}, \tag{10}$$

where $R_{t,m,n}$ is the normalized data; the μ_p and σ_p are the average and standard deviation of $\Delta BT_{t,m,n,p}$. The DBSCAN model could be used to detect anomalies. The Euclidean distance $D_{t_1,m_1,n_1,t_2,m_2,n_2}$ between two points ($P_1(t_1, m_1, n_1)$ and $P_2(t_2, m_2, n_2)$) is shown in Equation (11):

$$D_{t_1,m_1,n_1,t_2,m_2,n_2} = \sqrt{\sum_{p=7}^{14}\left(R_{t_1,m_1,n_1,p} - R_{t_2,m_2,n_2,p}\right)^2}, \tag{11}$$

For a given radius (E_0) and minimum integer (M_0), a point $P_c(t_c, m_c, n_c)$ is a core point if there are more than M_0 points within E_0 around it. It is denoted as $f_c(t_c, m_c, n_c) = 1$, as Equations (12) and (13) show [40]:

$$f_c(t_c, m_c, n_c) = \begin{cases}1, if \ \sum_{m=1}^{M}\sum_{n=1}^{N}\sum_{t=1}^{T} h(D_{t_c,m_c,n_c,t,m,n}) > M_0 \\ 0, otherwise\end{cases}, \tag{12}$$

$$h(x) = \begin{cases}1, if \ x \le E_0 \\ 0, otherwise\end{cases}. \tag{13}$$

A point $P_b(t_b, m_b, n_b)$ is a border point if it is not a core point and there is any core point within E_0 around it. It is denoted as $f_b(t_b, m_b, n_b) = 1$, as Equation (14) shows [40]:

$$f_b(t_b, m_b, n_b) = \begin{cases}1, if \ f_c(t_b, m_b, n_b) = 0 \ and \ \sum_{m=1}^{M}\sum_{n=1}^{N}\sum_{t=1}^{T} f_c(t, m, n) \times h(D_{t_b,m_b,n_b,t,m,n}) > 0 \\ 0, otherwise\end{cases}. \tag{14}$$

Other points $P_n(t_n, m_n, n_n)$ are noise points. In this paper, the noise points are regarded as abnormal points. It is denoted as $f_n(t_n, m_n, n_n) = 1$, as Equation (15) shows [40]:

$$f_n(t_n, m_n, n_n) = \begin{cases}1, if \ f_c(t_n, m_n, n_n) = 0 \ and \ f_b(t_n, m_n, n_n) = 0 \\ 0, otherwise\end{cases}. \tag{15}$$

2.3. Statistical Method

An anomaly detection algorithm needs to be validated statistically. We assume that the pre-seismic infrared anomalies commonly occurred near the epicenter. For the given predicted radius (R_0) and predicted time window (W_0), the region within R_0 around the anomaly and the time within W_0 after the anomaly determine the predicted range. The hit rate (HR) is defined in Equation (16) [41]:

$$HR = \frac{N_P}{N_E}, \quad (16)$$

where N_P is the number of earthquakes within the predicted range, and N_E is the number of all earthquakes within the study region. The two points ($P_1(t_1, m_1, n_1)$ and $P_2(t_2, m_2, n_2)$) belong to an anomaly sample, if they follow the condition of Equations (17)–(21):

$$f_n(t_1, m_1, n_1) = 1, \quad (17)$$

$$f_n(t_2, m_2, n_2) = 1, \quad (18)$$

$$|t_1 - t_2| \leq 1, \quad (19)$$

$$|m_1 - m_2| \leq 1, \quad (20)$$

$$|n_1 - n_2| \leq 1, \quad (21)$$

The correlation rate (CR) is defined in Equation (22) [42]:

$$CR = \frac{N_Y}{N_A}, \quad (22)$$

where N_Y is the number of anomaly samples related to earthquakes, and N_A is the number of all anomaly samples. The probability gain (Gain) is defined in Equation (23) [43]:

$$Gain = \frac{HR}{\rho}, \quad (23)$$

where ρ is the spatiotemporal occupancy of the predicted range.

3. Results

3.1. The Deviation of the LSTM Model

Both the pre-processed and predicted data are four-dimensional data. The data at 99.13°E, 29.68°N are shown in Figure 3. The pre-processed data are complicated and varies irregularly. The brightness temperatures in the 9th, 10th, and 14th channels are lower than those in the other channels because of the absorption of water vapors in the atmosphere. The deviation data reveal the difference between the observed and predicted data. As Figure 4 shows, it fluctuates around zero. The blue line is the deviation for the training dataset, and the orange line is the deviation for the test dataset. The images on a randomly selected date (30 January 2021) are used as examples to compare the original data with the deviation images. The images of the brightness temperature on 30 January 2021 are shown in Figure 5, and the deviation images are shown in Figure 6, where the digits denote the terrain in kilometers. The geospatial distribution of brightness temperature is related to the latitude and the topography. There were lower temperatures in the region with a high latitude and a high altitude. The spatiotemporal difference near the absorption band of water vapor was relatively less pronounced. The deviation images show the independence of the latitude. This indicates that the LSTM model could eliminate the background field of infrared brightness temperature, especially the difference caused by the latitude.

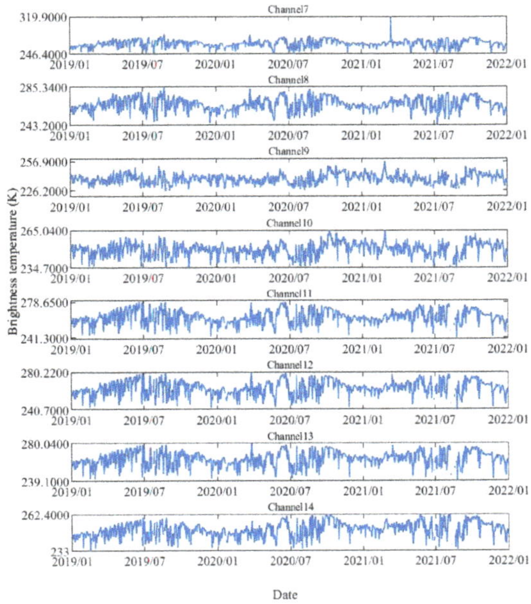

Figure 3. The pre-processed data at 99.13°E, 29.68°N.

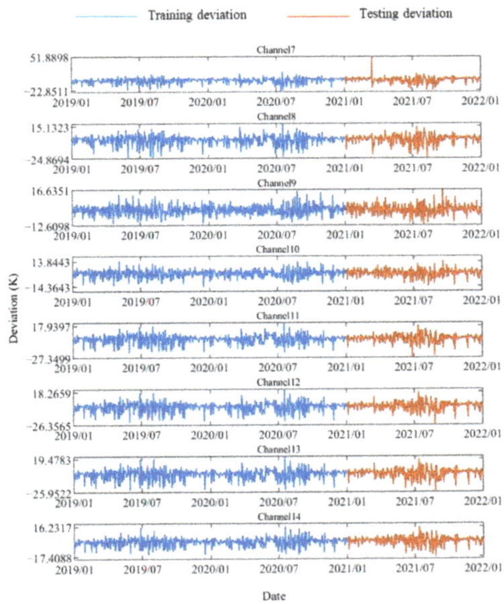

Figure 4. The deviation data at 99.13°E, 29.68°N.

The training and testing scores of the LSTM model are shown in Table 3. A smaller score, which means a smaller prediction deviation, indicates a better performance of the model. The RMSEs in the channels around the water vapor absorption band (channel 9, 10, and 14) are lower. Among the other channels, the RMSEs in the long-wave infrared channels (channels 11, 12, and 13) are slightly higher than those in the medium-wave infrared channels (channels 7 and 8). The frequency distributions of the deviation values are shown in Figure 7. Most of the deviations are around zero.

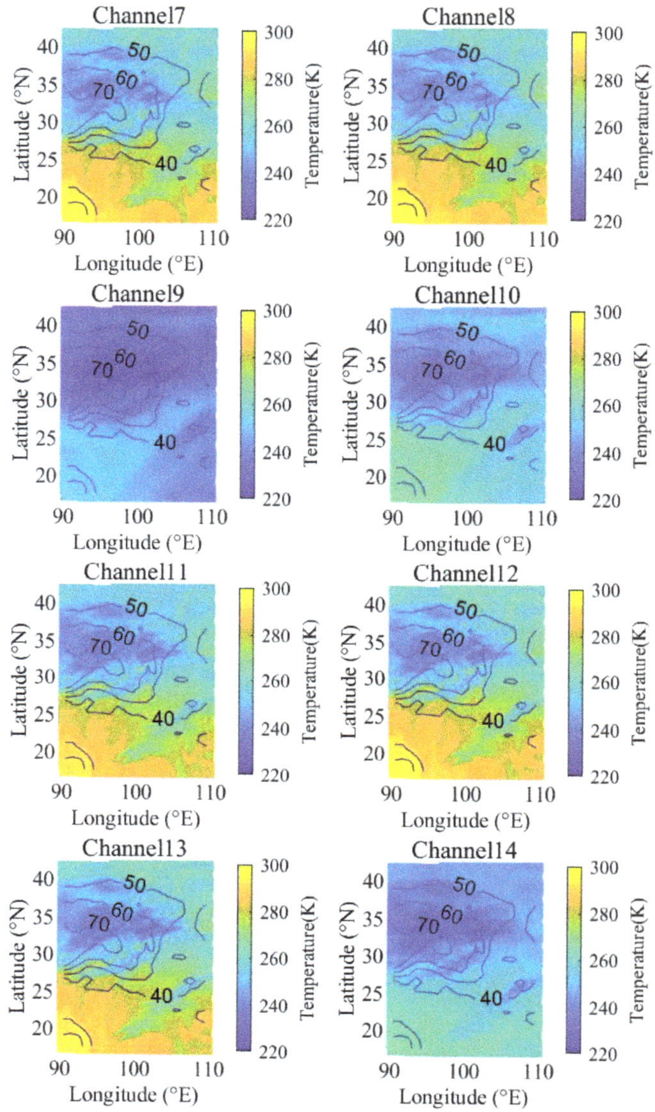

Figure 5. The images of brightness temperature on 30 January 2021.

Table 3. The training and testing scores of the LSTM model.

Channel	Training Score (μm)	Testing Score (μm)
7	7.4826	7.8116
8	7.2683	7.3682
9	4.3431	4.4512
10	4.9965	5.0628
11	7.7708	7.9001
12	8.4068	8.5414
13	8.624	8.7911
14	5.1109	5.4133

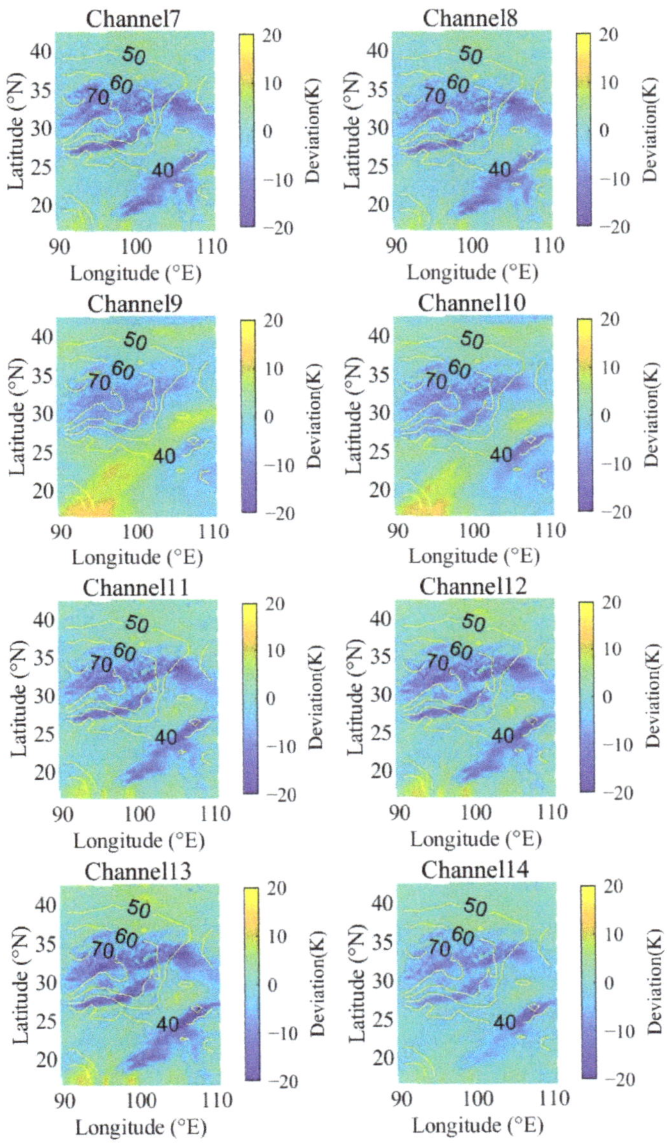

Figure 6. The deviation images on 30 January 2021.

3.2. Clustering Results

The clustering results of the DBSCAN model depend on its two parameters (E_0 and M_0). The abnormal rates for different parameters are shown in Table 4. These are the proportions of abnormal points in the whole study area. For the same radius, the larger the minimum integer is, the larger the abnormal rate is. A larger radius increases the amount of computation. Considering the algorithm efficiency and anomaly rate, the radius of 2 and the minimum integer of 100 are chosen in this study. For the chosen parameters ($E_0 = 2$, $M_0 = 100$), the deviation data were classified into two classes. About 4% of all the points were abnormal, and the others were normal.

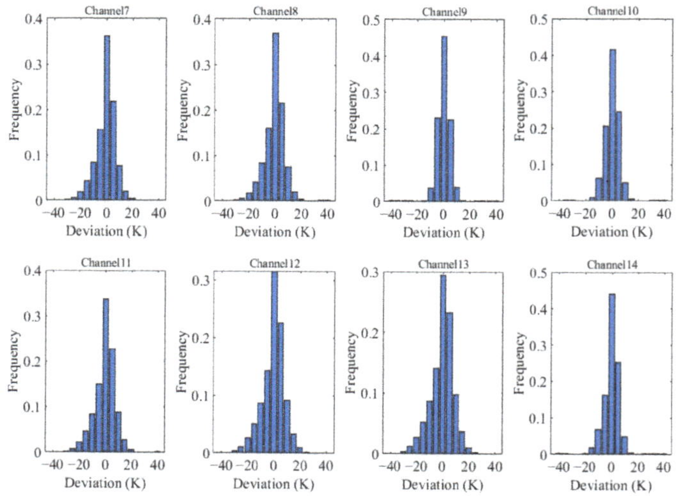

Figure 7. The frequency distributions of deviation values.

Table 4. The abnormal rate for different radii and minimum integers.

Radius E_0	Minimum Integer M_0	Abnormal Rate (%)
1	10	14.17
1	10^2	29.23
1	10^3	82.01
2	10	2.54
2	10^2	4.02
2	10^3	6.72
3	10	1.12
3	10^2	1.89
3	10^3	2.24

All anomalies within one month before the Madoi and Yangbi earthquakes in the study area are recorded in Table 5. The start time is the number of days relative to the Yangbi Earthquake time (21 May 2021, UTC+8). The duration is the number of days from the anomaly starting to the anomaly ending. Distance 1 is the distance between the anomaly and Yangbi epicenter, and Distance 2 is the distance between the anomaly and Madoi epicenter. The coverage area is the number of pixels covered by the anomaly. The longest anomaly occurred from 8 to 13 May 2021. It is about 420 km away from the Yangbi epicenter and 389 km away from the Madoi epicenter. The anomaly images are shown in Figure 8. The black points represent the abnormal points marked by the DBSCAN model. The two red stars represent the locations of the two mainshocks. The coverage area is the largest on 12 May 2021, which is 1157 pixels.

Table 5. The information on anomalies from 20 April 2021 to 20 May 2021 (UTC+8).

Start Time (Day)	Duration (Day)	Distance 1 (km)	Distance 2 (km)	Coverage Area (Pixel)
−29	2	1218.514	522.3128	335
−27	2	1614.612	1082.72	76
−27	3	694.1449	147.2815	233
−27	5	357.3739	1338.696	1310
−22	2	256.0721	989.042	174
−21	4	416.9308	308.354	777
−19	2	1818.656	871.9	65

Table 5. *Cont.*

Start Time (Day)	Duration (Day)	Distance 1 (km)	Distance 2 (km)	Coverage Area (Pixel)
−19	2	774.8023	1100.769	295
−17	4	11.84356	769.3796	809
−16	2	633.2236	515.8192	73
−14	2	622.0482	1264.55	138
−13	6	420.0436	388.9196	1157
−12	2	1438.045	704.7843	40
−11	4	297.1712	573.796	1261
−10	3	415.1837	1417.791	48
−10	2	629.9462	1580.574	85
−8	2	1046.354	79.24198	545
−6	2	682.544	1520.562	160
−5	5	412.239	1011.247	562
−4	2	557.2012	554.8826	349

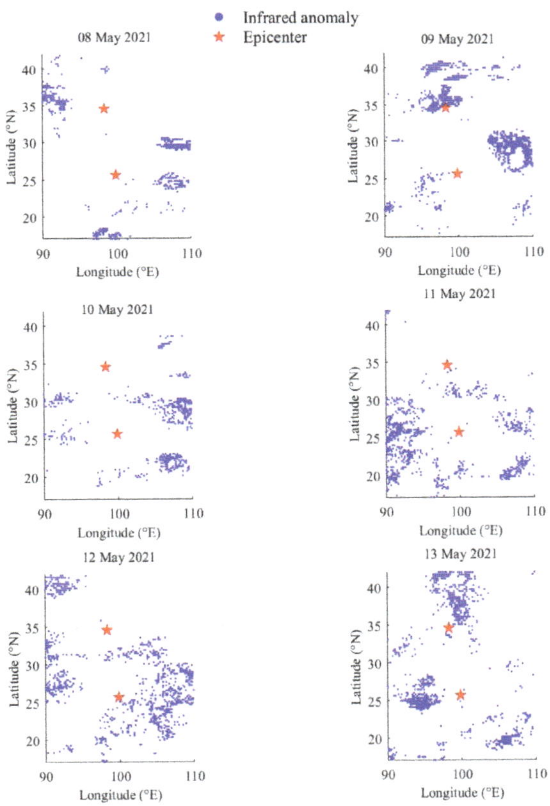

Figure 8. The anomaly images from 8 to 13 May 2021.

Figures 9 and 10 show the anomaly information of the Yangbi and Madoi earthquakes, respectively. The size of the scatter denotes the anomaly area (S), while the color denotes the anomaly duration. The coordinates denote the spatiotemporal interval between the anomaly and earthquake. The red dotted line is the radius of the seismogenic region calculated according to the so-called Dobrovolsky area, defined as Equation (24) [44]:

$$R = 10^{0.43M}, \tag{24}$$

where M is the magnitude of the earthquake. The large-area and long-duration anomalies are within the seismogenic zone (the region below the red dotted line in Figures 9 and 10).

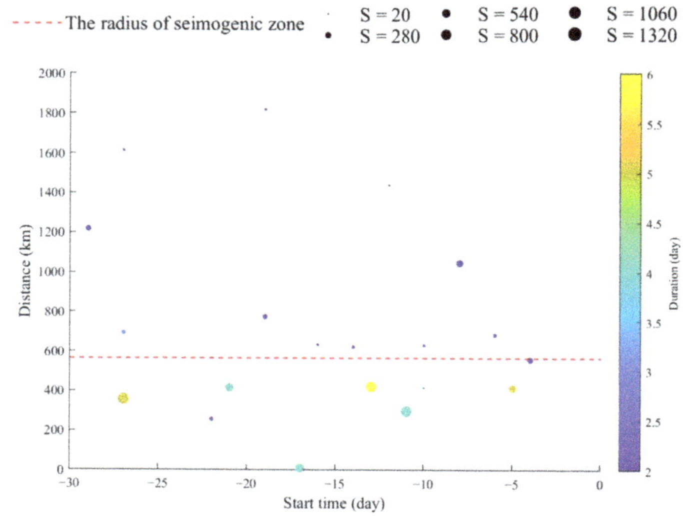

Figure 9. The anomaly information of Yangbi earthquake.

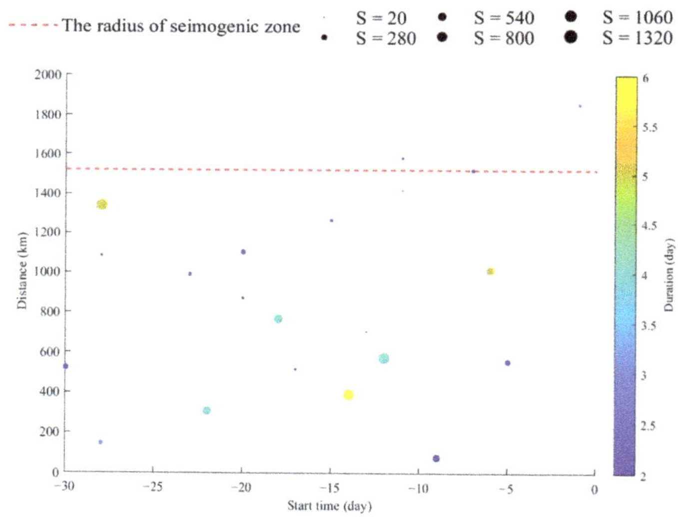

Figure 10. The anomaly information of Madoi earthquake.

3.3. Statistical Results

For the given predicted time window (W_0) of 14 days, Table 6 shows the correlation rate, the hit rate, and the probability gain with different predicted radii (R_0). The anomaly samples with a duration of over 5 days and a coverage area over 1000 pixels are used to determine the predicted range. There are 14 anomaly samples and 20 earthquakes with a magnitude over 5 in the study area, including 16 mainshocks, 1 foreshock, and 3 aftershocks. The magnitudes of the foreshock and aftershocks are no more than 5.6, as Table 2 shows. The foreshocks and aftershocks are excluded because they occupy a similar spatiotemporal area to the main shock. The highest probability gain is 1.9137, which is calculated with the predicted radius of 600 km. This means that multichannel infrared data could reduce the uncertainty of earthquake prediction. Meanwhile, the correlation is 64.29%, and the hit rate is 68.75%. This means that most of the anomaly samples are related to the earthquakes, and the method we proposed could detect the anomalies from multichannel infrared images before most of the earthquakes. The Molchan diagram is shown in Figure 11. The horizontal axis is the spatiotemporal occupancy of the predicted range, and the vertical axis is the miss rate (1-HR) [45]. The slope of the line through the star in Figure 12 is negative and is equal to the negative of the probability gain [46]. There are both the lowest miss rate and highest probability gain when the predicted radius is 600 km.

Table 6. The correlation rate, the hit rate, and the probability gain with different parameters.

Predicted Radius (km)	Correlation Rate	Hit Rate	Probability Gain
1000	0.6429	0.6875	1.6193
800	0.6429	0.6875	1.7358
600	0.6429	0.6875	1.9137
400	0.5714	0.5625	1.7971
200	0.3571	0.375	1.4576

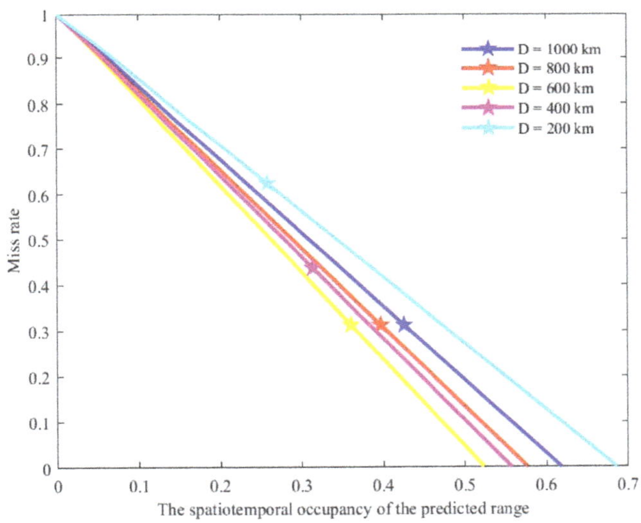

Figure 11. The Molchan diagram.

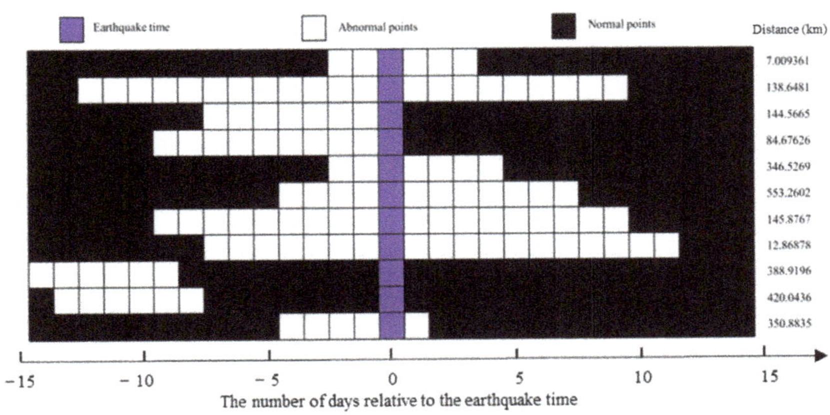

Figure 12. The spatiotemporal diagram of pre-seismic anomalies.

In this study, the anomalies were detected within 14 days before 11 earthquakes within 600 km around the epicenters. The information on these 11 earthquakes is listed in Table 7. The anomaly times and the distances between the anomaly and the epicenter are shown in Figure 12. Two earthquakes occurred after the anomaly disappeared. Two earthquakes occurred on the same day when the anomaly disappeared. In the other seven cases, the anomaly continued for several days after the earthquake.

Table 7. The considered earthquakes with a pre-seismic anomaly.

Time (UTC+8)	Magnitude	Latitude (°N)	Longitude (°E)	Depth (km)
16 September 2021, 04:33:31	6	29.2	105.34	10
26 August 2021, 07:38:18	5.5	38.88	95.5	15
13 August 2021, 12:21:35	5.8	34.58	97.54	8
29 July 2021, 16:39:27	5.7	22.7	96.04	20
7 July 2021, 14:43:48	5.2	19.65	101.2	10
16 June 2021, 16:48:58	5.8	38.14	93.81	10
12 June 2021, 18:00:46	5	24.96	97.89	16
10 June 2021, 19:46:07	5.1	24.34	101.91	8
22 May 2021, 02:04:11	7.4	34.59	98.34	17
21 May 2021, 21:48:34	6.4	25.67	99.87	8
19 March 2021, 14:11:26	6.1	31.94	92.74	10

4. Discussion

In this paper, the one-year infrared anomalies are detected and their correlations with the earthquakes are analyzed statistically. The maximum probability gain was 1.91 which was similar to the application result of the total electron content (TEC) anomaly [47]. Probability gains are influenced by the data used, the algorithms, and the regions studied. Filizzola et al. [48] studied the thermal anomalies in Turkey using robust satellite techniques the probability gain for earthquakes with a magnitude over 5 was 2.2 in their study. A probability gain greater than one means that the prediction method is better than a random guess. Jiao et al. [48] compared true earthquakes and random synthetic earthquakes to verify that the recognition ability of their method was higher than the random guess. At the same time, they proposed that it is expected to improve earthquake prediction performance by combining two–five kinds of anomalies.

As Figure 13 shows, the anomaly occurred in the southeast of the Madoi epicenter from 24 to 26 April and was 147 km away from the epicenter. It covered 233 pixels. The red stars indicate the locations of the epicenters, while the green points indicate the locations of the TEC anomaly. In our previous studies [47], there were TEC anomalies near the region on 23 April 2021. The abnormal region of TEC on 23 April 2021 overlaps with that of infrared data on 25 April 2021. In addition, TEC anomalies were observed near the two epicenters on 18 May 2021, and the infrared anomaly lasted from 17 May 2021 to 18 May 2021, as Figure 14 shows. During this period, the two types of anomalies were closest in location on 18 May 2021. This indicates that there is a certain spatial and temporal correlation between infrared and TEC anomalies before earthquakes.

Although the correlation rate of 64.29% and the hit rate of 68.75% were higher than that of the TEC anomalies, the probability gain was not improved. It may be that the spatiotemporal occupancy of the prediction range was too high because the abnormal coverage area in this paper is large, the prediction area radius is 600 km, and the prediction time window is 14 days. It is hard to pinpoint the exact location of an earthquake. Future research needs to combine multiple data sources to enhance the accuracy of earthquake prediction. Since both LSTM and DBSCAN models are suitable for multi-dimensional feature vectors, the method proposed in this paper can be used for pre-earthquake anomaly extraction from multi-source data, but normalizations of the spatiotemporal scale and amplitude are necessary for data pre-processing.

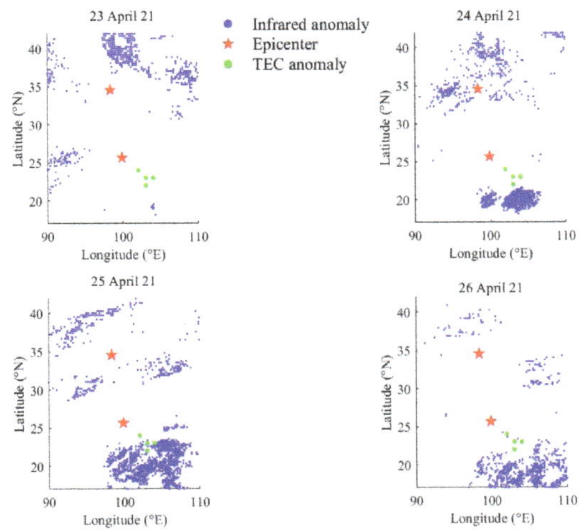

Figure 13. The infrared and TEC anomalies from 23 to 26 April 2021.

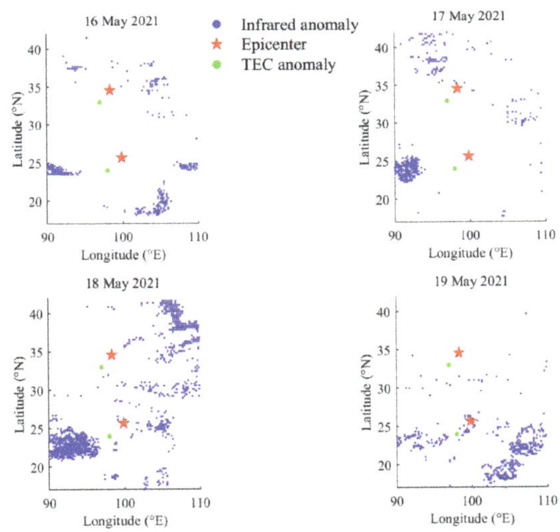

Figure 14. The infrared and TEC anomalies from 16 to 19 May 2021.

5. Conclusions

A new algorithm combining the LSTM and DBSCAN models was proposed to extract pre-seismic anomalies from the multichannel data of the Fengyun-4A satellite. The probability gain of 1.91 shows that this method is better than a random guess and can reduce the uncertainty of earthquake prediction. The correlation rate is 64.29%, and the hit rate is 68.75%. In this seismic case analysis, infrared and TEC anomalies exhibited a spatiotemporal correlation. This indicates that the method could be used to extract pre-seismic anomalies from multi-source data in future research for better prediction accuracy.

Supplementary Materials: The following supporting information can be downloaded at: https://www.mdpi.com/article/10.3390/rs15010259/s1.

Author Contributions: Conceptualization, G.C.; methodology, F.C.; validation, Y.Y.; formal analysis, Y.Y.; investigation, F.C.; writing—original draft preparation, Y.Y.; writing—review and editing, F.C.; visualization, Y.Y.; supervision, G.C.; funding acquisition, F.C. All authors have read and agreed to the published version of the manuscript.

Funding: This research received no external funding.

Institutional Review Board Statement: Not applicable.

Informed Consent Statement: Not applicable.

Data Availability Statement: Not applicable.

Acknowledgments: We acknowledge the data support of the National Satellite Meteorological Center and China earthquake networks center.

Conflicts of Interest: The authors declare no conflict of interest.

References

1. Varotsos, P.A.; Sarlis, N.V.; Skordas, E.S. Self-organized criticality and earthquake predictability: A long-standing question in the light of natural time analysis. *EPL (Europhys. Lett.)* **2020**, *132*, 29001. [CrossRef]
2. De Santis, A.; Perrone, L.; Calcara, M.; Campuzano, S.; Cianchini, G.; D'Arcangelo, S.; Di Mauro, D.; Marchetti, D.; Nardi, A.; Orlando, M.; et al. A comprehensive multiparametric and multilayer approach to study the preparation phase of large earthquakes from ground to space: The case study of the June 15 2019, M7.2 Kermadec Islands earthquake. *Remote Sens. Environ.* **2022**, *283*, 113325. [CrossRef]
3. Gu, G. Advantages of GNSS in Monitoring Crustal Deformation for Detection of Precursors to Strong Earthquakes. *Positioning* **2013**, *4*, 11–19. [CrossRef]
4. Wei, C.; Zhang, Y.; Guo, X.; Hui, S.; Qin, M.; Zhang, Y. Thermal Infrared Anomalies of Several Strong Earthquakes. *Sci. World J.* **2013**, *2013*, 208407. [CrossRef]
5. Bhardwaj, A.; Singh, S.; Sam, L.; Joshi, P.K.; Bhardwaj, A.; Martín-Torres, F.J.; Kumar, R. A review on remotely sensed land surface temperature anomaly as an earthquake precursor. *Int. J. Appl. Earth Obs. Geoinf.* **2017**, *63*, 158–166. [CrossRef]
6. Zhang, L.; Jiang, M.; Jing, F. Sea temperature variation associated with the 2021 Haiti Mw 7.2 earthquake and possible mechanism. *Geomat. Nat. Hazards Risk* **2022**, *13*, 2840–2863. [CrossRef]
7. Jing, F.; Shen, X.H.; Kang, C.L.; Xiong, P. Variations of multi-parameter observations in atmosphere related to earthquake. *Nat. Hazards Earth Syst. Sci.* **2013**, *13*, 27–33. [CrossRef]
8. Bao, Z.; Zhao, J.; Huang, P.; Yong, S.; Wang, X. A Deep Learning-Based Electromagnetic Signal for Earthquake Magnitude Prediction. *Sensors* **2021**, *21*, 4434. [CrossRef]
9. Uyeda, S.; Nagao, T.; Kamogawa, M. Short-term earthquake prediction: Current status of seismo-electromagnetics. *Tectonophysics* **2009**, *470*, 205–213. [CrossRef]
10. Adil, M.; Şentürk, E.; Pulinets, S.; Amory-Mazaudier, C. A Lithosphere–Atmosphere–Ionosphere Coupling Phenomenon Observed before M 7.7 Jamaica Earthquake. *Pure Appl. Geophys.* **2021**, *178*, 3869–3886. [CrossRef]
11. Huang, Q. Seismicity Pattern Changes Prior to the 2008 Ms7.3 Yutian Earthquake. *Entropy* **2019**, *21*, 118. [CrossRef] [PubMed]
12. Saraf, A.; Rawat, V.; Choudhury, S.; Dasgupta, S.; Das, J. Advances in understanding of the mechanism for generation of earthquake thermal precursors detected by satellites. *Int. J. Appl. Earth Obs. Geoinf.* **2009**, *11*, 373–379. [CrossRef]
13. Guo, Z.; Qiang, S.; Wang, C.; Liu, Z.; Gao, X.; Zhang, W.; Yu, Y.; Zhang, H.; Qiu, J. The mechanism of earthquake's thermal infrared radiation precursory on remote sensing images. In Proceedings of the IEEE International Geoscience and Remote Sensing Symposium, Toronto, ON, Canada, 24–28 June 2002; Volume 4, pp. 2036–2038. [CrossRef]
14. Qiang, Z.J.; Kong, L.C.; Zheng, L.Z.; Guo, M.-H.; Wang, G.-P.; Zhao, Y. An experimental study on temperature increasing mechanism of satellitic thermo-infrared. *Acta Seismol. Sin.* **1997**, *10*, 247–252. [CrossRef]
15. Jiao, Z.-H.; Zhao, J.; Shan, X. Pre-seismic anomalies from optical satellite observations: A review. *Nat. Hazards Earth Syst. Sci.* **2018**, *18*, 1013–1036. [CrossRef]
16. Genzano, N.; Filizzola, C.; Hattori, K.; Pergola, N.; Tramutoli, V. Statistical Correlation Analysis between Thermal Infrared Anomalies Observed from MTSATs and Large Earthquakes Occurred in Japan (2005–2015). *J. Geophys. Res. Solid Earth* **2021**, *126*, e2020JB020108. [CrossRef]
17. Pulinets, S.; Ouzounov, D. Lithosphere–Atmosphere–Ionosphere Coupling (LAIC) model—An unified concept for earthquake precursors validation. *J. Asian Earth Sci.* **2010**, *41*, 371–382. [CrossRef]
18. Venkatachalapathy, H.; Sreedharan, V.; Venkatanathan, N. Observation of Earthquake Precursors—A Study on OLR Scenario Prior to the Earthquakes of Indian and Neighboring Region Occurred in 2016. *Int. J. Earth Sci. Eng.* **2016**, *9*, 264–268.
19. Dong, N.; Liao, H. Characteristics of Thermal Infrared Anomalies during the Earthquakes in Wenchuan, Lushan in Ya'an and Jiuzhaigou. *IOP Conf. Ser. Earth Environ. Sci.* **2021**, *783*, 012132. [CrossRef]
20. Akhoondzadeh, M.; De Santis, A.; Marchetti, D.; Piscini, A.; Cianchini, G. Multi precursors analysis associated with the powerful Ecuador (M_W = 7.8) earthquake of 16 April 2016 using Swarm satellites data in conjunction with other multi-platform satellite and ground data. *Adv. Space Res.* **2017**, *61*, 248–263. [CrossRef]

21. Zhang, X.; Zhang, Y.; Guo, X.; Wei, C.; Zhang, L. Analysis of thermal infrared anomaly in the Nepal MS 8.1 earthquake. *Earth Sci. Front.* **2017**, *24*, 227–233. [CrossRef]
22. Yao, Q.-L.; Qiang, Z.-J. Thermal infrared anomalies as a precursor of strong earthquakes in the distant future. *Nat. Hazards* **2012**, *62*, 991–1003. [CrossRef]
23. Sun, D.; Zheng, H. Simulation Study of Infrared Transmittance under Different Atmospheric Conditions. *J. Phys. Conf. Ser.* **2022**, *2356*, 012045. [CrossRef]
24. Carolina, F.; Pergola, N.; Pietrapertosa, C.; Valerio, T. Robust satellite techniques for seismically active areas monitoring: A sensitivity analysis on September 7, 1999 Athens's earthquake. *Phys. Chem. Earth* **2004**, *29*, 517–527. [CrossRef]
25. Valerio, T.; Aliano, C.; Corrado, R.; Carolina, F.; Genzano, N.; Lisi, M.; Martinelli, G.; Pergola, N. On the possible origin of thermal infrared radiation (TIR) anomalies in earthquake-prone areas observed using robust satellite techniques (RST). *Chem. Geol.* **2013**, *339*, 157–168. [CrossRef]
26. Xie, T.; Kang, C.; Ma, W. Thermal infrared brightness temperature anomalies associated with the Yushu (China) Ms = 7.1 earthquake on 14 April 2010. *Nat. Hazards Earth Syst. Sci.* **2013**, *13*, 1105–1111. [CrossRef]
27. Akhoondzadeh, M. A comparison of classical and intelligent methods to detect potential thermal anomalies before the 11 August 2012 Varzeghan, Iran, earthquake (M_w = 6.4). *Nat. Hazards Earth Syst. Sci.* **2013**, *13*, 1077–1083. [CrossRef]
28. Saradjian, M.R.; Akhoondzadeh, M. Thermal anomalies detection before strong earthquakes (M > 6.0) using interquartile, wavelet and Kalman filter methods. *Nat. Hazards Earth Syst. Sci.* **2011**, *11*, 1099–1108. [CrossRef]
29. Akhoondzadeh, M. A MLP neural network as an investigator of TEC time series to detect seismo-ionospheric anomalies. *Adv. Space Res.* **2013**, *51*, 2048–2057. [CrossRef]
30. Akhoondzadeh, M. Support vector machines for TEC seismo-ionospheric anomalies detection. *Ann. Geophys.* **2013**, *31*, 173–186. [CrossRef]
31. Akhoondzadeh, M. Genetic algorithm for TEC seismo-ionospheric anomalies detection around the time of the Solomon (M_w = 8.0) earthquake of 6 February 2013. *Adv. Space Res.* **2013**, *52*, 581–590. [CrossRef]
32. Zhai, D.; Zhang, X.; Xiong, P. Detecting Thermal Anomalies of Earthquake Process within Outgoing Longwave Radiation Using Time Series Forecasting Models. *Ann. Geophys.* **2020**, *63*, PA548. [CrossRef]
33. Jing, F.; Zhang, L.; Singh, R. Pronounced Changes in Thermal Signals Associated with the Madoi (China) M 7.3 Earthquake from Passive Microwave and Infrared Satellite Data. *Remote Sens.* **2022**, *14*, 2539. [CrossRef]
34. Yang, X.; Zhang, T.-B.; Lu, Q.; Long, F.; Liang, M.-J.; Wu, W.-W.; Gong, Y.; Wei, J.-X.; Wu, J. Variation of Thermal Infrared Brightness Temperature Anomalies in the Madoi Earthquake and Associated Earthquakes in the Qinghai-Tibetan Plateau (China). *Front. Earth Sci.* **2022**, *10*, 823540. [CrossRef]
35. Guo, X.; Zhang, Y.-S.; Wei, C.-X.; Zhong, M.-J.; Zhang, X. Medium Wave Infrared Brightness Anomalies of Wenchuan 8.0 and Zhongba 6.8 Earthquakes. *Acta Geosci. Sin.* **2014**, *35*, 338–344. [CrossRef]
36. Zhang, P.; Zhu, L.; Tang, S.; Gao, L.; Chen, L.; Zheng, W.; Han, X.; Chen, J.; Shao, J. General Comparison of FY-4A/AGRI With Other GEO/LEO Instruments and Its Potential and Challenges in Non-meteorological Applications. *Front. Earth Sci.* **2019**, *6*, 224. [CrossRef]
37. Tramutoli, V.; Corrado, R.; Filizzola, C.; Genzano, N.; Lisi, M.; Pergola, N. From visual comparison to Robust Satellite Techniques: 30 years of thermal infrared satellite data analyses for the study of earthquake preparation phases. *Boll. Geofis. Teor. Appl.* **2015**, *56*, 167–202.
38. Yu, Y.; Si, X.; Hu, C.; Zhang, J. A Review of Recurrent Neural Networks: LSTM Cells and Network Architectures. *Neural Comput.* **2019**, *31*, 1235–1270. [CrossRef]
39. Hafeez, A.; Shah, M.; Shahzad, R. Machine Learning Based Thermal Anomaly Detection Associated with Three Earthquakes in Pakistan Using MODIS LST. In Proceedings of the 2021 Seventh International Conference on Aerospace Science and Engineering (ICASE), Islamabad, Pakistan, 14–16 December 2021; pp. 1–5. [CrossRef]
40. Schubert, E.; Sander, J.; Ester, M.; Kriegel, H.; Xu, X. DBSCAN revisited, revisited: Why and how you should (still) use DBSCAN. *ACM Trans. Database Syst. (TODS)* **2017**, *42*, 19. [CrossRef]
41. Piscini, A.; De Santis, A.; Marchetti, D.; Cianchini, G. A Multi-parametric Climatological Approach to Study the 2016 Amatrice–Norcia (Central Italy) Earthquake Preparatory Phase. *Pure Appl. Geophys.* **2017**, *174*, 3673–3688. [CrossRef]
42. Zhang, Y.; Meng, Q. A statistical analysis of TIR anomalies extracted by RSTs in relation to an earthquake in the Sichuan area using MODIS LST data. *Nat. Hazards Earth Syst. Sci.* **2019**, *19*, 535–549. [CrossRef]
43. Shebalin, P.N.; Narteau, C.; Zechar, J.D.; Holschneider, M. Combining earthquake forecasts using differential probability gains. *Earth Planet Sp.* **2014**, *66*, 37. [CrossRef]
44. Dobrovolsky, I.P.; Zubkov, S.I.; Miachkin, V.I. Estimation of the size of earthquake preparation zones. *Pure Appl. Geo-Phys.* **1979**, *117*, 1025–1044. [CrossRef]
45. Douglas Zechar, J.; Jordan, T.H. Testing alarm-based earthquake predictions. *Geophys. J. Int.* **2008**, *172*, 715–724. [CrossRef]
46. Filizzola, C.; Corrado, A.; Genzano, N.; Lisi, M.; Pergola, N.; Colonna, R.; Tramutoli, V. RST Analysis of Anomalous TIR Sequences in Relation with Earthquakes Occurred in Turkey in the Period 2004–2015. *Remote Sens.* **2022**, *14*, 381. [CrossRef]

47. Yue, Y.; Koivula, H.; Bilker-Koivula, M.; Chen, Y.; Chen, F.; Chen, G. TEC Anomalies Detection for Qinghai and Yunnan Earthquakes on 21 May 2021. *Remote Sens.* **2022**, *14*, 4152. [CrossRef]
48. Jiao, Z.; Shan, X. Pre-Seismic Temporal Integrated Anomalies from Multiparametric Remote Sensing Data. *Remote Sens.* **2022** *14*, 2343. [CrossRef]

remote sensing

MDPI

Article

Application of Model-Based Time Series Prediction of Infrared Long-Wave Radiation Data for Exploring the Precursory Patterns Associated with the 2021 Madoi Earthquake

Jingye Zhang [1], Ke Sun [1,*], Junqing Zhu [1], Ning Mao [1] and Dimitar Ouzounov [2]

[1] Institute of Earthquake Forecasting, CEA, Beijing 100036, China; zhangjy@ief.ac.cn (J.Z.); zhujq@ief.ac.cn (J.Z.); maoning@ief.ac.cn (N.M.)

[2] Institute for Earth, Computing, Human and Observing, Chapman University, Orange, CA 92866, USA; ouzounov@chapman.edu

* Correspondence: sunke@cea-ies.ac.cn

Abstract: Taking the Madoi M_S 7.4 earthquake of 21 May 2021 as an example, this paper proposes using time series prediction models to predict the outgoing long-wave radiation (OLR) anomalies and study short-term pre-earthquake signals. Five time series prediction models, including autoregressive integrated moving average (ARIMA) and long short-term memory (LSTM), were trained with the OLR time series data of the aseismic moments in the 5° × 5° spatial range around the epicenter. The model with the highest prediction accuracy was selected to retrospectively predict the OLR values during the aseismic period and before the earthquake in the area. It was found, by comparing the predicted time series values with the actual time series value, that the similarity indexes of the two time series before the earthquake were lower than the index of the aseismic period, indicating that the predicted time series before the earthquake significantly differed from the actual time series. Meanwhile, the temporal and spatial distribution characteristics of the anomalies in the 90 days before the earthquake were analyzed with a 95% confidence interval as the criterion of the anomalies, and the following was found: out of 25 grids, 18 grids showed anomalies—the anomalies of the different grids appeared on similar dates, and the anomalies of high values appeared centrally at the time of the earthquake, which supports the hypothesis that pre-earthquake signals may be associated with the earthquake.

Keywords: LSTM; OLR time series; Madoi earthquake; pre-earthquake signals

Citation: Zhang, J.; Sun, K.; Zhu, J.; Mao, N.; Ouzounov, D. Application of Model-Based Time Series Prediction of Infrared Long-Wave Radiation Data for Exploring the Precursory Patterns Associated with the 2021 Madoi Earthquake. *Remote Sens.* **2023**, *15*, 4748. https://doi.org/10.3390/rs15194748

Academic Editors: Dedalo Marchetti, Yunbin Yuan and Kaiguang Zhu

Received: 24 August 2023
Revised: 24 September 2023
Accepted: 25 September 2023
Published: 28 September 2023

1. Introduction

An earthquake is a process of the rapid release of the accumulated Earth's tectonic stress, which breaks through the critical value of the elastic rupture of rocks. In the past few decades, with the help of a large amount of pre-earthquake remote sensing data, researchers have found that large earthquakes are often accompanied by various anomalies before they occur, such as anomalies in the atmospheric temperature and humidity, the content of gases, including CH_4 and CO, at active ruptures, the surface latent heat flux (SLHF), total electron content (TEC), outgoing long-wave radiation (OLR), etc., which change significantly before an earthquake occurs [1–5]. Among them, the OLR data can characterize the radiative changes in the whole geophysical system, which contains the radiative information of the cloud top, the surface, and the atmosphere above it; this characterization helps to capture the anomalous changes in the geopathic system in the pre-earthquake area, and it is an essential parameter for the study of pre-earthquake signals [6,7].

The study of OLR anomalies linked to earthquakes first started in China: Sun et al. [8] from the Qinghai Seismological Bureau explored the relationship between long-wave radiation and earthquakes in 1990. Since then, researchers worldwide have begun to conduct extensive pre-seismic anomaly studies using OLR data. They have summarized several methods for OLR seismic information extraction, such as the vorticity background field method [9,10], the standard deviation thresholding method [11–13], and the method of the relative variance rate of power spectrum estimation [14,15]. The study of these methods proves that OLR has high potential for application in earthquake prediction.

The accumulation of OLR data over the years has significantly challenged traditional data processing algorithms. Rapid advancements in machine learning have shown promising results across various industries in recent years. The field of seismology has also benefited from the thriving development of machine learning. Experts and scholars can more precisely explore the essential features and patterns hidden in seismic data via intelligent data analysis and pattern recognition. Currently, machine learning applications in seismology focus mainly on two aspects. Firstly, they are used for preliminary earthquake magnitude, timing, and location predictions. Wang et al. [16] constructed a two-dimensional input long short-term memory (LSTM) network capable of learning the correlations between earthquakes at different locations and times and used it for predictions. Their results indicated that this system could make accurate predictions at different temporal and spatial scales. Using multiple seismic activity parameters, Bikash Sadhukhan et al. [17] utilized deep-learning techniques to establish a correlation model between calculated seismic indicators and potential seismic events. This model may predict earthquake magnitudes at different locations and has shown significant and positive results for earthquakes ranging from 3.5 to 6.0 magnitude. Secondly, machine learning can be applied to address complex problems, such as analyzing and interpreting precursory information related to earthquakes. Erman et al. [18] proposed a multi-network-based hybrid long short-term memory (N-LSTM) for ionospheric anomaly detection. This model had good prediction accuracy and stability and successfully detected two total electron content (TEC) anomalies before the Nepal earthquake. Xiong et al. [19] utilized machine learning to identify electromagnetic precursory disturbances in DEMETER data, comparing machine learning algorithms and selecting LightGBM for optimal performance in identifying electromagnetic precursory disturbances before earthquakes. The results indicated that the electromagnetic precursory data within the seismogeneic zone, calculated using the Dobrovolsky formula and a time window of approximately a few hours before the earthquake, provided an effective discrimination of electromagnetic precursory disturbances. Draz et al. [20] investigated multi-parameter precursors with different physical properties, including sea surface temperature (SST), air temperature (AT), relative humidity (RH), outgoing long-wave radiation (OLR), and TEC using standard deviation (STDEV), wavelet transformation, and LSTM networks to identify potential pre- and post-earthquake anomalies. Each method identified noticeable abnormal changes in atmospheric and ionospheric precursors before and after earthquakes. Their research demonstrates the significant relevance of machine learning techniques in detecting seismic anomalies, supporting further studies on the lithosphere–atmosphere–ionosphere coupling (LAIC) mechanism.

Based on previous experience and existing foundations, we chose five commonly used time series forecasting models in this study: one traditional parametric model, autoregressive integrated moving average; two machine learning models, support vector machine (SVM) and extreme gradient-boosting (XGBoost); and two deep learning models, the LSTM and the bi-directional long short-term memory (BILSTM) model. The prediction performance of the models was evaluated with the help of the root mean square error (RMSE) as a metric of accuracy. The results show that the LSTM model exhibited the best performance. Therefore, in this paper, the LSTM model was chosen to predict and analyze the OLR time series anomalies in the 90 days before the Madoi M_s 7.4 earthquake as an example of studying the precursor information of an earthquake. The experimental results were satisfactory, indicating that the pre-earthquake OLR time series analysis based on the LSTM prediction model had good predictive capability and broad application prospects.

The rest of this paper is organized as follows: Section 2 outlines the tectonic context of the seismogenic region and the HIRS/4 OLR-18 data product used in this paper. Section 3 describes the principles of the different time series prediction models and the anomaly discrimination and extraction methods. Section 4 tests the prediction performance of the various models, based on which the anomaly discrimination and extraction results are obtained. These results are discussed in Section 5, and Section 6 presents the conclusions and outlook for future work in this area.

2. Data and Data Preprocessing

2.1. Study Earthquakes

The Madoi earthquake occurred in the Bayanhar block in the north-central part of the Qinghai–Tibet Plateau, one of the most representative active blocks of lateral extrusion. The seismogenic fault is the NW-striking Jiangcuo fault, dominated by sinistral strike–slip movement.

The "SERIES OF EARTHQUAKE CASES IN CHINA" is important scientific material for studying Chinese earthquakes and exploring earthquake predictions [21–25]. In the compilation of the series, the statistics of fixed-point anomaly data are distributed as follows: $M_S \geq 7.0$ earthquake, within 500 km; $6 \leq M_S < 7.0$ earthquake, within 300 km; and $5.0 \leq M_S < 6$ earthquakes, within 200 km. Therefore, to fully consider the impact range of the Madoi earthquake, 25 pixels centered on the epicenter at 5° × 5° (32°–37°N, 96°–101°E) were selected as the focus area (Figure 1) to study the preseismic OLR anomalies of the Madoi earthquake.

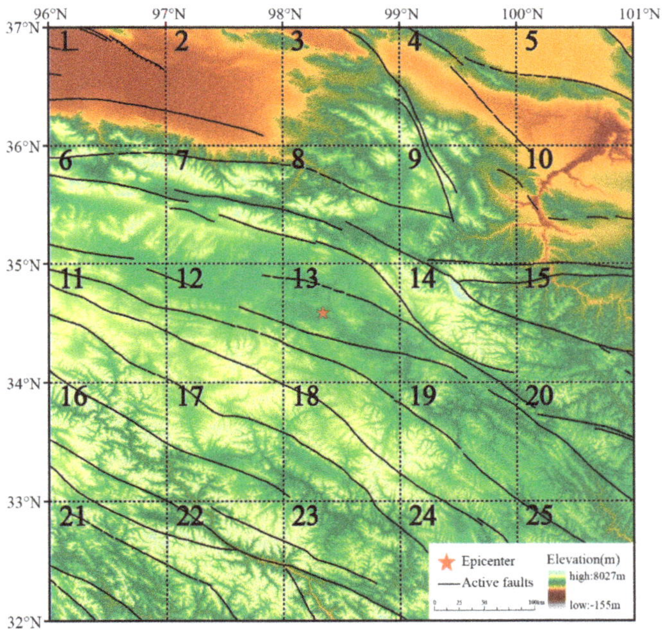

Figure 1. Sketch map of geological structure in the study area. The study area consists of 25 grids, denoted by the numbers 1 to 25.

2.2. Data

The OLR data were calculated by matching the infrared channel (10.5–12.5 μm) data from NOAA satellites with the total measurements in the broadband (4–50 μm) acquired via the Large Meteorological Experiment Satellite (NIMBUS). The OLR is closely related to factors, such as the temperature and humidity in the atmosphere, and mainly reflects radiation information from cloud tops. Since its launch in 1974, several decades of OLR data have been accumulated, which are of high completeness and continuity and can be downloaded free of charge via NCEP's FTP server (ftp://ftp.cpc.ncep.noaa.gov/precip/noaa18_1x1/) (accessed on 20 March 2023). In this paper, the night-time OLR data from the NOAA-18 satellite were selected to minimize the influence of solar radiation and human activities on the experimental results, with a temporal resolution of 1 day, a spatial resolution of 1° × 1°, and a data unit of W/m^2. Each data file comprises 360 × 180 global grid points (Figure 2).

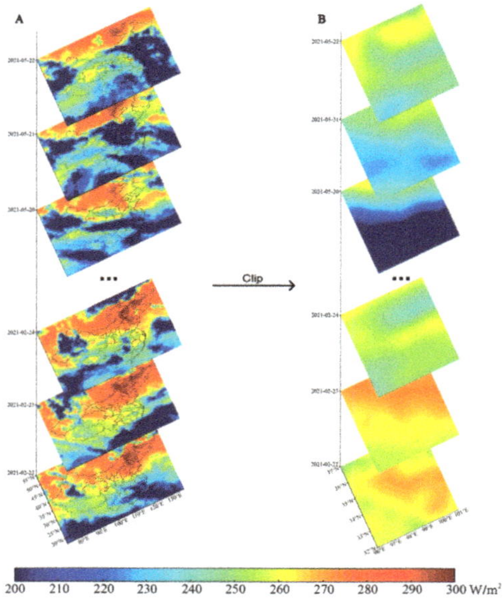

Figure 2. (**A**) OLR values at the China scale 90 days before the Madoi earthquake (22 February 2021–22 May 2021). (**B**) OLR values within the 5° × 5° spatial extent of the epicenter of the Madoi earthquake cropped out from A.

2.3. Data Preprocessing

Limited by the physical cycle of solar–terrestrial activity, the OLR emitted by the Earth shows periodic changes, especially seasonal changes [26]. To thoroughly study these characteristic variations and minimize the influence of factors such as climate variations, weather patterns, human activities, and vegetation coverage, a five-year OLR time series from 22 May 2016 to 22 May 2021 was selected. At the same time, we noticed these conclusions: Lu et al. [27] studied the variation characteristics of the thermal infrared anomalies of 20 moderately strong earthquakes in Tibet and found that the longest time of long-wave radiation anomalies before the earthquake was 90 days, and the post-earthquake anomalies corresponded well with earthquakes of magnitude five or above; Song et al. [28] found that the thermal anomaly began to gather in space about three months before the Wenchuan earthquake; and Eleftheriou et al. [29] studied dozens of earthquakes with magnitudes ranging from 4.0 to 7.9, and they believed that earthquake anomalies could only be considered to be related to earthquakes if they occurred within 30 days after the earthquake. Therefore, to accurately train the time series prediction model when there is no earthquake, the data of 90 days before and 30 days after the two earthquakes of magnitude five or above that occurred during this period (the M_S 5.3 earthquake in Qinghai Province on 6 May 2018, and the M_S 5.6 earthquake in Shiqu, Sichuan Province on 1 April 2020), totaling 240 days, were deleted. The remaining data were divided into a training set and a test set. The training set included the data from the start time to 90 days before the earthquake, which was used to learn the OLR time series change law and verify the model's performance without an earthquake. The test set was the data obtained 90 days before the earthquake. The singular values in the data were replaced by the average of the same period over the years.

3. Methods

Based on the OLR data, a time series prediction model was selected to predict the OLR values in the pre-earthquake epicenter's 5° × 5° spatial range. After preprocessing the OLR data, the training and selection of the model was the key to obtaining the subsequent analysis data, and it was also the core of the whole experiment. The trained model was used to predict the OLR values for ninety days when there was no earthquake and ninety days before the earthquake, respectively, and the similarity test was performed with the actual values to detect whether there was any anomalous period before the earthquake. Finally, 95% confidence intervals were set to extract and analyze the temporal and spatial characteristics of the anomalies to detect the possible precursor information of earthquakes. Figure 3 shows the whole process of the experiment.

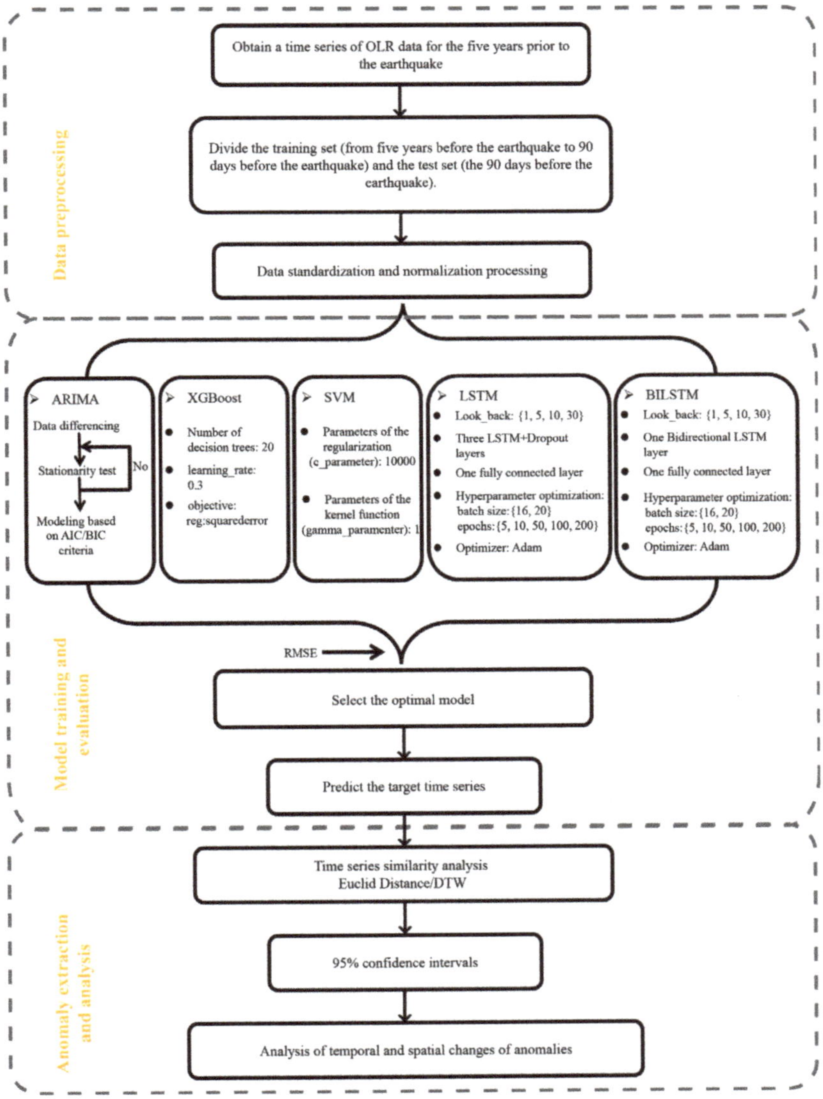

Figure 3. Flow chart of the experiment.

3.1. Time Series Forecasting Models

3.1.1. ARIMA

The ARIMA model is a famous time series forecasting method proposed by Box and Jenkins in the early 1970s [30]. The model is relatively simple. It treats the data series of the forecasting object over time as a random sequence, describes this sequence approximately by fitting a mathematical model, and predicts the future values based on the time series of past and present values. The ARIMA (p, d, q) model is described as follows:

$$y_t = \varphi_1 y_{t-1} + \cdots + \varphi_p y_{t-p} + \varepsilon_t + \theta_1 \varepsilon_{t-1} + \cdots + \theta_q \varepsilon_{t-q} \tag{1}$$

where p is the order of the autoregressive model, d is the number of differences that make the series smooth, q is the order of the sliding average model, y_t is the value of the OLR time series, φ_i is the autoregressive coefficient, θ_i is the moving average coefficient, and ε_t is the white noise process.

3.1.2. SVM

SVM is a supervised learning method in machine learning that transforms multiple features nonlinearly via the kernel function, aiming to extract more information from multiple feature inputs and obtain more accurate prediction results [31]. The specific mathematical description is as follows:

The hyperplane is $y = \omega^T x + b$, where ω and b represent the weight vector and bias, and T denotes the matrix transposition operation. The sample distance of the hyperplane to different classes was maximized according to the SVM principle, which is expressed as

$$arg\ max\left\{\frac{1}{\| \omega \|} \cdot min[(\omega^T x + b) \cdot y_i]\right\} \tag{2}$$

where $\| \omega \|$ represents the L2 norm of the weight vector used to balance the model's complexity and classification accuracy.

According to the distance formula from the point in space to the plane, the distance d from the sample point to the hyperplane can be obtained:

$$d = (\omega^T x + b) \cdot \frac{1}{\| \omega \|} \tag{3}$$

At this point, the optimization objective of the SVM algorithm is obtained after the introduction of the penalty factor C and the relaxation variable ξ:

$$\min_{\omega, b} \frac{1}{2} \| \omega \|^2 + \mathtt{C} \sum_{i=1}^{l} \xi_i \tag{4}$$

$$s.t. y_i \left(\omega^T x + b\right) \gg 1 - \xi_i \tag{5}$$

The traditional solution is to introduce Lagrange multipliers:

$$\frac{1}{2} \sum_{i=1}^{l} \sum_{j=1}^{l} \propto_i \propto_j y_i y_j k(x_i x_j) - \sum_{i=1}^{l} \propto_i \tag{6}$$

$$s.t. \sum_{i=1}^{l} \propto_i y_i = 0,\ 0 \leq \propto_i \leq \mathtt{C},\ i = 1, 2, \cdots, l \tag{7}$$

where k () is the kernel function of the SVM algorithm.

3.1.3. XGBoost

XGBoost is an optimized distributed gradient boost library that can train the model quickly and more efficiently [32]. The objective function during training consists of two parts: the first part is the gradient boost algorithm loss (Formula (8)), and the second part is the regularization term (Formula (9)).

$$L(\varnothing) = \sum_{i=1}^{n} l((y_i', y_i) + f_k(x_i)) + \Omega(f_k) \tag{8}$$

where n is the number of training function samples, l is the loss for a single sample, assuming it is a convex function, y_i' is the model's predicted value for the training sample, y_i is the true labeled value of the training sample, and $f_k(x_i)$ represents the predicted value of the decision tree k for sample i.

The regularization term defines the complexity of the model:

$$\Omega(f_k) = \gamma T + \frac{1}{2}\lambda \sum_{j=1}^{T} \omega_j^2 \tag{9}$$

where γ and λ are manually set parameters, ω is the vector formed by the values of all leaf nodes of the decision tree, and T is the number of leaf nodes.

3.1.4. LSTM/BILSTM

LSTM is a popular RNN structure that can learn long-term dependent information [33]. The LSTM network generally consists of three gate units (forgotten gate f_t, input gate i_t, and output gate o_t) and a memory cell (Figure 4). The forgetting gate is responsible for receiving the h_{t-1} output from the hidden layer at the previous time, the x_t newly input, and determining the information to be forgotten, the input gate is responsible for controlling the update and storage of the information, the output gate determines the information to be output in the current state, and the memory cell controls the transmission of the information.

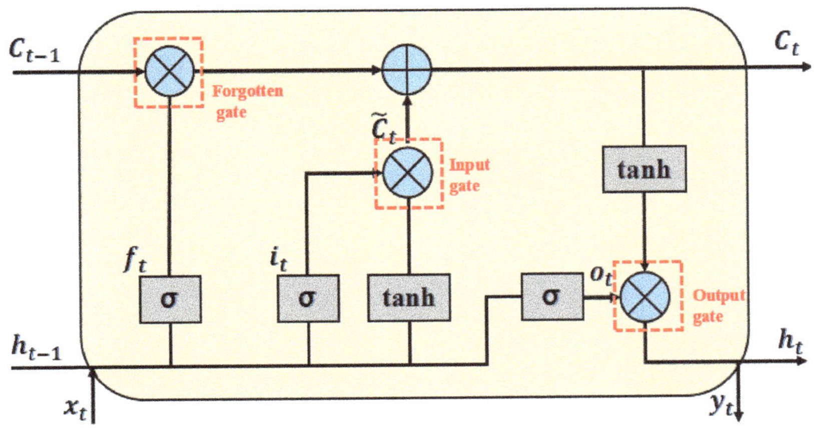

Figure 4. LSTM unit structure diagram.

Based on LSTM, BILSTM combines the information of the input sequence in both forward and backward directions. For the output at time t, the forward LSTM layer has information for time t and before in the input sequence, and the backward LSTM layer has information for time t and after in the input sequence (Figure 5) [34]. The vectors output by the two LSTM layers can be added, averaged, or concatenated.

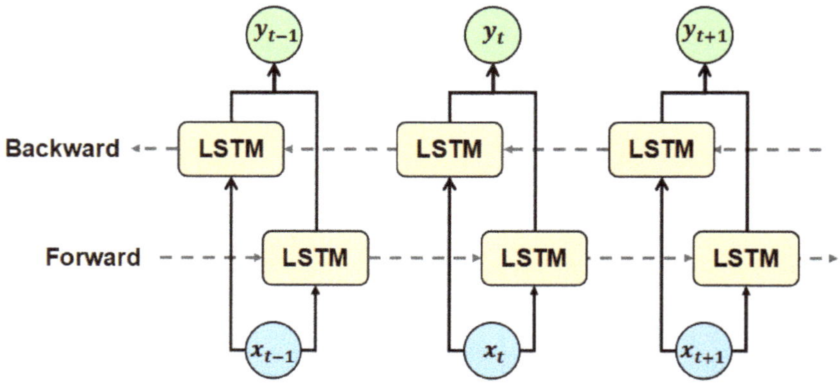

Figure 5. BILSTM unit structure diagram.

3.2. Model Performance Comparison

This study evaluates the model's predictive performance using the root mean square error (RMSE), which measures the deviation between the observed and predicted values. RMSE, a commonly used error test, has been widely used in artificial intelligence earthquake prediction research [18,35,36]. The formula for calculating RMSE is as follows

$$RMSE = \sqrt{\frac{1}{N}\sum_{t=1}^{N}(OLR_P - OLR_A)^2} \tag{10}$$

where N is the number of observations involved in evaluating the model, OLR_P is predicted OLR values, and OLR_A is the actual OLR value.

3.3. Time Series Similarity Analysis

The historical data of 90 days without the earthquake (1 September 2020–29 November 2020) and 90 days before the earthquake (21 February 2021–22 May 2021) were selected to predict the two OLR time series, respectively, using the model with the highest prediction accuracy after comparison with Section 3.2. The similarity test index uses the Euclidean distance (ED) and dynamic time warping (DTW). The ED and DTW between the predicted and actual values in the two periods were calculated, respectively, to analyze whether there was an abnormal period before the earthquake.

3.3.1. ED

ED is the most common distance measure, measuring the absolute distance between two points in a multi-dimensional space. When the ED is used to compare two time series, a one-to-one correspondence is established between each point in the sequence, and the distance between every two points is calculated and then summed. The smaller the distance, the higher the similarity. The formula for calculating ED is as follows:

$$ED = \sqrt{\sum_{i=1}^{n}(x_{i1} - x_{i2})^2} \tag{11}$$

where x_{i1} represents the *i*-dimensional coordinate of the first point and x_{i2} represents the i-dimensional coordinate of the second point.

3.3.2. DTW

DTW is a new similarity measurement method that can find the best matching path between data in two arbitrary long time series by adjusting the corresponding relationship between time points. It is robust to noise and can effectively measure the similarity of time series. Suppose first there are two time series, X and Y, of length m and n, respectively. Define a distance matrix D of size $m \times n$, where $D(i,j)$ denotes the distance between the ith element of X and the jth sum element of Y. Then, define a cumulative matrix C, also of size $m \times n$, where $C(i,j)$ represents the minimum cumulative distances from the first element of X to the ith element and from the first element of Y to the jth element. The formula for DTW is derived as follows:

$$C(1,1) = D(1,1) \tag{12}$$

$$C(i,1) = D(i,1) + C(i-1,1),\ 2 \leq i \leq m \tag{13}$$

$$C(1,j) = D(1,j) + C(1,j-1)\ ,\ 2 \leq j \leq n \tag{14}$$

$$C(i,j) = D(i,j) + min(C(i-1,j), C(i,j-1), C(i-1,j-1)),\ 2 \leq i \leq m,\ 2 \leq j \leq n \tag{15}$$

Ultimately, the DTW distance or similarity can be derived from $C(m,n)$. The smaller values of $C(m,n)$ indicate higher similarity between two time series, while larger values indicate lower similarity.

3.4. *Abnormal Identification and Extraction*

A confidence interval refers to the estimated interval of the population parameter constructed by the sample statistics, which shows the degree of confidence that the parameter's actual value falls around the measured value. Besides being often used in statistical research, it has also been effectively applied in discriminating pre-earthquake anomalies [37–39]. Based on these results, we calculated residuals from the raw data and obtained 95% confidence intervals for the predicted values. The difference between the predicted value and the actual value of the time series prediction model is calculated, and the difference between the predicted and actual values is counted as an anomaly when the difference exceeds 95% of the confidence interval. When the difference is higher (or lower) than the upper (or lower) limit of the confidence interval, the actual value is lower (or higher) than the predicted value, and it is regarded as a cold anomaly (or a hot anomaly). The temporal and spatial variation characteristics of the anomalies are then analyzed.

4. Results

4.1. *Model Performance*

After determining the optimal parameters for each method, we compared the performance of the different methods. Table 1 presents the prediction performance of each model for 25 grid OLR time series. As shown in Table 1, there were significant differences between the different prediction models. The two deep-learning algorithms performed the best in this experiment, where the LSTM algorithm had the smallest sum of RMSE and the best prediction performance, and the BILSTM algorithm performed second only to the LSTM algorithm. The two machine learning algorithms (SVM and XGBoost) performed equally well next to the deep-learning algorithm, while the traditional parametric model ARIMA had the worst prediction performance. In critical situations, such as detecting OLR data anomalies, relying on models with high uncertainty and low stability does not make sense, so LSTM was chosen as the research algorithm for this experiment.

Table 1. Error results based on time series prediction model at different grid points.

Grid	ARIMA	SVM	XGBoost	LSTM	BILSTM
1	34.2619	35.1532	30.8762	30.5294	33.6217
2	53.1949	36.1017	34.9021	30.7129	32.3657
3	58.2728	33.0449	31.8505	26.7471	27.9391
4	44.444	30.2318	28.8857	26.7325	27.6086
5	44.444	30.2318	28.8857	26.7325	27.6086
6	50.9688	40.6984	35.6925	33.1588	34.6644
7	53.0276	34.4316	36.2105	32.0643	33.0467
8	47.2451	34.6952	35.5667	30.8911	31.451
9	53.2237	32.278	33.349	30.1235	29.7703
10	54.5762	35.9806	38.6714	32.6847	32.0692
11	53.5756	40.6919	39.8142	36.5470	37.1474
12	53.5756	40.6919	39.8142	36.5470	37.1474
13	70.4604	38.8847	36.4824	33.1946	34.5191
14	41.412	40.2266	37.6286	36.5276	36.032
15	41.7221	41.2979	41.9942	36.7073	36.7855
16	44.5678	40.9163	37.4803	35.4564	34.5752
17	58.5795	42.6388	37.4096	35.6641	37.3662
18	52.4613	39.7556	37.5302	34.8795	37.4875
19	47.4852	37.6338	37.0222	36.4709	37.1824
20	39.4942	39.1195	40.4514	37.6639	36.3644
21	61.4502	41.8658	40.5014	38.3213	37.3602
22	61.4502	41.8658	40.5014	38.3213	37.3602
23	52.6675	35.1916	34.3085	31.4399	33.413
24	42.6541	32.3148	32.4968	31.4682	32.2569
25	38.0416	32.2304	34.0134	31.7403	31.5389
Total	1253.2563	928.1726	902.3391	831.3261	846.6816

4.2. Similarity Comparison

4.2.1. ED Comparison

The results of the ED values between the actual and predicted time series of OLR for the 25 grids around the epicenter for the 90 days when there was no earthquake and the 90 days before the earthquake are shown in Table 2. As shown in Table 2, the ED values for the 25 grids when there was no earthquake were all lower than those of the pre-earthquake ED values. The two sets of results were processed as differences to visualize the similarity between the two sets of time series and explore the possible regional range of anomalies. The results show that the difference in ED was more significant in the east (grids 14 and 15) and south (grids 17–25) of the epicenter and smaller in the west (grids 11 and 12) and north (grids 1 and 3–12) of the epicenter. The results of the ED comparison reflect abnormal changes in the actual time series before the earthquake. At the same time, the more significant differences in the eastern and southern parts of the epicenter indicate that anomalies were more likely to occur in this region.

Table 2. ED between actual and predicted time series.

Gird	ED in the Aseismic Period	ED in the Seismic Period	Difference Value
1	186.8928	215.2030	28.3102
2	182.2871	243.9496	61.6625
3	198.2106	222.8911	24.6805
4	193.0780	222.0978	29.0198
5	193.0780	222.0978	29.0198
6	200.2466	218.4609	18.2143

Table 2. *Cont.*

Gird	ED in the Aseismic Period	ED in the Seismic Period	Difference Value
7	193.9372	231.1959	37.2587
8	200.1804	239.3742	39.1938
9	200.1804	239.3742	39.1938
10	210.2695	231.7173	21.4478
11	205.3519	218.5727	13.2208
12	216.1626	243.7730	27.6104
13	210.8625	254.2664	43.4039
14	193.5884	244.6240	51.0356
15	193.5884	244.6240	51.0356
16	193.7733	216.9230	23.1497
17	199.9631	259.6670	59.7039
18	188.1746	253.0382	64.8636
19	188.1746	253.0382	64.8636
20	214.1401	287.3047	73.1646
21	222.7757	282.6755	59.8998
22	217.8504	294.7557	76.9053
23	180.2219	262.8554	82.6335
24	180.2219	262.8554	82.6335
25	210.8567	296.3231	85.4664

4.2.2. DTW Comparison

Table 3 shows the DTW values of the actual and predicted time series of the OLR for the 25 grids around the epicenter for 90 days with no earthquake and 90 days before an earthquake. Table 3 shows that the DTW values in the 90 days at the time of no earthquake were all smaller than the DTW of 90 days before the earthquake. Again, the difference between the two was processed, and the results showed that the difference results were more prominent for the grids south of the epicenter. The DTW comparison results show that the similarity between the actual time sequence and the predicted time sequence of the aseismic period was better than that before the earthquake. This further indicates that pre-earthquake signals may occur within the 90 days before an earthquake.

Table 3. DTW between actual and predicted time series.

Gird	DTW in the Aseismic Period	DTW in the Seismic Period	Difference Value
1	1048.3101	1122.6331	74.323
2	1192.8760	1458.1691	265.2931
3	1233.5800	1460.4359	226.8559
4	1201.0691	1287.7728	86.7037
5	1201.0691	1287.7728	86.7037
6	1252.1896	1292.0052	39.8156
7	1218.4646	1584.6326	366.168
8	1159.1498	1524.3408	365.191
9	1159.1498	1524.3408	365.191
10	1328.1454	1423.4520	95.3066
11	1306.5641	1342.0472	35.4831
12	1258.7495	1451.5585	192.809
13	1262.1207	1613.9854	351.8647
14	1150.4503	1752.4018	601.9515
15	1150.4503	1752.4018	601.9515
16	1213.2118	1244.7485	31.5367
17	1120.0947	1756.5435	636.4488

Table 3. *Cont.*

Gird	DTW in the Aseismic Period	DTW in the Seismic Period	Difference Value
18	983.54445	1500.6782	517.13375
19	983.54445	1500.6782	517.13375
20	1331.7225	1870.4373	538.7148
21	1252.8426	2052.9131	800.0705
22	1285.1753	2056.8329	771.6576
23	990.76198	1797.7720	807.01002
24	990.76198	1797.7720	807.01002
25	1326.1863	1954.9361	628.7498

4.3. Abnormal Temporal and Spatial Characteristics

4.3.1. Temporal Scale Characteristics

Figure 6 shows the anomalies in the temporal dimension. According to our definition of hot and cold anomalies in Section 3.4, it is clear that all the pre-seismic anomalies of this predicted Madoi earthquake were "hot" anomalies. The anomalies appeared in 13 periods during the 90 days before the earthquake. The earliest anomaly appeared on 13 March, with a small amplitude. The first concentration of anomalies occurred on 18 March, covering six grids, and another anomaly of smaller magnitude occurred on 31 March. There were three anomalies on 14 April, 18 April, and 24 April, with three grids simultaneously having anomalies on 18 April. After entering May, the frequency of anomalies increased, and the anomaly amplitude increased. On May 6, three grids were abnormal, and the maximum radiation energy exceeded the lower limit of 17.05. On 9 May, the anomaly range was the largest, 14 grids were abnormal at the same time, and the maximum radiation energy exceeded the lower limit of 24.17. Four days later, the concentrated anomaly occurred again, 11 grids were abnormal, the intensity of the anomaly increased further, and the maximum radiation energy of six grids exceeded the lower limit of 30. On the next day, the concentrated anomaly still existed, but the scope and intensity decreased; seven grids were abnormal, and the maximum radiation energy was 21.62. There were three more minor anomalies on 16 May, 17 May, and 20 May, respectively. In general, the closer to the time of earthquake occurrence, the greater the anomaly intensity, the wider the coverage area, and the anomaly development trend is appearing–strengthening–weakening–disappearing.

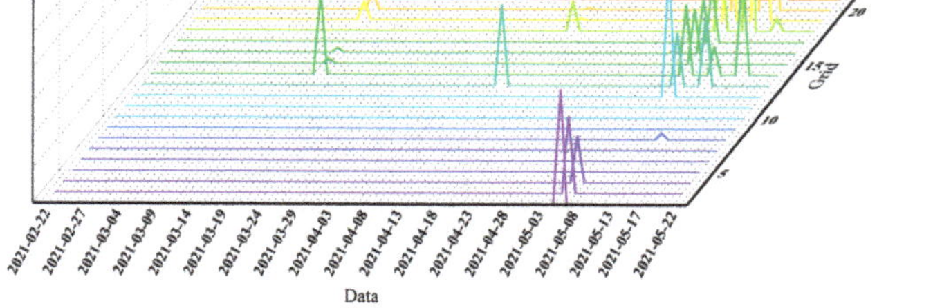

Figure 6. OLR anomaly prediction before the Madoi earthquake—temporal scale. The anomalies in different grids are represented by different colors.

4.3.2. Spatial Scale Characteristics

Figure 7 shows the anomalies of 25 grids in the spatial dimension, 7 of which did not exceed the confidence interval, and these grids were mainly distributed in the northern and northeastern regions of the epicenter. Before the earthquake, some slight anomalies occurred northwest of the epicenter. Anomalies with strong intensity and large areas were mainly concentrated in the south and southeast of the epicenter. Combined with Figure 1, we found more fault zones in the south of the epicenter, while fewer fault zones were in the north. Therefore, within one month before the Madoi earthquake, the anomalies were mainly distributed in the south and southeast of the epicenter with more fault zones, and we speculate that the location of frequent anomalies may have had a particular relationship with the activity of the fault zone.

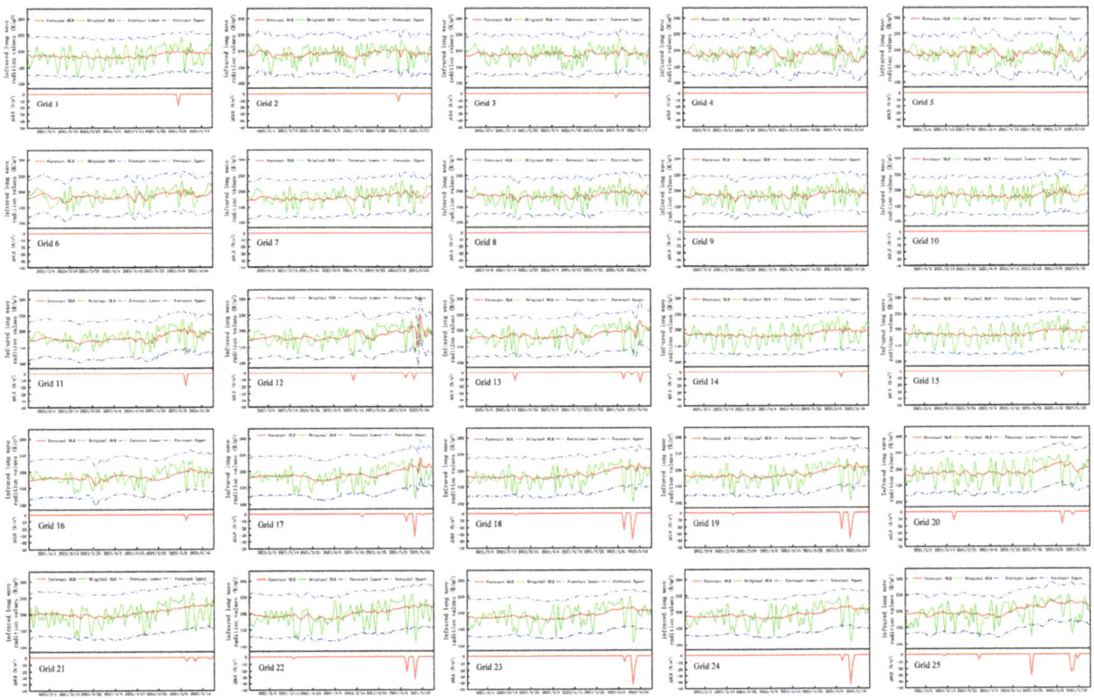

Figure 7. OLR anomaly prediction before the Madoi earthquake—spatial scale.

5. Discussion

5.1. Consistency of LSTM Results with Traditional Methods

Tramutoli introduced the Absolutely Local Index of Change of the Environment (ALICE) method [40], which relies on the robust satellite technology (RST) approach for detecting pre-seismic thermal infrared anomalies. This method constructs a background field based on multi-year observational data to depict the state of thermal infrared anomalies during periods of seismic quiescence. It has demonstrated valuable applications in several earthquake events [41,42]. To validate the effectiveness of the LSTM model in detecting preseismic OLR anomalies, we conducted a comparative analysis between the predicted anomalies and those extracted using the conventional ALICE method. The ALICE formula is defined as

$$ALICE(x,y,t) = \frac{V(x,y,t) - \mu(x,y)}{\sigma(x,y)} \tag{16}$$

In the equation, $ALICE(x,y,t)$ represents the anomaly value at time t for the coordinate position (x, y) of a pixel; $V(x,y,t)$ represents the pixel value at time t for the coordinate position (x, y); $\mu(x,y)$ represents the multi-year average at the same location and time; and $\sigma(x,y)$ represents the corresponding standard deviation.

The ALICE method was employed to select a research area encompassing a 10° × 10° region with the epicenter of the Madoi earthquake at its center. The chosen time frame spanned the 90 days preceding the earthquake, aligning with the time window used in the LSTM method for predicting preseismic anomalies. This approach was adopted to acquire insights into the evolution of OLR anomalies leading up to the Madoi earthquake, subsequently generating spatial distribution maps of OLR anomalies (Figure 8). According to a previous study, an ALICE value greater than 2 was considered to fall within the category of anomalies [43].

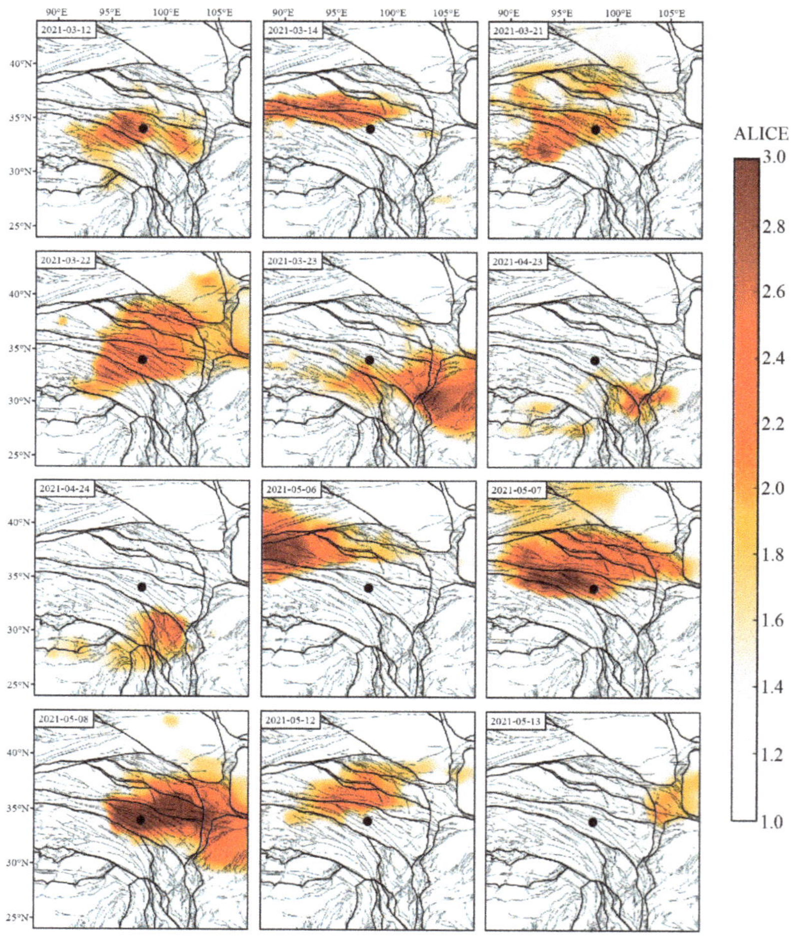

Figure 8. Daily scale thermal infrared anomaly of the Madoi earthquake. The black dots represent the location of the epicenter of the Madoi earthquake, the thick black lines represent different blocks, and the thin black lines represent active faults.

From Figure 8, it is evident that the ALICE method successfully detected significant preseismic OLR anomalies. In total, there were 12 instances of anomalies observed. The first anomaly occurred on 12 March near the epicenter. The anomaly reappeared on March 14th in the form of strips covering the seismic fault zone. Starting from 21 March, the anomaly shifted eastward, expanding its coverage during this process and lasting for three days. Subsequently, a period of calm persisted for a month. Within the last 30 days preceding the earthquake, the anomalies' frequency, extent, and intensity increased. Seven anomalies occurred within this time frame. On 23 April, an anomaly moved from the southeast to the south of the epicenter and lasted for two days. On 6 May, an anomaly appeared northwest of the epicenter and then moved eastward along the seismogenic fault zone, approaching the epicenter. On 8 May, the anomaly peaked, covering the epicenter with an ALICE value of 3 before disappearing. On 12 May, an anomaly emerged in the northwest of the epicenter, lasting for two days, and disappeared from the eastern part of the epicenter. Comparing the temporal characteristics of anomalies extracted using the ALICE method with the predictions from the LSTM method shown in Figure 6, we observe a strong correspondence regarding the occurrence time, frequency, and anomaly intensity. This suggests that utilizing the LSTM method to detect preseismic OLR anomalies is trustworthy to a certain extent and holds research value for practical applications.

5.2. Robustness of the LSTM Model

To verify whether the model still maintains good prediction performance for earthquake precursor signals under different spatial and temporal conditions, we tested the model using information from six earthquakes and two hypothetical random earthquakes (A and B) that occurred in mainland China in recent years. Table 4 summarizes and organizes this information. Figure 9 illustrates the preseismic OLR anomaly extraction results for these earthquake examples.

Table 4. Earthquake information used in our study.

Date (y-m-d)	Longitude (°E)	Latitude (°N)	Depth (km)	Magnitude (M_S)	Location
2022-01-08	101.26	37.77	10	6.9	Menyuan, Qinghai
2021-05-21	99.87	25.67	8	6.4	Yangbi, Yunnan
2021-03-19	92.74	31.94	10	6.1	Biru, Tibet
2020-06-26	82.33	35.73	10	6.4	Yutian, Xinjiang
2019-04-24	94.61	28.40	10	6.3	Medog, Tibet
2017-08-09	82.89	44.27	11	6.6	Jinghe, Xinjiang
2019-02-28	112	44	—	—	A
2019-01-31	83	40	—	—	B

The following information can be obtained from Figure 9: Before the Menyuan earthquake, six anomalies were observed. The first anomaly occurred on the 79th day before the earthquake; then, there were two centralized anomalies occurring on the 80th day and the 54th day before the earthquake, and the last anomaly centrally appeared on the 32nd day before the earthquake and then tended to be calm until the earthquake occurred. There were eight days of anomalies before the Yangbi earthquake, with the earliest anomaly occurring on the 60th day before the earthquake and the last anomaly occurring on the 9th day before the earthquake, and almost every anomaly appeared on multiple grids at the same time. There were also eight days of anomalies before the Biru earthquake, with the earliest anomaly occurring on the 85th day before the earthquake, the last anomaly occurring on the 5th day before the earthquake, and anomalies occurring on 13 grids at the same time. For the Yutian earthquake, anomalies first appeared 87 days before and continued until the 2nd day before, totaling 22 days with anomalies. On the 50th day before, 18 grid cells exhibited anomalies simultaneously; at other times, anomalies were mostly concentrated in multiple grid cells. There were 14 days of anomalies before the Medog earthquake, with the first anomaly occurring on the 84th day before the earthquake and the last anomaly occurring on the 2nd day before the earthquake. The fluctuation in anomalies before the earthquake was obvious, with concentrated anomalies occurring for many days. The earliest anomaly of the Jinghe earthquake occurred on the 82nd day before the earthquake, and several strong and concentrated anomalies appeared from the 80th to the 70th day before the earthquake, with the earthquake occurring on the 4th day after the last concentrated anomaly ended.

Overall, the model successfully detected significant OLR anomaly changes within 90 days before all six actual earthquakes. The anomalies exhibited several distinctive features: firstly, the anomalies had a widespread impact, often simultaneously affecting multiple grid cells. Secondly, the anomalies displayed substantial intensity, frequently exceeding the confidence interval by more than 30. Thirdly, the anomalies were recurrent, with multiple anomalies observed within the 90 days before the earthquakes. On the other hand, the anomalies extracted for hypothetical earthquakes A and B were very quiet, with occasional minor anomalies, but these fluctuations did not correspond to the three characteristic phenomena mentioned above. Therefore, we believe that this model demonstrates robustness in exploring seismic precursors. It performed well in extracting and tracking pre-earthquake abnormal signals.

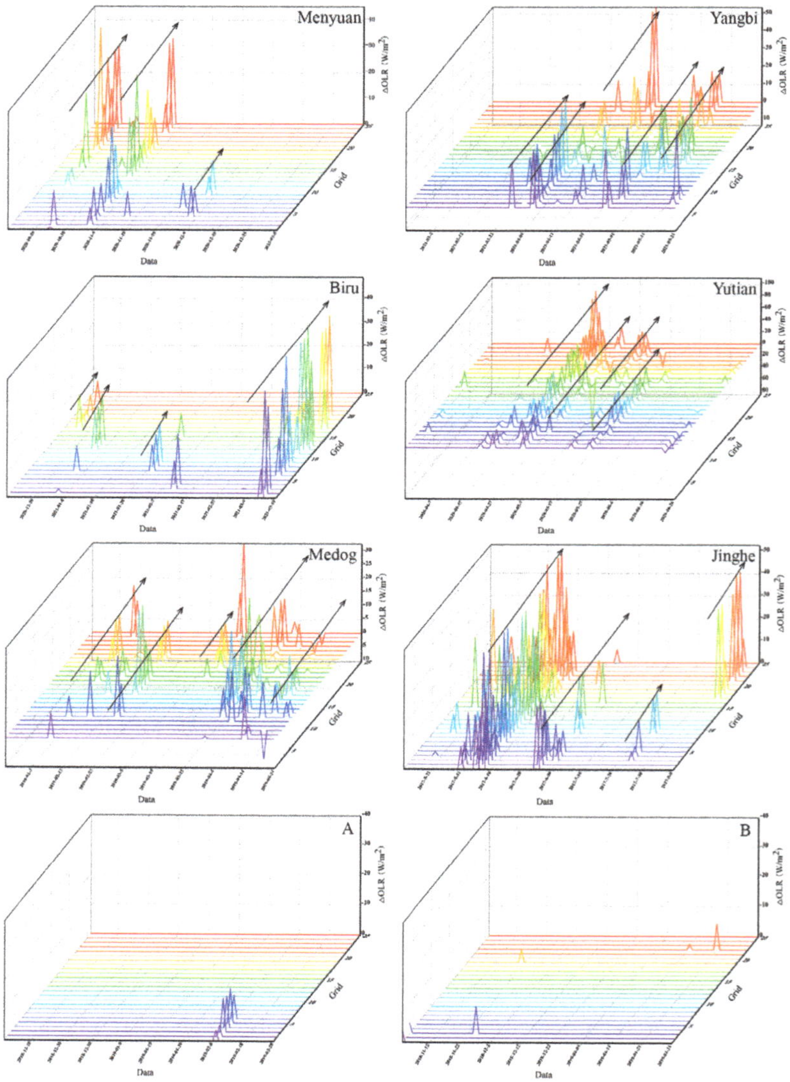

Figure 9. Extraction results of preseismic OLR anomalies for actual earthquakes and hypothetical random earthquakes. The black arrow indicates a concentrated anomaly. The anomalies in different grids are represented by different colors.

5.3. Advantages of the LSTM Model

Finding anomalies in any time series data is a very critical task. Based on statistical theory, some anomaly discrimination methods have proven the existence of infrared long-wave anomalies before earthquakes: Kang et al. [44] analyzed the thermal and infrared change characteristics before and after 14 November 2001, west of the Kunlun Mountain Pass magnitude 8.1 earthquake, based on the OLR data of NOAA satellites and the brightness temperature data of the Japanese geostationary meteorological satellite GMS-5. The results showed that the warming anomaly in OLR occurred one month before the earthquake, the brightness temperature anomaly occurred one week before the earthquake, and the anomaly characteristics were consistent with the tectonic distribution of the earthquake region in the spatial pattern. Zhang and Meng [43] conducted long-term statistical studies on the relationship between thermal infrared anomalies and earthquakes in different regions using multi-year nightly OLR data from NOAA satellites and Molchan plots. The results show that TIR anomalies were highly correlated with magnitude 4.0 normal fault or reverse fault earthquakes and that TIR anomalies induced by earthquakes should be spatially and temporally persistent. Although these methods have achieved good results in earthquake monitoring and forecasting, the results may be potentially different due to the setting of anomaly discrimination rules affected by a variety of factors [45,46], and it is difficult to summarize a universal method for studying different seismic cases. The LSTM algorithm belongs to the data-driven method, which fully considers the time correlation of the time-series data itself and utilizes historical data for training and prediction based on the data of the current time step so that LSTM can achieve end-to-end training and prediction on time series data without manually extracting features or performing other preprocessing steps, which can reduce the error caused by human or environmental factors' interference, and the applicability is high. In addition, the traditional OLR anomaly extraction algorithm mainly focuses on hot anomalies [14,47]. Still, this experiment subdivides the types of anomalies by setting confidence intervals, which can capture cold and hot anomalies simultaneously, to facilitate the study of the generation mechanism of anomalies. For example, Han et al. [48] studied the strain evolution characteristics around the epicenter for many years before the Madoi earthquake, and the results showed that, in the early stages of the earthquake, the block where the epicenter was located was in the deformation gradient zone of tension and pressure conversion, which made it easy to cause energy accumulation, and was a seismic risk area. At the same time, considering that, according to rock stress experiment results, a rock is heated via extrusion and cooled via tension [49], we speculate that the anomaly type may be able to predict the regulation of active faults on rock stress, which provides a new method for exploring the relationship between anomaly generation and tectonic activity.

5.4. Multi-Parameter Coupling Relationship and Anomaly Generation Mechanism Analysis before Earthquake

According to the LAIC model, complex variations occur before an earthquake. A review of the research literature related to the Madoi earthquake reveals several significant findings: Du and Zhang [50] analyzed ionospheric disturbances in the seismogenic area before and after the Madoi earthquake. They noticed that anomalous ionospheric disturbances began to appear about 40 days before the earthquake. About 20 days before the earthquake, the electron density of Ne changed dramatically, and these anomalies disappeared after the earthquake. Zhang et al. [51] discovered multiple occurrences of electromagnetic anomalies within the last two weeks before the earthquake. These anomalies had short intervals between adjacent events and a wide-ranging impact. They suggested a potential connection between these anomalies and the occurrence of the Madoi earthquake. Wang et al. [52], using the Benioff strain as a response parameter, studied the evolution of load–unload response ratio (LURR) anomalies within 400 km of the Madoi earthquake epicenter. They observed that the LURR values gradually increased during the two months leading up to the earthquake, reaching their highest point approximately one month before the event. Afterward, they gradually decreased, indicating that the rock medium in the seismogenic area had reached the end of its yielding stage. Su et al. [53] observed that the water level near the epicenter's observation station dropped significantly on February 25 and the HCO_3^- concentration in the observation well water had continued to rise since March. Considering that it was caused by increased carbon dioxide dissolution in the water, they thought that it might indicate an increase in fault activity in this area. Jing et al. [54] analyzed the spatial and temporal evolution of the preseismic thermal variations in the Madoi earthquake using the Index of Microwave Radiation Anomaly (IMRA) based on eight years of microwave brightness temperature (MWBT) data. Changes in MWBT are directly dependent on the surface and subsurface temperature and emissivity. Their results showed that enhanced IMRA had been distributed along the seismogenic fault since mid-February, with longer-lasting and stronger anomalies reoccurring in March and April 2021. They also analyzed near-surface CO variations detected via AIRS sensors, noting two significant CO anomalies on 23 February and 23 March. They suggested that these might be indicative of accelerated gas release due to changes in tectonic stress. Luo et al. [55] studied the water radon values observed at the Huangyuan station near the Madoi earthquake epicenter. Their findings revealed a distinctive pattern of the water radon values from April to June, characterized by a "decline (during the seismic event)–sustained low values–rebound–recovery" anomaly. This pattern was similar to the short-term anomalies observed in water radon values before and after several recent earthquakes. They also noted a sharp drop in water radon measurements three days before the earthquake, suggesting that water radon value changes could serve as short-term precursory anomalies for the Madoi earthquake.

The chronological sequence of these anomalies is illustrated in Figure 10. Building upon the findings from these multi-parameter studies and following the LAIC concept [56], an attempt was made to construct a coupled parameter model for the Madoi earthquake, as depicted in Figure 11. The accumulation of stresses in the crustal medium of the seismogenic zone before the earthquake could have some effects: (1) Some of the rocks composing the crust carry electrical charges. These charges exist as electron vacancies in the valence band and typically remain stable. Under the influence of stress, positive holes (p-holes) in the valence band are activated, impacting the rock's conductivity and altering its geophysical and chemical properties, generating electromagnetic radiation, thermal infrared radiation, and more. (2) Stress accumulation induces thermoelastic effects in rocks, causing them to heat up due to friction. (3) Stress accumulation leads to the fracturing of crustal rocks, creating a connected network for subsurface gases' (such as radon) transportation, resulting in a continuous seepage of gases from the rocks. (4) Stress accumulation also induces changes in the preseismic geophysical environment. These alterations significantly impact the atmospheric electric field, ultimately leading to abnormal variations in electron density (e.g., Ne) within the ionosphere. Furthermore, both positive charge vacancies and escaping radon can trigger air ionization, which has an important effect on long-wave radiation. Therefore, our interpretation of the preseismic OLR anomalies related to the Madoi earthquake suggests that they may result from multiple mechanisms acting in concert, including processes like greenhouse gas degassing and electromagnetic field anomalies. Moreover, because changes in long-wave radiation anomalies represent the integral of all thermal effects related to the earthquake, from the Earth's surface to the top of the troposphere, encompassing variations in surface temperature, air temperature, and latent heat flux, OLR exhibits more extensive anomaly changes compared with other parameters.

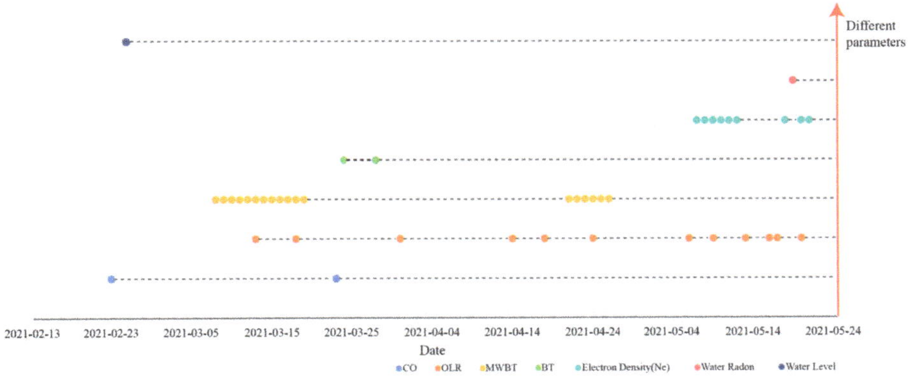

Figure 10. Time distribution of multi-parameter anomalies in Madoi earthquake.

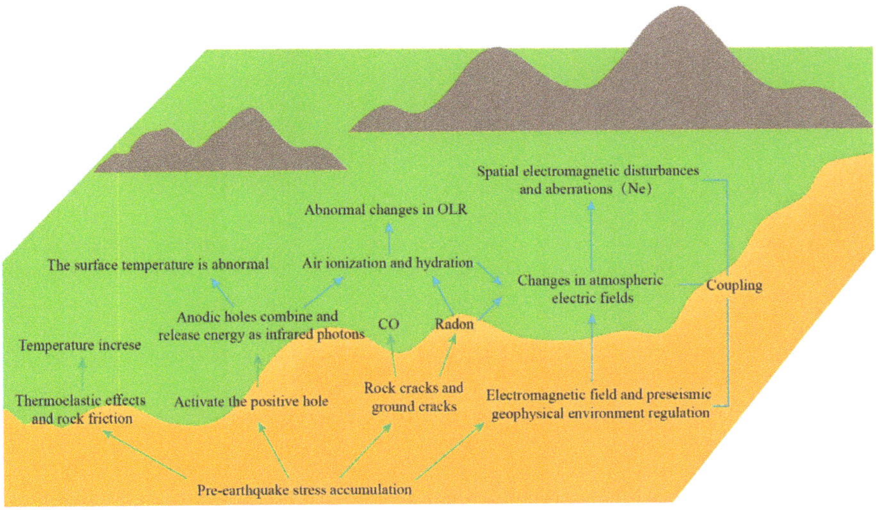

Figure 11. Diagram of Madoi earthquake multi-parameter coupling model. The green arrows represent underground processes and the blue arrows represent surface and atmospheric processes.

6. Conclusions

Based on the OLR data, the precursory anomalies of the Madoi earthquake are preliminarily explored in this paper. The LSTM model with the best performance is selected to learn the variation law of the OLR time series without an earthquake, and the OLR values in two periods (90 days in the aseismic period and 90 days before the earthquake) before the Madoi earthquake are predicted, respectively. Based on the evaluation of ED and DTW time series similarity methods, the similarity degree between the predicted value and the actual value in the two periods is quantified, which is used to determine whether there is an abnormal period within the 90 days before the earthquake. Finally, the 95% confidence interval is used to extract and track the spatiotemporal features of the anomalies. The following conclusions are drawn:

(1) The two time series similarity analysis methods proved that the similarity between the time series of the actual values at the time of the aseismic and the predicted values were better than those before the earthquake, indicating an anomalous period.

(2) The results of the RMSE-based accuracy evaluation show that, among the five selected time series prediction models, the prediction performance of the LSTM model is better than several other time series models, which is attributed to the strong learning ability

of the deep-learning model on data features. After accepting the input time series data, the LSTM layer, utilizing its ability to process the contextual information and the memory capability, has the potential to dig in and capture the potential nonlinear features in the time-series data.

(3) The anomalies are extracted, and the temporal and spatial change characteristics of the anomalies are analyzed using 95% confidence intervals. On the time scale, the earliest anomalies appeared seventy days before the earthquake. The closer to the moment of the earthquake, the higher the frequency of anomalies, the wider the coverage area, and the higher the intensity; the anomalies exceeded the lower limit of the confidence interval of 40 W/m^2 on the ninth day before the earthquake. On the spatial scale, the anomalies are concentrated in the southern and southeastern parts of the epicenter, where the distribution of fault zones is denser, and are less frequent in the northern part of the epicenter, where the fault zones are sparse, which suggests that tectonic activity is potentially related to the occurrence of earthquakes and the emergence of radiation anomalies. In addition, it is worth noting that the selection of the confidence interval affects the extraction of anomalies, and as the threshold setting of the confidence interval increases, some small amplitude chance anomalies unrelated to earthquakes will be filtered out, and the results will be more reliable.

(4) The method of exploring the pre-earthquake signals of the Madoi earthquake with deep-learning prediction OLR time series is feasible. The application of the method achieved the expected purpose. After the occurrence of anomalies, we should focus on the rupture zones and faults in a locking state around the relevant geological bodies, which may be advantageous locations for the origin of future strong earthquakes. At present, only a complete experiment has been carried out for the Madoi earthquake, and this method will be applied to the processing and analysis of other moderate and strong earthquakes in the future to further verify the feasibility of the method of exploring pre-earthquake signals by predicting OLR time series with deep learning.

Author Contributions: Conceptualization, J.Z. (Jingye Zhang) and K.S.; methodology, J.Z. (Jingye Zhang); software, J.Z. (Jingye Zhang) and J.Z. (Junqing Zhu); validation, J.Z. (Jingye Zhang) and K.S.; formal analysis, J.Z. (Jingye Zhang); investigation, N.M.; resources, J.Z. (Jingye Zhang); writing—original draft preparation, J.Z. (Jingye Zhang); writing—review and editing, J.Z. (Jingye Zhang), D.O. (Dimitar Ouzounov) and N.M.; visualization, J.Z. (Junqing Zhu); supervision, K.S.; project administration, K.S.; funding acquisition, K.S. All authors have read and agreed to the published version of the manuscript.

Funding: The work of J.Z., K.S., J.Z. and N.M. was funded by the National Natural Science Foundation of China under Grant No. U2039202, the National Key Research and Development Program of China under Grant No. 2019YFC1509202, and the Central Public-Interest Scientific Institution Basal Research Fund No. CEAIEF2022050500. The work of D.O. for this manuscript has been supported by the Institute of ECHO at Chapman University research funds.

Institutional Review Board Statement: Not applicable.

Informed Consent Statement: Not applicable.

Data Availability Statement: This study uses NOAA-18 satellite 1° × 1° OLR data, which can be downloaded via NCEP's FTP server (ftp://ftp.cpc.ncep.noaa.gov/precip/noaa18_1x1/) (accessed on 20 March 2023).

Acknowledgments: We thank the National Natural Science Foundation of China and the National Key Research and Development Program of China for funding this study. We thank NOAA for providing the OLR data and Sha Yin for their guidance in producing the figures.

Conflicts of Interest: The authors declare no conflict of interest.

References

1. Zhang, Y.; Meng, Q.; Wang, Z.; Lu, X.; Hu, D. Temperature Variations in Multiple Air Layers before the M_W 6.2 2014 Ludian Earthquake, Yunnan, China. *Remote Sens.* **2021**, *13*, 884. [CrossRef]
2. Cui, Y.; Ouzounov, D.; Hatzopoulos, N.; Sun, K.; Zou, Z.; Du, J. Satellite observation of CH_4 and CO anomalies associated with the Wenchuan M_S 8.0 and Lushan M_S 7.0 earthquakes in China. *Chem. Geol.* **2017**, *469*, 185–191. [CrossRef]
3. Singh, R.; Cervone, G.; Kafatos, M.; Prasad, A.K.; Sahoo, A.K.; Sun, D.; Tang, D.L.; Yang, R. Multi-sensor studies of the Sumatra earthquake and tsunami of December 26 2004. *Int. J. Remote Sens.* **2007**, *28*, 2885–2896. [CrossRef]
4. Xiong, B.; Li, X.; Wang, Y.Q.; Zhang, H.M.; Liu, Z.J.; Ding, F.; Zhao, B.Q. Prediction of ionospheric TEC over China based on long and short-term memory neural network. *Chin. J. Geophys. -Chin. Ed.* **2022**, *65*, 2365–2377.
5. Kang, C.L.; Han, Y.B.; Liu, D.F.; Cao, Z.Q. The OLR anomaly and mechanism before Tibet earthquake(M6.9). *Prog. Geophys.* **2008**, *23*, 1703–1708.
6. Ouzounov, D.; Liu, D.; Kang, C.; Cervone, G.; Kafatos, M.; Taylor, P. Outgoing long wave radiation variability from IR satellite data prior to major earthquakes. *Tectonophysics* **2007**, *431*, 211–220. [CrossRef]
7. Pulinets, S.; Ouzounov, D. *The Possibility of Earthquake Forecasting: Learning from Nature*; Institute of Physics Books; IOP Publishing: Bristol, UK, 2018; p. 168.
8. Sun, H.B.; Zhang, X.D. Primary exploration in relating to earthquake and Outgoing long-wave Radiation. *Plateau Earthq. Res.* **1990**, *24*, 6873.
9. Liu, D.F.; Kang, C.L. Predicting heavy disasters by outgoing longwave radiation (OLR) of the earth. *Earth Sci. Front.* **2003**, *2*, 427–435.
10. Jing, F.; Shen, X.H.; Kang, C.L.; Xiong, P. Variations of multi-parameter observations in atmosphere related to earthquake. *Nat. Hazards Earth Syst. Sci.* **2013**, *13*, 27–33. [CrossRef]
11. Sun, K.; Shan, X.J.; Ouzounov, D.; Shen, X.H.; Jing, F. Analyzing long wave radiation data associated with the 2015 Nepal earthquakes based on Multi-orbit satellite observations. *Chin. J. Geophys.-Chin. Ed.* **2017**, *60*, 3457–3465.
12. Tramutoli, V.; Aliano, C.; Corrado, R.; Filizzola, C.; Genzano, N.; Lisi, M.; Martinelli, G.; Pergola, N. On the possible origin of thermal infrared radiation (TIR) anomalies in earthquake-prone areas observed using robust satellite techniques (RST). *Chem. Geol.* **2013**, *339*, 157–168. [CrossRef]
13. Jing, F.; Shen, X.H.; Wang, H.; Kang, C.L.; Xiong, P. Infrared characteristics analysis of the 2015 Nepal M_S8.1 earthquake. *Acta Seismol. Sin.* **2016**, *38*, 429–437.
14. Guo, X.; Zhang, Y.S.; Zhong, M.J.; Shen, W.R.; Wei, C.X. Variation characteristics of OLR for the Wenchuan earthquake. *Chin. J. Geophys.* **2010**, *53*, 2688–2695.
15. Al Banna, M.H.; Abu Taher, K.; Kaiser, M.S.; Mahmud, M.; Rahman, M.S.; Hosen, A.S.M.S.; Cho, G.H. Application of Artificial Intelligence in Predicting Earthquakes: State-of-the-Art and Future Challenges. *IEEE Assess* **2020**, *08*, 192880–192923. [CrossRef]
16. Wang, Q.L.; Guo, Y.F.; Yu, L.X.; Li, P. Earthquake Prediction Based on Spatio-Temporal Data Mining: An LSTM Network Approach. *IEEE Trans. Emerg. Top. Comput.* **2020**, *8*, 148–158. [CrossRef]
17. Sadhukhan, B.; Chakraborty, S.; Mukherjee, S. Predicting the magnitude of an impending earthquake using deep learning techniques. *Earth Sci. Inform.* **2023**, *16*, 803–823. [CrossRef]
18. Senturk, E.; Saqib, M.; Adil, M.A. A Multi-Network based Hybrid LSTM model for ionospheric anomaly detection: A case study of the M_W 7.8 Nepal earthquake. *Adv. Space Res.* **2022**, *70*, 440–455. [CrossRef]
19. Xiong, P.; Long, C.; Zhou, H.Y.; Battiston, R.; Zhang, X.M.; Shen, X.H. Identification of electromagnetic pre-earthquake perturbations from the DEMETER data by machine learning. *Remote Sens.* **2020**, *12*, 3643. [CrossRef]
20. Draz, M.U.; Shah, M.; Jamjareegulgarn, P.; Shahzad, R.; Hasan, A.M.; Ghamry, N.A. Deep machine learning based possible atmospheric and Ionospheric precursors of the 2021 M_W 7.1 Japan Earthquake. *Remote Sens.* **2023**, *15*, 1904. [CrossRef]
21. Chen, Q.F.; Zheng, D.L.; Che, S. *Earthquake Cases in China (1992–1994)*; Seismological Press: Beijing, China, 2002.
22. Chen, Q.F.; Zheng, D.L.; Liu, G.P.; Li, M. *Earthquake Cases in China (1995–1996)*; Seismological Press: Beijing, China, 2002.
23. Chen, Q.F.; Zheng, D.L.; Gao, R.S. *Earthquake Cases in China (1997–1999)*; Seismological Press: Beijing, China, 2002.
24. Jiang, H.K.; Fu, H.; Yang, M.L.; Ma, H.S. *Earthquake Cases in China (2003–2006)*; Seismological Press: Beijing, China, 2014.
25. Jiang, H.K.; Yang, M.L.; Fu, H. *Earthquake Cases in China (2011–2012)*; Seismological Press: Beijing, China, 2018.
26. Jing, F.; Shen, X.H.; Kang, C.L.; Meng, Q.Y.; Xiong, P. Preliminary Analysis of the Background Features of Outgoing Longwave Radiation in China. *Earthquake* **2009**, *29*, 90–97.
27. Lu, X.; Meng, Q.; Gu, X.; Zhang, X.; Xie, T.; Geng, F. Thermal infrared anomalies associated with multi-year earthquakes in the Tibet region based on China's FY-2E satellite data. *Adv. Space Res.* **2016**, *58*, 989–1001. [CrossRef]
28. Song, D.M.; Zang, L.; Shan, X.J.; Yuan, Y.; Cui, J.Y.; Shao, H.M.; Shen, C.; Shi, H.T. A study on the algorithm for extracting earthquake thermal infrared anomalies based on the yearly trend of LST. *Seismol. Geol.* **2016**, *38*, 680–695.
29. Eleftheriou, A.; Filizzola, C.; Genzano, N.; Lacava, T.; Lisi, M.; Paciello, R.; Pergola, N.; Vallianatos, F.; Tramutoli, V. Long-Term RST Analysis of Anomalous TIR Sequences in Relation with Earthquakes Occurred in Greece in the Period 2004–2013. *Pure Appl. Geophys.* **2016**, *173*, 285–303. [CrossRef]
30. Box, G.E.P.; Jenkins, G.M. *Time Series Analysis: Forecasting and Control*; Holden Day: San Francisco, CA, USA, 1976.
31. Zhu, J.B.; Song, J.D.; Li, S.Y. Magnitude estimation for the February 13, 2021 M_J7.3 earthquake near the coast of Fukushima Japan based on support vector machine. *World Earthq. Eng.* **2021**, *37*, 74–81.

32. Wang, T.T.; Bian, Y.J.; Zhang, Y.X.; Hou, X.L. Classification of earthquakes, explosions and mining-induced earthquakes based on XGBoost algorithm. *Comput. Geosci.* **2023**, *170*, 105242. [CrossRef]
33. Hsu, T.Y.; Pratomo, A. Early Peak Ground Acceleration Prediction for On-Site Earthquake Early Warning Using LSTM Neural Network. *Front. Earth Sci.* **2022**, *10*, 911947. [CrossRef]
34. Lin, H.; Zhang, S.; Li, Q.; Li, Y.; Li, J.; Yang, Y. A new method for heart rate prediction based on LSTM-BiLSTM-Att. *Measurement* **2023**, *207*, 112384. [CrossRef]
35. Xiong, P.; Zhai, D.L.; Long, C.; Zhou, H.Y.; Zhang, X.M.; Shen, X.H. Long Short-Term Memory Neural Network for Ionospheric Total Electron Content Forecasting Over China. *Space Weather* **2021**, *19*, 2365–2377. [CrossRef]
36. Nicolis, O.; Plaza, F.; Salas, R. Prediction of intensity and location of seismic events using deep learning. *Spat. Stat.* **2021**, *42*, 100442. [CrossRef]
37. Chen, S.P.; Yin, L.; Liang, S.M.; Hu, X.Y.; Yu, X.Y. Application of Deep Learning to Predict GPS Time Series in Exploring Precursors of Menyuan M_S6.4 Earthquake. *J. Geod. Geodyn.* **2020**, *40*, 1248–1253.
38. Zhai, D.L.; Zhang, X.M.; Xiong, P. Detecting Thermal Anomalies of Earthquake Process Within Outgoing Longwave Radiation Using Time Series Forecasting Models. *Ann. Geophys.* **2020**, *63*, 548. [CrossRef]
39. Zhang, X.H.; Ren, X.D.; Wu, F.B.; Chen, Y.Y. A new method for detection of pre-earthquake ionospheric anomalies. *Chin. J. Geophys.-Chin. Ed.* **2013**, *56*, 441–449.
40. Tramutoli, V. Robust AVHRR techniques (RAT) for environmental monitoring: Theory and applications. *Proc. SPIE-Int. Soc. Opt. Eng.* **1998**, *3496*, 101–113.
41. Di Bello, G.; Filizzola, C.; Lacava, T.; Marchese, F.; Pergola, N.; Pietrapertosa, C.; Piscitelli, S.; Scaffidi, I.; Tramutoli, V. Robust satellite techniques for volcanic and seismic hazards monitoring. *Ann. Geophys.* **2004**, *47*, 49–64.
42. Genzano, N.; Aliano, C.; Filizzola, C.; Pergola, N.; Tramutoli, V. A robust satellite technique for monitoring seismically active areas: The case of Bhuj-Gujarat earthquake. *Tectonophysics* **2007**, *431*, 197–210. [CrossRef]
43. Zhang, Y.; Meng, Q. A statistical analysis of TIR anomalies extracted by RSTs in relation to an earthquake in the Sichuan area using MODIS LST data. *Nat. Hazards Earth Syst. Sci.* **2019**, *19*, 535–549. [CrossRef]
44. Kang, C.L.; Chen, Z.W.; Chen, L.Z.; Tian, Q.J.; Liu, D.F. Analysis on the satellite infrared anomaly feature before west to Kunlun Mountain Pass *M*8.1 earthquake. *China Earthq. Eng. J.* **2003**, *25*, 12–15.
45. Cui, J.; Shen, X.H.; Zhang, J.F.; Ma, W.Y.; Chu, W. Analysis of spatiotemporal variations in middle-tropospheric to upper-tropospheric methane during the Wenchuan M_S = 8.0 earthquake by three indices. *Nat. Hazards Earth Syst. Sci.* **2019**, *19*, 2841–2854. [CrossRef]
46. Meng, Q.; Zhang, Y. Discovery of spatial-temporal causal interactions between thermal and methane anomalies associated with the Wenchuan earthquake. *Eur. Phys. J.-Spec. Top.* **2021**, *230*, 247–261. [CrossRef]
47. Guo, X.; Zou, Y.; Zhang, X.; Wang, Y. Anomalies in Outgoing Long-Wave Radiation for Several Strong Earthquakes in Mainland China. *China Earthq. Eng. J.* **2019**, *41*, 1221–1227.
48. Han, S.S.; Tan, K.; Lu, X.F.; Zhang, C.H.; Li, Q. Analysis of Strain Evolution Characteristics before the 2021 Maduo 7.4 Earthquake in Qinghai. *J. Geod. Geodyn.* **2023**, *43*, 641–645.
49. Deng, M.D.; Geng, N.G.; Cui, C.Y.; Zhi, Y.Q.; Fan, Z.F.; Ji, Q.Q. The Study on the Variation of Thermal State of Rocks Caused by the Variation of Stress State of Rocks. *Earthq. Res. China* **1997**, *02*, 85–91.
50. Du, X.H.; Zhang, X.M. Ionospheric Disturbances Possibly Associated with Yangbi M_S6.4 and Maduo M_S7.4 Earthquakes in China from China Seismo Electromagnetic Satellite. *Atmosphere* **2022**, *13*, 438. [CrossRef]
51. Zhang, X.M.; Liu, J.; Zhou, Y.L.; Du, X.H. The ionospheric Perturbations Around Yunnan Yangbi M_S6.4 and Qinghai Maduo M_S7.4 Earthquakes in 2021. *Earthquake* **2022**, *42*, 1–20.
52. Wang, S.Y.; Tian, H.; Ma, K.X.; Yu, C.; Ma, W.Y.; Yu, H.Z. Correlation between short-term and imminent anomalies of LURR and OLR before the Maduo M_S7.4 earthquake in Qinghai Province, 2021. *China Earthq. Eng. J.* **2021**, *43*, 847–852.
53. Su, W.G.; Liu, L.; Sun, X.H. Characteristics of underground fluid anomalies in Zuoshu station before Maduo M_S7.4 and Menyuan M_S6.9 earthquakes. *China Earthq. Eng. J.* **2022**, *44*, 700–706, 712.
54. Jing, F.; Zhang, L.; Singh, R.P. Pronounced Changes in Thermal Signals Associated with the Madoi (China) *M* 7.3 Earthquake from Passive Microwave and Infrared Satellite Data. *Remote Sens.* **2022**, *14*, 2539. [CrossRef]
55. Luo, B.S.; Bai, Z.X.; Li, Y.F. Analysis on abnormal characteristic of water radon in HuangYuan station before and after the M_S 7.4 Maduo earthquake in 2021. *Plateau Earthq. Res.* **2022**, *34*, 53–57.
56. Pulinets, S.; Ouzounov, D. Lithosphere-Atmosphere-Ionosphere Coupling (LAIC) model—An unified concept for earthquake precursors validation. *J. Asian Earth Sci.* **2011**, *41*, 371–382. [CrossRef]

remote sensing

MDPI

Article

Study of the OLR Anomalies before the 2023 Turkey M7.8 Earthquake

Jun Liu [1], Jing Cui [2,*], Ying Zhang [3], Jie Zhu [4], Yalan Huang [2], Lin Wang [1] and Xuhui Shen [5]

1 School of Ecological Environment, Institute of Disaster Prevention, Sanhe 065201, China; liujun.3s@cidp.edu.cn (J.L.); 21661332@sr.cidp.edu.cn (L.W.)
2 National Institute of Natural Hazards, Ministry of Emergency Management of the People's Republic of China, Beijing 100085, China; huangyalan22@mails.ucas.ac.cn
3 Department of Geophysics, School Earth and Space Sciences, Peking University, Beijing 100091, China; zhangying2016@radi.ac.cn
4 China Earthquake Networks Center, Beijing 100045, China; zhujie@seis.ac.cn
5 National Space Science Center, Chinese Academy of Sciences, Beijing 100190, China; xuhuishen@ninhm.ac.cn
* Correspondence: jingcui@ninhm.ac.cn; Tel.: +86-01062846770

Abstract: Using the model of the additive tectonic stress from celestial tide-generating force, we studied the relationship between the seismogenic structure and celestial tide-generating stress in the M7.8 Turkey earthquake on 6 February 2023. We analyzed the daily continuous variation characteristics of OLR before and after the Turkey earthquake and discussed the correlation characteristics of tidal stress, OLR, and the earthquake. The results showed that the observed OLR anomaly according to the tidal stress variation cycle "C" (1–15 February) presented a phase change in time, which was synchronized with a continuous trough-to-peak change in the additional tectonic main pressure stress. The spatial distribution of OLR anomalies was mainly concentrated in the southwest section of the East Anatolian Fault Zone, which indicates that seismic tectonic movements were the main causes of OLR anomaly variation during this earthquake. An OLR anomaly change was related to this M7.8 "Swarm Type" of earthquake in Turkey. Impending earthquake OLR anomalies represent that the stress of the seismogenic structure in the seismogenic region has entered a critical state, which can provide stress monitoring and a seismogenic region indication for earthquakes induced by tidal force. The change cycle of the celestial tide-generating force provides a time indication for the identification of seismic thermal anomalies, and it indicates that the combination of the additional tectonic stress of the tidal force and the change of OLR anomaly has value for the research on the short-impending earthquake precursor.

Keywords: earthquake; OLR; short-impending anomaly; tidal stress; Turkey

Citation: Liu, J.; Cui, J.; Zhang, Y.; Zhu, J.; Huang, Y.; Wang, L.; Shen, X. Study of the OLR Anomalies before the 2023 Turkey M7.8 Earthquake. *Remote Sens.* **2023**, *15*, 5078. https://doi.org/10.3390/rs15215078

Academic Editor: Yunbin Yuan

Received: 23 August 2023
Revised: 19 October 2023
Accepted: 20 October 2023
Published: 24 October 2023

1. Introduction

Earthquakes are the result of the tectonic deformation and rupture of faults under tectonic stress, but in situ stress cannot be directly observed at present [1]. By observing the change in heat-related physical quantities, tectonic stress information can be obtained indirectly [2,3]. The occurrence of atmospheric thermal anomalies before earthquakes has been widely recognized by scholars [4,5]. Scholars from the former Soviet Union [3] found that the phenomenon of short-impending infrared radiation enhancement occurs at the intersection of faults before earthquakes, which introduced a new way to study earthquakes using satellite thermal infrared remote sensing technology. In the last 30 years, many scholars have explored the application of satellite remote sensing technology in seismic activity monitoring and earthquake precursor research, and many achievements have been made in the studies on the generation mechanism of thermal anomalies [3,6–13] and the extraction algorithm of seismic thermal anomalies [14–21]. Pre-earthquake thermal anomalies have also been verified in many case studies [22,23].

In order to scientifically elucidate the physical mechanism of infrared thermal anomalies in tectonic earthquakes, both domestic and foreign scholars have conducted extensive exploration and research using various methods such as air field experiments and rock loading experiments. Consequently, several hypotheses including the gas-thermal [3,6], stress-induced heat [7–9], "P-hole" [11], and latent heat release due to radon decay theories [13] have been proposed, but remain inconclusive. However, considering the sources of heat during the earthquake preparation process, it is evident that with the progressive accumulation of tectonic stress, mechanical energy, along with other forms of physical and chemical energy, are generated deep within the Earth's interior. Subsequently, a significant portion of this energy is converted into heat, which is then released through various mechanisms in the form of infrared electromagnetic radiation [14]. At present, the methods for pre-earthquake thermal anomaly extraction using satellite infrared information include the background field difference analysis method [15], the robust satellite data analysis technique (RST) [16,17,24,25], the eddy field calculation mean algorithm [26,27], the wavelet power spectrum method [21], the spatio-temporally weighted two-step method [28], etc. These methods require many years of historical data. Moreover, the normal background field established by the multi-year data-averaging algorithm based on statistical principles will mask small fluctuations in the data and miss the weak thermal anomaly signal before the earthquake. At the same time, due to the uncertainty of the selection time (5 years, 10 years, etc.), the different background fields will present different results [29–31].

The occurrence of earthquakes is a manifestation of internal tectonic movements within the Earth. However, as the Earth is not an isolated celestial body, its movement is inevitably influenced by macroscopic celestial movements. One external mechanical factor that affects the Earth's movement is sun–moon tidal forces. This force generates solid tides in the Earth's interior, leading to periodic tidal stresses with magnitudes in the order of 10^3 Pa. Although these tidal stresses are much smaller than seismic stress drops, their cyclic loading rate exceeds that of tectonic stress accumulation by two orders of magnitude [32,33]. The stress in the Earth's crust results from both tectonic and tidal forces acting upon it. When seismic tectonic stress reaches a critical state, causing rock sliding, and if a rapidly changing tidal stress is superimposed in an appropriate direction, it may trigger an earthquake [32,34–37]. In order to further study the dynamic change of tidal stress in the process of time series and its relationship with the seismic process, Ma et al. [38] proposed a model of the additive tectonic stress from celestial tide-generating force (abbreviated as ATSCTF). This model calculates the components of tidal stress (∂P, ∂T, and ∂N) generated by the celestial tide-generating force at the epicenter along the direction of the main pressure stress P-axis, the tension stress T-axis, and the vertical stress N-axis. The induced effects of additional tectonic stresses on seismic faults are classified into three types [39,40]. By drawing time-series variation curves for each component according to the ATSCTF models, this provides a clear temporal context with mechanical significance for studying the thermal anomalies associated with earthquakes and enables the monitoring of tectonic stress states through the observation of thermal anomalies caused by tide-induced earthquakes [41,42].

At present, when infrared remote sensing data are used to study seismic and tectonic activities, the cloud interference often hinders the observation of temperature anomalies before earthquakes. Outgoing long-wave radiation (OLR) represents the energy density of electromagnetic waves emitted into outer space via Earth's atmospheric system, and the radiation physical quantity most directly reflects the underlying surface properties and energy variation parameters. Moreover, its band (10.5–12.5 μm) is concentrated in the atmospheric window and minimally interferes with clouds [31]. Therefore, OLR data were selected for this research. Aiming to study the 2023 "earthquake swarm type" M7.8 earthquake in Turkey, we combined the celestial tide-generating force and OLR data to study the correlation between two physical parameters and earthquakes and discussed the process of earthquake breeding and occurrence.

2. Study Area

According to the U.S. National Earthquake Information Center (NEIC), at 01:17 (UTC) on 6 February 2023 (4:18 local time on 6 February), a 7.8 magnitude earthquake with a focal depth of 17.9 km occurred 26 km east of Nurdaği in Gaziantep province, south-central Turkey, near the northern Syrian border. The epicenter was located at 37.17°N and 37.03°E, and aftershocks continued. Nine hours later, at 10:24 (UTC) (13:25 local time), a 7.5 magnitude earthquake occurred again about 96 km northeast of the epicenter of the 7.8 magnitude earthquake. The epicenter was located at 38.02°N, 37.20°E. The focal depth of this earthquake was 10 km. These two powerful earthquakes caused many casualties and property losses in Turkey, and this is the worst disaster to hit the country in more than 80 years, arousing the attention of the international community.

Most of Turkey is located on the Anatolian block where the Eurasian plate, the Arabian plate, and the African plate meet, and its geological structure is very complex. Two huge strike-slip fault zones, the East Anatolian Fault and the North Anatolian Fault [36], developed in Turkey under the influence of the tectonic compression of multiple plates. The two fault zones converge in eastern Turkey (Figure 1). Both earthquakes occurred near the northeasterly extending East Anatolian Fault zone, which is one of the major intracontinental conversion faults in the Eastern Mediterranean region. Together with the North Anatolian right-rotation Fault zone, it accommodated the westward compression tectonic movement of the Anatolian block. This fault zone is also controlled by the northwest subduction of the Arabian plate. The rupture of this earthquake was more than 400 km long.

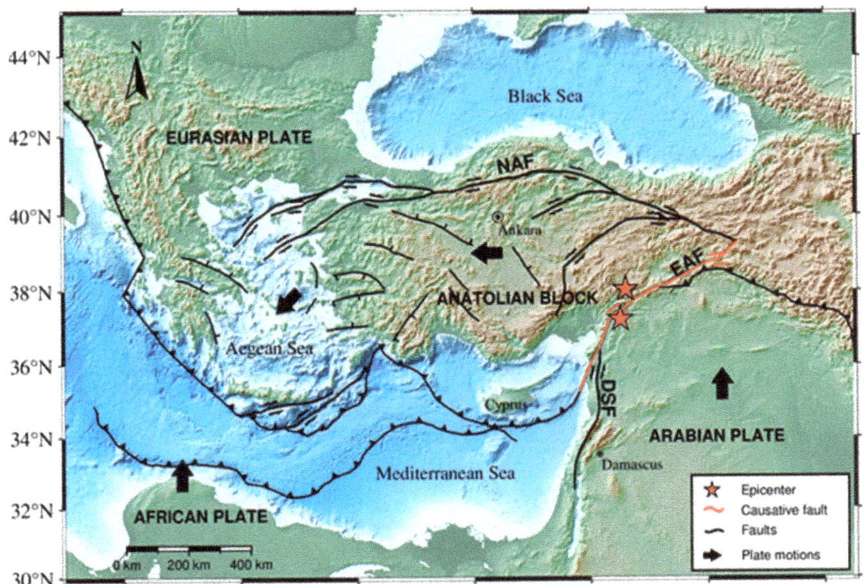

Figure 1. Distribution of major active faults in Turkey earthquake epicenter and study area [43,44]. NAF: North Anatolian Fault; EAF: East Anatolian Fault; DSF: Dead Sea Fault.

3. Materials and Methods

3.1. Calculation of the Additive Tectonic Stress from the Celestial Tide-Generating Force

To analyze the influence of additive tectonic stress from the celestial tide-generating force on the seismogenesis fault, the tidal stress component in the axis coordinate system of seismic stress is calculated. It is generated by the celestial tide-generating force of the sun and the moon at the epicenter along the directions of the main pressure stress

P-axis, the tension stress T-axis, and the vertical stress N-axis. This is performed to explore the relationship between the size, direction, and time of seismic occurrence and the region of seismic occurrence [39] and provide timely guidance for the selection of the OLR abnormal background.

The calculation of tidal stress components at the epicenter is divided into three steps. Firstly, the tidal stress components separately generated by the moon and the sun at the source in the spherical coordinate system are calculated according to Kelvin's method [45]. Then, the tidal stress components from the spherical coordinate system are converted to the rectangular coordinate system. Finally, the tidal stress component in the rectangular coordinate system of the earthquake source generated by the sun and the moon is converted into the stress component in the main axis coordinate system of the earthquake stress, and then the tidal stress components at the source along the P-axis, T-axis, and N-axis, ∂_P, ∂_T, and ∂_N, respectively, are obtained [40]. The converted formula is as follows:

$$\begin{cases} \partial_P = \sigma_{xx}B_{PX}^2 + \sigma_{yy}B_{PY}^2 + \sigma_{zz}B_{PZ}^2 + 2\sigma_{xy}B_{PX}B_{PY} + 2\sigma_{yz}B_{PY}B_{PZ} + 2\sigma_{zx}B_{PZ}B_{PX} \\ \partial_T = \sigma_{xx}B_{TX}^2 + \sigma_{yy}B_{TY}^2 + \sigma_{zz}B_{TZ}^2 + 2\sigma_{xy}B_{TX}B_{TY} + 2\sigma_{yz}B_{TY}B_{TZ} + 2\sigma_{zx}B_{TZ}B_{TX} \\ \partial_N = \sigma_{xx}B_{NX}^2 + \sigma_{yy}B_{NY}^2 + \sigma_{zz}B_{NZ}^2 + 2\sigma_{xy}B_{NX}B_{NY} + 2\sigma_{yz}B_{NY}B_{NZ} + 2\sigma_{zx}B_{NZ}B_{NX} \end{cases} \quad (1)$$

In Equation (1), $\sigma_{ij}(i,j = x,y,z)$ represents the 9 tidal stress components generated by the joint action of the moon and the sun in the source coordinate system; $B_{i,j}(i = P,T,N, j = X,Y,Z)$ is the coordinate transformation matrix consisting of the rotation angles H, Q, and V in the transformation between the source rectangular coordinate system and the seismic stress main axis coordinate system, which are expressed as:

$$\begin{cases} B_{PX} = -\sin H \cos Q \sin V - \sin Q \cos V \\ B_{PY} = -\sin H \sin Q \sin V + \cos Q \cos V \\ B_{PZ} = \sin V \cos H \\ B_{TX} = -\sin H \cos Q \cos V + \sin Q \sin V \\ B_{TY} = -\sin H \sin Q \cos V - \cos Q \sin V \\ B_{TY} = \cos H \cos V \\ B_{NX} = \cos H \cos Q \\ B_{NY} = \cos H \sin Q \\ B_{NZ} = \sin H \end{cases} \quad (2)$$

In Equation (2), H, Q, and V are the rotation angles of the coordinate transformation, which can be obtained from the source parameters (inclination and strike of the P-axis and T-axis).

3.2. OLR Data Processing

OLR is the energy density of electromagnetic waves emitted into outer space via Earth's atmospheric system, also known as the thermal radiation flux density, which is measured in W/m^2. It is measured using an NOAA polar orbit satellite load radiation measuring instrument, which scans the Earth and the atmosphere in the infrared window channel (10.5–12.5 μm) and detects long-wave radiation emitted from the ground.

This research adopted OLR data taken from http://www.emc.ncep.noaa.gov (accessed on 5 May 2023), which have a temporal resolution of 24 h and a spatial resolution of $1 \times 1°$, covering a total of 360×181 grids worldwide. Since OLR is based on remote sensing output results in the infrared band, its band approximates the long-wave atmospheric window, for which atmospheric decay is weak, and its band is close to the surface long-wave radiation; additionally, it is sensitive to temperature changes in the sea surface and near the ground. Therefore, it is suitable for monitoring some geo-hazard signs associated with "thermal" genesis phenomena.

The OLR values of the grid points (inter-daily scale) in the study area are calculated using Formula (3) to extract the changing characteristics of the OLR data before and every day after the earthquake. Subsequently, the numerical distribution of each grid point representing information in the radiation enhancement area is obtained.

$$\Delta T_i(lon, lat) = T_i(lon, lat) - T_{background}(lon, lat) \tag{3}$$

$\Delta T_i(lon, lat)$ represents the incremental OLR value in a Gaussian grid (lon, lat); $T_i(lon, lat)$ represents the OLR value in a Gaussian grid (lon, lat); $T_{background}$ (lon, lat) represents the OLR value in a Gaussian grid (lon, lat) on a fixed background day; lon is the geodetic longitude, $lon = 1, 2, \ldots, 360$; lat is the geodetic latitude, $lat = 1, 2, \ldots, 180$; i is the date. The background day is determined based on the inflection point of the period for additional tectonic stress caused by tidal-generating forces.

4. Results

4.1. Analysis of the Additive Tectonic Stress from the Celestial Tide-Generating Force Change in Turkey Earthquake

The tidal stress generated by the sun and the moon exhibits periodic and continuous changes, with its magnitude being influenced by the focal mechanism of earthquakes. The turning point in the period of tidal stress change represents a position where the impact of tidal forces on the tectonic environment alters, and this period is typically divided based on such turning points (wave crests or troughs) [41]. According to the focal mechanism solution provided by USGS (https://earthquake.usgs.gov/) (accessed on 19 February 2023), we calculated the additional tectonic stress component (∂P, ∂T, and ∂N) at the epicenter of the Turkey earthquake for the period from 1 January to 1 March 2023 using the ATSCTF model [38]. The results were then plotted as a time-dependent curve in Figure 2. The figure illustrates that the additional stress components ∂T and ∂N generated by tidal forces from celestial bodies were relatively small along the tension stress T-axis and the vertical stress N-axis, whereas the component ∂P generated in the direction of the main pressure stress P-axis was significant. Consequently, the period of tidal stress variation can be divided into four consecutive periods (marked as A, B, C, and D) based on the inflection point (trough) of the change curve for the ∂P value. Period A corresponds to 3–17 January, period B corresponds to 18–31 January, period C corresponds to 1–15 February, and period D corresponds to 16 February–1 March. The initial day of each cycle serves as the foundational background day. The earthquake occurred in close proximity to the relative high point of the principal compressive stress ∂P during its transition from trough to peak [40], suggesting that the celestial tide-generating force continuously amplified the positive pressure on the fault surface. When the active tectonic stress reaches a critical threshold, the tide-generating force may trigger destabilization and the rupture of seismic structures under stress, leading to earthquakes.

The M7.8 earthquake in Turkey occurred in cycle "C", but no earthquake occurred (http://www.ceic.ac.cn) (accessed on 20 June 2023) near the epicenter during similar phases of cycles A, B, and D, indicating that the celestial tide-generating force is only one of the important external factors that trigger and induce earthquakes. Only when the active tectonic stress reaches the critical point of rupture can the earthquake be induced [41]. To judge the stress accumulation in the seismogenic region and the size of the change in tectonic stress in the corresponding period, it is necessary to further analyze the spatial and temporal evolution characteristics of OLR against the background of the tidal force change period.

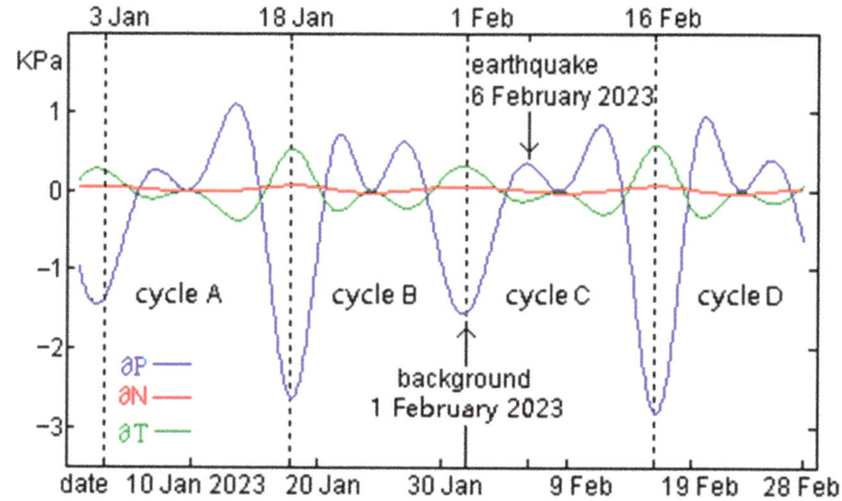

Figure 2. The additive tectonic stress components ∂P, ∂T, and ∂N changed over time before and after the earthquake in Turkey.

4.2. Spatial and Temporal Evolution Characteristics of OLR Anomaly in Turkey "Swarm Type" Earthquake

The choice of background date is critical when extracting OLR anomalies using Equation (3). Therefore, to avoid uncertainties in the calculation results due to differences in the duration of the background time in the statistical processing method of multi-year data [26,27], this paper provides a temporal guide with a clear physical meaning for the selection of the background date based on the cycle of the additive tectonic stress from the celestial tide-generating force variation. Taking the phase low point of the main stress value ∂P, with the variation curve as the reference (Figure 2), 3 January, 18 January, 1 February and 16 February were selected as the background dates for the OLR data, respectively. According to Equation 3, the nighttime surface long-wave radiation values in the spatial range (20–50°N, 20–50°E) from 3 January to 1 March were subtracted from the OLR background values to obtain continuous day-by-day images of the OLR changes before and after the earthquakes in Turkey.

As shown in Figure 3c, within the study area, the epicenter of the Turkey earthquake and its adjacent areas showed a significantly anomalous OLR change before and after the earthquake after the low peak phase of the additive tectonic main stress from the celestial tide-generating force ∂P. Obvious OLR anomalies appeared south and southwest of the epicenter on 2 February, spanning the intersection of the Anatolian, African, and Arabian plates. The OLR anomalies were mainly distributed along both sides of the southwest end of the NE direction East Anatolia Fault Zone, indicating that the fault had a controlling effect on the OLR anomaly increase. On 3–4 February, the OLR anomaly declined, and its extent narrowed, but it was still symmetrically distributed on both sides of the NE direction East Anatolia Fault Zone. On 5 February, the OLR anomaly declined significantly in intensity and was spatially scattered in a northeast direction along the East Anatolian Fault Zone. The earthquake occurred on 6 February, when the additive tectonic stress from the celestial tide-generating main pressure stress ∂P reached its sub-peak phase, the positive pressure on the fault surface increased, and this promoted the fault stress state to reach a critical condition for rupture sliding and seismogenesis. The OLR anomaly was enhanced in the images again at the southwest end of the East Anatolian Fault Zone. After the mainshock, aftershock activity was frequent between 7 February and 15 February, and densely distributed at the southwest end of the East Anatolian Fault Zone. The frequent vibration of the earthquake formed numerous collapsed areas or large-scale surface ruptures, which became thermal channels and released large amounts of heat, manifesting as the continuous expansion of the area and the magnitude of the OLR anomaly.

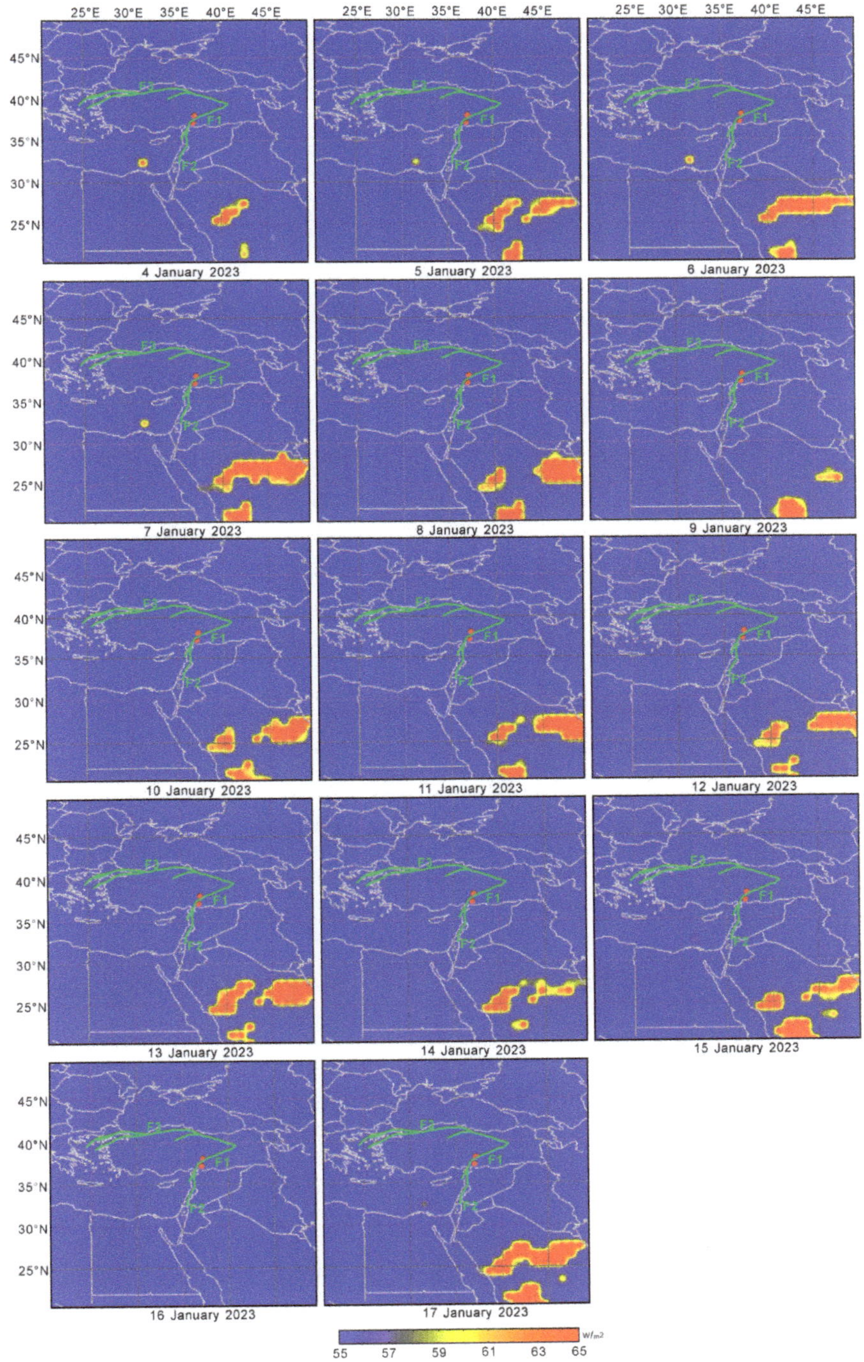

(**a**) A period

Figure 3. *Cont.*

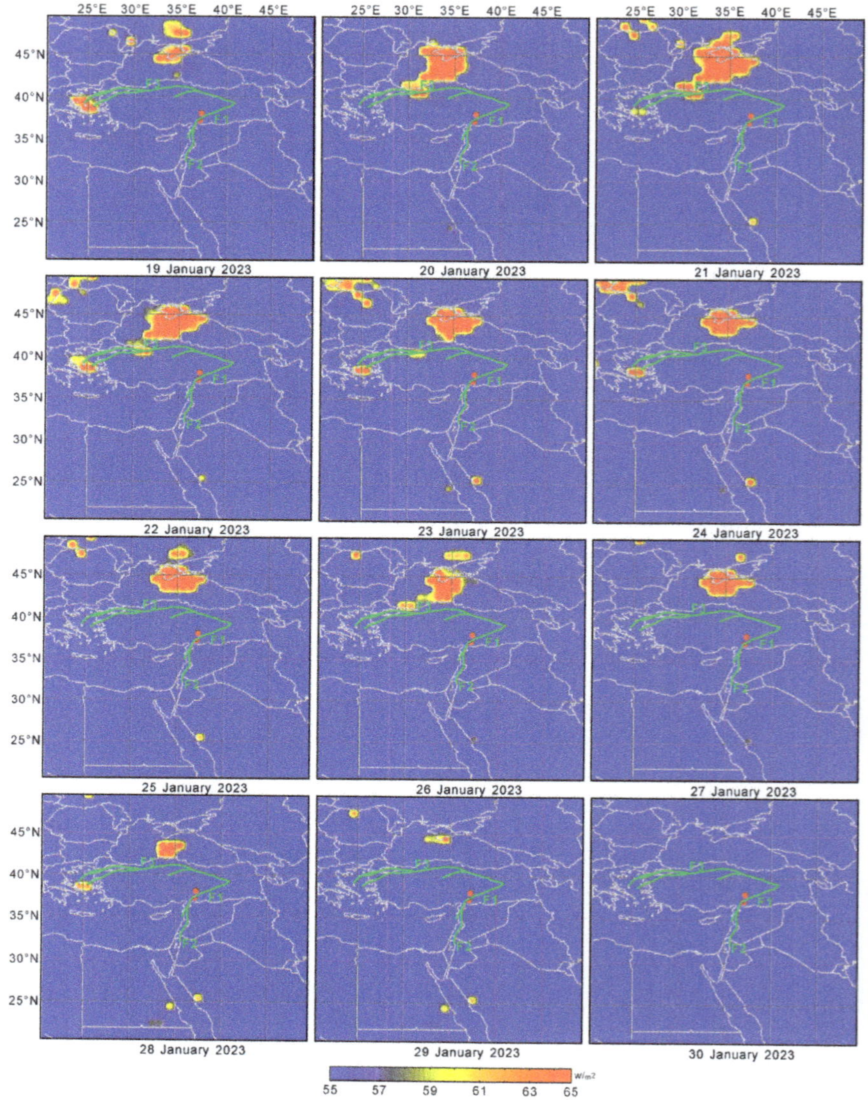

(**b**) B period

Figure 3. *Cont.*

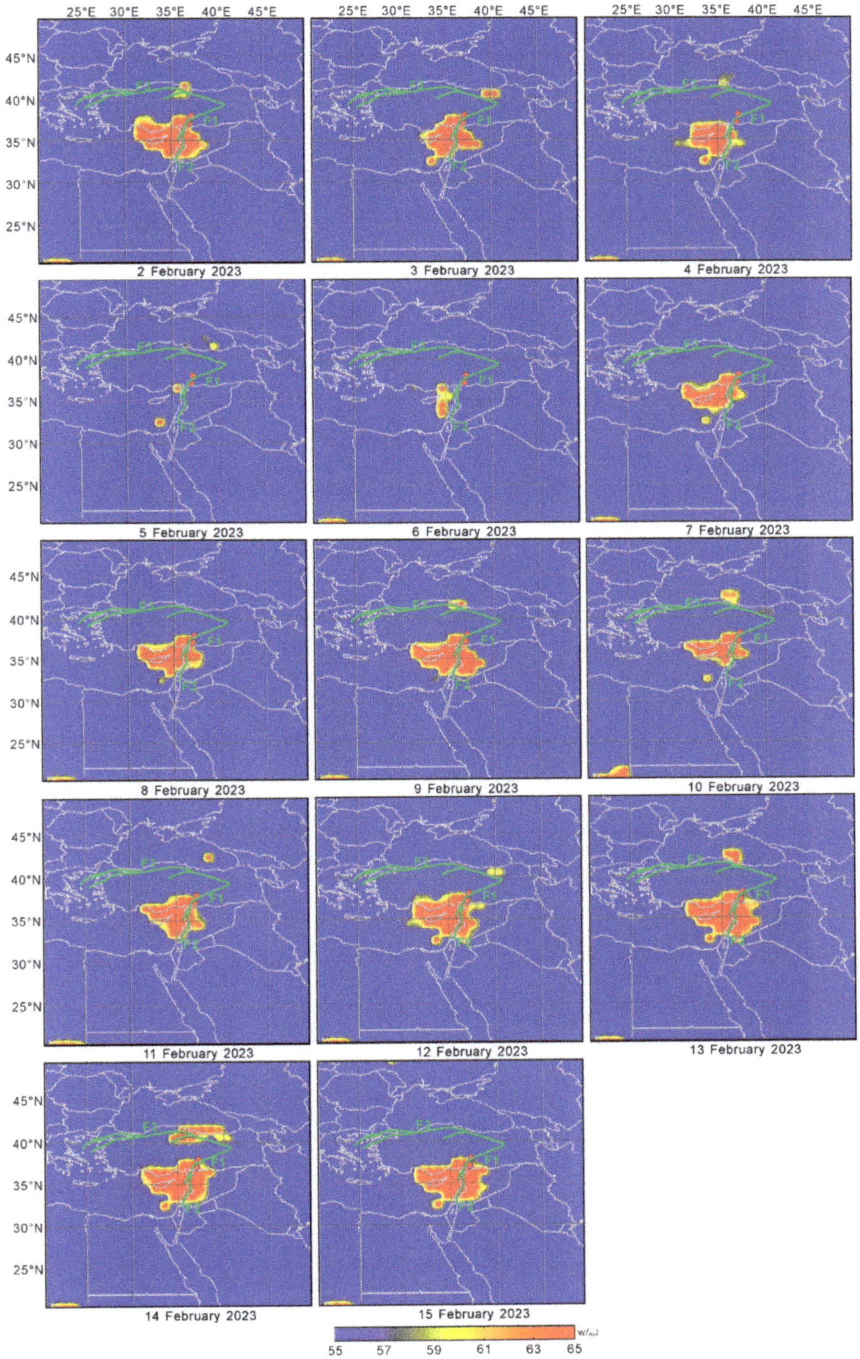

(c) C period

Figure 3. *Cont.*

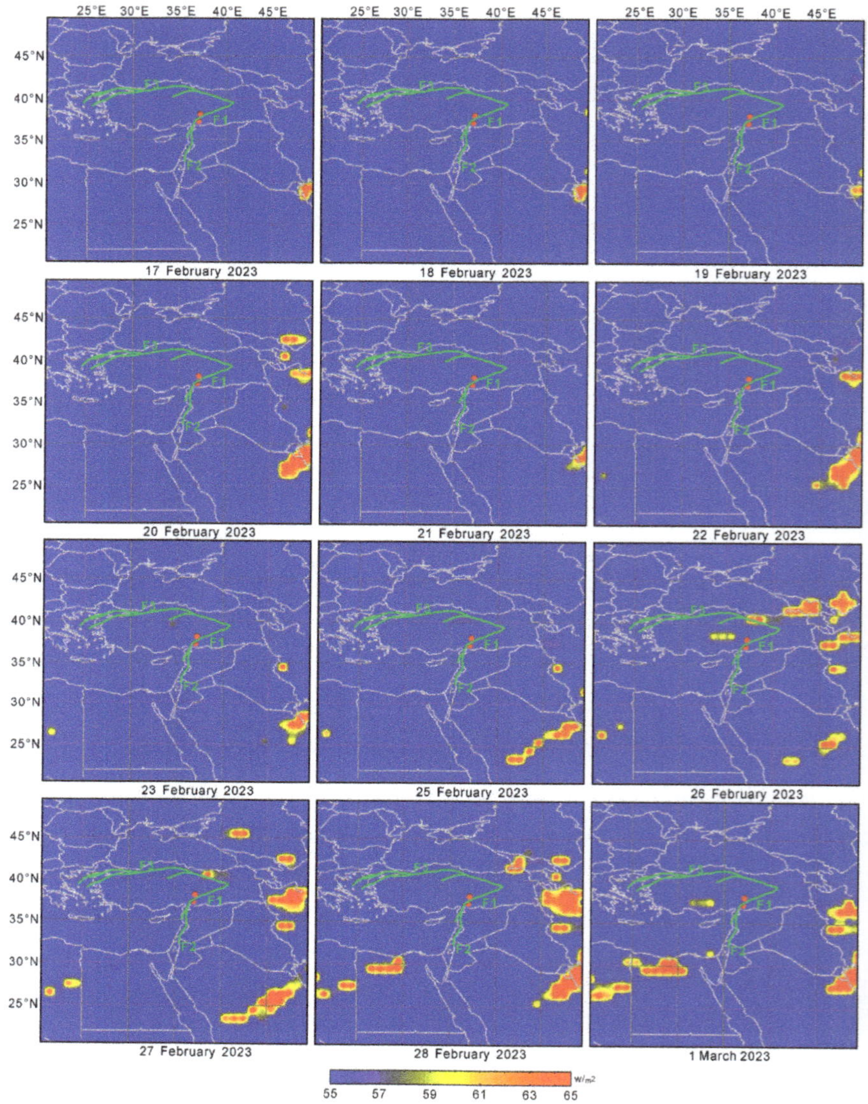

(**d**) D period

Figure 3. Images of the spatial and temporal evolution of ground-based long-wave radiation anomalies from the 7.8 magnitude earthquake in Turkey. The red dots are the epicenters. F1: East Anatolian Fault; F2: North Anatolian Fault; F3: Dead Sea Fault.

As shown in Figure 3a,b,d, there was no significant enhancement of OLR in and around the epicenter of the Turkey earthquake during the pre-earthquake tidal change cycle "A", "B", and the post-earthquake tidal change cycle "D", and no earthquake occurred. It should be noted that the OLR image for 24 February is missing from cycle "D" (Figure 3d) due to missing OLR data for that day.

5. Discussion

During the seismogenic cycle (cycle "C"), the earthquakes in Turkey occurred when the trough-to-peak main pressure stress change occurred, which was when the magnitude and direction of the tidal stress suddenly changed, prompting the destabilizing rupture of the southwest section of the East Anatolian Fault in a high-stress state. However, no earthquake occurred near the epicenter in similar phases of other cycles of tidal stress change, indicating that tidal stress cannot directly trigger earthquakes and that the relationship between tidal stress and earthquakes cannot be studied in isolation from the regional tectonic setting.

From the analysis of Figure 3c, it can be seen that the OLR short-impending anomalies before and after the M7.8 earthquake in Turkey, which were observed synchronously according to the cycle changes in additive tectonic stress from the celestial tide-generating force, showed a change over time, undergoing the continuous evolution of initial warming (2 February)→anomalous decay (3–5 February)→mainshock (6 February)→aftershock→continuous warming (7–15 February). This is consistent with the phase change in infrared radiation temperature in the rock stress rupture experiment [46]. The above process basically reflects the seismogenic process of micro-rupture→closure→major rupture (seismogenesis)→the rupture of active structures in all directions→tectonic adjustment (aftershock) during the seismogenesis. Combined with the theory of tidal force-induced earthquakes, when tectonic stresses have accumulated to enable rock rupture and sliding, earthquakes may be triggered by the superimposition of rapidly changing celestial tidal forces and additive tectonic stresses. It can be assumed that the OLR anomaly, which is synchronized with the cycle of tidal force change, is a thermal image representation of the seismic tectonic stress approaching the critical state of rupture, and that the tidal force induces earthquakes by changing the tectonic stress environment in the subsurface when the ground stress is sufficient. These "Swarm Type" earthquakes in Turkey are characterized by a high seismic frequency, a long duration of activity, and slow energy decay. After these two powerful earthquakes, the transition zone between the Dead Sea Fault Zone and the East Anatolian Fault Zone was characterized by high aftershock activity, with a total of 6000 aftershocks until 19 February, 40 of which were of magnitude 5–6, and 436 of which were of magnitude 4–5. Ongoing seismic activity induced the release of a substantial amount of thermal radiation through surface ruptures, which were visually depicted in the OLR change image as a persistent thermal radiation anomaly located in close proximity to the southwestern region of the East Fault belt. It has been shown that with the decay of the earthquake magnitude, the seismotectonics of the region will stabilize, and OLR anomalies will gradually disappear [41]. It has been observed in the rock-loading test that a transient cooling phenomenon occurred prior to rock instability, and this cooling anomaly served as a significant precursor to rock failure and instability, indicating the imminent macro failure of the rock [12,47,48]. Furthermore, the infrared seismic monitoring results revealed that apart from warming anomalies, certain earthquake events [49–51] exhibited a preceding cooling phenomenon before the occurrence of earthquakes. A similar process was also observed during the earthquakes in 2023 in Turkey. Therefore, we posit that the abnormal attenuation phenomenon of OLR prior to the earthquake on 5 February may signify a lock-in period before changes in rock stress lead to a rupture and could potentially serve as short-term imminent precursor information for earthquakes. The thermal anomalies associated with the Turkish earthquake were primarily distributed along the southwest section of the East Anatolian Fault Zone and northern section of the Dead Sea Fault, which aligned with the spatial distribution patterns of asperities identified in the East Anatolian Fault Zone [52]. These findings indicate that the OLR anomaly was influenced by the spatial distribution of seismogenic faults and tectonic activities, while representing thermal image manifestations of tectonic stress changes during earthquake preparation. The spatial and temporal evolutions of OLR shown in Figure 3c were consistent with the characteristics of seismic thermal anomaly expression, and it is suggested that there was a correlation between this OLR anomaly change and seismic activity.

To help verify the method's recognition rate, we achieved a coverage of $40 \times 40°$ for another earthquake of magnitude M7.2 that occurred on 23 October 2011 at 10:41:22 UTC close to the city of Van in eastern Turkey at a shallow depth (10 km) below the shore of Lake Van (38.8°N, 43.5°E). The relationship between the Van earthquake and ATSCTF is shown in Figure 4. The earthquake occurred during cycle "B" at the time when the ∂P value increased to the maximum amplitude and the ∂T value decreased to the minimum amplitude. The changes in ∂P resulted in an increase in the positive stress of the fault plane, and the changes in ∂T resulted in a decrease in the rupture slide intensity of the fault plane. However, unlike the observations noted for the 2023 earthquake, the changes in ∂P and ∂T were too small to trigger an earthquake (Figure 4). The ∂N value of the ATSCTF was much stronger, and the earthquake occurred at a relatively high value of ∂N. This indicates that the ATSCTF was strong enough to enhance the tectonics under conditions of high tectonic stress by disturbing and breaking the balance of the tectonic stress, thereby triggering an earthquake. However, no earthquakes occurred during the other three cycles, not even during cycle "D", for which the ∂N value was higher than it was in cycle "B".

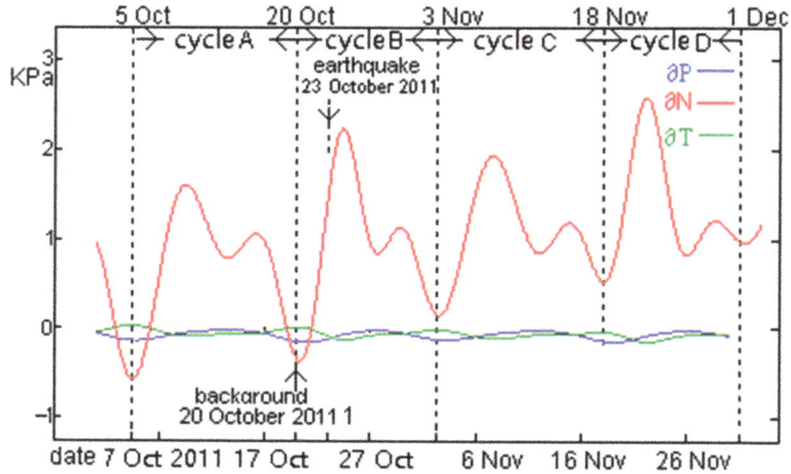

Figure 4. The periodic changes in ∂P, ∂T, and ∂N of the ATSCTF for Turkey's Van earthquake on 23 October 2011.

Figures 2 and 4 show that the 6 February, 2023 M7.8-strong earthquake and 23 October 2011 M7.2-strong earthquake both occurred at the high points (Figures 2 and 4), indicating that the tidal force had a certain triggering effect on the occurrence of earthquakes.

We obtained the spatial and temporal evolution of OLR before, during, and after the earthquake, which correspond to cycles "A", "B", and "C" (Figure 5). The results show that there was no significant enhancement of OLR in and around the epicenter of the Turkish earthquake during the pre-earthquake tidal change cycle "A" and the post-earthquake tidal change cycle "C", and no earthquake occurred. In cycle "B", the OLR changes during the ATSCTF period of the earthquake were clearly shown. On 21 October, the OLR value increased in the north of the epicenter. There was a distinct increase in the OLR value on 22 October, and the OLR area extended mainly to the north of the epicenter. After 22 October, the OLR area shrank around the epicenter. The main shock occurred on 23 October. After the shock was over, the higher OLR value regions were still concentrated at the center of the epicenter. The distribution of anomalous areas was approximately parallel to the geological structure associated with the North Anatolian Fault. The OLR value decreased gradually along the East Anatolian Fault until 25 October (Figure 5). The enhancement demonstrated a similar evolution process (increase→earthquake→shrink around the epicenter) for the 2023 earthquake, and the evolution process was consistent with the rock failure process.

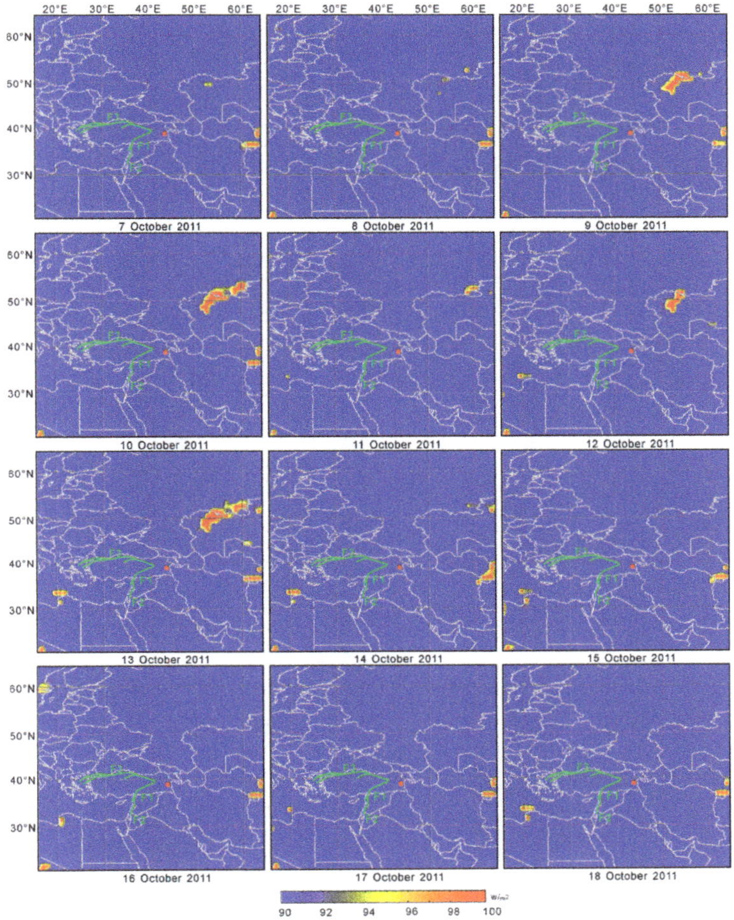

(**a**) A period

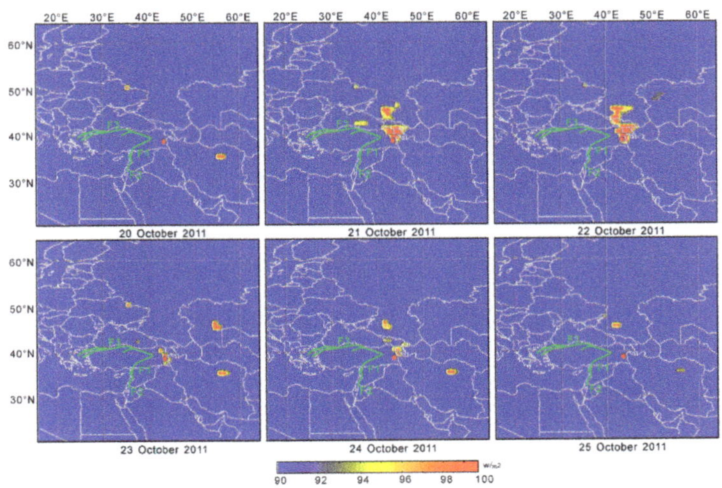

(**b**) B period

Figure 5. *Cont.*

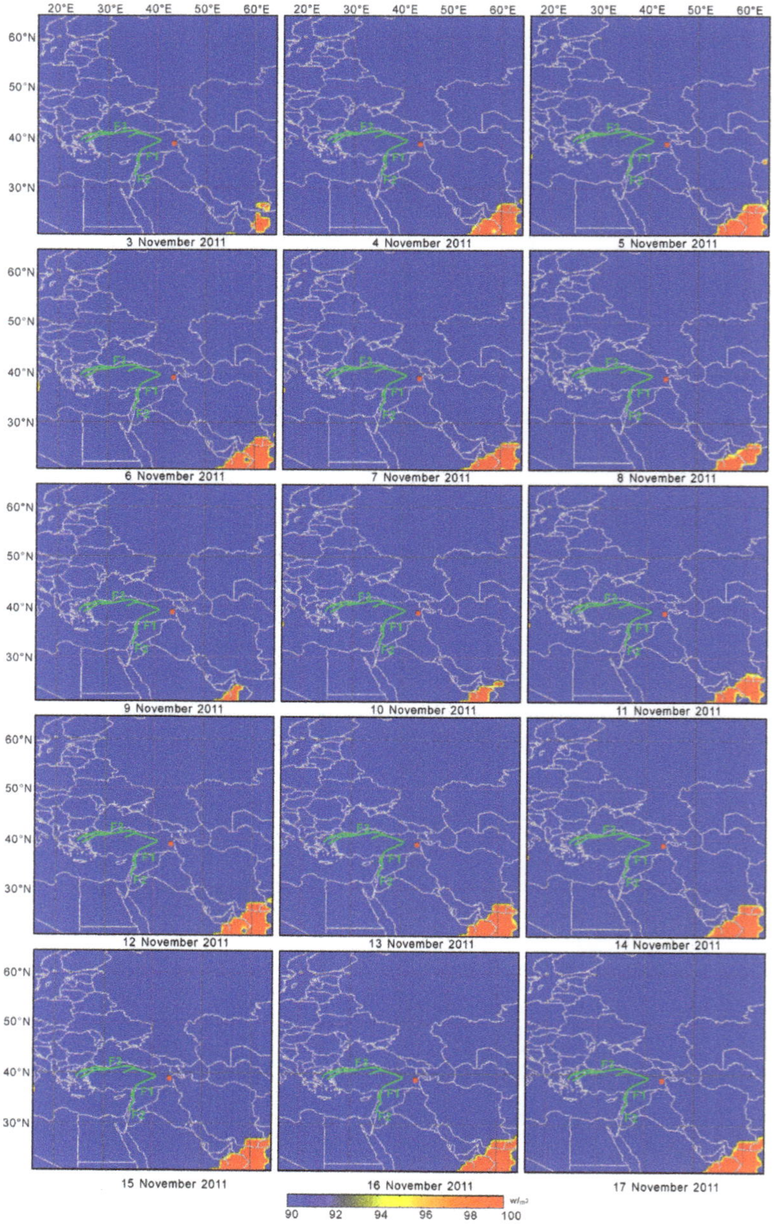

(**c**) C period

Figure 5. Incremental field distribution of OLR during the Turkey earthquake on 23 October 2011. The red dots are the epicenters. F1: East Anatolian Fault; F2: North Anatolian Fault; F3: Dead Sea Fault.

6. Conclusions

This paper used celestial tide-generating force and OLR data to jointly carry out research on seismic thermal anomaly identification. On the one hand, the change cycle of additive tectonic stress caused by the celestial tide-generating force can provide a clear time indication for the selection of remotely sensed anomaly-monitoring background data for short-impending earthquakes. At the same time, a mechanical basis for the identification of pre-earthquake thermal anomalies was added to eliminate the uncertainty of the conclusion caused by the randomness of background time selection in statistical processing methods that have existed for many years. On the other hand, the temporal and spatial evolution of OLR anomalies also provide an opportunity to monitor the stress variation of seismogenic structures for tidal force-induced earthquakes. Therefore, the temporal and spatial evolution characteristics of OLR anomalies induced by the celestial tide-generating force and tectonic stress change provide an approach for the research on a short-impending earthquake precursor. It is necessary to further study the physical mechanism of this model of tidal force-induced earthquakes and to explore the regular characteristics of thermal anomalies, so as to improve the capability of monitoring short-impending anomalies of strong earthquakes in specific regions in the future.

Author Contributions: Conceptualization, J.L. and J.C.; methodology, J.L., J.C. and Y.Z.; software, J.L.; validation, Y.Z., J.Z., Y.H. and L.W.; formal analysis, J.L.; investigation, J.L.; resources, J.C.; data curation, J.C.; writing—original draft preparation, J.L; writing—review and editing, J.C., Y.Z. and X.S.; visualization, J.C.; supervision, J.C.; project administration, J.C.; funding acquisition, J.C. All authors have read and agreed to the published version of the manuscript.

Funding: The work is supported by the National Key Research and Development Project (No. 2021YFB3901203) and the APSCO Earthquake Research Project Phase II.

Data Availability Statement: The OLR data of this study can be downloaded from http://www.emc.ncep.noaa.gov (accessed on 5 May 2023).

Acknowledgments: Thanks to the NOAA for the OLR data. We also thank all reviewers for their comments and suggestions.

Conflicts of Interest: The authors declare no conflict of interest.

References

1. Chen, S.Y.; Liu, P.X.; Chen, L.C.; Liu, Q.Y. Evidence from seismological observation for detecting dynamic change in crustal stress by bedrock temperature. *Chin. Sci. Bull.* **2020**, *65*, 2395–2405. [CrossRef]
2. Milne, J. *Earthquakes and Other Earth Movements*; Kegan Paul, Trench, & Company: London, UK, 1913; p. 213.
3. Gorny, V.I.; Salman, A.G.; Tronin, A.A.; Shilin, B.V. The Earth's outgoing IR radiation as an indicator of seismic activity. *Proc. Acad. Sci. USSR* **1988**, *301*, 67–69.
4. Mogi, K. Fundamental Studies on Earthquake Prediction. In *A Collection of Papers of International Symposium on ISC-SEP*; Seismological Press: Beijing, China, 1984; pp. 619–652.
5. Qiang, Z.J.; Kong, L.C.; Zheng, L.Z.; Guo, M.H.; Wang, G.P.; Zhao, Y. An experimental study on temperature increasing mechanism of satellitic thermo-infrared. *Acta Seismol. Sin.* **1997**, *10*, 101–106.
6. Qiang, Z.J.; Kong, L.C.; Wang, G.P.; Li, Q.Z.; Lin, C.G.; Xu, X.D. Earth Gas Emis-Sion, Infrared Thermo-Amomaly and Seismicity. *Chin. Sci. Bull.* **1992**, *37*, 2259–2262.
7. Geng, N.G.; Cui, C.Y.; Deng, M.D.; Zhi, Y.Q. Remote sensing rock mechanics and its application prospects. *Prog. Geophys.* **1993**, *8*, 1–7.
8. Deng, M.D.; Geng, N.G.; Cui, C.Y.; Zhi, Y.Q.; Fan, Z.F.; Ji, Q.Q. The study on the variation of thermal state of rocks caused by the variation of stress state of rock. *Earthq. Res. China* **1997**, *13*, 179–185.
9. Wu, L.X.; Cui, C.Y.; Geng, N.G.; Wang, J.Z. Remote sensing rock mechanics (RSRM) and associated experimental studies. *Int. J. Rock Mech. Min. Sci.* **2000**, *37*, 879–888. [CrossRef]
10. Wu, L.X.; Liu, S.J.; Li, J.P.; Dong, Y.Q.; Xu, X.D. Theoretical analysis to impending tectonic earthquake warning based on satellite infrared anomaly. In Proceedings of the 2007 IEEE International Geoscience and Remote Sensing Symposium, Barcelona, Spain, 23–28 July 2007; pp. 3723–3727.
11. Freund, F. Charge generation and propagation in igneous rocks. *J. Geodyn.* **2002**, *33*, 543–570.
12. Ma, J.; Liu, L.Q.; Liu, P.X.; Ma, S.L. Thermal precursory pattern of fault unstable sliding: An experimental study of en echelon faults. *Chin. J. Geophys.* **2007**, *50*, 1141–1149.
13. Pulinets, S.A. Natural radioactivity, earthquakes, and the ionosphere. *EOS Trans. Am. Geophys. Union.* **2007**, *88*, 217–218.
14. Lisi, M.; Filizzola, C.; Genzano, N.; Grimaldi, C.S.L.; Lacava, T.; Marchese, F.; Mazzeo, G.; Pergola, N.; Tramutoli, V. A study on the Abruzzo 6 April 2009 earthquake by applying the RST approach to 15 years of AVHRR TIR observations. *Nat. Hazards Earth Syst. Sci.* **2010**, *10*, 395–406. [CrossRef]
15. Saraf, A.K.; Choudhury, S. Thermal remote sensing technique in the study of pre-earthquake thermal anomalies. *J. Ind. Geophys. Union.* **2005**, *9*, 197–207.

16. Tramutoli, V.; Cuomo, V.; Filizzola, C.; Pergola, C.; Pietrapertosa, C. Assessing the potential of thermal infrared satellite surveys for monitoring seismically active areas: The case of Kocaeli (Izmit) earthquake, August 17, 1999. *Remote Sens. Environ.* **2005**, *96*, 409–426. [CrossRef]
17. Pergola, N.; Aliano, C.; Coviello, I.; Filizzola, C.; Genzano, N.; Lacava, T.; Lisi, M.; Mazzeo, G.; Tramutoli, V. Using RST approach and EOS-MODIS radiances for monitoring seismically active regions: A study on the 6 April 2009 Abruzzo earthquake. *Nat. Hazards Earth Syst. Sci.* **2010**, *10*, 239–249. [CrossRef]
18. Saradjian, M.R.; Akhoondzadeh, M. Thermal anomalies detection before strong earthquakes (M > 6.0) using interquartile, wavelet and Kalman filter methods. *Nat. Hazards Earth Syst. Sci.* **2011**, *11*, 1099–1108. [CrossRef]
19. Jing, F.; Singh, R.P.; Shen, X. Land–atmosphere–meteorological coupling associated with the 2015 Gorkha (M 7.8) and Dolakha (M 7.3) Nepal earthquakes. Geomatics. *Nat. Hazards Risk.* **2019**, *10*, 1267–1284. [CrossRef]
20. Ghosh, S.; Chowdhury, S.; Kundu, S.; Sasmal, S.; Politis, D.Z.; Potirakis, S.M.; Hayakawa, M.; Chakraborty, S.; Chakrabarti, S.K. Unusual Surface Latent Heat Flux Variations and Their Critical Dynamics Revealed before Strong Earthquakes. *Entropy.* **2022**, *24*, 23. [CrossRef] [PubMed]
21. Zhang, L.F.; Zhang, X.; Guo, Y.X.; Hu, W.Y.; Sun, X.H. Characteristics of thermal infrared anomalies in the middle-eastern segmentof Qilian seismic belt guided by Menyuan M6.9 earthquake. *China Earthq. Eng. J.* **2022**, *44*, 398–407.
22. Wu, L.X.; Mao, W.F.; Liu, S.J.; Xu, Z.Y.; Li, Z.W.; Qi, Y.; Miao, Z.L. Mechanisms of altering infrared-microwave radiation from stressed rock and key issues on crust stress remote sensing. *J. Remote Sens.* **2018**, *22*, 146–161.
23. Jing, F.; Singh, R.P.; Cui, Y.J.; Sun, K. Microwave Brightness Temperature Characteristics of Three Strong Earthquakes in Sichuan Province, China. *IEEE J. Sel. Top. Appl. Earth Observ. Remote Sens.* **2020**, *13*, 513–522. [CrossRef]
24. Song, D.M.; Zang, L.; Shan, X.J.; Yuan, Y.; Cui, J.Y.; Shao, H.M.; Shen, C.; Shi, H.T. A study on the algorithm for extracting earthquake thermal infrared anomalies based on the yearly trend of LST. *Seismol. Geology.* **2016**, *38*, 680–695.
25. Meng, Y.F.; Meng, Q.Y.; Zhang, Y.; Zhou, S.J.; Liu, W.B. Correlation between thermal anomalies and earthquakes based on nighttime multi-temporal remote sensing data. *Acta Seismol. Sin.* **2021**, *43*, 124–135.
26. Kang, C.L.; Zhang, Y.M.; Liu, F.D.; Jing, F. Long-Wave-Radiation Patterns prior to the Wenchuan M8.0 Earthquake. *Earthquake* **2009**, *29*, 116–120.
27. Chakraborty, S.; Sasmal, S.; Chakrabarti, S.K.; Bhattacharya, A. Observational signatures of unusual outgoing longwave radiation (OLR) and atmospheric gravity waves (AGW) as precursory effects of May 2015 Nepal earthquakes. *J. Geodyn.* **2018**, *113*, 43–51. [CrossRef]
28. Qi, Y.; Wu, L.X.; He, M.; Mao, W.F. Spatio-temporally weighted two-step method for retrieving seismic MBT anomaly: May 2008 Wenchuan earthquake sequence being a case. *IEEE J. Sel. Top. Appl. Earth Observ. Remote Sens.* **2020**, *13*, 382–391. [CrossRef]
29. Ren, J.; Kang, C.L.; Ma, W.Y.; Yang, T.Q.; Lin, X.Y. Preliminary study on the tidal force and Outgoing Longwave Radiation change. *Earthq. Res. China* **2015**, *31*, 447–455.
30. Wu, L.X.; Qin, K.; Liu, S.J. Progress in Analysis to Remote Sensed Thermal Abnormity with Fault Activity and Seismogenic Process. *Acta Geodaetica et Cartographica Sin.* **2017**, *46*, 1470–1481.
31. Su, B.; Li, H.; Ma, W.Y.; Zhao, J.; Yao, Q.; Cui, J.; Yue, C.; Kang, C.L. The Outgoing Longwave Radiation Analysis of Medium and Strong Earthquakes. *IEEE J. Sel. Top. Appl. Earth Observ. Remote Sens.* **2021**, *14*, 6962–6973. [CrossRef]
32. Heaton, T.H. Tidal triggering of earthquakes. *Geophys. J. Int.* **1975**, *43*, 307–326. [CrossRef]
33. Emter, D. Tidal triggering of earthquakes and volcanic events. In *Lecture Notes in Earth Sciences*; Springer Verlag: Berlin, Germany, 1997; Volume 66, pp. 293–309.
34. Tanaka, S.; Ohtake, M.; Sato, H. Tidal triggering of earthquakes in Japan related to the regional tectonic stress. *Earth Planets Space* **2004**, *56*, 511–515. [CrossRef]
35. Sun, C.Q.; Yan, C.H.; Wu, X.P.; Xu, Y.J. The effect of Tidal Triggering on Seismic Fault in Eastern Tibetan Plateau and Its Neighboring Areas. *Chin. J. Geophys.* **2014**, *57*, 2054–2064.
36. Xie, C.D.; Lei, X.L.; Wu, X.P.; Fu, H.; Xiong, Z.Y.; Hu, X.L.; Li, S. Effect of tidal stress on fault nucleation and failure of the 2007 M s 6.4 Ning'er earthquake. Sci. *China Earth Sci.* **2016**, *59*, 397–407. [CrossRef]
37. Chen, X.Z. The tidal triggering of earthquakes. *Prog. Earthq. Sci.* **2021**, *51*, 145–160.
38. Ma, W.Y.; Xu, X.D.; Zhang, X.C.; Wu, H.F.; Xu, B.H. A preliminary study on the use of NCEP temperature images and astro-tidal-triggering to forecast short-impending earthquake. *Seismol. Geol.* **2006**, *28*, 447–455.
39. Ma, W.Y.; Ma, W.M.; Zhao, H.; Li, H. Temperature changing process of the Hokkaido (Japan) earthquake on 25 September 2003. *Nat. Hazards Earth Syst. Sci.* **2008**, *8*, 985–989. [CrossRef]
40. Ma, W.Y.; Wang, H.; Li, F.S.; Ma, W.M. Relation between the celestial tide-generating stress and the temperature variations of the Abruzzo M=6.3 earthquake in April 2009. *Nat. Hazards Earth Syst. Sci.* **2012**, *12*, 819–827. [CrossRef]
41. Ma, W.Y.; Zhang, X.D.; Liu, J.; Yao, Q.; Zhou, B.; Yue, C.; Kang, C.L.; Lu, X. Influences of multiple layers of air temperature differences on tidal forces and tectonic stress before, during and after the Jiu jiang earthquake. *Remote Sens. Environ.* **2018**, *210*, 159–165.
42. Zhang, Y.; Meng, Q.Y.; Wang, Z.; Lu, X.; Hu, D. Temperature Variations in Multiple Air Layers before the Mw 6.2 2014 Ludian Earthquake, Yunnan, China. *Remote Sens.* **2021**, *13*, 884. [CrossRef]
43. Duman, T.Y.; Çan, T.; Emre, Ö.; Kadirioğlu, F.T.; Baştürk, N.B.; Kılıç, T.; Arslan, S.; Özalp, S.; Kartal, R.F.; Kalafat, D.; et al. Seismotectonic database of Turkey. *Bull. Earthq. Eng.* **2018**, *16*, 3277–3316. [CrossRef]

44. Meng, J.N.; Sinoplu, O.; Zhou, Z.P.; Tokay, B.; Kusky, T.; Bozkurt, E.; Wang, L. Greece and Turkey Shaken by African tectonic retreat. *Sci. Rep.* **2021**, *11*, 6486. [CrossRef]
45. Wu, Q.P. *Gravitology and Solid Earth Tide*; Earthquake Publisher: Beijing, China, 1997; pp. 152–212.
46. Wu, L.X.; Liu, S.J.; Wu, Y.H.; Wang, C.Y. Precursors for rock fracturing and failure-Part II: IRR T-Curve abnormalities. *Int. J. Rock Mech. Min. Sci.* **2006**, *43*, 483–493. [CrossRef]
47. Liu, S.J.; Wu, L.X.; Zhang, Y.B. Temporal-Spatial Evolution Features of Infrared Thermal Images Before Rock Failure. *J. Northeast. Univ. Nat. Sci.* **2009**, *30*, 1034–1038.
48. Ren, Y.Q.; Liu, P.X.; Ma, J.; Chen, S.Y. Experimental study on evolution of thermal field of en echelon fault during the meta-instability stage. *Chin. J. Geophys.* **2013**, *56*, 2348–2357.
49. Wei, L.J.; Guo, J.F.; Cui, H.; Li, H.B.; Qiang, Z.J. Satellite Thermal Infrared Anomaly: A Short-term and Impending Earthquake Precursor before the Wenchuan Ms8.0 Earthquake in Sichuan, China. *Acta Geosci. Sin.* **2008**, *29*, 583–591.
50. Ma, W.Y.; Kang, C.L.; Xie, T.; Ren, J.; Zhong, X.H. The changes of the tidal force and the outgoing long-wave radiation of Lushan (China) MS 7.0 earthquake. *Prog. Geophys.* **2014**, *29*, 2047–2050.
51. Ma, W.Y.; Kang, C.L.; Liu, J.; Yue, C.; Lu, X. Evolution characteristics of multiple stratification air temperature in vertical of Kunlun Mountains Ms8.1 earthquake. *J. Remote Sens.* **2018**, *22*, 174–180.
52. Xu, L.; Aoki, Y.; Wang, J.; Cui, Y.; Chen, Q.; Yang, Y.; Yao, Z. The 2023 Mw 7.8 and 7.6 Earthquake Doublet in Southeast Türkiye: Coseismic and Early Postseismic Deformation, Faulting Model, and Potential Seismic Hazard. *Seismol. Res. Lett.* **2023**, 1–12. [CrossRef]

remote sensing

Article

A Spatially Self-Adaptive Multiparametric Anomaly Identification Scheme Based on Global Strong Earthquakes

Zhonghu Jiao *, Yumeng Hao and Xinjian Shan

State Key Laboratory of Earthquake Dynamics, Institute of Geology, China Earthquake Administration, Beijing 100029, China; haoyumeng21@mails.ucas.ac.cn (Y.H.); xjshan@ies.ac.cn (X.S.)
* Correspondence: jzh@ies.ac.cn

Abstract: Earthquake forecasting aims to determine the likelihood of a damaging earthquake occurring in a particular area within a period of days to months. This provides ample preparation time for potential seismic hazards, resulting in significant socioeconomic benefits. Surface and atmospheric parameters derived from satellite thermal infrared observations have been utilized to identify pre-earthquake anomalies that may serve as potential precursors for earthquake forecasting. However, the correlation between these anomalies and impending earthquakes remains a significant challenge due to high false alarm and missed detection rates. To address this issue, we propose a spatially self-adaptive multiparametric anomaly identification scheme based on global strong earthquakes to establish the optimal recognition criteria. Each optimal parameter exhibits significant spatial variability within the seismically active region and indicates transient and subtle anomaly signals with a limited frequency of occurrences (<10 for most regions). In comparison to the fixed criterion for identifying anomalies, this new scheme significantly improves the positive Matthew's correlation coefficient (MCC) values from ~0.03 to 0.122–0.152. Additionally, we have developed a multi-parameter anomaly synthesis method based on the best MCC value of each parameter anomaly. On average, the MCC increased from 0.143 to 0.186, and there are now more earthquake-prone regions with MCC values > 0.5. Our research emphasizes the critical importance of a multiparametric system in earthquake forecasting, where each geophysical parameter can be assigned a distinct weight, and the findings specifically identify OLR, including all-sky and clear-sky ones, as the most influential parameter on a global scale, highlighting the potential significance of OLR anomalies for seismic forecasting. Encouraging results imply the effectiveness of utilizing multiparametric anomalies and provide some confidence in advancing our knowledge of operational earthquake forecasting with a more quantitative approach.

Keywords: earthquake anomalies; earthquake forecasting; statistical analysis; AIRS; thermal infrared remote sensing

Citation: Jiao, Z.; Hao, Y.; Shan, X. A Spatially Self-Adaptive Multiparametric Anomaly Identification Scheme Based on Global Strong Earthquakes. *Remote Sens.* **2023**, *15*, 3803. https://doi.org/10.3390/rs15153803

Academic Editors: Dedalo Marchetti, Yunbin Yuan and Kaiguang Zhu

Received: 6 July 2023
Revised: 28 July 2023
Accepted: 29 July 2023
Published: 31 July 2023

1. Introduction

Earthquakes are among the most devastating natural phenomena, capable of causing severe loss of life, along with economic and social damages. Short-term earthquake forecasting is deemed a viable approach to mitigate the impact of catastrophic earthquakes [1,2]. Considerable efforts have been undertaken to establish a statistical correlation between surficial and atmospheric anomalies and earthquakes, and various pre-seismic anomalies have been detected and analyzed. Notably, surface temperature [3–5], atmospheric temperature [6–8], water vapor [6,9,10], and outgoing longwave radiation [6,7,11] exhibit significant fluctuations and a coupling effect before strong earthquakes, indicating their potential association with the earthquake preparatory phase. Notwithstanding notable advancements, reliable precursor parameters or anomaly detection methods remain challenging in providing operational earthquake forecasts [2,12]. Hence, it is imperative to find any evidence of pre-seismic anomalies associated with earthquakes and plausible earthquake precursors through satellite- or simulation-based technologies in a comprehensive manner.

The seismogenic phase is a complex process involving the progressive localization of volumetric deformation, rock damage, and slow slip transients at multiple spatial and temporal scales [13]. Despite some proposed hypotheses [14–19], the underlying geophysical mechanism remains unclear, particularly regarding the potential correlation between anomalous phenomena and earthquake activity. This presents a significant challenge because there is a lack of understanding regarding which geophysical or geochemical parameters indicate earthquake preparation and the spatiotemporal evolution characteristics of anomalous fluctuations triggered by seismogenic processes. Meanwhile, broad evaluations of pre-earthquake anomalies indicate the widespread distribution of various anomalous signals in spatial and temporal domains [5,6,20].

Most studies on pre-seismic anomalies are based on case studies, which often exhibit unique spatiotemporal characteristics of specific anomalies, making it challenging to extend their findings to other seismic events or regions. These studies utilize varying shock events, data sources, and methods for detecting anomalies. Consequently, their results only reflect local and limited anomalous characteristics preceding major earthquakes. Therefore, some studies have attempted to establish statistical correlations between pre-earthquake anomalies and strong earthquakes, but have focused solely on seismically active regions such as Taiwan and Japan [21,22]. This narrow regional focus restricts their ability to gain a comprehensive global perspective and impedes their capacity to identify the fundamental characteristics and evolution of earthquake-triggered anomalous phenomena. Moreover, given the absence of a quantitative index for evaluating the forecasting capacity of pre-seismic anomalies under a standardized baseline, it is challenging to conduct longitudinal comparisons of different anomaly parameters and detection methods in order to identify the most effective precursors or techniques [20]. The complex and challenging nature of understanding pre-earthquake anomalies and their relationship with seismic activity, therefore, highlight the need for a global perspective and standardized evaluation methods.

Several studies have attempted to address the gap in understanding earthquake precursors. Eleftheriou et al. [23] conducted a long-term correlation analysis of $M \geq 4$ earthquakes in Greece from 2004 to 2013 to identify thermal anomalies. They compared their findings using the Molchan diagram to validate the refined Robust Satellite Technique (RST) data analysis approach. Zhang and Meng [24] built on this work and used additional assessment metrics to evaluate land surface temperature anomalies before 3615 earthquakes in the Sichuan region of China. These works demonstrate the importance of using appropriate data analysis techniques to identify and validate pre-earthquake anomalies. Fu et al. [22] analyzed OLR anomalies for $M \geq 6.0$ earthquakes in the Taiwan region from 2009 to 2019, with a high percentage of earthquakes ($\sim$77%) showing pre-seismic anomalies. The results of a statistical correlation analysis conducted by Genzano et al. [21] indicated a noncausal correlation between brightness temperature anomalies and $M \geq 6$ earthquakes in Japan from 2005 to 2015, as evidenced by the Molchan diagram.

However, it is important to note that their studies were confined to local seismically active regions and spanned only an approximately 10-year period, potentially limiting the generalizability of their findings. Consequently, there is a pressing need for more extensive and long-term investigations encompassing diverse tectonic settings and seismic regions, while considering a wide range of geophysical parameters. Such endeavors are crucial in bolstering earthquake forecasting capabilities. One notable limitation in their approach is its reliance on a single geophysical parameter (e.g., thermal anomalies) to detect pre-seismic signals. This singular focus may not adequately capture the complexity and diversity of earthquake precursors. Additionally, the utilization of various anomaly methods and criteria for defining and identifying anomalies introduces the risk of yielding inconsistent or incomparable results across different regions and time periods. Moreover, their research lacks comprehensive and quantitative metrics for retrospectively evaluating the forecast capability of pre-seismic anomalies. The absence of standardized evaluation methods hinders the ability to gauge the effectiveness of their forecasting approach consistently. To overcome these challenges and advance earthquake forecasting methodologies, it is imperative to conduct quantitative evaluations of pre-earthquake anomalies, encompassing relevant geophysical parameters. Embracing a multiparametric approach and exploring multidisciplinary effects can lead to comprehensive and robust forecasting techniques, providing valuable insights for earthquake preparedness and risk mitigation.

Our previous studies also focused on this topic, and a statistical evaluation framework was proposed to address the limitations associated with earthquake forecasting [20,25]. The proposed framework facilitates the development of a standardized and quantitative approach for evaluating earthquake forecasting capability based on remotely sensed data products in raster format. The current study aims to improve this framework in two ways. Firstly, given the significant differences in regional tectonic background and seismic dynamics in seismically active regions across the globe, the anomaly recognition criteria cannot remain constant. To overcome this issue, a spatially self-adaptive scheme was proposed, which allows for the identification of local optimal anomaly recognition criteria. Secondly, a straightforward method for enhancing earthquake forecasting capability was implemented by utilizing the optimal forecasting abilities of multiple parameters derived from the Atmospheric Infrared Sounder (AIRS) product. The present study establishes a new baseline for evaluating potential precursory parameters and anomaly detection techniques. This development is particularly relevant for advancing the operational earthquake forecasting system that utilizes thermal anomalies from satellite platforms. Such a system is crucial for effective environmental management and planning in seismically active regions.

2. Data

2.1. Earthquake Catalog Data

The United States Geological Survey (USGS) Earthquake Hazards Program is committed to investigating the spatial and temporal characteristics and the tectonic mechanisms of earthquakes. For this study, earthquake catalog data comprising 1825 earthquakes with a magnitude of $\geq$6 and focal depths of $\leq$70 km between 2006 and 2020 were utilized from USGS. Shallow earthquakes were selected owing to their greater damage potential and significant anomalous variations, as highlighted by previous studies [26,27]. The spatial distribution and magnitudes of the chosen earthquakes over a 15-year period are presented in Figure 1. They are observed to occur predominantly along the main plate boundaries and mid-ocean ridges. These regions are characterized by heightened plate movements and higher frequencies of seismic and volcanic activities, as reported by Jenkins et al. [28].

2.2. AIRS/Aqua Level 3 Product

The AIRS instrument, which is a new generation of high spectral resolution infrared sounder with 2378 channels, is onboard the Aqua satellite of the Earth Observation System. It was launched on 4 May 2002 and can observe the emitted infrared radiation of the earth–atmosphere system in three spectral bands: 3.74–4.61 μm (2169–2674 cm^{-1}), 6.20–8.22 μm (1217–1613 cm^{-1}), and 8.8–15.4 μm (649–1136 cm^{-1}). With a horizontal spatial resolution of 13.5 km at the nadir, increasing to 41×21.4 km^2 at the maximum zenith angle, AIRS is capable of providing a temporal resolution twice per day with a swath width of 1650 km, which covers over 95% of the Earth's surface per day [29]. The radiances observed by AIRS can be used to estimate geophysical and geochemical parameters, including surface temperature, atmospheric temperature, water vapor content, trace gases, cloud parameters, and dust parameters. The AIRS/Aqua L3 daily standard physical retrieval (AIRS3STD) V7.0 product was utilized to compute pre-seismic multiparametric anomalies, which has a spatial resolution of 1°, and only nighttime data were considered in a daily temporal resolution. The used parameters consisted of skin temperature (ST; K), air temperature (AT; K), total integrated column water vapor burden (CWV; kg/m^2), outgoing longwave radiation flux (OLR; W/m^2), and clear-sky OLR (COLR; W/m^2).

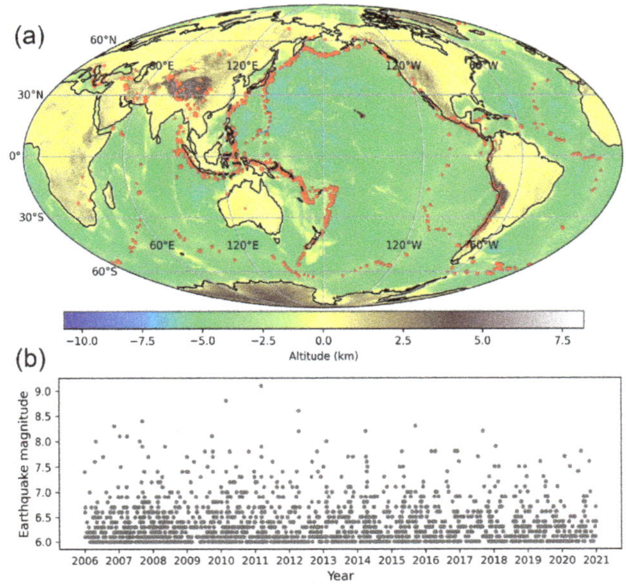

Figure 1. A total of 1825 global earthquakes with magnitudes ≥ 6 and focal depths ≤ 70 km, that occurred between 2006 and 2020, from the United States Geological Survey (USGS) Earthquake Hazards Program catalog. (**a**) Earthquake locations depicted as red dots. The altitude data used are sourced from the ETOPO Global Relief Model 2022, available at https://www.ncei.noaa.gov/products/etopo-global-relief-model (accessed on 28 July 2023). (**b**) Earthquake magnitudes varied with time.

3. Methods

3.1. Anomaly Detection Method

Anomaly detection is the process of identifying outliers from the normal or long-term average reference values of specific geophysical and geochemical parameters, such as surface temperature and OLR, in a spatial or temporal domain. These outliers serve as indicators of potential pre-seismic anomalies, which are anomalous fluctuations that are triggered to some extent by seismogenic processes. These anomalies can be detected through diverse methods for the first step of earthquake-related anomaly recognition [27]. In this study, a z-score (*ZS*) normalization method was utilized, represented as follows:

$$ZS(x,y,t) = \frac{v(x,y,t) - \mu(x,y)}{\delta(x,y)}, \tag{1}$$

where $v(x,y,t)$ represents the current value of a given parameter (e.g., OLR) at a specific location (x,y) and time t. The average $\mu(x,y)$ and standard deviation $\delta(x,y)$ of the reference field are calculated at the same or similar time intervals over multiple years at the location (x,y). The widely used ZS method has a simple formula that provides a clear mathematical interpretation—it is the multiple of the value from the average condition with respect to the amplitude of variation. The anomaly represents the difference between $v(x,y,t)$ retrieved from satellite data and $\mu(x,y)$ of the reference field over a long-term average trend. The anomalous results calculated by the ZS method are only meaningful when $\delta(x,y)$ is not zero or is not significantly close to zero. If we consider the occurrence of anomalies as a normal distribution, the anomalies calculated by Equation (1) with values of 1, 2, and 3 represent probabilities of 68.17%, 99.45%, and 99.73%, respectively. Larger anomalies imply a higher likelihood of rare events occurring. From another perspective, the difference between $v(x,y,t)$ and $\mu(x,y)$ of the reference field represents the signal, and $\delta(x,y)$ of the reference field is the noise. The higher the signal-to-noise ratio, the greater the possibility of anomalies prior to an earthquake.

3.2. Anomaly Recognition Criteria

The anomalous signals associated with earthquakes can be subtle and transient due to various natural and anthropogenic processes that significantly influence the calculated anomalies based on Equation (1) [30]. Therefore, it is essential to have a set of anomaly recognition criteria to determine whether the anomalies are caused by seismogenic processes as much as possible. The availability of anomalous perturbances prior to upcoming earthquakes largely depends on the quality of remote sensing products and anomaly detection approaches. To enhance reliability or reduce false positive rates indicated by pre-seismic anomalies, a variety of conditions or limits have been proposed [10,21]. In previous work, Jiao and Shan [25] fixed the anomaly recognition criteria at the global scale, ignoring the spatial differentiation of pre-seismic anomalies due to various tectonic environments and earthquake preparatory conditions. Therefore, to improve the earthquake forecasting capability, different values of each parameter in anomaly recognition criteria can be chosen and adapted to our framework through comparison with previous works. The following parameters and their respective potential values have been selected:

1. The temporal range for warning about impending earthquakes prior to the current day is 30, 60, 90, 120, or 150 days;
2. A pixel with an absolute anomaly value, which is calculated by the ZS method, of $\geq$1, 1.5, 2.0, 2.5, 3.0, or 3.5 will be considered a valid anomaly pixel;
3. Within the sliding window, the number of valid anomaly pixels must be $\geq$5%, 10%, 20%, 30%, 40%, or 50% inside a 1-day resolution anomaly window;
4. Within an observation period (e.g., 90 days), if there are days where $\geq$5%, 10%, 20%, 25%, 30%, 40%, or 50% satisfy the above conditions, a pre-seismic anomaly is identified based on daily AIRS data. For instance, if there are more than 9 days (i.e., 90 days $\times$ 10%) within a 90-day period that meet the specified conditions, an anomalous event is confirmed.

Additionally, a sliding spatial window of $7^\circ \times 7^\circ$ surrounding a central grid of 1° spatial resolution was used in this study. When a pre-seismic anomaly is identified at a specific time and position, an alarm warns that a $M \geq 6$ earthquake is likely to occur within the next 30 days within the sliding spatial window. For each pixel with 1° spatial resolution at a global scale, a set of specified parameters is optimally determined, using the approach described in Section 3.4.

3.3. Metrics to Evaluate Earthquake Forecasting Capability

To evaluate the earthquake forecasting capability of pre-seismic anomalies from AIRS data, a comprehensive indicator, Matthew's correlation coefficient (MCC), was used. MCC is a statistical index that is more reliable than accuracy and F1 score for binary classification evaluation. It is relatively balanced and provides a high score only when predictions have good results in all four confusion matrix categories (i.e., true positives, true negatives, false positives, and false negatives), thus avoiding overoptimistic assessment results found in other statistical rates [31]. When using equivalent terms that are directly related to earthquake forecasting [25], MCC is redefined as

$$\text{MCC} = \frac{Acc \times Nor - Fal \times Mis}{\sqrt{(Acc + Fal)(Acc + Mis)(Nor + Fal)(Nor + Mis)}} \tag{2}$$

where *Acc* denotes the accuracy forecasting count, *Nor* is the normal count, *Fal* represents the false alarm count, and *Mis* is the missed detection count. The MCC value ranges from -1 to 1, with 1 indicating perfect earthquake forecast, 0 meaning that the proposed model is no better than a random guess, and -1 denoting exact opposite prediction results.

MCC values from −1 to 0 are considered invalid because they denote a normal condition (i.e., no earthquakes) when anomalies occur. This contradicts the assumption that various anomalous phenomena observed during the seismogenic stage are generated by underlying fault activity that triggers impending earthquakes. Additionally, when calculating the global mean condition, positive and negative MCC values counteract each other, leading to problematic statistical outcomes. Furthermore, if we are to consider the negative MCC of a pixel in seismically active regions, we could alter the anomaly recognition criteria in the opposite direction to achieve valid earthquake forecasting. While this approach may be statistically feasible, it may impede the exploration of the underlying geophysical mechanism through satellite observations and ground measurements. This raises concerns about the prevalence of anomalous fluctuations and the calm phase before earthquakes that are perceived as accidental occurrences. Consequently, from this perspective, anomaly detection methods relying on a reference field for long-term averages become devoid of meaningful basics. The consideration of negative MCC values as invalid still remains an open question that necessitates further investigation to ascertain its reasonability and implications.

3.4. Determination of Spatially Self-Adaptive Anomaly Identification Parameters

To obtain the optimal anomaly recognition criteria for each pixel with a 1-degree spatial resolution, a local self-adaptive scheme was proposed. Firstly, nighttime data of multiparameter anomalies within a 7 × 7 sliding window for a single day are extracted from the period of 2006 to 2020, generating a spatiotemporal data cube for the central pixel. Secondly, the possible values of four parameters described in Section 3.2 are iteratively applied to the data cube, and the combination of parameter values with the highest MCC value is selected as the final optimal parameters for that pixel. Finally, the calculations are applied to each pixel through the sliding window, and optimal parameters are gridded to indicate their spatial distributions. There are 1800 loops (i.e., 5 base days × 6 absolute anomaly values × 6 anomaly ratios × 6 anomaly counts) for each pixel globally (180 × 360 pixels), and, due to the overwhelming computation cost, limited discrete values of parameters of the anomaly recognition criteria are used in this study to relieve computational constraints. It should be noted that anomalies of the five used geophysical parameters (ST, AT, CWV, COLR, and OLR) of the AIRS product require independent optimal parameters.

To prevent overfitting, a 3-fold cross-validation was conducted on the training set to fine-tune the anomaly identification parameters and determine the optimal parameter configuration. The spatiotemporal data cube was divided into three consecutive folds, consisting of a training set and a test set (Table 1). Each fold was utilized as a validation set once, while the remaining two folds constituted the training set. The optimal parameters were determined using all the data in the training set, and the earthquake forecasting capability was evaluated using the test data set. Note that this study utilized data from only 1825 earthquakes occurring over a 15-year period. A higher number of folds in cross-validation leads to fewer earthquakes being used in the test dataset. For instance, in the case of 5-fold cross-validation, only earthquakes from a 3-year period were used to assess earthquake forecasting capability. In seismically active regions, calculating the MCC may encounter challenges due to the rarity of strong shocks occurring at the same location (i.e., one pixel) within a relatively short period. From a statistical standpoint, this can lead to unstable results. As a tradeoff, we selected 3-fold cross-validation to increase the number of earthquake cases in the test dataset, and to also have the ability to assess the reliability of our proposed approach with different folds.

Table 1. The time division for the 3-fold cross-validation implemented to acquire the optimal anomaly identification parameters for each geophysical parameter.

No.	Years for Training Data	Years for Test Data
1	2011–2020	2006–2010
2	2006–2010, and 2016–2020	2011–2015
3	2006–2015	2016–2020

4. Results

4.1. Optimal Parameters of Anomaly Recognition Criteria

The optimal parameters for anomaly recognition criteria in Section 3.2 were determined using a spatially self-adaptive anomaly identification scheme, as depicted in Figure 2, for AIRS OLR anomalies using training data within 2011–2020. Each optimal parameter exhibited significant spatial variation in seismically active regions, and their values tended to become lower, indicating that most pre-seismic anomalies are short-term and subtle signals, which makes it harder to identify seismic-related anomalies prior to major earthquakes. The largest number of base days was for 30 days, while the 60- and 90-day values dominated seismically active regions in China (Figure 2a). A previous study about the multiple anomalies before the 2015 Nepal earthquake sequence summarized by Wu et al. [30] showed similar characteristics. The anomaly count was calculated using a fixed ratio of base days, and, due to the lower values of base days, anomaly counts over most areas were less than 30 (Figure 2b). Therefore, recognizable anomalies that forecast impending earthquakes are not very frequent, the same as was seen for the identified thermal anomalies in relation to earthquakes by Filizzola et al. [32]. In Figure 2c, the values of anomaly ratio (i.e., the ratio between the number of pixels that have anomalies occurred and total pixels in a $7^\circ \times 7^\circ$ window) have relatively equal counts except for the anomaly ratio value of 0.05. The absolute anomaly values were always less than 3.0 (Figure 2d), indicating generally weak anomalous signs that are possibly associated with imminent earthquakes. This threshold value is always adopted by studies when using analogical anomaly detection methods [27]. Given that the anomalous values follow a normal distribution, the valid absolute anomaly values account for a 99.7% probability, according to Equation (1), meaning that it is a non-trivial probability event. The statistical results also reveal that the recognition capability of a temporal integrated anomaly method can outperform random guessing [33].

Although the spatiotemporal evolution patterns of earthquake-related anomalies have not been definitively ascertained (i.e., no widespread consensus yet), our research provides a quantitative description of the characteristics of pre-earthquake anomalies or seismic precursory signals through the use of four criteria (see Section 3.2). The optimal results obtained from these criteria represent the most likely possibility for forecasting earthquakes (Figure 2), and their respective values offer insights into the subtlety and short-term nature of seismic precursory signals, thereby emphasizing the complexities involved in identifying seismic-related anomalies. To further establish the connection between these signals and seismic activity, we analyzed the MCC results, as shown in the subsequent sections. MCC serves as a metric to represent the possibility of earthquake forecasting based on anomalous phenomena, with higher MCC values suggesting a stronger likelihood of a robust connection between the identified precursory signals and seismic events. This analytical approach allows us to gain a more comprehensive understanding of the efficacy and potential of using seismic precursors for earthquake forecasting, further validating the relevance and significance of our research findings.

In addition, we employed 3-fold cross-validation to estimate the optimal parameters of the anomaly recognition criteria. The results obtained for different training periods are presented in Figures A1 and A2 in the Appendix A. The spatial distributions of the optimal parameters exhibited minimal variations among each other, and their histograms demonstrated a high level of consistency. This indicates that the calculation method used to obtain the optimal parameters for the anomaly recognition criteria is relatively reliable.

Therefore, in this study, we present the results only for the first fold of cross-validation. The results for the anomalies of AIRS ST, AT, CWV, and COLR displayed similar characteristics and are, therefore, not shown.

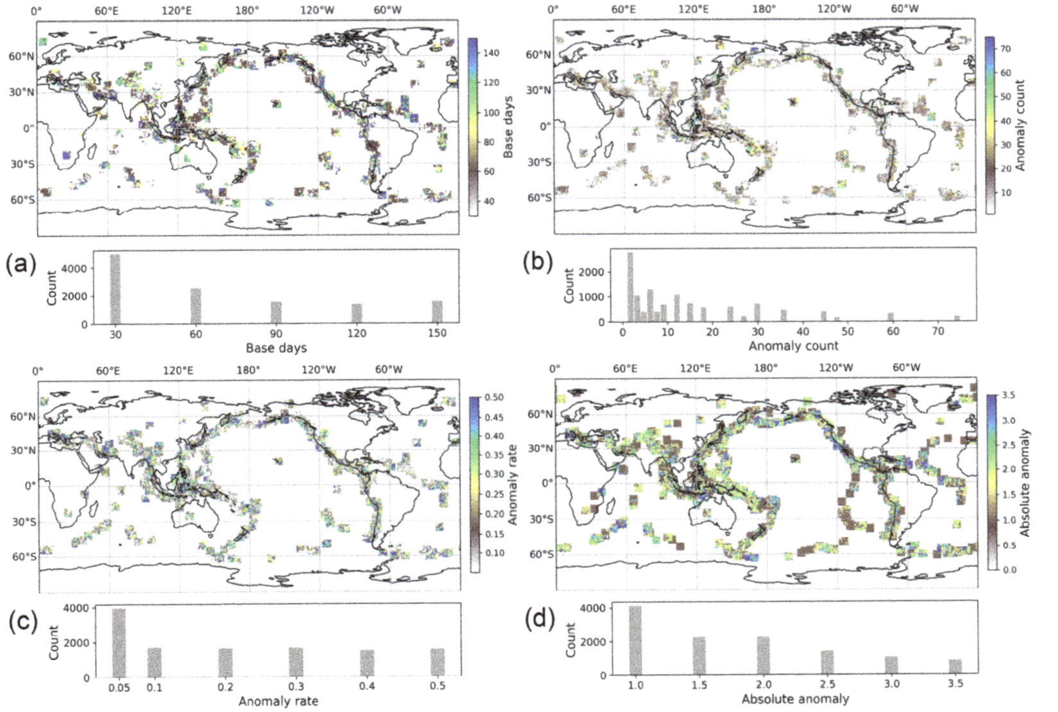

Figure 2. The optimal parameters of the anomaly recognition criteria employed for forecasting imminent earthquakes using the outgoing longwave radiation anomaly based on training data within 2011–2020, as indicated by their spatial distributions and histograms in the top and bottom subfigures, respectively. The parameters include (**a**) the base days preceding earthquakes; (**b**) the anomaly count that occurred during the base days; (**c**) the anomaly rate within the spatial window, which is 7 × 7 pixels in this study; and (**d**) the absolute anomaly value used to identify potential earthquake-related anomalies.

4.2. Optimal Parameters of Multiparameter Composed Anomalies

To enhance earthquake forecasting capacity, a simple composition method was utilized to select the optimal anomaly parameter for a pixel based on the best MCC value at the test stage for the years 2006–2010. This approach ensures that a region not only possesses the best anomaly recognition criteria, but also the dominant geophysical parameter that generates the most effective earthquake forecasting capability based on its anomalous variations prior to impending earthquakes. The spatial distributions and histograms of optimal parameters for composed (COM) anomalies are illustrated in Figure 3. Anomaly recognition criteria exhibited analogous statistical characteristics to those in Figure 2. Optimal parameters at some pixels obtained at the training stage cannot generate valid forecasting (i.e., MCC ≤ 0) at the test stage, causing less predictable areas, as shown in Table 2. The results with different training periods under a 3-fold cross validation are shown in Figures A3 and A4.

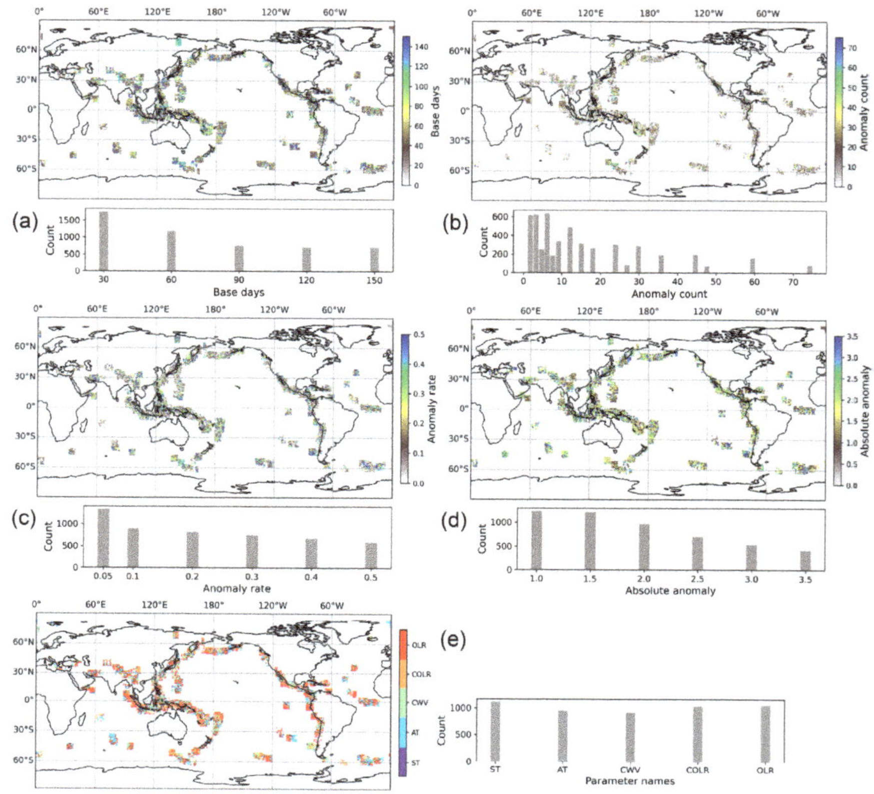

Figure 3. The optimal parameters for anomaly recognition criteria that are utilized for forecasting impending earthquakes in multiparameter composed anomalies based on training data within 2011–2020. These anomalies are analyzed based on their spatial distributions and histograms, as shown in the top and bottom subfigures, respectively. The parameters include (**a**) the base days preceding earthquakes; (**b**) the anomaly count that occurred during the base days; (**c**) the anomaly rate within the spatial window, which is 7 × 7 pixels in this study; (**d**) the absolute anomaly value used to identify potential earthquake-related anomalies; and (**e**) the geophysical parameters used in the composed anomalies. ST: skin temperature; AT: air temperature; CWV: total integrated column water vapor burden; OLR: outgoing longwave radiation flux; COLR: clear-sky OLR.

Table 2. The statistical results of the Matthew's correlation coefficient (MCC) conducted for earthquake forecasting utilizing anomalies derived from selected geophysical parameters. The seismically active region, as depicted in Figure 4 of [25], encompasses a total of 12,034 pixels. The ratio refers to the number of valid anomaly pixels divided by the aforementioned total count. ST: skin temperature; AT: air temperature; CWV: total integrated column water vapor burden; OLR: outgoing longwave radiation flux; COLR: clear-sky OLR; COM: composed anomalies.

Stage	Training				Test			
Parameter	**Min**	**Max**	**Mean**	**Ratio**	**Min**	**Max**	**Mean**	**Ratio**
ST	0.008	1	0.272	0.836	0.001	1	0.122	0.200
AT	0.029	1	0.299	0.836	0.002	1	0.150	0.170
CWV	0.024	1	0.305	0.836	0.002	1	0.145	0.165
COLR	0.024	1	0.303	0.836	0.002	1	0.150	0.174
OLR	0.034	1	0.338	0.836	0.002	0.701	0.152	0.164
COM	/	/	/	/	0.002	1	0.186	0.424

The key information from optimal parameters for such composed anomalies was that the dominant geophysical parameter at the global scale is OLR, encompassing all-sky and clear-sky OLRs, as displayed in Figure 3e with red and orange areas. The secondary parameter is ST, whereas CWV and AT have a minimal percentage. The implication is that OLR anomalies could prove to be a more viable candidate for global earthquake forecasting under the background of the consideration that a multi-parametric strategy holds promise in advancing the development of seismic forecasting [21]. The findings presented by Fu et al. [22] offer compelling evidence that OLR may serve as a potential indicator for forecasting earthquakes in the Taiwan region. OLR refers to the total longwave radiation emitted from the Earth–atmosphere system at the top of the atmosphere in a spectral range of 4 to 100 μm. It is primarily influenced by surface temperature and emissivity, atmospheric profile of temperature and water vapor, clouds, and greenhouse gases. From the perspective of the lithosphere–atmosphere–ionosphere coupling mechanism [30,34], OLR represents an integrated effect of anomalous fluctuations that occur within the seismogenic zone, spanning from the lithosphere to the atmosphere. Therefore, OLR anomalies could be a delegate of the gross effect beneath the top of the atmosphere (e.g., air temperature anomalies).

4.3. MCCs of AIRS Multiparameter Anomalies Using Optimal Parameters

During the training stage, most pixels within the seismically active regions demonstrated valid forecasting capability, as indicated by a MCC greater than 0, shown in Figure 4 and summarized in Table 2. This finding indicates that the optimal parameters of anomaly recognition criteria effectively forecast earthquakes within the period from 2011 to 2020 for most seismically active regions. Interestingly, certain regions exhibited particularly high MCC values exceeding 0.5, as depicted by the bright yellow and green areas. These regions were widely distributed without following any specific spatial pattern. The histograms of MCC values for each parameter anomaly demonstrated a similar characteristic, with most areas having MCC values below 0.4, an average MCC of approximately 0.2, and only a few regions with MCC values exceeding 0.5. Moreover, the spatial distribution of MCC values varied among different geophysical parameters, as illustrated in Figure 4. This suggests the potential for a complementary effect, where different anomalies may combine to improve earthquake forecasting capabilities. These observations provide valuable insights into the spatial patterns and distribution of forecasting skill, paving the way for further research and potential enhancements to earthquake forecasting methodologies.

Figure 5 presents the spatial distribution and statistical histograms of the MCC, which depict the seismic forecasting performance of five geophysical parameters derived from AIRS during the test stage spanning from 2006 to 2010. In comparison to the results obtained during the training stage (Figure 4), fewer regions demonstrated valid forecasting capabilities. This can be attributed to the limited occurrence of earthquakes with a magnitude of $\geq$6 and focal depths of $\leq$70 km between 2006 and 2010, as the calculation of MCC requires an adequate number of earthquake cases for reliable evaluation. Furthermore, it is noteworthy that similar spatial distributions of MCCs as well as the statistical results in Table 2 were observed across different anomalies of geophysical parameters. The average MCC values ranged from 0.122 for ST anomalies to 0.152 for OLR anomalies. The valid ratios at the test stage ranged from 0.164 for OLR anomalies to 0.200 for ST anomalies. These results exhibited a significant decrease compared to those at the training stage (Table 2), where the average MCC values ranged between 0.272 and 0.338, with a ratio of 0.836. However, despite the decrease in performance during the test stage, it is important to recognize that our proposed scheme is still capable of capturing to some extent the underlying mechanism of earthquake occurrences. The challenging nature of earthquake forecasting in the time period, characterized by a scarcity of relevant events, contributes to the observed decrease in forecasting capability. Nonetheless, the insights gained from these results contribute to a deeper understanding of the complexities involved in earthquake forecasting and offer valuable directions for further research and improvements to our methodology.

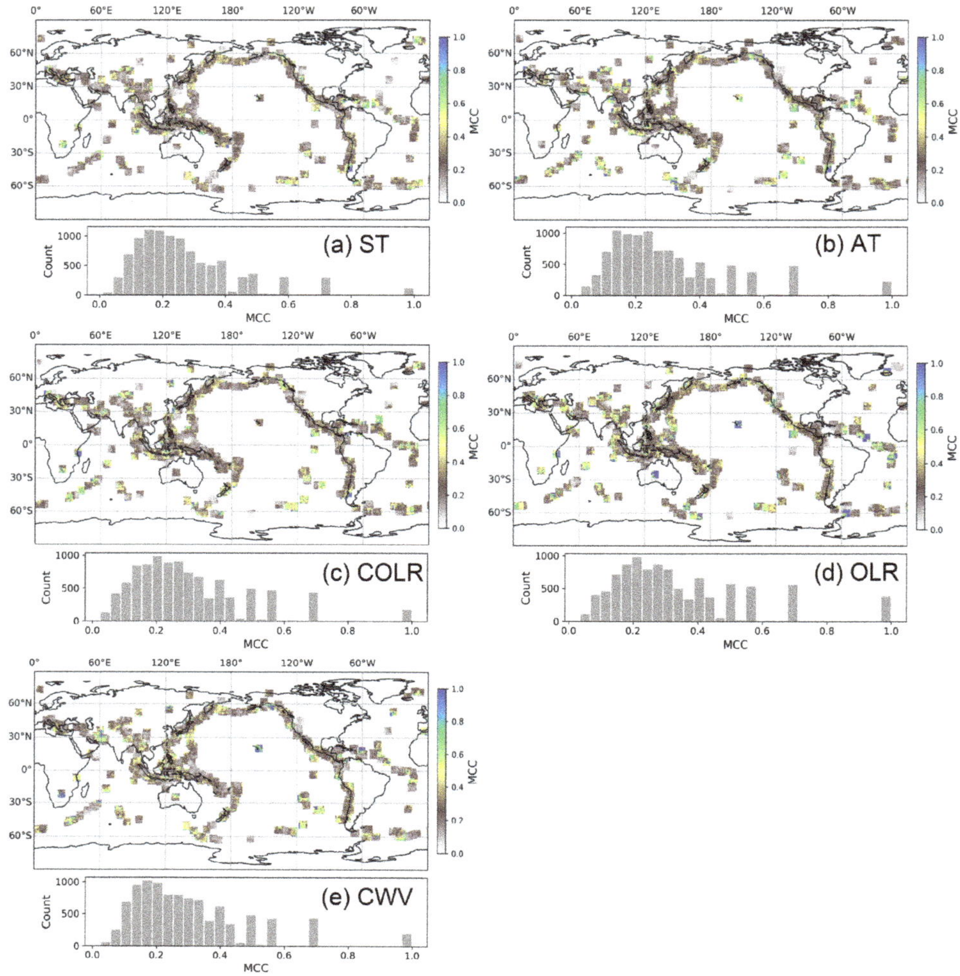

Figure 4. The spatial distributions and histograms of Matthew's correlation coefficients (MCC) obtained using the optimal parameters of anomaly recognition criteria at the training stage based on training data within 2011–2020 for (**a**) skin temperature (ST), (**b**) air temperature (AT), (**c**) clear-sky outgoing longwave radiation (COLR), (**d**) outgoing longwave radiation (OLR), and (**e**) total integrated column water vapor burden (CWV) anomalies.

The MCCs of composed anomalies showed significant improvement due to the increase in the number of predictable regions from ~0.175 to 0.424 in Table 2. On average, the MCCs increased from 0.143 (average of all geophysical parameters) to 0.186 when composed anomalies were applied. These values may not be high enough to meet the requirements of an operational earthquake forecasting system. Machine learning-based methods have demonstrated the capability to achieve higher MCC values [35], despite employing different earthquake cases and distinct data processing approaches. Nevertheless, the simple composite scheme demonstrates the effectiveness of applying multiparametric anomalies and is feasible for improving the earthquake forecasting capability. Previous studies have suggested the integration of different precursors of lithosphere, atmosphere, and ionosphere to promote reliable earthquake forecasting [6,7,36]. Our proposed scheme provides a probable and sustainable path towards achieving this goal.

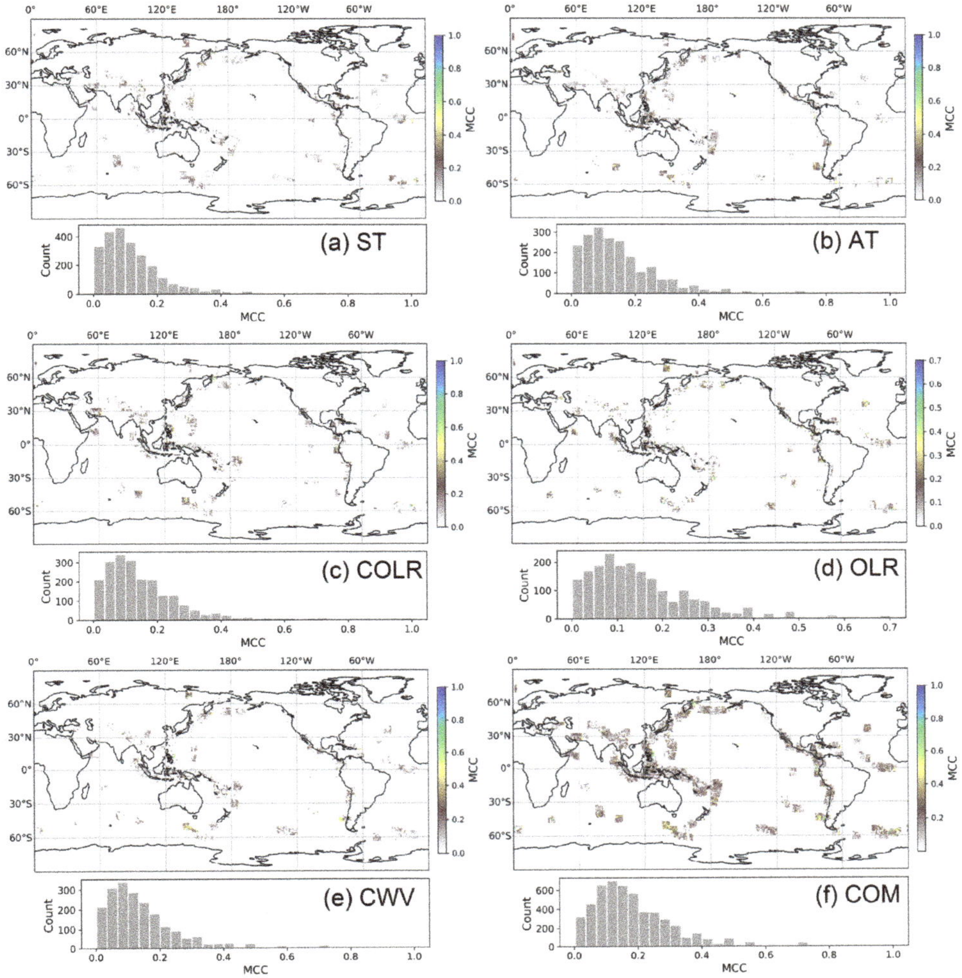

Figure 5. The spatial distributions and histograms of Matthew's correlation coefficients (MCC) obtained using the optimal parameters of anomaly recognition criteria at the test stage for 2006–2010, based on training data within 2011–2020 for (**a**) skin temperature (ST), (**b**) air temperature (AT), (**c**) clear-sky outgoing longwave radiation (COLR), (**d**) outgoing longwave radiation (OLR), (**e**) total integrated column water vapor burden (CWV), and (**f**) composed (COM) anomalies.

5. Discussion

5.1. Comparison with the Results Obtained Using Fixed Anomaly Recognition Criteria

In contrast to fixed anomaly recognition criteria, the spatially self-adaptive anomaly identification scheme demonstrates significant improvements. Figure 6 illustrates that the fixed criterion lacks adaptability, resulting in relatively low MCC values (averaging less than 0.05) for a significant portion of seismically active regions. These low values indicate inaccurate predictions, where pre-seismic anomalies are not followed by earthquakes. Consequently, the fixed criteria fail to identify pre-seismic anomalies in these areas, undermining the reliability of earthquake forecasting. Conversely, the optimal parameters of the local self-adaptive scheme generate higher MCC values (at least greater than 0.12), as shown in Table 2 for the majority of seismically active regions (Figure 5). This crucial advancement represents a significant improvement in producing reliable earthquake forecasts.

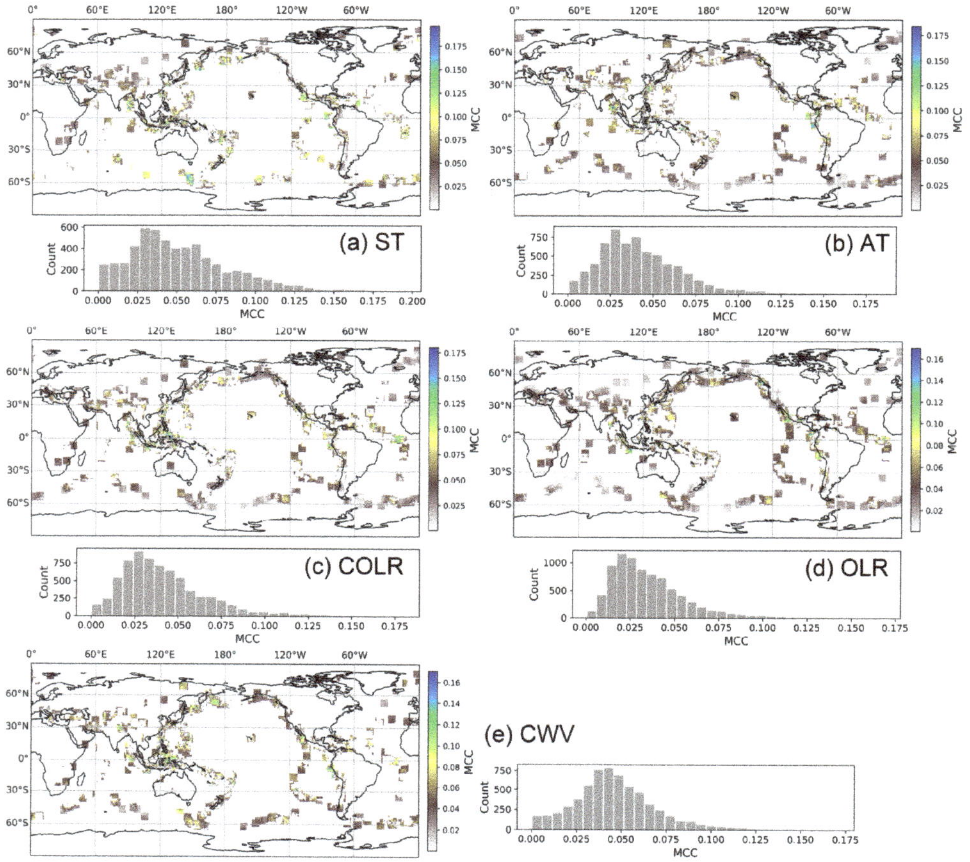

Figure 6. The spatial distributions and histograms of Matthew's correlation coefficients (MCC) when using a single set of anomaly recognition criterion for (**a**) skin temperature (ST), (**b**) air temperature (AT), (**c**) clear-sky outgoing longwave radiation (COLR), (**d**) outgoing longwave radiation (OLR), and (**e**) total integrated column water vapor burden (CWV) anomalies within 2006–2010. The temporal range for predicting impending earthquakes was 120 days. The pixel with absolute anomaly value of $\geq$2 is considered as a valid anomaly pixel. Within the sliding window, the number of valid anomaly pixels is $\geq$10%. Within the observation period (i.e., 120 days), if there are $\geq$12 days that satisfy the above conditions, a pre-seismic anomaly is identified.

5.2. MCCs of Inland, Coastal, and Oceanic Earthquakes

The earthquakes investigated in this study were categorized into three types based on their location: inland, coastal, and oceanic events. The MCC histograms presented in Figure 7 reveal that the location of earthquakes has a limited influence on the earthquake forecasting performance of each AIRS parameter anomaly. Specifically, for inland earthquakes, there is a scarcity of samples with higher MCC values. However, for all geophysical parameters, except for OLR, the maximum MCC value can reach 1.0. Across all five geophysical parameters, the majority of MCC values are below 0.5. Nevertheless, when considering the combined anomalies, the overall average MCC can reach 0.186 (Table 2). The MCC histograms of the same geophysical anomaly for the three types of earthquakes exhibit similar patterns, indicating that earthquake location does not affect the forecasting capability too much. The statistical results are presented in Table 3, revealing that oceanic earthquakes consistently exhibit slightly higher mean MCC values for each parameter, albeit with limited increments. This demonstrates that the use of composed anomalies results in a significant improvement of approximately 0.043 in MCC on average, compared to the use of a single geophysical parameter. This improvement is independent of the type of earthquake location, including inland, oceanic, and coastal earthquakes (Figure 7). Our previous research [33] also reveals that statistical characteristics based on earthquake focal mechanisms do not exhibit significant differences, despite the use of different approaches and earthquake events. This suggests that the geophysical mechanism underlying the generation of anomalous signals prior to upcoming earthquakes is possibly consistent across earthquake types and locations [34,37].

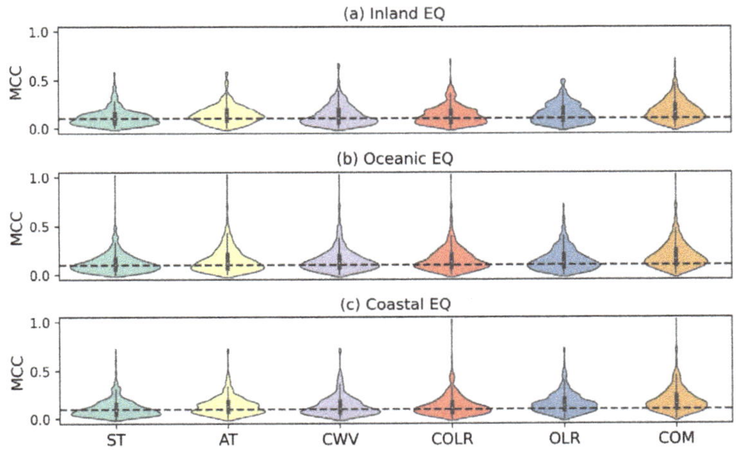

Figure 7. Histograms of Matthew's correlation coefficients (MCC), which were obtained for (**a**) inland, (**b**) oceanic, and (**c**) coastal earthquakes (EQ) for five geophysical and composed parameters. They were computed using the optimal parameters of anomaly recognition. The dash line denotes the MCC of 0.1. ST: skin temperature; AT: air temperature; CWV: total integrated column water vapor burden; OLR: outgoing longwave radiation flux; COLR: clear-sky OLR; COM: composed anomalies.

Table 3. The statistical results of the Matthew's correlation coefficient (MCC) for inland, oceanic, and coastal earthquakes (EQ). ST: skin temperature; AT: air temperature; CWV: total integrated column water vapor burden; OLR: outgoing longwave radiation flux; COLR: clear-sky OLR; COM: composed anomalies.

Type	Parameter	Min	Max	Mean
Inland EQ	ST	0.002	0.567	0.118
	AT	0.002	0.567	0.149
	CWV	0.002	0.649	0.139
	COLR	0.004	0.695	0.140
	OLR	0.002	0.487	0.152
	COM	0.002	0.695	0.176
Oceanic EQ	ST	0.001	1.000	0.124
	AT	0.002	1.000	0.153
	CWV	0.002	1.000	0.149
	COLR	0.002	1.000	0.157
	OLR	0.002	0.701	0.150
	COM	0.002	1.000	0.189
Coastal EQ	ST	0.002	0.701	0.116
	AT	0.002	0.701	0.142
	CWV	0.002	0.701	0.137
	COLR	0.002	1.000	0.136
	OLR	0.003	0.701	0.157
	COM	0.002	1.000	0.183

5.3. Earthquake Forecasting Capability of Pre-Seismic Anomalies in China Mainland

China Mainland is a region known for its susceptibility to seismic activity, characterized by strong and shallow earthquakes that can cause substantial damage. The frequent occurrence of earthquakes in China necessitates the ability to forecast impending seismic events in the short term, typically within a span of several weeks. This forecasting capability is crucial for minimizing potential risks associated with earthquakes and their secondary hazards, such as landslides and debris flow [38]. Figure 8, which specifically focuses on China and its immediate vicinity, reveals that the MCC of individual geophysical parameters typically ranges between 0 and 0.4. Valid anomalies primarily occur along the Qinghai–Tibet Plateau and the south–north seismic belt, both being prominent regions with significant seismic activity [39]. When considering the combination of all geophysical parameters, a greater number of areas in midwestern China, the region with the highest seismic activity, exhibit higher MCC values, averaging around 0.15.

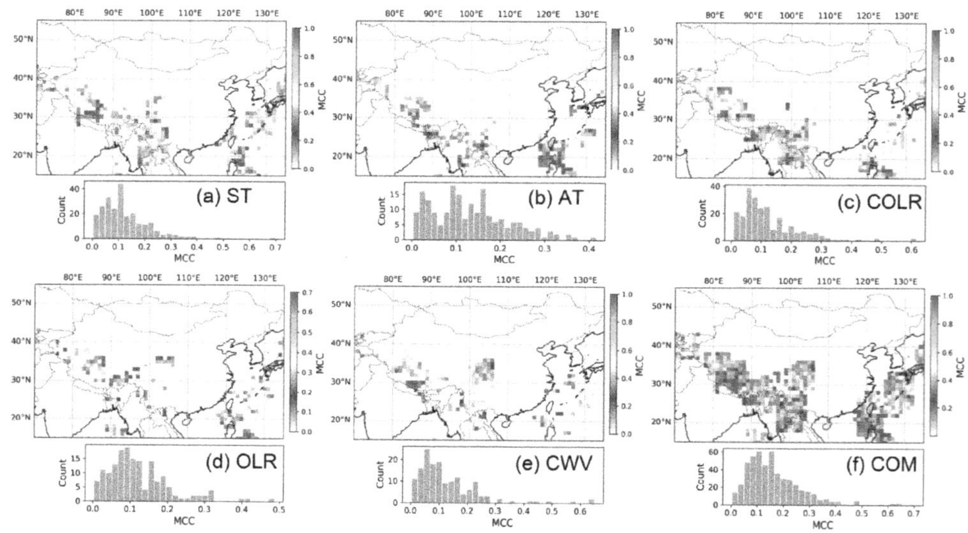

Figure 8. The spatial distributions and histograms (top and bottom sub-figures, respectively) of Matthew's correlation coefficients (MCC) of five geophysical and composed (COM) parameters in China and the surrounding region. ST: skin temperature; AT: air temperature; CWV: total integrated column water vapor burden; OLR: outgoing longwave radiation flux; COLR: clear-sky OLR.

5.4. A Statistical Evaluation Framework for Earthquake Forecasting

In this study, we have developed a spatially self-adaptive multiparametric anomaly identification scheme based on global strong earthquakes, building upon recent advancements in earthquake forecasting research. Our scheme represents a notable improvement over conventional methods that employ fixed anomaly recognition criteria. Additionally, we have incorporated a simple composed approach, further enhancing the earthquake forecasting capability of the proposed scheme. These findings underscore the significance of our approach in advancing earthquake forecasting methodologies. When considering the application of these results to operational earthquake forecasting, the generated global MCC map, in conjunction with new observation data, can serve as a valuable reference for assessing the reliability or probability of anomalous events to indicate upcoming earthquakes. This integrated approach can enhance the credibility and confidence in identifying significant anomalies in earthquake forecasting endeavors.

This scheme plays a pivotal role within our proposed statistical evaluation framework for earthquake forecasting, as illustrated in Figure 9 [20,25]. It aims to enhance the recognition of pre-seismic anomalies that are potentially associated with impending strong earthquakes (Parts 3, 4, and 5 in Figure 9), specifically focusing on earthquakes with a magnitude greater than six in this study. This focus on strong earthquakes is supported by several recent studies [6,40], which highlight the significance of early detection and accurate forecasting for such events. To evaluate the earthquake forecasting capability in a more comprehensive manner, we introduced the use of random synthetic earthquakes in the prospective statistics. This approach enables us to assess earthquake forecasting capabilities across stochastic locations worldwide and over different time periods. The effectiveness of this methodology has been demonstrated in a previous study [33], where it proved valuable in examining the performance of anomaly detection methods.

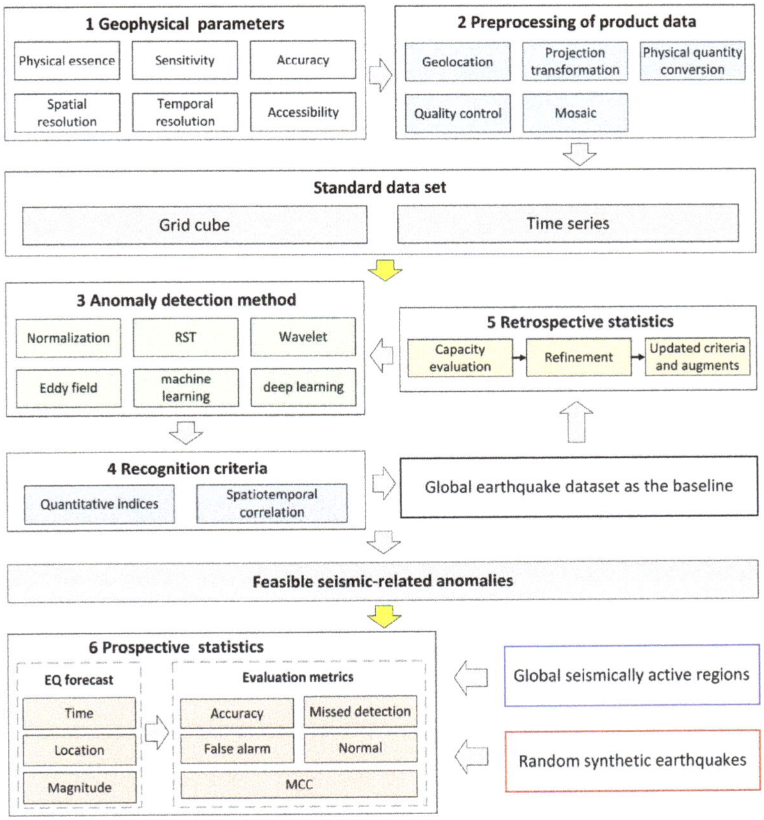

Figure 9. A statistical evaluation framework for earthquake forecasting.

The primary objective of our framework is to provide a quantitative and consistent evaluation of the earthquake forecasting capability, encompassing various geophysical and geochemical parameters. These parameters are primarily obtained through state-of-the-art remote sensing technologies (Parts 1 and 2 in Figure 9), coupled with advanced anomaly detection methods (Part 3). This holistic approach sets our study apart from traditional case studies prevalent in the literature. Recent publications [25,30] emphasize the need for comprehensive evaluation frameworks that incorporate diverse parameters and anomaly detection techniques. Through this simple yet robust framework, we aim to identify the optimal parameters and anomaly detection methods for earthquake forecasting. While the accurate forecast of earthquakes remains a significant challenge, the evaluation metrics, such as the MCC, allow us to estimate the most plausible lower limits of forecasting performance. Any advancements achieved exceeding these lower limits are crucial steps toward advancing operational earthquake forecasting.

5.5. Challenges and Possibilities of Current Scheme

We employed five geophysical parameters derived from AIRS observations to compute pre-seismic anomalies and forecast impending strong earthquakes. As a high spectral resolution infrared sounder, cloud cover significantly affects the availability of data in the AIRS3STD product. Consequently, missing data during the observation period can lead to gaps in anomalous results and altered historical data that affect the calculation of the reference field, subsequently impacting the forecasting state (i.e., accurate forecast, normal condition, false alarm, or missed detection), as specified in Section 3.2. This, in turn, can influence the MCC of a given geophysical parameter. To address this issue, we will incorporate reanalysis data from sources such as the European Centre for Medium-Range Weather Forecasts Reanalysis v5 (ERA5). ERA5 datasets offer seamless spatiotemporal coverage for the relevant geophysical parameters, enabling us to validate and refine our proposed forecasting scheme. Furthermore, this will allow us to investigate the influence of gaps in the product data, providing valuable insights into their potential impact on earthquake forecasting capability.

The quality of the utilized AIRS product is a crucial aspect, given that these data products have undergone extensive development and validation over the years, resulting in their increased maturity (https://airs.jpl.nasa.gov/data/guides-docs, accessed on 28 July 2023). While the validation of various geophysical parameters of the AIRS product is beyond the scope of this study and constitutes a vast topic on its own, it remains an essential consideration. Nevertheless, it is worth noting that, if the data product exhibits no systematic biases, the ZS method employed in this study can mitigate certain retrieval errors by employing the manipulation of the difference and ratio, as presented in Equation (1). This issue also holds significance for other retrieved parameters, as retrieval algorithms typically involve simplifying assumptions to model complex physical processes, inevitably introducing retrieval errors. The uncertainties stemming from these assumptions can significantly impact anomaly detections and even earthquake forecasting, yet their precise extent remains unclear.

The optimal parameters for anomaly recognition criteria were determined using 3-fold cross-validation. This approach was adopted to mitigate the significant overfitting issue that arises when utilizing all 1825 earthquakes to estimate the optimal parameters. The overfitting problem adversely affects earthquake forecasting for the years 2021 and 2022, resulting in low MCC values, which are not presented in this paper. The application of cross-validation in our study effectively improved the overfitting problem, as demonstrated in Section 4.3. However, achieving a set of unified optimal parameters still requires the implementation of specific optimization methods. The primary challenge in this context stems from the limited number of available earthquakes for analysis. Consequently, we must derive the best model based on a small number of samples. This limitation becomes even more critical when employing deep learning methods that demonstrate its advantage in forecasting earthquakes [35], as they typically necessitate a large number of training samples to yield meaningful results [41]. As a result, it is imperative to carefully address this constraint when considering the use of deep learning approaches in our research.

In recent years, probabilistic and physical earthquake forecasting systems have emerged as cutting-edge research in the field [42,43]. To enhance earthquake forecasting capabilities, a simple composed approach was adopted in this study, incorporating anomalies of multiple geophysical parameters. However, a limitation of this method is the inability to quantify the individual contribution of each geophysical parameter anomaly to the final probabilistic forecasting outcome, given the integration of multiparametric anomalies. To address this challenge, we are exploring statistical models that enable the calculation of forecasting probabilities for specific geophysical parameters, leveraging both long-term historical data and short-term observations. Subsequently, employing a multiparametric integration model becomes feasible, allowing for the calculation of earthquake occurrence probabilities in the near future. The development of operational earthquake forecasts serves as a foundation for enhancing global resilience to earthquakes, facilitating preparedness and mitigation efforts worldwide.

6. Conclusions

We have developed a spatially self-adaptive scheme to identify pre-seismic multiparametric anomalies using data from global strong earthquakes recorded between 2006 and 2020. This scheme represents a significant improvement over fixed anomaly recognition criteria, which often lead to invalid forecasts in many seismically active regions. Through our research, we have obtained optimal parameters for the anomaly recognition criteria in global seismically active regions, revealing substantial spatial variation on a global scale. The identified optimal values indicate that most anomalies associated with impending earthquakes are characterized by short-term and subtle signals with a low occurrence frequency. Moreover, we have found that the locations of earthquakes have limited impact on the earthquake forecasting capability of this scheme. Additionally, we have proposed a simple multiparametric composition approach to enhance the earthquake forecasting capability, where each geophysical parameter can contribute a distinct role to overall capability, which has demonstrated superior effectiveness compared to single-parameter approaches, as indicated by improved MCCs. The MCCs of composed anomalies have shown significant improvement in a larger number of regions, with MCC values exceeding 0.5 and an average increase from 0.143 to 0.186. Furthermore, our study has identified OLR including all-sky and clear-sky ones as the dominant geophysical parameter on a global scale, and OLR anomalies exhibit promising potential as a viable contender for earthquake forecasting.

In this study, earthquake forecasting relies on data-driven anomaly detection models that evaluate the probability and future state of imminent strong earthquakes. The results obtained provide valuable insights into our ability to forecast catastrophic earthquake events with greater accuracy. Although we have not yet achieved demonstrably successful predictions of large earthquakes, this study presents a promising approach to the development of remote sensing-based earthquake monitoring and forecasting systems.

Author Contributions: Conceptualization, Z.J.; Formal analysis, Z.J.; Funding acquisition, Z.J. and X.S.; Methodology, Z.J.; Writing—original draft, Z.J.; Writing—review & editing, Z.J., Y.H. and X.S. All authors have read and agreed to the published version of the manuscript.

Funding: This research was funded by the Basic Science Research Plan of the Institute of Geology, China Earthquake Administration, grant number IGCEA2002, and National Key Research and Development Program of China, grant number 2019YFC1509202, 2018YFC1503602.

Data Availability Statement: AIRS AIRS3STD data are available from the Goddard Earth Sciences Data and Information Services Center (GES DISC) at https://disc.gsfc.nasa.gov/datasets/AIRS3STD_7.0/summary (accessed on 2 July 2023). Earthquake catalog data are available from the United States Geological Survey (USGS) at https://www.usgs.gov/natural-hazards/earthquake-hazards (accessed on 2 July 2023).

Acknowledgments: The authors would like to acknowledge the reviewers and editors for their valuable comments and suggestions, which helped improve the quality of this work.

Conflicts of Interest: The authors declare no conflict of interest.

Appendix A

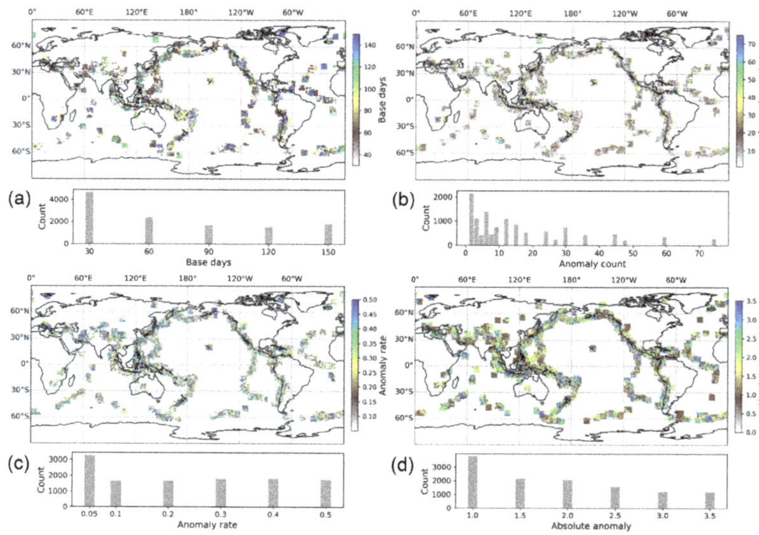

Figure A1. The optimal parameters of the anomaly recognition criteria employed for forecasting imminent earthquakes using the outgoing longwave radiation anomaly based on training data within 2006–2010 and 2016–2020, as indicated by their spatial distributions and histograms in the top and bottom subfigures, respectively. The parameters include (**a**) the base days preceding earthquakes; (**b**) the anomaly count that occurred during the base days; (**c**) the anomaly rate within the spatial window, which is 7 × 7 pixels in this study; and (**d**) the absolute anomaly value used to identify potential earthquake-related anomalies.

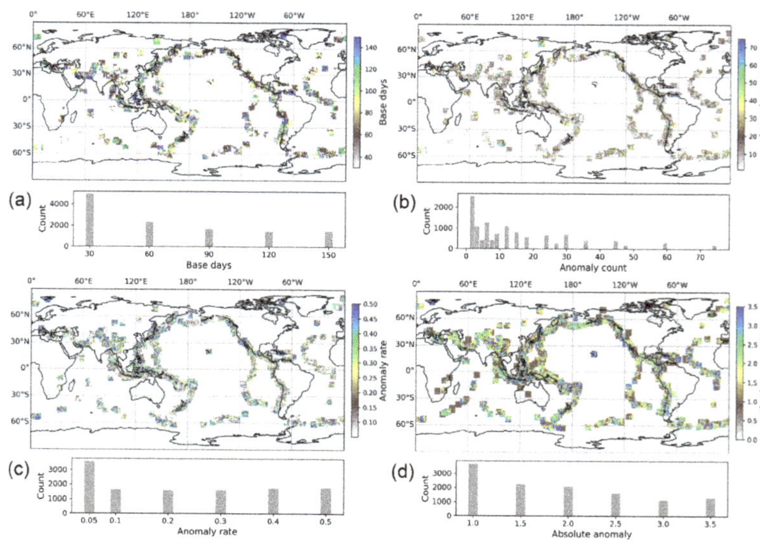

Figure A2. The optimal parameters of the anomaly recognition criteria employed for forecasting imminent earthquakes using the outgoing longwave radiation anomaly based on training data within 2006–2015, as indicated by their spatial distributions and histograms in the top and bottom subfigures, respectively. The parameters include (**a**) the base days preceding earthquakes; (**b**) the anomaly count that occurred during the base days; (**c**) the anomaly rate within the spatial window, which is 7 × 7 pixels in this study; and (**d**) the absolute anomaly value used to identify potential earthquake-related anomalies.

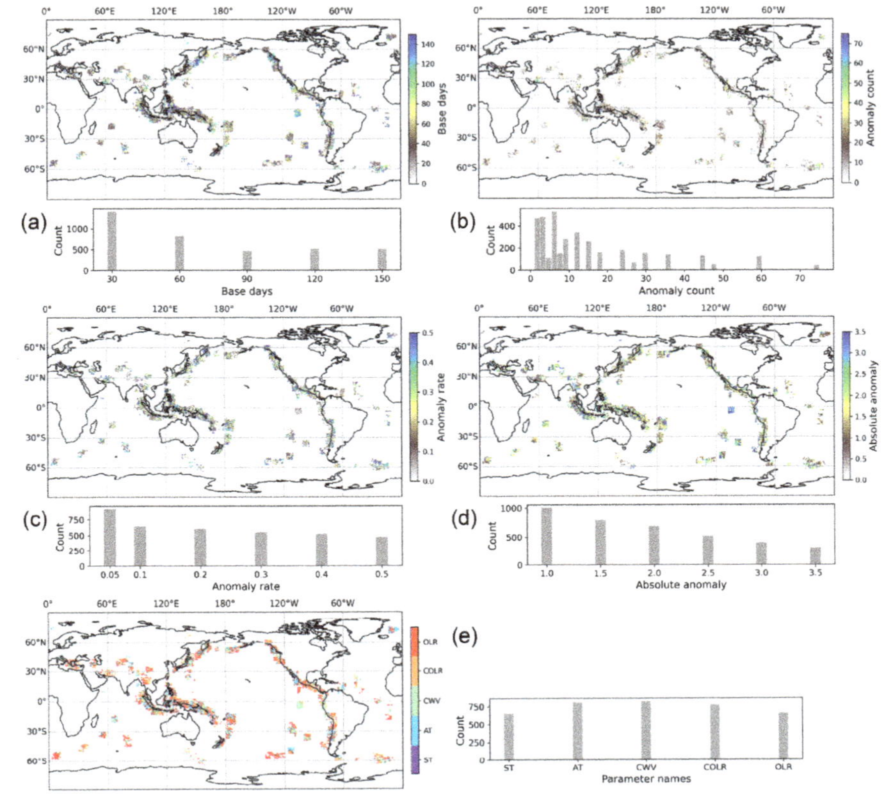

Figure A3. The optimal parameters for anomaly recognition criteria that are utilized for forecasting impending earthquakes in multiparameter composed anomalies based on training data within 2006–2010, and 2016–2020. These anomalies are analyzed based on their spatial distributions and histograms, as shown in the top and bottom subfigures, respectively. The parameters include (**a**) the base days preceding earthquakes; (**b**) the anomaly count that occurred during the base days; (**c**) the anomaly rate within the spatial window, which is 7×7 pixels in this study; (**d**) the absolute anomaly value used to identify potential earthquake-related anomalies; and (**e**) the geophysical parameters used in the composed anomalies. ST: skin temperature; AT: air temperature; CWV: total integrated column water vapor burden; OLR: outgoing longwave radiation flux; COLR: clear-sky OLR.

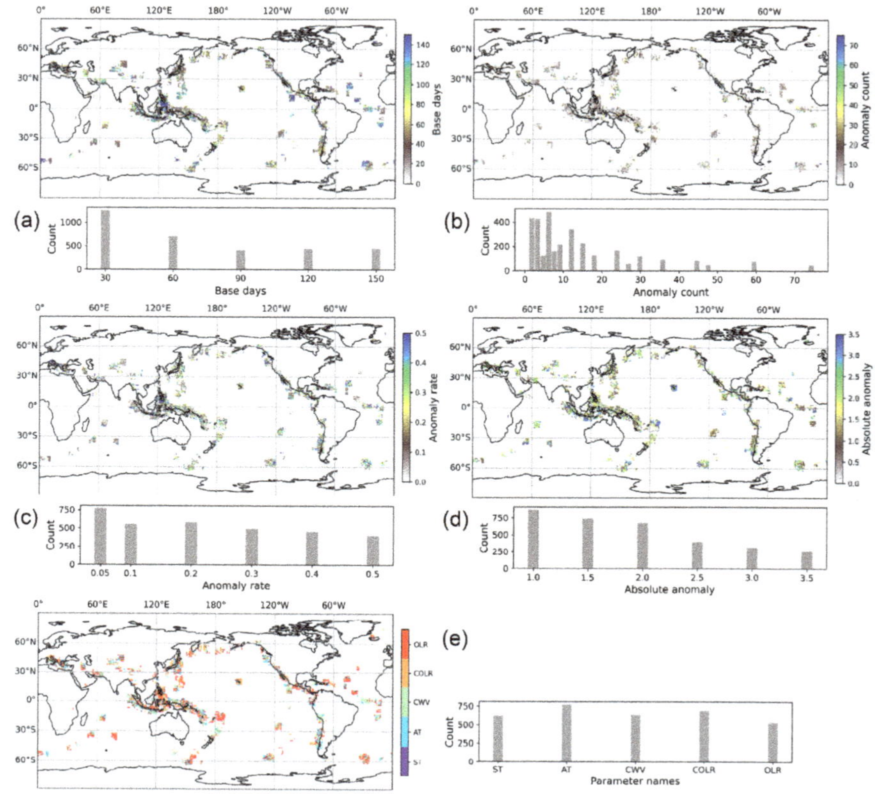

Figure A4. The optimal parameters for anomaly recognition criteria that are utilized for forecasting impending earthquakes in multiparameter composed anomalies based on training data within 2006–2015. These anomalies are analyzed based on their spatial distributions and histograms, as shown in the top and bottom subfigures, respectively. The parameters include (**a**) the base days preceding earthquakes; (**b**) the anomaly count that occurred during the base days; (**c**) the anomaly rate within the spatial window, which is 7 × 7 pixels in this study; (**d**) the absolute anomaly value used to identify potential earthquake-related anomalies; and (**e**) the geophysical parameters used in the composed anomalies. ST: skin temperature; AT: air temperature; CWV: total integrated column water vapor burden; OLR: outgoing longwave radiation flux; COLR: clear-sky OLR.

References

1. Marzocchi, W.; Lombardi, A.M.; Casarotti, E. The Establishment of an Operational Earthquake Forecasting System in Italy. *Seismol. Res. Lett.* **2014**, *85*, 961–969. [CrossRef]
2. Huang, Q. Forecasting the epicenter of a future major earthquake. *Proc. Natl. Acad. Sci. USA* **2015**, *112*, 944–945. [CrossRef]
3. Jiao, Z.H.; Shan, X. Short-Term Responses of Land Surface Temperature Anomalies to Earthquakes in China. In Proceedings of the 2021 IEEE International Geoscience and Remote Sensing Symposium IGARSS, Brussels, Belgium, 11–16 July 2021; pp. 8412–8415.
4. Chen, S.; Liu, P.; Feng, T.; Wang, D.; Jiao, Z.; Chen, L.; Xu, Z.; Zhang, G. Exploring Changes in Land Surface Temperature Possibly Associated with Earthquake: Case of the April 2015 Nepal Mw 7.9 Earthquake. *Entropy* **2020**, *22*, 377. [CrossRef]
5. Khalili, M.; Alavi Panah, S.K.; Abdollahi Eskandar, S.S. Using Robust Satellite Technique (RST) to determine thermal anomalies before a strong earthquake: A case study of the Saravan earthquake (April 16th, 2013, MW = 7.8, Iran). *J. Asian Earth Sci.* **2019**, *173*, 70–78. [CrossRef]
6. Satti, M.S.; Ehsan, M.; Abbas, A.; Shah, M.; de Oliveira-Júnior, J.F.; Naqvi, N.A. Atmospheric and ionospheric precursors associated with M $\geq$ 6.5 earthquakes from multiple satellites. *J. Atmos. Sol.-Terr. Phys.* **2022**, *227*, 105802. [CrossRef]
7. Shah, M.; Aibar, A.C.; Tariq, M.A.; Ahmed, J.; Ahmed, A. Possible ionosphere and atmosphere precursory analysis related to Mw > 6.0 earthquakes in Japan. *Remote Sens. Environ.* **2020**, *239*, 111620. [CrossRef]
8. Jing, F.; Singh, R.P.; Shen, X. Land–Atmosphere–Meteorological coupling associated with the 2015 Gorkha (M 7.8) and Dolakha (M 7.3) Nepal earthquakes. *Geomat. Nat. Hazards Risk* **2019**, *10*, 1267–1284. [CrossRef]

9. Marchetti, D.; De Santis, A.; Shen, X.; Campuzano, S.A.; Perrone, L.; Piscini, A.; Di Giovambattista, R.; Jin, S.; Ippolito, A.; Cianchini, G.; et al. Possible Lithosphere-Atmosphere-Ionosphere Coupling effects prior to the 2018 Mw = 7.5 Indonesia earthquake from seismic, atmospheric and ionospheric data. *J. Asian Earth Sci.* **2020**, *188*, 104097. [CrossRef]
10. Wu, L.; Zheng, S.; De Santis, A.; Qin, K.; Di Mauro, R.; Liu, S.; Rainone, M.L. Geosphere coupling and hydrothermal anomalies before the 2009 Mw 6.3 L'Aquila earthquake in Italy. *Nat. Hazards Earth Syst. Sci.* **2016**, *16*, 1859–1880. [CrossRef]
11. Sun, K.; Shan, X.; Ouzounov, D.; Shen, X.; Jing, F. Analyzing long wave radiation data associated with the 2015 Nepal earthquakes based on Multi-orbit satellite observations. *Chin. J. Geophys.* **2017**, *60*, 3457–3465.
12. Jordan, T.H.; Chen, Y.-T.; Gasparini, P.; Madariaga, R.; Main, I.; Marzocchi, W.; Papadopoulos, G.; Sobolev, G.; Yamaoka, K.; Zschau, J. Operational earthquake forecasting: State of Knowledge and Guidelines for Utilization. *Ann. Geophys.* **2011**, *54*, 316–391.
13. Kato, A.; Ben-Zion, Y. The generation of large earthquakes. *Nat. Rev. Earth Environ.* **2020**, *2*, 26–39.
14. Molchanov, O.; Fedorov, E.; Schekotov, A.; Gordeev, E.; Chebrov, V.; Surkov, V.; Rozhnoi, A.; Andreevsky, S.; Iudin, D.; Yunga, S.; et al. Lithosphere-atmosphere-ionosphere coupling as governing mechanism for preseismic short-term events in atmosphere and ionosphere. *Nat. Hazards Earth Syst. Sci.* **2004**, *4*, 757–767. [CrossRef]
15. Denisenko, V.V.; Boudjada, M.Y.; Lammer, H. Propagation of Seismogenic Electric Currents Through the Earth's Atmosphere. *J. Geophys. Res. Space Phys.* **2018**, *123*, 4290–4297.
16. Tramutoli, V.; Aliano, C.; Corrado, R.; Filizzola, C.; Genzano, N.; Lisi, M.; Martinelli, G.; Pergola, N. On the possible origin of thermal infrared radiation (TIR) anomalies in earthquake-prone areas observed using robust satellite techniques (RST). *Chem. Geol.* **2013**, *339*, 157–168.
17. Freund, F.; Ouillon, G.; Scoville, J.; Sornette, D. Earthquake precursors in the light of peroxy defects theory: Critical review of systematic observations. *Eur. Phys. J. Spec. Top.* **2021**, *230*, 7–46.
18. Ren, Y.; Ma, J.; Liu, P.; Chen, S. Experimental Study of Thermal Field Evolution in the Short-Impending Stage Before Earthquakes. *Pure Appl. Geophys.* **2017**, *175*, 2527–2539. [CrossRef]
19. Liu, P.; Chen, S.; Liu, Q.; Guo, Y.; Ren, Y.; Zhuo, Y.; Feng, J. A Potential Mechanism of the Satellite Thermal Infrared Seismic Anomaly Based on Change in Temperature Caused by Stress Variation: Theoretical, Experimental and Field Investigations. *Remote Sens.* **2022**, *14*, 5697. [CrossRef]
20. Jiao, Z.-H.; Shan, X. Statistical framework for the evaluation of earthquake forecasting: A case study based on satellite surface temperature anomalies. *J. Asian Earth Sci.* **2021**, *211*, 104710. [CrossRef]
21. Genzano, N.; Filizzola, C.; Hattori, K.; Pergola, N.; Tramutoli, V. Statistical Correlation Analysis Between Thermal Infrared Anomalies Observed From MTSATs and Large Earthquakes Occurred in Japan (2005–2015). *J. Geophys. Res. Solid Earth* **2021**, *126*, e2020JB020108.
22. Fu, C.-C.; Lee, L.-C.; Ouzounov, D.; Jan, J.-C. Earth's Outgoing Longwave Radiation Variability Prior to M $\geq$ 6.0 Earthquakes in the Taiwan Area During 2009–2019. *Front. Earth Sci.* **2020**, *8*, 364. [CrossRef]
23. Eleftheriou, A.; Filizzola, C.; Genzano, N.; Lacava, T.; Lisi, M.; Paciello, R.; Pergola, N.; Vallianatos, F.; Tramutoli, V. Long-term RST analysis of anomalous TIR sequences in relation with earthquakes occurred in Greece in the period 2004–2013. *Pure Appl. Geophys.* **2016**, *173*, 285–303. [CrossRef]
24. Zhang, Y.; Meng, Q. A statistical analysis of TIR anomalies extracted by RSTs in relation to an earthquake in the Sichuan area using MODIS LST data. *Nat. Hazards Earth Syst. Sci.* **2019**, *19*, 535–549. [CrossRef]
25. Jiao, Z.-H.; Shan, X. Consecutive statistical evaluation framework for earthquake forecasting: Evaluating satellite surface temperature anomaly detection methods. *J. Asian Earth Sci. X* **2022**, *7*, 100096. [CrossRef]
26. Picozza, P.; Conti, L.; Sotgiu, A. Looking for Earthquake Precursors From Space: A Critical Review. *Front. Earth Sci.* **2021**, *9*, 676775. [CrossRef]
27. Jiao, Z.-H.; Zhao, J.; Shan, X. Pre-seismic anomalies from optical satellite observations: A review. *Nat. Hazards Earth Syst. Sci.* **2018**, *18*, 1013–1036. [CrossRef]
28. Jenkins, A.P.; Biggs, J.; Rust, A.C.; Rougier, J.C. Decadal Timescale Correlations Between Global Earthquake Activity and Volcanic Eruption Rates. *Geophys. Res. Lett.* **2021**, *48*, e2021GL093550. [CrossRef]
29. Chahine, M.T.; Pagano, T.S.; Aumann, H.H.; Atlas, R.; Barnet, C.; Blaisdell, J.; Chen, L.; Divakarla, M.; Fetzer, E.J.; Goldberg, M.; et al. AIRS: Improving weather forecasting and providing new data on greenhouse gases. *Bull. Am. Meteorol. Soc.* **2006**, *87*, 911–926. [CrossRef]
30. Wu, L.; Qi, Y.; Mao, W.; Lu, J.; Ding, Y.; Peng, B.; Xie, B. Scrutinizing and rooting the multiple anomalies of Nepal earthquake sequence in 2015 with the deviation–time–space criterion and homologous lithosphere–coversphere–atmosphere–ionosphere coupling physics. *Nat. Hazards Earth Syst. Sci.* **2023**, *23*, 231–249. [CrossRef]
31. Chicco, D.; Jurman, G. The advantages of the Matthews correlation coefficient (MCC) over F1 score and accuracy in binary classification evaluation. *BMC Genom.* **2020**, *21*, 6. [CrossRef]
32. Filizzola, C.; Corrado, A.; Genzano, N.; Lisi, M.; Pergola, N.; Colonna, R.; Tramutoli, V. RST Analysis of Anomalous TIR Sequences in Relation with Earthquakes Occurred in Turkey in the Period 2004–2015. *Remote Sens.* **2022**, *14*, 381. [CrossRef]
33. Jiao, Z.; Shan, X. Pre-Seismic Temporal Integrated Anomalies from Multiparametric Remote Sensing Data. *Remote Sens.* **2022**, *14*, 2343. [CrossRef]

34. Pulinets, S.; Ouzounov, D. Lithosphere–Atmosphere–Ionosphere Coupling (LAIC) model—An unified concept for earthquake precursors validation. *J. Asian Earth Sci.* **2011**, *41*, 371–382. [CrossRef]
35. Xiong, P.; Tong, L.; Zhang, K.; Shen, X.; Battiston, R.; Ouzounov, D.; Iuppa, R.; Crookes, D.; Long, C.; Zhou, H. Towards advancing the earthquake forecasting by machine learning of satellite data. *Sci. Total Environ.* **2021**, *771*, 145256. [CrossRef]
36. Genzano, N.; Filizzola, C.; Lisi, M.; Pergola, N.; Tramutoli, V. Toward the development of a multi parametric system for a short-term assessment of the seismic hazard in Italy. *Ann. Geophys.* **2020**, *63*, 550. [CrossRef]
37. Freund, F. Toward a unified solid state theory for pre-earthquake signals. *Acta Geophys.* **2010**, *58*, 719. [CrossRef]
38. Quigley, M.C.; Hughes, M.W.; Bradley, B.A.; van Ballegooy, S.; Reid, C.; Morgenroth, J.; Horton, T.; Duffy, B.; Pettinga, J.R. The 2010–2011 Canterbury Earthquake Sequence: Environmental effects, seismic triggering thresholds and geologic legacy. *Tectonophysics* **2016**, *672–673*, 228–274. [CrossRef]
39. Han, Y.; Zang, Y.; Meng, L.; Wang, Y.; Deng, S.; Ma, Y.; Xie, M. A summary of seismic activities in and around China in 2021. *Earthq. Res. Adv.* **2022**, *2*, 100157. [CrossRef]
40. Xu, X.; Chen, S.; Zhang, S.; Dai, R. Analysis of Potential Precursory Pattern at Earth Surface and the Above Atmosphere and Ionosphere Preceding Two Mw $\geq$ 7 Earthquakes in Mexico in 2020–2021. *Earth Space Sci.* **2022**, *9*, e2022EA002267. [CrossRef]
41. Reichstein, M.; Camps-Valls, G.; Stevens, B.; Jung, M.; Denzler, J.; Carvalhais, N.; Prabhat, F. Deep learning and process understanding for data-driven Earth system science. *Nature* **2019**, *566*, 195–204. [CrossRef]
42. Hardebeck, J.L. Spatial Clustering of Aftershocks Impacts the Performance of Physics-Based Earthquake Forecasting Models. *J. Geophys. Res. Solid Earth* **2021**, *126*, e2020JB020824. [CrossRef]
43. Rundle, J.B.; Stein, S.; Donnellan, A.; Turcotte, D.L.; Klein, W.; Saylor, C. The complex dynamics of earthquake fault systems: New approaches to forecasting and nowcasting of earthquakes. *Rep. Prog. Phys.* **2021**, *84*, 076801. [CrossRef] [PubMed]

remote sensing

MDPI

Article

Studying the Impact of the Geospace Environment on Solar Lithosphere Coupling and Earthquake Activity

Dimitar Ouzounov [1,*] and Galina Khachikyan [2]

1 Institute for Earth, Computing, Human and Observing (Institute for ECHO), Chapman University, Orange, CA 92866, USA
2 Institute of Seismology, Almaty 050060, Kazakhstan; galina.khachikyan@seismology.kz
* Correspondence: ouzounov@chapman.edu

Abstract: In solar–terrestrial physics, there is an open question: does a geomagnetic storm affect earthquakes? We expand research in this direction, analyzing the seismic situation after geomagnetic storms (GMs) accompanied by the precipitation of relativistic electrons from the outer radiation belt to form an additional radiation belt (RB) around lower geomagnetic lines. We consider four widely discussed cases in the literature for long-lived (weeks, months) RBs due to GMs and revealed that the 1/GMs 24 March 1991 with a new RB at L~2.6 was followed by an M7.0 earthquake in Alaska, 30 May 1991, near footprint L = 2.69; the 2/GMs 29 October 2003 (Ap = 204) with new RB first in the slot region at L = 2–2.5 cases followed by an M7.8 earthquake on 17 November 2003 at the Aleutian Islands near footprint L = 2.1, and after forming an RB at L~1.5 which lasted for ~26 months, two mega quakes, M9.1 in 2004 and M8.6 in 2005, occurred at the globe; the 3/GMs 3 September 2012 with a new RB at L= 3.0–3.5 was followed by an M7.8 earthquake in Canada near footprint L = 3.2; and the 4/GMs 21 June 2015 with a new RB at L = 1.5–1.8 was followed by an M6.3 earthquake on 7 September 2015 in New Zealand, near footprint L = 1.58. The obtained results suggest that (1) major earthquakes occur near the footprints of geomagnetic lines filled with relativistic electrons precipitating from the outer radiation belt due to geomagnetic storms, and (2) the time delay between geomagnetic storm onset and earthquake occurrence may vary from several weeks to several months. The results may expand the framework for developing mathematical magnetosphere–ionosphere coupling models.

Keywords: earthquakes; cosmic rays; geomagnetic storms; radiation belt; LAIC

Citation: Ouzounov, D.; Khachikyan, G. Studying the Impact of the Geospace Environment on Solar Lithosphere Coupling and Earthquake Activity. *Remote Sens.* **2024**, *16*, 24. https://doi.org/10.3390/rs16010024

Academic Editors: Dedalo Marchetti, Yunbin Yuan and Kaiguang Zhu

Received: 30 October 2023
Revised: 6 December 2023
Accepted: 8 December 2023
Published: 20 December 2023

1. Introduction

The idea that space weather—as measured by sunspots, cosmic rays, solar wind, interplanetary magnetic fields, geomagnetic activity, and precipitation of charged particles from the radiation belt—may play a role in triggering earthquakes has a long history but has not received physical justification for the time being. It was noted in [1] that the problem's difficulty is that the unknown is a physical mechanism of action involving relatively weak fields of cosmic origin on tectonic processes. The authors of [2] suggested that geomagnetic variations affect rocks, e.g., by inducing eddy currents in the Earth's crust via an alternating magnetic field, leading to rocks' additional heating. They modeled the distribution of heat flows generated by 98 magnetic storms on the surface and in the interior of the Northern Tien Shan. They showed that the average amount of energy released in the Earth in the form of heat is about 7.4×10^{11} J, which is comparable to the amount of energy released in this territory over the same time interval. Simultaneously, they noted that the heat flow transferred by storms into the ground is 2–3 orders of magnitude lower than the heat flow from the ground and, therefore, cannot play a significant role in the earthquake preparation phase but can most likely accelerate the process of releasing the accumulated tectonic stress if it has already started.

It was also assumed that the action could be forceful, owing to the movement of telluric currents in the Earth's magnetic field. However, [3] showed that these ponderomotive forces produce only the weakest stresses and deformations in the Earth's crust. As the authors of [4,5] suggested, the Alfvèn waves, which can drive currents in the ionosphere and reradiate the energy as electromagnetic waves that propagate to the ground [6], could mediate between space weather and seismic activity.

Over the last 30 years, international community cooperation has suggested the concept of the lithosphere–atmosphere–ionosphere coupling (LAIC), which indicates that electromagnetic disturbances in the lithosphere accompany the earthquake preparation process and can modify the electric field in the global electric circuit (GEC) (see [7–11] and references herein). The GEC links the electric fields and currents flowing in different parts of the atmosphere [12,13]. There are three main generators of fields and currents in the global electric circuit [13]: (1) the troposphere generator (continuous thunderstorm activity of the Earth, with about 46 lightning strikes per second); (2) the ionospheric wind dynamo; and (3) the solar wind/magnetosphere dynamo. The troposphere generator may provide a 200 kV to 600 kV potential difference between the ground and the ionosphere. The ionosphere generator (ionospheric wind dynamo) may provide about 5 kV to 15 kV of potential difference between the high and low latitudes. The magnetospheric generator (solar wind/magnetospheric dynamo) can provide a potential drop of 40 kV to 130 kV across the polar cap. Thunderstorms in the troposphere are the strongest source of voltages/currents in this circuit. The electric current flows upward from the thunderstorm areas and spreads out all over the globe through the ionosphere and through the magnetosphere along magnetic field lines to the opposite hemisphere. The current returns to the surface of the Earth as the fair-weather troposphere-to-Earth current. The density of the vertical electric conduction current in the GEC depends on conductivity and is about 10^{-12} A/m^2 on average.

The LAIC concept considers that radon emanation may accompany an earthquake preparation process [7]. The radon concentration shows relatively strong spatial variation and differs widely above continental and oceanic areas. The average radon flux density for the entire ocean is equal to 0.0382 mBq m^{-2} s^{-1} [14] and is much smaller than the typical estimates for the average flux density from land, which is in the approximate range of 20 to 35 mBq m^{-2} s^{-1} [15]. Overland, the radon concentration depends on the tectonic conditions. For example, in Mysore city (India), it is of order 20 Bq/m^3 [16]; simultaneously, in the vicinity of active faults, radon concentration may reach 2000 kBq/m^3, five orders of magnitude larger [17]. Thus, Radon emanation from an area of earthquake preparation (active faults) may increase the conductivity and decrease the resistance of the surface-layer air (the lower tropospheric part of the GEC). A decrease in boundary layer resistance reduces the total ground–ionosphere electrical resistance. This, in turn, increases the vertical fair-weather current in the GEC and lowers the ionosphere (to maintain the continuity of the electron flow). Hence, the effects of earthquake preparation modify the ionosphere parameters. The concept of the GEC was successfully used in [18,19] to explain the results from the DEMETER satellite observations [20], which showed a statistically significant decrease in the natural VLF (~1.7 kHz) wave intensity related to 8400 nighttime earthquakes of $M \geq 5.0$ within 440 km of the epicenters.

In [21,22], a modified configuration of the GEC is discussed. Its external element is located on the magnetopause with the electromotive force generator driven by solar wind energy; its internal element is in the solid Earth, with the electromotive force generator driven by the tectonic processes [22]. The modified GEC may be considered a mediator in transmitting solar wind energy into the Earth's crust. The functioning of a GEC will depend on electrical conductivity along the entire path from the magnetopause to the Earth's crust. It is mentioned above that in the near-ground troposphere, ionization may be produced by the isotopes of radon [14–17], whose emanation is expected to be strong at overactive faults.

In the upper troposphere and lower stratosphere, the galactic cosmic rays influence the ionization and, therefore, the electrical parameters of the atmosphere [23,24]. Cosmic rays are energetic particles (primarily protons and α-particles), which, according to their origin, are categorized as either galactic cosmic rays (GCRs) or solar cosmic rays (solar energetic particles, SEPs) [23]. GCRs are always present in the Earth's vicinity, while SEPs occur sporadically, usually in conjunction with the Sun's giant eruptive events (solar flares or coronal mass ejections). The energy inputted by cosmic rays in the Earth's atmosphere is about 10^{-9} times that of solar energy [25]. However, cosmic rays are the only source of ion production in the lower stratosphere and upper troposphere [23,24]. The cosmic rays are inputted to the electrical properties of the atmosphere and the global electric circuit by manipulating atmospheric conductivity, vertical electric current, and vertical ionospheric potential gradients. It is shown in [26,27] that changes in the surface potential and ionospheric potential gradients correlate well with changes in GCR. About 95% of cosmic ray particles have an energy of 0.1–15 GeV, which contains more than 60% of all cosmic ray particle energy [28,29]. The cosmic ray particles with energy $\leq$15 GeV undergo an 11-year modulation, and the flux of cosmic ray particles with this energy may decrease more than twofold when the solar activity period changes from the minimum to the maximum [28]. Thus, the effect of cosmic ray ionization in the lower stratosphere and upper troposphere should be more dominant during the period of minimum solar activity. In addition to continuous ionization in the Earth's atmosphere caused by a galactic cosmic ray, sporadic ionization occurs during solar energetic particle events, potentially affecting the Earth's environment [30,31], possibly contributing to GEC operation and earthquake occurrence. In the pure conductive mesosphere, the ionization increases due to the precipitation of energetic electrons from the radiation belt up to the stratopause (~55 km altitude). This process is more effective in times of geomagnetic storms.

In the present paper, we analyze the response of seismic activity to the effects of precipitation of relativistic electrons from the outer radiation belt during geomagnetic storms (Section 3.1). Furthermore, we compare global seismic activity's temporal variations with solar activity's temporal variations and galactic cosmic rays' flux (Section 3.2).

2. Data and Methods

For analysis, we used data from the USGS Global Seismological Catalog (https://earthquake.usgs.gov/earthquake, accessed on 25 October 2018) for 1973–2017 for earthquakes with a magnitude of $M \geq 4.5$ (more than 220,000 events). For the epicenters of all analyzed earthquakes, we calculated the values of the McIlvaine parameter L [32] (indicating the distance of geomagnetic lines to the center of the Earth at the equator, expressed in terms of the radii of the Earth) using the IGRF model and the computer codes of the GEOPACK [33]. The McIlwain parameter (L)—is named after the scientist Carl McIlwain and has two more names: (L-shell), magnetic shell; (L-value), number of the magnetic field line)—is taken to be equal to the ratio of the average distance of the geomagnetic shell (geomagnetic field line) from the center of the Earth in the plane from the geomagnetic equator to the radius of the Earth. The equation for calculating the value of L at a point with geographic coordinates (λ, φ) is $L = Re(\lambda,\varphi)/Re\cos^2\Lambda$, where λ is geographic latitude, φ is geographic longitude, Λ is geomagnetic latitude, and *Re* is the radius of the Earth at the equator. The equation for calculating the radius of the Earth at a point with geographic coordinates (λ, φ) is $Re(\lambda, \varphi) = \alpha + b \cos^2 \lambda + c \cos^4 \lambda$, where α = 6356.912 km is the radius of the Earth at the geographic pole, b = 21.3677 km, and c = 0.108 km. The value of geomagnetic latitude at a point with geographic coordinates (λ, φ) can be obtained from the equation $\sin\Lambda = \sin \lambda \sin \lambda_p + \cos \lambda \cos \lambda_p \cos (\varphi - \varphi_p)$, where λ_p, φ_p are the geographic latitude and longitude of the geomagnetic pole; these change over time (secular variations of the geomagnetic field), which is reflected in the international reference model IGRF, which describes the main geomagnetic field and its secular variations produced by an internal source located in the liquid outer core of the Earth.

A geomagnetic storm is an interval of several days during which there is a significant reduction in the horizontal component of the geomagnetic field at the Earth's surface [34]. Following [35], geomagnetic storms may be small (Dst from −30 to −50 nT), moderate (Dst from −50 to −100 nT), strong (Dst from −100 to −200 nT), powerful (Dst from −200 to −350 nT), and extra strong (giant) (Dst lower than −350 nT). We investigate the response of seismic activity to geomagnetic storms, and the data for the disturbance storm (Dst index) were obtained from NASA/GSFC's Space Physics Data Facility's CDAWeb service and OMNI data (https://cdaweb.gsfc.nasa.gov/index.html, accessed on 30 September 2023). Geomagnetic storms, the bright manifestation of space weather variations, influence the Earth's radiation belts (toroids of very high-energy magnetically trapped charged particles), as discovered in 1958 by Van Allen [36]. Radiation belts typically comprise two distinct zones (inner and outer) spatially separated by the slot region (Figure 1 from [37]).

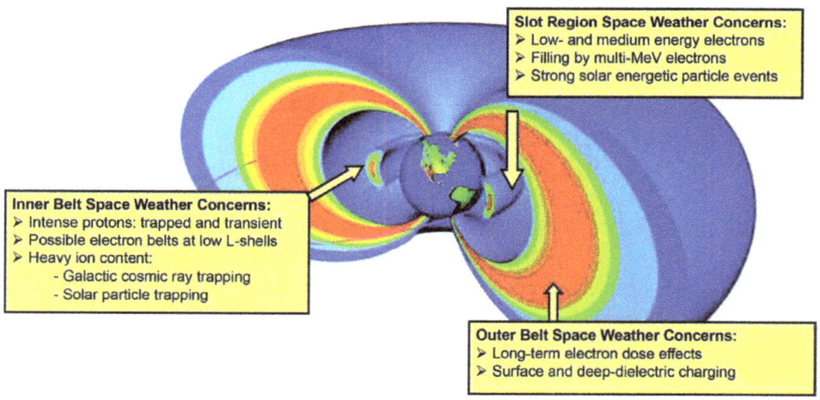

Figure 1. Schematic diagram of Earth's radiation belt, reprinted from Ref. [37].

The inner Van Allen zone, mainly populated by energetic protons, extends from just above the dense atmosphere to an equatorial altitude of about 10,000 km above the Earth's surface. The left label in Figure 1 shows several space weather concerns for the inner radiation belt zone [37]: the intense, high-energy trapped protons, the variable, trapped solar energetic particles, the trapped galactic cosmic rays, and the trapped energetic electrons. The upper label in Figure 1 shows that the slot region can present several space-weather concerns, including low- and medium-energy electron enhancements, multi-MeV electrons (on rare occasions), and strong solar energetic particle events (again on relatively rare occasions). The slot region extends from L~2.0 to L~3.0. The outer zone of the radiation belt extends from L~3.0 to L~6.5. It comprises mildly to highly relativistic electrons and varies widely in time and particle intensity. Due to geomagnetic storms, the high-energy electrons can spill down from the outer radiation belt and form a new radiation belt (storage ring) around certain geomagnetic lines in the inner magnetosphere [38–46]. The new radiation belts normally last for several days but very rare, they may last for up to several weeks, months, or even years. We analyze the response of seismic activity to four widely discussed geomagnetic storms in the scientific literature, which were all followed by a newly formed storage ring in the inner magnetosphere. Three of them lasted for up to several months; they were created by geomagnetic storms on 24 March 1991 [41,42], 1 September 2012 [43], and 21 June 2015 [44]; one of them, created by a giant storm on 31 October 2003, lasted for more than two years [40]. To compare seismic data variations with solar activity variations, we used sunspot numbers from the World Data Center SILSO, Royal Observatory of Belgium, Brussels (https://wwwbis.sidc.be/silso/datafiles, accessed on 30 September 2023). The sunspot number is one of the space weather parameters usually used to investigate long-term solar activity variations, including 11-year variations.

3. Results

3.1. Increasing Seismic Activity near Magnetic Field Line Footprints of Newly Created Radiation Belts Arising from Geomagnetic Storms

3.1.1. Case Study in the Years 1991, 2012, and 2015

On 24 March 1991, a powerful geomagnetic storm started at 04:30 UT and reached its negative extremum in the primary phase on 25 March 1991 at 00:30 UT, with Dst = −298 nT. At that time, the CRRES satellite was near L~2.6, and its instruments recorded powerful fluxes of electrons with E~15 MeV and protons with E~20–110 MeV [41]. The MIR orbital station also observed the newly created radiation belt for about 2 years [42].

On 1 September 2012, a moderate geomagnetic storm started at 22:30 UT and reached its negative extremum in the primary phase on 3 September 2012 at 10:30 UT, with Dst = −69 nT. At that time, the Relativistic Electron Proton Telescope (REPT) on NASA's Van Allen probes board recorded a flux of energetic electrons (3.6 MeV, 4.5 MeV, and 5.6 MeV) at geomagnetic lines $3.0 \leq L \leq 3.5$. A newly formed belt of relativistic electrons was observed for about a month [43] and was then destroyed by the next strong geomagnetic storm on 1 October 2012 [38].

On 21 June 2015, a powerful geomagnetic storm started at 18:30 UT and reached its negative extremum in the primary phase on 23 June 2015 at 04:30 UT, with Dst = −204 nT. At that time, the Magnetic Electron Ion Spectrometer (*MagEIS*) on NASA's Van Allen probes board recorded a new belt of relativistic electrons with energy E= ~1.06 MeV at geomagnetic lines L~1.5–1.8, which persisted for ~11 months [44].

Figure 2 presents monthly earthquakes of magnitude $M \geq 4.5$ that occurred in 1991, 2012, and 2015 near the footprints of the newly created belts of relativistic electrons.

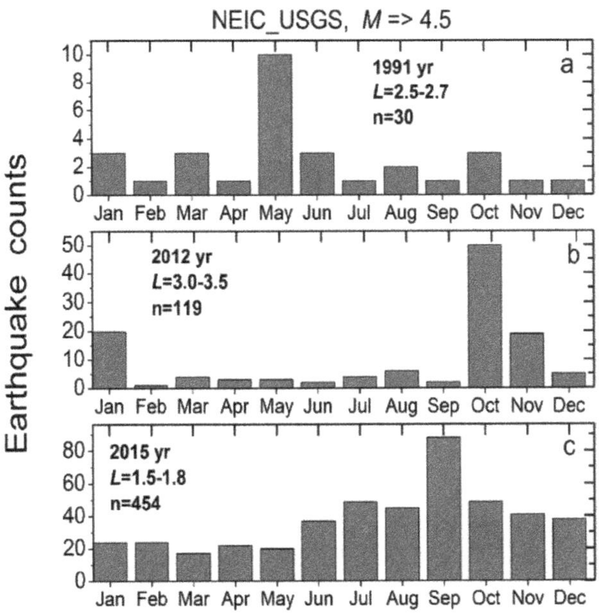

Figure 2. Histograms of the monthly number of earthquakes of $M \geq 4.5$ occurring near the footprints of geomagnetic lines belonging to the newly created belts of relativistic electrons: (**a**) 1991, L = 2.5–2.7; (**b**) 2012, L = 3.0–3.5; and (**c**) 2015, L = 1.5–1.8.

Figure 2a shows the histogram of the number of earthquakes of $M \geq 4.5$ near the footprint of the geomagnetic lines L = 2.5–2.7 in different months of 1991. It is seen that the most significant number of earthquakes occurred in May 1991. The strongest was the M7.0 earthquake in Alaska on 30 May 1991, with epicenter coordinates of 54.57°N, 161.61°E, near the footprint of the geomagnetic line L~2.69 and closely adjacent to the new radiation belt created around L~2.6 after a magnetic storm on 24 March 1991 [41,42]. We see that an increase in seismic activity at the base L = 2.5–2.7 in May 1991 occurred ~2 months after the geomagnetic storm onset.

Figure 2b shows the distribution by month across 2012 for the number of earthquakes with a magnitude of M $\geq$ 4.5 that occurred near the footprint of the geomagnetic lines L = 3.0–3.5, around which a belt of high-energy electrons was formed due to a geomagnetic storm on 1 September 2012 [43]. For the base of L = 3.0–3.5, the number of earthquakes strongly increased in October 2012; the largest here was an earthquake of M = 7.8, which occurred off the coast of Canada on 28 October 2012, with coordinates 52.79°N, 132.1°W, near the footprint of L = ~3.3. This earthquake also occurred ~2 months after the geomagnetic storm onset on 1 September 2012, which created a storage ring of relativistic electrons.

Figure 2c presents the monthly number of earthquakes with M $\geq$ 4.5 in 2015 near the footprint of geomagnetic lines L = 1.5–1.8. A belt of high-energy electrons was formed due to a geomagnetic storm on 21 June 2015 [44]. The increase of seismic activity here started in June 2015, just after a strong geomagnetic storm, but it peaked in September 2015. The largest earthquake (M = 6.3) occurred on 7 September 2015, near New Zealand, with epicenter coordinates of 32.82 °S, 177.86 °W in the footprint of L~1.58. The M6.3 events again persisted over 2 months following the magnetic storm onset.

Figure 3 visualizes the correlation between geomagnetic storm onset and the strongest earthquake in the footprints of magnetic lines belonging to newly created radiation belts.

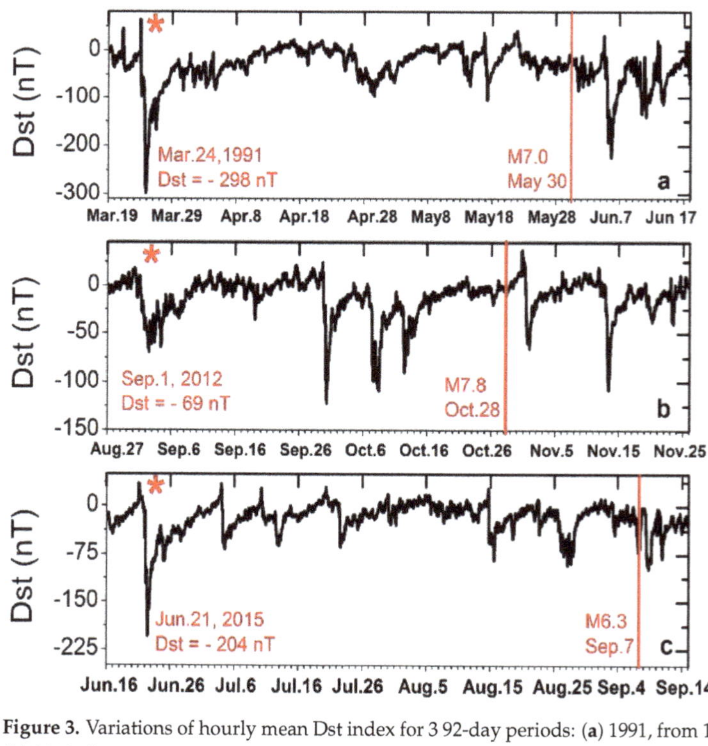

Figure 3. Variations of hourly mean Dst index for 3 92-day periods: (**a**) 1991, from 19 March, 04:30 UT; (**b**) 2012, from 27 August, 22:30 UT; and (**c**) 2015, from 16 June, 18:30 UT. Upper red star marks dates of geomagnetic storm onsets in March 1991, September 2012, and June 2015, and red vertical lines indicate the dates of the strongest earthquakes that occurred in the footprint of magnetic lines belonging to the radiation belts newly created in the inner magnetosphere due to indicated magnetic storms: M7.0 in Alaska on 30 May 1991, near the footprint of L~2.69; M7.8 near Canada on 28 October 2012, near the footprint of L~3.3; and M6.3 in New Zealand on 7 September 2015, near the footprint of L~1.58.

It is seen from Figure 3 that the time delay between storm onset and earthquake occurrence was about 67 days in 1991, 57 days in 2012, and 78 days in 2015; its mean value equals ~67 days. Figure 3b shows that the geomagnetic storm on 1 September 2012 was only moderate, but the strong one occurred on 1 October 2012, with Dst = −122 nT. One may suggest that this stronger storm induced the M7.8 earthquake. As we see it, finding a correlation between a particular magnetic storm and an induced earthquake is only possible under some initial idea. Our idea was to investigate seismicity in the footprint of magnetic lines belonging to the new radiation belts created by magnetic storms in the inner magnetosphere, and this allowed us to obtain similar results for all three considered cases. It was mentioned in [46] that long-living radiation belts are rarely created due to magnetic storms; more often, the newly created belts survive for a few days. The moderate geomagnetic storm on 1 September 2012 (Dst = −69 nT) produced a relatively long-lasting radiation belt [43], but the next strong magnetic storm (Dst = −122) destroyed it [38].

To test whether the relationship of seismic activity with the magnetic storm was significant in the spatial domain (near the footprints of magnetic lines belonging to a new radiation belt created due to the storm), we present Figure 4. It shows the distribution by month in 2012 for the number of earthquakes of M $\geq$ 4.5 that occurred near the footprint of L = 3.0–3.5 (red bars), as well as in neighboring geomagnetic lines L = 2.5–2.99 (green) and L = 3.51–5.0 (blue).

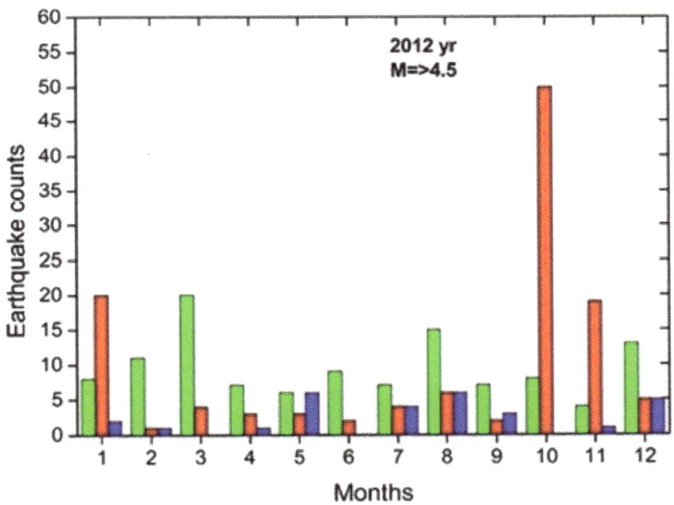

Figure 4. Distribution by month in 2012 for the number of earthquakes of magnitude M $\geq$ 4.5 that occurred at the base of geomagnetic lines L = 2.5–2.99 (green, 115 events), L = 3.0–3.5 (red, around which a belt of high-energy electrons was formed after the geomagnetic storm on 1 September 2012, 119 events), and L = 3.51–5.0 (blue, 29 events).

It is seen from Figure 4 that the distribution of earthquakes by month was uniform at the base of the lower (L = 2.5–2.99) and higher (L = 3.51–5.0) lines. For the base of L = 3.0–3.5 (around which the belt of high-energy particles was formed due to a magnetic storm in September 2012), the number of earthquakes strongly increased ~2 months after the onset of the magnetic storm. On the other hand, we present Figure 4 only as an example to demonstrate the situation for the year 2012. One cannot expect the picture to be the same for all possible cases because there are many reasons for earthquake occurrence. Sometimes, earthquakes may occur at the base of the nearest magnetic lines; these do not relate to the lines of the newly formed radiation belt. This could possibly be the next step in a more detailed investigation.

3.1.2. Case Study in the Year 2003

The deepest, most recent high-energy electron radiation belt (around L~1.5) was detected after a giant magnetic storm at the end of October (beginning of November 2003 (Halloween storm), when the Dst-index reached up to −400 nT [37]. At this time, the Solar Anomalous and Magnetospheric Particle Explorer (SAMPEX) detected multi-MeV electron transport from the outer zone of the radiation belt through the slot region into the inner magnetosphere up to L~1.5 [39,40]. Figure 5 shows the distribution of the flux intensity of relativistic electrons with energies of 2–6 MeV, according to the SAMPEX satellite data from [40] (fragment of their Figure 1).

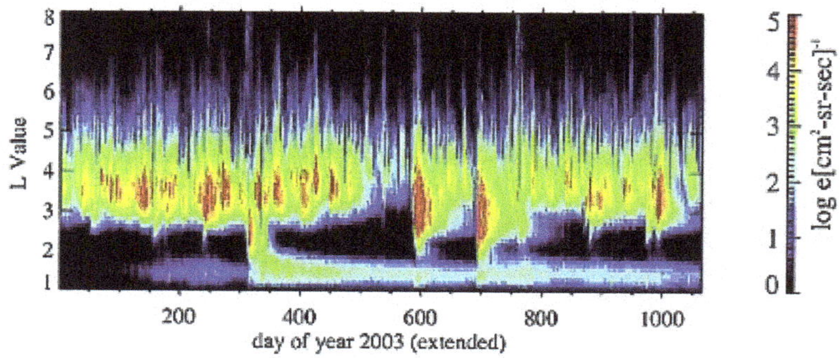

Figure 5. Distribution on geomagnetic lines in 2003–2005 the flux intensity of relativistic electrons with energies of 2–6 MeV according to the SAMPEX satellite data adapted from Ref. [40] (fragment of their Figure 1).

It is seen from Figure 5 that the precipitation of high energy electrons from the outer radiation belt (which has a maximum around L~4.0) started on 31 October (304 days from the beginning of the year 2003) and lasted for about 30 days (during the entirety of November 2003). In this time interval, the relativistic electrons strongly filled the geomagnetic field lines L~2.0–2.5, as shown in Figure 5. Usually, this range of field lines is purely filled by charged particles, provided it belongs to a slot region (L~2.0–3.0, Figure 1). In analogy with the previous cases, which were presented in Figure 2, one may expect an increase in seismic activity in the regions located near the footprint of magnetic lines L~2.0–2.5 some weeks after the onset of the magnetic storm on 31 October 2023, and this does indeed occur. Hence, Figure 6 presents monthly earthquakes of magnitude M $\geq$ 4.5 that occurred in January–December of 2003 near the footprints of L = 2.0–2.5. It is seen that for the base of L = 2.0–2.5, the number of earthquakes strongly increased in November 2003; the largest here was an earthquake with M = 7.8, which occurred at the Rat Islands, Aleutian Islands, Alaska, on 17 November 2003, with coordinates 51.146°N, 178.650°E near the footprint of L~2.1. This earthquake happened ~17 days (over 2 weeks) after the geomagnetic storm onset on 31 October 2003.

The SAMPEX instruments detected a newly formed belt of relativistic electrons in the inner magnetosphere around L~1.5 for ~26 months (from November 2003 to December 2005 [39,40]). It is reasonable to compare the amount of global seismic energy released before and after this 26-month interval. To do this, we used the USGS Seismological Catalog for representative magnitudes M $\geq$ 4.5 (https://earthquake.usgs.gov/earthquakes/search, accessed on 25 October 2018) from July 1999 to April 2010 (65,684 earthquakes). For each of the events, the released seismic energy was estimated using the Gutenberg and Richter relation [47] ($E_s = 10^{(11.8+1.5M)}$), where M is the magnitude, and the Es value is in ergs. Then, the sum of Es in each of the five 26-month intervals was estimated. The intervals were July 1999–August 2001 (10,427 events), September 2001–October 2003 (10,709 events), November 2003–December 2005 (15,106 events), January 2006–February 2008 (15,233 events), and March 2008–April 2010 (14,209 events). In Figure 7, the released seismic energy is given in Joules. It is seen in the 26-months bin

(November 2003–December 2005) (when a belt of relativistic electrons was present in the inner magnetosphere due to the giant storm of 31 October 2003) that the released seismic energy peaked strongly (~3.7×10^{18} Joule) according to results in the 26-month intervals before and after. According to statistics, one earthquake with a magnitude of M8 or more occurs annually. However, when an intense belt of relativistic electrons existed in the lower magnetosphere in November 2003–December 2005 (after a giant magnetic storm in late October 2003), three such events occurred around the globe relatively rapidly (slightly more than 3 months). These were M8.1 (north of Macquarie Island on 23 December 2004), M9.1 (Sumatra, the Andaman Islands on 26 December 2004), and M8.6 (northern Sumatra on 28 March 2005).

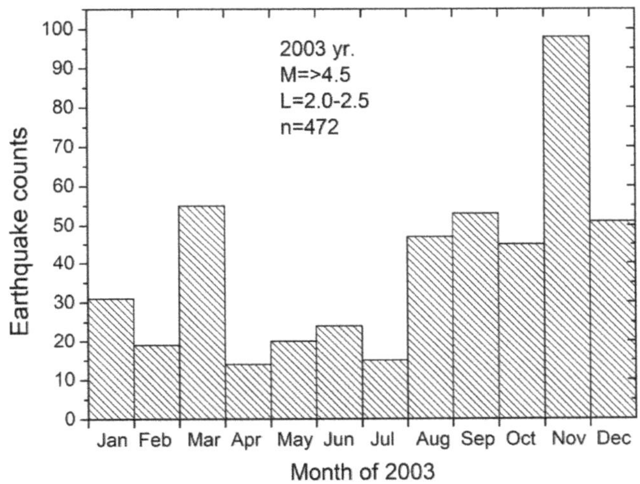

Figure 6. Histograms of the monthly number of earthquakes of M $\geq$ 4.5 occurring near the footprints of geomagnetic lines L = 2.0−2.5 from January to December of 2003.

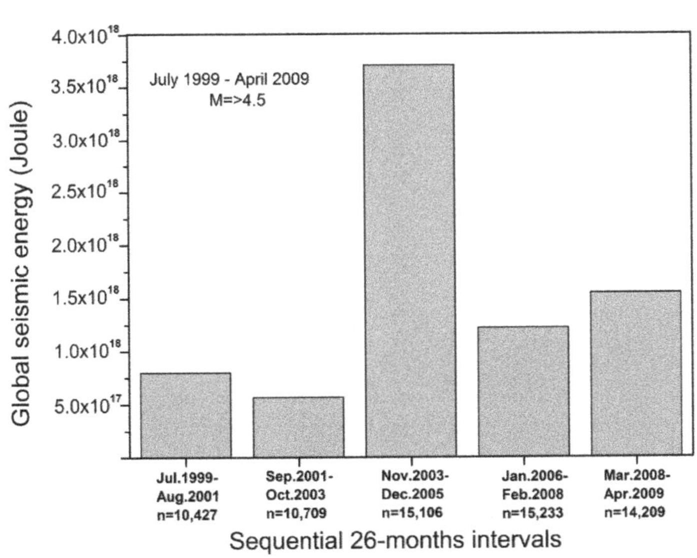

Figure 7. Histograms of the global seismic energy released into the five 26-month intervals: July 1999−August 2001, September 2001−October 2003, November 2003−December 2005, January 2006−February 2008, and March 2008−April 2010, based on the USGS seismological catalog for M $\geq$ 4.5.

The results in Figures 2–7 clearly show that the geomagnetic field lines, when they are filled with high-energy particles, may take part in seismic processes (earthquakes tend to occur near the footprints of these magnetic lines). Here, we need to mention another result showing that the magnetic field lines are somehow involved in the earthquake preparation process [48]. For example, Figure 8 shows the geomagnetic field lines connecting 38 geomagnetically conjugated points at the Earth's surface in the southern and northern hemispheres.

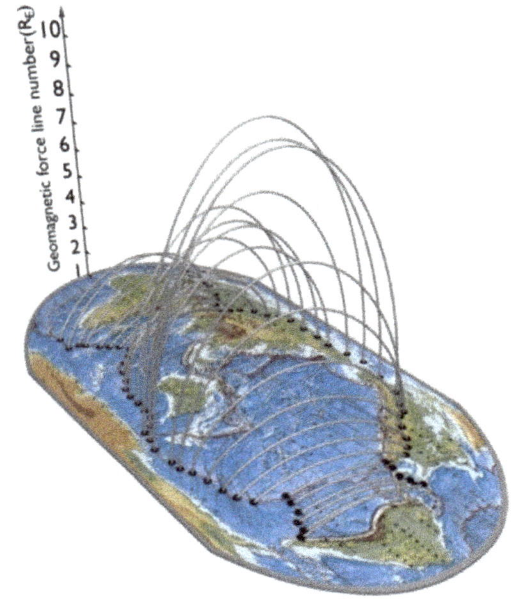

Figure 8. Geomagnetic field lines connecting magnetically conjugated sites, adapted with permission from Ref. [48].

In the southern hemisphere, the coordinates of 38 points were determined, distributed relatively evenly along one of the seismic belts (the boundary of the Antarctic lithospheric plate). By using the GEOPACK computational package [33], we calculated the *L*-value (McIlwain parameter) for each of the 38 points, and the coordinates of a magnetically conjugate point in the northern hemisphere were determined. Figure 8 shows that the Antarctic lithospheric plate border, which is traced by the earthquake epicenters, is in magnetic conjugation with the zone at the junction of orogenic and platform structures in the northern hemisphere. The effect of geomagnetic conjugation is most clearly manifested between the southern boundary of the Nazca tectonic Plate and the northern boundaries of the Cocos and Caribbean tectonic plates. It is well known that many processes in the upper geospheres (ionosphere, magnetosphere) are magnetically conjugated (occur at the same geomagnetic L-line); the same is evidently true for the solid Earth.

3.2. On Temporal Variations of Solar Activity, Galactic Cosmic Rays, and Global Seismicity

The results in Figures 2–7 support the idea that seismic activity increases when the atmosphere conductivity increases. In the considered cases, the precipitation of high-energy electrons from the outer radiation belt due to the magnetic storms increased conductivity at mesosphere heights [45], which is, on average, purely conductive. It was considered above that in a purely conductive stratosphere and troposphere, the galactic cosmic rays (GCRs) are mainly input to ionization and conductivity [23–31]. The flux of GCR increases in a solar minimum [28,29]; thus, one may expect increasing seismic activity in a solar minimum. Nowadays, earthquake rows are insufficient (especially for strong events [49]) to carry out such analysis with a high statistical significance; nevertheless, a tendency to increase seismic activity with decreasing solar activity is evident.

Hence, in [50], an out-of-phase relationship was found between secular variations in solar activity and seismic energy released on the planet in 1690–2002. The same result was obtained in [51,52]. It was shown in [53] that out of the 12 strongest earthquakes (with a magnitude of more than 7.5) that occurred in Japan in 1700–2005, nine events (~70%) were confined to periods of low solar activity when the intensity of galactic cosmic rays was increased.

Figure 9 presents the daily counts of earthquakes with a magnitude of $M \geq 4.5$ that occurred on the Earth from 1 January 1973 to 31 December 2017, estimated based on the USGS Global Seismological Catalog. Over 226,000 earthquakes with a magnitude of $M \geq 4.5$ occurred during these years.

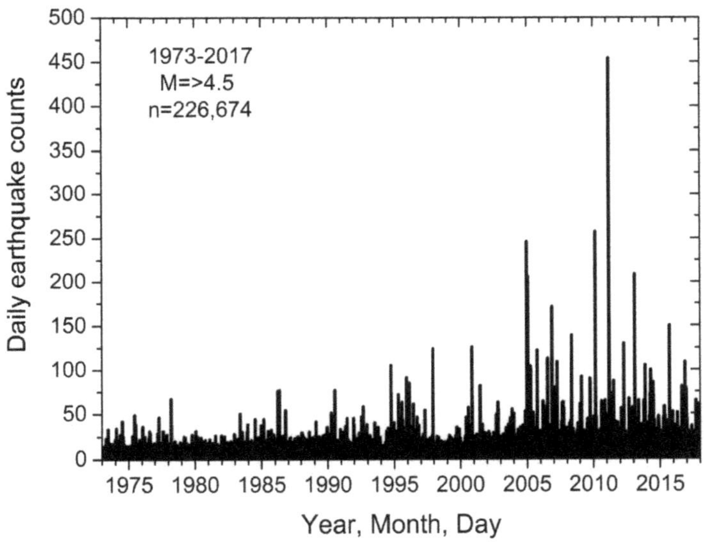

Figure 9. The daily counts of earthquakes with a magnitude of $M \geq 4.5$ occurred on the Earth from 1 January 1973 to 31 December 2017.

Figure 9 shows that the number of seismic events around the globe has shown a sharp increase since the beginning of the 21st century. To underline that this is not a result of increasing global seismic network, which could register more effectively low magnitude earthquakes, we present in Figure 10 daily values of the released at the Earth seismic energy ($\log Es = 11.8 + 1.5\, M$) for the same period, from 1 January 1973 to 31 December 2017. The thick red line is a linear trend, which shows a monotonic, slightly increasing seismic energy from 1973 to 2017; the dashed square marks the cluster of the five strongest earthquakes (magnitudes more than 8.5) that occurred on the globe at the beginning of the 21st century: M9.1 (Sumatra, the Andaman Islands on 26 December 2004), M8.6 (northern Sumatra on 28 March 2005), M 8.8 (Chile on 27 February 2010), M 9.0 (Great Tohoku Earthquake, Japan on 11 March 2011), and M 8.6 (northern Sumatra on 11 April 2012).

Because of the hypothesis that seismic activity increases with increasing galactic cosmic ray (GCR) flux—which increases conductivity in the lower stratosphere and upper troposphere and thus increases the effectivity of the global electric circuit functioning, of which one element is an earthquake [8]—we show in Figure 11 the monthly time variation of galactic cosmic ray count (pink curve) and sunspot number (black curve) for the period 1979–2020, as obtained from [54]. Figure 11 shows out-of-phase variations in galactic cosmic rays and sunspot numbers. The black dashed line in Figure 11 is the time mean value of GCR counts. In contrast, the gold and green dashed lines indicate the one-half standard deviation above and below the mean, defining the maximum and minimum GCR groups, respectively. It can be seen that during the two decades of the 21st century, the intensity of GCR increased compared to previous decades. This supports the suggestion that the functioning of the GEC may have been more effective at the beginning of the 21st century, which could be responsible for a monotonic increase in global seismic activity from 1973 to 2017 (red trend in Figure 10).

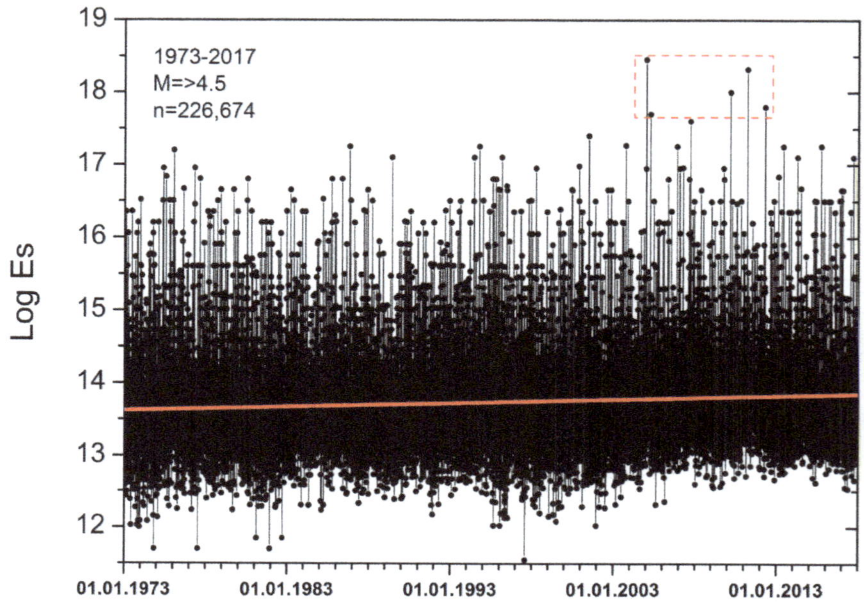

Figure 10. The logarithm of the daily amount of released global seismic energy (Joules) from 1 January 1973 to 31 December 2017 in earthquakes of $M \geq 4.5$; the red line is a linear trend; the dashed square marks the cluster of the five strongest earthquakes (magnitudes more than 8.5).

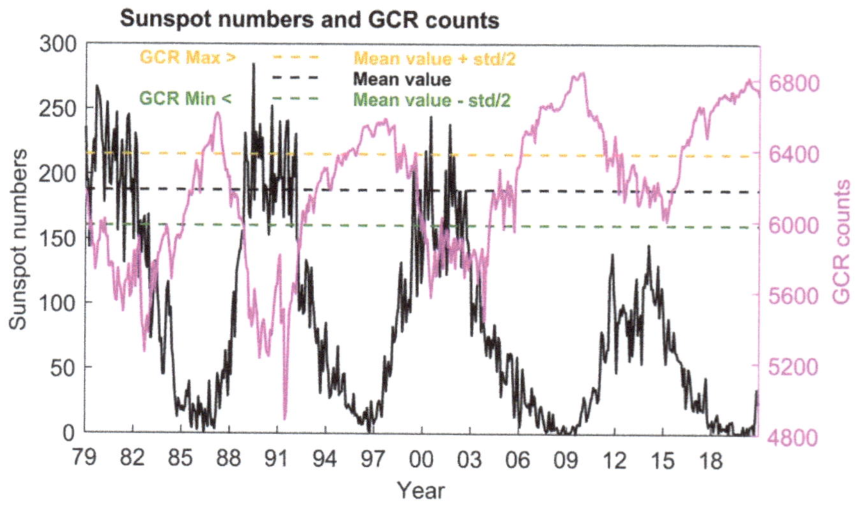

Figure 11. Monthly time variation of sunspot number (black curve) and galactic cosmic ray (GCR) count (pink curve) for the period 1979–2020, adapted from Ref. [54] (fragment of their Figure 1).

4. Discussion

In [4], it was suggested that the Sun may be considered a significant agent provoking earthquakes during substantial increases in solar wind activity. One of Earth's most striking manifestations of solar wind activity is geomagnetic storms. It is shown in many papers that a geomagnetic storm may precede the occurrence of an earthquake. The authors in [55] investigated seismicity variations in the Northern Tien-Shan region from 1975 to 1996, before and after geomagnetic storms with a sudden storm commencement (SSC). They showed that in the territories underlain by rocks with low electrical resistivity, the seismic activity increased on average 2–6 days after the onset of a geomagnetic storm. In [56], it was found for 1994–2017 that 17 earthquakes with magnitudes $M \geq 6.5$ occurred within 2 days after 50 magnetic storms, when Kp-indices were above 7; all these events took place in the eastern hemisphere but were absent in the western hemisphere. In [57], 4666 earthquakes with magnitudes of $Mw \geq 6$ that occurred between 1932 and 2016 in the Pacific Rim region with longitudes of 120°–160°E and 70°–130°W were investigated, and it was concluded that external forces acting at the time of geomagnetic storms might trigger an earthquake with a time delay of ~12–14 days. In [58], the authors investigated the correlation between geomagnetic storms and $Mw \geq 7.0$ global earthquakes from 1957 to 2020; their results gave statistically significant evidence that the probability of geomagnetic storms increased around 26–27 days before earthquakes. In [59], the response of global seismicity to St. Patrick's Day (17 March) geomagnetic storms in 2013 and 2015 was investigated. It was found that after a time delay of ~30 and ~39 days after storm onsets in 2013 and 2015, respectively, strong crust earthquakes occurred in relatively close continental areas: in Iran (M7.7, 16 April 2013) and Nepal (M7.8, 25 April 2015). Besides this, after a time delay of ~68 and ~74 days after storm onsets in 2013 and 2015, respectively, strong deep-focused earthquakes occurred at relatively close ocean areas: beneath the Sea of Okhotsk (M8.3, 24 May 2013, Russia) and beneath the Pacific Ocean (M7.8, 30 May 2015, Japan). The authors of [60] analyzed the magnetic field data from the Swarm three-identical satellite constellation concerning global Mw5.5+ shallow earthquakes from November 2013 to November 2021. They revealed that after the magnetic storm, the anticipation time of large earthquakes (Mw7.5+) may reach up to several years before an event occurrence.

Many years of searching for statistical connections between solar, geomagnetic, and seismic processes finally led to the first mathematical model [61] to consider the hypothesis of electromagnetic earthquakes being triggered by a sharp rise of telluric currents in the lithosphere, including crust faults due to the interaction of solar flare X-ray radiation with the ionosphere–atmosphere–lithosphere system producing an increase in telluric currents in the crust faults. The results of their numerical estimations showed that after an X-class solar flare, the telluric current density in the conductive layer of the lithosphere might be 2–3 orders higher than the average telluric current density in the lithosphere, and that it may be comparable with the parameters of the electric current pulses generated in the lithosphere by artificial pulsed power sources [62]. As they noted, these artificial electrical pulses injected into the Earth's crust in seismic-prone areas resulted in the electromagnetic triggering of weak earthquakes and the spatiotemporal re-distribution of the regional seismicity of Pamir and Northern Tien Shan. Our results in Figures 2–7 show that due to the impact of the geospace environment on the lithosphere, relatively strong earthquakes may occur (magnitude 7^+). Moreover, these earthquakes may look as addressed (targeted) because they occurred exactly near the footprint of geomagnetic lines filled with relativistic electrons precipitating from the outer radiation belt due to geomagnetic storms.

Moreover, the time delay between the onset of a geomagnetic storm and the occurrence of an earthquake may vary from weeks to months. This means the solar wind energy is somehow delivered to some regions of the lithosphere. For several weeks or months, it somehow prepares rock media in those regions for destruction in a kind of earthquake. These results may be helpful for the further development of mathematical models of the solar–terrestrial coupling stated in [61].

Figures 9 and 10 compare temporal variations in global seismicity, sunspot numbers and galactic cosmic rays. Understandably, the rows of seismic and cosmic ray data were relatively short in comparison with sunspot data, which have been known about since 1755; nevertheless, a tendency of increasing global seismic activity with decreasing sunspot numbers and increasing galactic cosmic rays is evident and agrees with the findings of other authors, e.g., [50–53].

The dashed red square in Figure 10 marks the cluster of five mega-earthquakes in 2004–2012. The clustering started shortly after the giant magnetic storm in late October 2003, which provoked the long-lived storage of relativistic electrons forming around L~1.5. This storage persisted for more than 2 years (~26 months). It could have been repeatedly replenished with high-energy charged particles during other geomagnetic storms, as seen in Figure 5. Such an assumption does not seem very irrational because in [60], it was found that the delay time can be as long as several years for very strong earthquakes (Mw7.5+).

It is necessary to mention that after a cluster of strong earthquakes at the beginning of the 21st century (and especially after the M9.0 event in Japan), at the meeting of the Seismological Society of America in 2011, a heated debate ensured regarding whether the observed temporary clustering of strong earthquakes had some physical cause [49,63]. It was noted [49], "If global clustering of earthquakes is so important that it should be taken into account when assessing seismic hazard, then seismological data should reject the "null hypothesis" that the temporal distribution of earthquakes on the planet is described by the distribution Poisson (uniform distribution of random independent events that occur at a constant speed)." However, questions immediately arose about the reliability of the results of such testing conducted on a time-limited sample of sporadic events (which are the strongest earthquakes). In [64], the effectiveness of traditional statistical tests to unambiguously answer the question of the existence or absence of the clustering of earthquakes in catalogs with a small number of events was evaluated. It was concluded that the test results cannot be considered reliable when analyzing a relatively short sample with rare events. Thus, it was concluded [49], "... even if there is a global process leading to the generation and clustering of earthquakes, the length of the instrumental seismological series is currently too small to discover reliably. This situation may change either with the accumulation of seismological data in the future or with the advent of the physically justified hypothesis of earthquake generation, which will positively affect the test results." We can consider a new hypothesis that the Sun could be a significant earthquake-provoking agent [4]. One of these hypotheses [61] suggests that electromagnetic earthquakes are triggered by a sharp rise of telluric currents in the lithosphere due to the interaction of solar flare X-ray radiation with the ionosphere–atmosphere–lithosphere system. Another hypothesis [8,65] suggests that the process of earthquake preparation and realization could be related to the functioning of the global electric circuit (GEC), and earthquake statistics support this suggestion. Hence, one of the main characteristics of the GEC is a unitary variation—a dependence of the fair-weather electric field on universal time, called the Carnegie curve [66,67]—which demonstrates a steady increase of the electric field in fair-weather regions at ~19:00 UT. It was revealed in [68,69] that the global seismic activity also shows a unitary variation, which has a rather strong correlation with the Carnegie curve.

5. Conclusions

We analyzed seismic activity for four well-known cases of the appearance of long-lived (several weeks, months, or years) radiation belts (storage rings) newly created by geomagnetic storms. In addition, some short-lived storage rings (several days' duration) were also generated by each storm. Thus, it seems appropriate to conduct similar studies on short-lived storage rings using data from, for example, NASA's "Van Alllen Probes" satellite (2012–2019), which carried high-tech equipment that did not fail during magnetic storms and intensified solar wind streams. Since it has been discovered that the delay time between a geomagnetic storm and the occurrence of an earthquake can be several weeks or even months, and since an earthquake is an intra-terrestrial phenomenon, it seems appropriate to begin retrospective studies of variations in the parameters of the solid Earth in the vicinity of the epicenters of those (targeted) earthquakes that appeared after geomagnetic storms.

Author Contributions: D.O. and G.K. provided the concepts for the manuscript. G.K. organized and wrote the manuscript. All authors provided critical feedback and helped shape the research, analysis, and manuscript. All authors have read and agreed to the published version of the manuscript.

Funding: This research is partly funded by the Science Committee of the Ministry of Science and Higher Education of the Republic of Kazakhstan under the grant #IRN-AP19677977.

Institutional Review Board Statement: Not applicable.

Informed Consent Statement: Not applicable.

Data Availability Statement: The original contributions presented in the study are included in the article; further inquiries can be directed to the corresponding author.

Acknowledgments: We thank the US Geological Survey and European–Mediterranean Seismological Centre for providing earthquake information services and data. We acknowledge the use of the NASA/GSFC's Space Physics Data Facility's CDAWeb service, OMNIdata. We especially thank the CRRES and Van Allen Probes scientific missions for providing observations of the radiation belt dynamics, and J. B. Blake et al., 1992, [41]; D. N. Baker et al., 2004, [39], 2007 [40]; R. M. Thorne et al., 2013, [43]; and S. G. Claudepierre et al., 2017, [44], whose results we used in our investigations. Lastly, we thank our reviewers.

Conflicts of Interest: The authors declare that the research was conducted without any commercial or financial relationships that could be construed as a potential conflict of interest.

References

1. Guglielmi, A.V.; Klain, B.I. On the influence of the Sun on the seismicity of the Earth. *Sol. Terr. Phys.* **2020**, *6*, 111–115. [CrossRef]
2. Fainberg, E.B.; Avagimov, A.A.; Zeigarnik, V.A.; Vasil'eva, T.A. Generation of heat flows in the Earth's interior by global geomagnetic storms. *Phys. Solid Earth* **2004**, *40*, 315–322. (In Russian)
3. Guglielmi, A.V. Ponderomotor forces in the crust and in the Earth's magnetosphere. *Earth Phys.* **1992**, *7*, 35–39.
4. Anagnostopoulos, G.; Spyroglou, I.; Rigas, A.; Preka-Papadema, P.; Mavromichalaki, H.; Kiosses, I. The Sun as a Significant Agent Provoking Earthquakes. *Eur. Phys. J. Spec. Top.* **2021**, *230*, 287–333. [CrossRef]
5. Guglielmi, A.V.; Zotov, O.D. On geoseismic noise and helioseismic oscillations. *Izvestiya Phys. Solid. Earth* **2021**, *57*, 567–572. [CrossRef]
6. McPherron, R. Magnetic Pulsations: Their Sources and Relation to Solar Wind and Geomagnetic Activity. *Surv. Geophys.* **2005**, *26*, 545–592. [CrossRef]
7. Pulinets, S.A.; Boyarchuk, K. *Ionospheric Precursors of Earthquakes*; Springer: Berlin/Heidelberg, Germany, 2004.
8. Pulinets, S.A. Physical mechanism of the vertical electric field generation over active tectonic faults. *Adv. Space Res.* **2009**, *44*, 767–773. [CrossRef]
9. Pulinets, S.; Ouzounov, D. Lithosphere-Atmosphere-Ionosphere Coupling (LAIC) model—An unified concept for earthquake precursors validation. *J. Asian Earth Sci.* **2011**, *41*, 371–382. [CrossRef]
10. Ouzounov, D.; Pulinets, K.S.; Hattori, P.; Taylor, P. (Eds.) *Pre-Earthquake Processes: A Multi-disciplinary Approach to Earthquake Prediction Studies, American Geophysical Union*; John Wiley & Sons: Hoboken, NJ, USA, 2018; 385p.
11. Pulinets, S.; Ouzounov, D.; Karelin, A.; Davidenko, D. Lithosphere–Atmosphere–Ionosphere–Magnetosphere Coupling—A Concept for Pre-Earthquake Signals Generation. In *Pre Earthquake Processes: A Multidisciplinary Approach to Earthquake Prediction Studies*; John Wiley & Sons: Hoboken, NJ, USA, 2018; pp. 79–98.
12. Roble, R.G.; Tzur, I. The Global Atmospheric Electrical Circuit. In *Study in Geophysics—The Earth's Electrical Environment*; National Academy Press: Washington, DC, USA, 1986; pp. 206–231.
13. Roble, R.G. On Modeling Component Processes in the Earth's Global Electric Circuit. *J. Atmos. Terr. Phys.* **1991**, *53*, 831–847. [CrossRef]
14. Schery, S.D.; Huang, S. An estimate of the global distribution of radon emissions from the ocean. *Geophys. Res. Lett.* **2004**, *31*, L19104. [CrossRef]
15. Schery, S.D.; Wasiolek, M.A. Modeling ^{222}Rn flux from the Earth's surface. In *Radon and Thoron in the Human Environment*; Katase, A., Shimo, M., Eds.; World Scientific River Edge: River Edge, NJ, USA, 1998; pp. 207–217.
16. Chandrashekara, M.S.; Sannappa, J.; Paramesh, L. Studies on atmospheric electrical conductivity related to radon and its progeny concentrations in the lower atmosphere at Mysore. *Atmos. Envir.* **2006**, *40*, 87–95. [CrossRef]

17. Kobeissi, M.A.; Gomez, F.; Tabet, C. Measurement of anomalous radon gas emanation across the Yammouneh Fault in southern Lebanon: A possible approach to earthquake prediction. *Int. J. Disaster Risk Sci.* **2015**, *6*, 250–266. [CrossRef]
18. Harrison, R.; Aplin, K.; Rycroft, M. Atmospheric Electricity Coupling Between Earthquake Regions and the Ionosphere. *J. Atmos. Sol. Terr. Phys.* **2010**, *72*, 376–381. [CrossRef]
19. Rycroft, M.J.; Harrison, G. Electromagnetic Atmosphere-Plasma Coupling: The Global Atmospheric Electric Circuit. *Space Sci. Rev.* **2012**, *168*, 363–384. [CrossRef]
20. Němec, F.; Santolík, O.; Parrot, M. Decrease of Intensity of ELF/VLF Waves Observed in the Upper Ionosphere Close to Earthquakes: A Statistical Study. *J. Geophys. Res.* **2009**, *114*, A04303. [CrossRef]
21. Bering, E.A.; Few, A.A.; Benbrook, J.R. The global electric circuit. *Phys. Today* **1998**, *51*, 24–30. [CrossRef]
22. Makarova, L.N.; Shirochkov, A.V. A New Approach to the Global Electric Circuit Conception. 1998. Available online: http://www.sgo.fi/SPECIAL/Contributions/Makarova.pdf (accessed on 10 November 2021).
23. Bazilevskaya, G.A.; Usoskin, I.G.; Flückiger, E.O.; Harrison, R.G.; Desorgher, L.; Bütikofer, R.; Krainev, M.B.; Makhmutov, V.S.; Stozhkov, Y.I.; Svirzhevskaya, A.K.; et al. Cosmic ray induced ion production in the atmosphere. *Space Sci. Rev.* **2008**, *137*, 149–173. [CrossRef]
24. Phillips, T.; Johnson, S.; Koske-Phillips, A.; White, M.; Yarborough, A.; Lamb, A.; Schultz, J. Space weather ballooning. *Space Weather* **2016**, *14*, 697–703. [CrossRef]
25. Frohlich, C.; Lean, J. Total solar irradiance variation in new eyes to see inside the sun and stars. In Proceedings of the IAU Symposium, Kyoto, Japan, 26–30 August 1997; p. 185.
26. Markson, R. Modulation of the Earth's electric field by cosmic radiation. *Nature* **1981**, *291*, 304–308. [CrossRef]
27. Harrison, R.G. The Global Atmospheric Electrical Circuit and Climate. *Surv. Geophys.* **2004**, *25*, 441–484. [CrossRef]
28. Carslaw, K.S.; Harrison, R.G.; Kirkby, J. Cosmic rays, clouds, and climate. *Science* **2002**, *298*, 1732–1737. [CrossRef] [PubMed]
29. Kirkby, J. Cosmic Rays and Climate. *SurvGeophys* **2007**, *28*, 333–375. [CrossRef]
30. Usoskin, I.G.; Desorgher, L.; Velinov, P.; Storini, M.; Flückiger, E.O.; Bütikofer, R.; Kovaltsov, G.A. Ionization of the Earth's atmosphere by solar and galactic cosmic rays. *Acta Geophys.* **2009**, *57*, 88–101. [CrossRef]
31. Vainio, R.; Desorgher, L.; Heynderickx, D.; Storini, M.; Flückiger, E.; Horne, R.B.; Kovaltsov, G.A.; Kudela, K.; Laurenza, M.; McKenna-Lawlor, S.; et al. Dynamics of the Earth's particle radiation environment. *Space Sci. Rev.* **2009**, *147*, 187–231. [CrossRef]
32. McIlwain, C.E. Coordinates for mapping the distribution of geomagnetically trapped particles. *J. Geophys. Res.* **1961**, *66*, 3681–3691. [CrossRef]
33. Tsyganenko, N.A. Geopack: A Set of Fortran Subroutines for Computations of the Geomagnetic Field in the Earth's Magnetosphere. 2008. Available online: https://geo.phys.spbu.ru/~tsyganenko/Geopack-2008.html (accessed on 1 October 2012).
34. Gonzalez, W.D.; Joselyn, J.A.; Kamide, Y.; Kroehl, H.W.; Rostoker, G.; Tsurutani, B.T.; Vasyliunas, V.M. What is a Geomagnetic Storm? *J. Geophys. Res.* **1994**, *99*, 5771–5792. [CrossRef]
35. Loewe, C.A.; Prölss, G.W. Classification and mean behavior of magnetic storms. *J. Geophys. Res. Space Phys.* **1997**, *102*, 14209–14213. [CrossRef]
36. Van Allen, J.A.; Ludwig, G.H.; Ray, E.C.; McIlwain, C.E. Observation of high intensity radiation by satellites 1958 alpha and gamma. *J. Jet Propuls.* **1958**, *28*, 588–592. [CrossRef]
37. Baker, D.N.; Erickson, P.J.; Fennell, J.F.; Foster, J.C.; Jaynes, A.N.; Verronen, P.T. Space Weather Effects in the Earth's Radiation Belts. *Space Sci. Rev.* **2018**, *214*, 17. [CrossRef]
38. Baker, D.N.; Kanekal, S.G.; Hoxie, V.C.; Henderson, M.G.; Li, X.; Spence, H.E.; Elkington, S.R.; Friedel, R.H.W.; Goldstein, J.; Hudson, M.K.; et al. A long-lived relativistic electron storage ring embedded in Earth's outer Van Allen belt. *Science* **2013**, *340*, 186–190. [CrossRef]
39. Baker, D.N.; Kanekal, S.G.; Li, X.; Monk, S.P.; Goldstein, J.; Burch, J.L. An extreme distortion of the Van Allen belt arising from the Halloween solar storm in 2003. *Nature* **2004**, *432*, 878–881. [CrossRef] [PubMed]
40. Baker, D.N.; Kanekal, S.G.; Horne, R.B.; Meredith, N.P.; Glauert, S.A. Low-altitude measurements of 2–6 MeV electron trapping lifetimes at $1.5 \leq L \leq 2.5$. *Geophys. Res. Lett.* **2007**, *34*, L20110. [CrossRef]
41. Blake, J.B.; Kolasinski, W.A.; Fillius, R.W.; Mullen, E.G. Injection of electrons and protons with energies of tens of MeV into $L < 3$ on March 24 1991. *Geophys. Res. Lett.* **1992**, *19*, 821–824. [CrossRef]
42. Galper, A.M. Earth radiation belt. *Soros Educ. J.* **1999**, 75–81. Available online: http://cosmic-rays.ru/Resource/9906_075.pdf (accessed on 10 November 2021).
43. Thorne, R.M.; Li, W.; Ni, B.; Ma, Q.; Bortnik, J.; Baker, D.N.; Spence, H.E.; Reeves, G.D.; Henderson, M.G.; Kletzing, C.A.; et al. Evolution and slow decay of an unusual narrow ring of relativistic electrons near L~3.2 following the September 2012 magnetic storm. *Geophys. Res. Lett.* **2013**, *40*, 3507–3511. [CrossRef]
44. Claudepierre, S.G.; O'Brien, T.P.; Fennell, J.F.; Blake, J.B.; Clemmons, J.H.; Looper, M.D.; Mazur, J.E.; Roeder, J.L.; Turner, D.L.; Reeves, G.D.; et al. The hidden dynamics of relativistic electrons (0.7-1.5 MeV) in the inner zone and slot region. *J. Geophys. Res. Space Phys.* **2017**, *122*, 3127–3144. [CrossRef]
45. Kavanagh, A.J.; Cobbett, N.; Kirsch, P. Radiation Belt Slot Region Filling Events: Sustained Energetic Precipitation Into the Mesosphere. *J. Geophys. Res. Space Phys.* **2018**, *123*, 7999–8020. [CrossRef]
46. Li, W.; Hudson, M.K. Earth's Van Allen Radiation Belts: From Discovery to the Van Allen Probes Era. *J. Geophys. Res. Space Phys.* **2019**, *124*, 8319–8351. [CrossRef]

47. Gutenberg, B.; Richter, C. Magnitude and energy of earthquakes. *Nature* **1955**, *176*, 795. [CrossRef]
48. Khachikyan, G.; Zhakupov, N.; Kadyrkhanova, N. Geomagnetic conjugacy of modern tectonic structures. *Geodyn. Tectonophys.* **2013**, *4*, 187–195. [CrossRef]
49. Michael, A.J. Random variability explains apparent global clustering of large earthquakes. *Geophys. Res. Lett.* **2011**, *38*, L21301. [CrossRef]
50. Shestopalov, I.P.; Kharin, E.P. Secular variations of solar activity and seismicity of the Earth. *Geophys. J.* **2006**, *28*, 59–70.
51. Zhang, G.Q. Relationship between global seismicity and solar activities. *Acta Seismol. Sin.* **1998**, *11*, 495–500.
52. Huzaimy, J.M.; Yumoto, K. Possible correlation between solar activity and global seismicity. In Proceeding of the 2011 IEEE International Conference on Space Science and Communication (IconSpace), Penang, Malaysia, 12–13 July 2011; pp. 138–141.
53. Ebisuzaki, T.; Miyahara, H.; Katoaka, R.; Sato, T.; Ishimine, Y. Explosive volcanic eruptions triggered by cosmic rays: Volcano as a bubble chamber. *Gondwana Res.* **2011**, *19*, 1054–1061. [CrossRef]
54. Kumar, V.; Dhaka, S.K.; Hitchman, M.H.; Yoden, S. The influence of solar-modulated regional circulations and galactic cosmic rays on global cloud distribution. *Sci. Rep.* **2023**, *13*, 3707. [CrossRef]
55. Sobolev, G.A.; Zakrzhevskaya, N.A.; Kharin, E.P. On the relation between seismicity and magnetic storms. *Phys. Solid. Earth.* **2001**, *37*, 917–927.
56. Sobolev, G.A. The effect of strong magnetic storms on the occurrence of large earthquakes. *Izv. Phys. Solid. Arth.* **2021**, *57*, 20–36. [CrossRef]
57. Urata, N.; Duma, G.; Freund, F. Geomagnetic Kp Index and Earthquakes. *Open J. Earthq. Res.* **2018**, *7*, 39–52. [CrossRef]
58. Chen, H.; Wang, R.; Miao, M.; Liu, X.; Ma, Y.; Hattori, K.; Han, P. A Statistical Study of the Correlation between Geomagnetic Storms and M $\geq$ 7.0 Global Earthquakes during 1957–2020. *Entropy* **2020**, *22*, 1270. [CrossRef]
59. Ouzounov, D.; Khachikyan, G. A similarity in patterns of global seismicity after St. Patrick's Day geomagnetic storms of 2013 and 2015. *arXiv* **2023**, arXiv:2302.14173. [CrossRef]
60. Marchetti, D.; De Santis, A.; Campuzano, S.A.; Zhu, K.; Soldani, M.; D'arcangelo, S.; Orlando, M.; Wang, T.; Cianchini, G.; Di Mauro, D.; et al. Worldwide Statistical Correlation of Eight Years of Swarm Satellite Data with M5.5+ Earthquakes: New Hints about the Preseismic Phenomena from Space. *Remote Sens.* **2022**, *14*, 2649. [CrossRef]
61. Sorokin, V.; Yaschenko, A.; Mushkarev, G.; Novikov, V. Telluric Currents Generated by Solar Flare Radiation: Physical Model and Numerical Estimations. *Atmosphere* **2023**, *14*, 458. [CrossRef]
62. Zeigarnik, V.A.; Bogomolov, L.M.; Novikov, V.A. Electromagnetic Earthquake Triggering: Field Observations, Laboratory Experiments, and Physical Mechanisms—A Review. *Izv. Phys. Solid. Earth* **2022**, *58*, 30–58. [CrossRef]
63. Bufe, C.G.; Perkins, D. The 2011 Tohoku earthquake: Resumption of temporal clustering of Earth's megaquakes. *Seismol. Res. Lett.* **2011**, *82*, 455.
64. Ammon, C.J.; Aster, R.C.; Lay, T.; Simpson, D.W. The Tohoku earthquake and a 110-year spatio-temporal record of global seismic strain release. *Seismol. Res. Lett.* **2011**, *82*, 454.
65. Namgaladze, A.A. Earthquakes, and global electrical circuit. *Russ. J. Phys. Chem.* **2013**, *7*, 589–593. [CrossRef]
66. Whipple, F.J.W. On the association of the diurnal variation of electric potential gradient in fine weather with the distribution of thunderstorms over the globe. *Q. J. R. Meteorol. Soc.* **1929**, *55*, 1–18. [CrossRef]
67. Harrison, R.G. The Carnegie Curve. *Surv. Geophys.* **2013**, *34*, 209–232. [CrossRef]
68. Pulinets, S.A.; Khachikyan, G.Y. Unitary Variation in the Seismic Regime of the Earth: Carnegie-Curve Matching. *Geomagn. Aeron.* **2020**, *60*, 787–792. [CrossRef]
69. Pulinets, S.; Khachikyan, G. The Global Electric Circuit and Global Seismicity. *Geosciences* **2021**, *11*, 491. [CrossRef]

remote sensing

MDPI

Article

Worldwide Statistical Correlation of Eight Years of *Swarm* Satellite Data with M5.5+ Earthquakes: New Hints about the Preseismic Phenomena from Space

Dedalo Marchetti [1], Angelo De Santis [2], Saioa A. Campuzano [3], Kaiguang Zhu [1,*], Maurizio Soldani [2], Serena D'Arcangelo [2,4], Martina Orlando [2], Ting Wang [1], Gianfranco Cianchini [2], Domenico Di Mauro [2], Alessandro Ippolito [2], Adriano Nardi [2], Dario Sabbagh [2], Wenqi Chen [1], Xiaodan He [1], Xuhui Shen [5], Jiami Wen [1], Donghua Zhang [1], Hanshuo Zhang [1], Yiqun Zhang [1] and Zhima Zeren [5]

1 College of Instrumentation and Electrical Engineering, Jilin University, Changchun 130061, China; dedalomarchetti@jlu.edu.cn (D.M.); tingwang20@mails.jlu.edu.cn (T.W.); wqchen21@mails.jlu.edu.cn (W.C.); hexd19@mails.jlu.edu.cn (X.H.); wenjm20@mails.jlu.edu.cn (J.W.); zhangdh6518@mails.jlu.edu.cn (D.Z.); zhanghs6518@mails.jlu.edu.cn (H.Z.); yiqun21@mails.jlu.edu.cn (Y.Z.)

2 Istituto Nazionale di Geofisica e Vulcanologia, 00143 Rome, Italy; angelo.desantis@ingv.it (A.D.S.); maurizio.soldani@ingv.it (M.S.); serena.darcangelo@ingv.it (S.D.); martina.orlando@ingv.it (M.O.); gianfranco.cianchini@ingv.it (G.C.); domenico.dimauro@ingv.it (D.D.M.); alessandro.ippolito@ingv.it (A.I.); adriano.nardi@ingv.it (A.N.); dario.sabbagh@ingv.it (D.S.)

3 Instituto de Geociencias IGEO (CSIC-UCM), 28040 Madrid, Spain; saioa.arquero@igeo.ucm-csic.es

4 Facultad de Ciencias Físicas, Universidad Complutense de Madrid (UCM), 28040 Madrid, Spain

5 Space Observation Research Center, National Institute of Natural Hazards, MEMC, Beijing 100085, China; xuhuishen@ninhm.ac.cn (X.S.); zerenzhima@ninhm.ac.cn (Z.Z.)

* Correspondence: zhukaiguang@jlu.edu.cn

Citation: Marchetti, D.; De Santis, A.; Campuzano, S.A.; Zhu, K.; Soldani, M.; D'Arcangelo, S.; Orlando, M.; Wang, T.; Cianchini, G.; Di Mauro, D.; et al. Worldwide Statistical Correlation of Eight Years of *Swarm* Satellite Data with M5.5+ Earthquakes: New Hints about the Preseismic Phenomena from Space. *Remote Sens.* **2022**, *14*, 2649. https://doi.org/10.3390/rs14112649

Academic Editor: Michael E. Gorbunov

Received: 7 May 2022
Accepted: 30 May 2022
Published: 1 June 2022

Abstract: Nowadays, the possibility that medium-large earthquakes could produce some electromagnetic ionospheric disturbances during their preparatory phase is controversial in the scientific community. Some previous works using satellite data from DEMETER, *Swarm* and, recently, CSES provided several pieces of evidence supporting the existence of such precursory phenomena in terms of single case studies and statical analyses. In this work, we applied a Worldwide Statistical Correlation approach to **M5.5+ shallow earthquakes** using the first **8 years of *Swarm*** (i.e., from November 2013 to November 2021) **magnetic field and electron density signals** in order to improve the significance of previous statistical studies and provide some new results on how earthquake features could influence ionospheric electromagnetic disturbances. We implemented new methodologies based on the hypothesis that the anticipation time of anomalies of larger earthquakes is usually longer than that of anomalies of smaller magnitude. We also considered the **signal's frequency** to introduce a new identification criterion for the anomalies. We find that taking into account the frequency can improve the statistical significance (up to 25% for magnetic data and up to 100% for electron density). Furthermore, we noted that the **frequency** of the *Swarm* **magnetic field signal** of possible precursor anomalies seems to slightly **increase as the earthquake is approaching**. Finally, we checked a possible relationship between the frequency of the detected anomalies and earthquake features. The earthquake focal mechanism seems to have a low or null influence on the frequency of the detected anomalies, while the epicenter location appears to play an important role. In fact, **land** earthquakes are more likely to be preceded by **slower** (lower frequency) magnetic field signals, whereas **sea** seismic events show a higher probability of being preceded by **faster** (higher frequency) magnetic field signals.

Keywords: electromagnetic; frequency; *Swarm*; earthquake; precursors; ionosphere

1. Introduction

In this paper, we will deal with the delicate research on the possible precursors of earthquakes, i.e., those anomalous phenomena that might appear before earthquakes because they are caused by their preparation phase and thus anticipate their imminent occurrence. The earthquake precursors could be simply classified as seismic precursors and non-seismic precursors: the former are eventual modifications of seismicity or seismological parameters before the occurrence of an earthquake, while the latter, the object of our present research, are alterations of some geophysical parameters, such as atmospheric chemical and physical characteristics, radon emission, and, furthermore, possible electromagnetic disturbances up to the ionosphere [1–3]. Historically, some evidence of electromagnetic precursors has been reported in the data from ground magnetic observatories—for example, before the M6.9 Spitak (Armenia) 1988 and M7.1 Loma-Prieta 1989 earthquakes by Fraser-Smith et al. [4] and Molchanov et al. [5]. A model to explain the generation of Ultra Low Frequency (ULF) electromagnetic waves responsible for such pre-earthquake disturbance was proposed by Molchanov and Hayakawa [6] and based on electromagnetic pulses produced by the separation of electric charges at the fault level due to micro-fracturing activity during the preparation phase of the earthquake. Variations of this explanation have been proposed; e.g., [7], introducing a better model for microfracturing rate variation on the time and improving the geometry of the fault. However, one of the most important theories was put forward by Freund [8,9], which proposed the generation of positive holes (called p-holes) due to the peroxy defects on the rocks also induced by stress on the rock. Freund proposed such theory as a "universal" explanation of all other precursors: for example, he proposed that the detected Thermal-InfraRed anomalies could be explained by photon emission in such wavelengths due to the recombination at the Earth's surface of the peroxy bonds, as well as their location atop topographical profiles by the electrical migration of p-holes in the lithosphere due to the "tip electric effect" [9]. Even if p-holes are a different source with respect to the one initially proposed, the propagation mechanism of the eventual pre-earthquake ULF waves could, in principle, remain unchanged. Other theories have also been developed and warrant consideration—for example, supposing that the release of p-holes at the Earth's surface can produce an accumulation of electric charges that could gradually create a perturbation in the atmosphere and then propagate up to the ionosphere in the form of plasma bubble, interacting and following the magnetic field, as numerically simulated by Kuo et al. [10,11]. Such simulation opened a scientific discussion, as Prokhorov and Zolotov [12] found that some assumptions in the model were too simplified, and Kuo and Lee replied with an update of the original work, showing that the model seems, in any case, valid [13]. Furthermore, Denisenko et al. [14,15] found some possible inaccuracies in the model of Kuo et al., proposing another model that improved the calculus of the conductivity in the ionosphere, but the value on the ground of the electric field strength signal necessary to create a small perturbation of the ionosphere was found to be very large, even if it is still possible. As stated by the same authors of all these papers, more pieces of evidence are necessary to understand if such phenomena exist and which could be the most reliable model to describe them. This paper aims to provide some further observational points for such discussion.

According to Pulinets and Ouzounov [16], the ionization of the air could be also produced by the release of radon instead of (or in addition to) the p-holes. Additionally, other mechanisms for electric (or electromagnetic) disturbances have been proposed based on the change of the resistivity of rocks caused by the movements of fluids (the highly conductive water that fills the new cracks) in the crust before the occurrence of the earthquake under the so-called "Dilatancy Model" by Scholz et al. [17]. For instance, the movement of fluids in Central Italy prior to the occurrence of significant earthquakes has been detected from seismological analyses [18]. Changes in resistivity have been supposed and calculated to describe the atmospheric electric and magnetic field anomalies detected by a ground antenna [19]. Furthermore, ground and atmospheric observations, together with *Swarm* magnetic field satellite anomalous data, have been proposed as good candidates to describe a pre-earthquake chain of anomalies [20]. A comprehensive review of several models of the Lithosphere–Atmosphere–Ionosphere (or, in some cases, just the Lithosphere–Ionosphere) Coupling models has been provided by Liperovsky et al. [21], where other possible physical couplings are also proposed such as pre-earthquake acoustic gravity waves, atmospheric electric currents possibly generated by the release of charged aerosol, variations in the D and E layers in the ionosphere induced by radon release, a direct resonator between the lithosphere and the ionosphere that is supposed to form a spherical capacitor in this model, and even more challenging possible explanations. Previous statistical works have provided pieces of evidence of the correlation of the ionospheric disturbances in terms of electron density measured by the DEMETER satellite, finding a significant increase from 10 to 6 days prior to the seismic M4.8+ events [22–24], and an increase in the electric field (typically around 10/20 mV/m) from a few minutes to some days prior to several M5+ earthquakes was reported by Zolotov [25] using DEMETER and INTERCOSMOS-BULGARIA-1300 satellites. Previous studies by some of the authors of this paper analyzed the *Swarm* and China Seismo Electromagnetic Satellite (CSES) data prior to the occurrence of about 20 earthquakes in recent years with a magnitude "M" in the range $6.0 \leq M \leq 8.3$ [26–28]. The worldwide systematic association of detected ionospheric disturbances in the magnetic field and electron density with earthquakes has been statistically proved by De Santis et al. [29], and a preliminary investigation of CSES-01 electron density data has also shown promising results [30].

The present work is the natural extension of the first statistical and systematic analysis of correlation using satellite magnetic field data from *Swarm* satellites with global M5.5+ shallow earthquakes [29]. Here, a temporal extension of the analysis is provided in terms of analyzed time (i.e., 8 years instead of 4.7 years) and the preparation time before the earthquake (from 1000 days before, instead of 500 days before). Furthermore, some new methodologies are also provided to corroborate the results.

In Section 2, we provide a brief introduction about the investigated dataset and the data analysis method. Section 3 illustrates the results, which are deeply discussed in Section 4. In Section 5, we present some conclusions and future perspectives.

2. Materials and Methods

2.1. Investigated Datasets

This study investigates satellite datasets from the *Swarm* satellite mission and earthquakes from the United States Geological Survey (USGS) catalog, as described in the following sections.

2.1.1. Satellite Datasets

In this paper, we analyze the ionospheric disturbances retrieved from satellite in situ measurements possibly related to earthquakes. In particular, we analyze data from the *Swarm* three-identical satellite constellation launched by the European Space Agency (ESA) on 22 November 2013. At the present time (April 2022), all three satellites are still in orbit, with most instruments in good condition. The *Swarm* constellation is the "state of the art" to monitor with the best accuracy the spatial structure and the temporal evolution of the Earth's magnetic field from space [31]. All three *Swarm* satellites include identical payloads onboard. In particular, in this work, the Y-East component of the magnetic field and electron density (*Ne*) are analyzed as originally measured by fluxgate magnetometers (Vector Field Magnetometer, VFM) and Langmuir probes (Electric Field Instrument, EFI), respectively. The Y-East magnetic field component has been selected, as it is expected to be more sensitive to the internal perturbations [32] and, in the case of Lithosphere–Atmosphere–Ionosphere Coupling (LAIC) coupling, to the Field Aligned Currents (FAC) induced by air ionization from radon release or p-holes. At the ionospheric heights, these currents are expected to be almost horizontal, while at the ground/atmospheric level they are expected to be almost vertical [21]. Among the two X-North and Y-East horizontal components, the Y-East component should be more sensitive to such kinds of perturbations, as it is quasi-orthogonal to the main geomagnetic field generated by the Earth's outer core, so we expect that the same absolute perturbation is more evident on the Y-East component. The used magnetic field data are at "Low-Resolution" and are provided at a sample rate of 1 Hz in a daily file that contains 86,400 samples associated with 31 or 32 tracks (i.e., half orbits) per day. *Swarm Ne* is sampled at 2 Hz and is provided in the Electric Field Instrument (EFI) product. There is an open debate on the discrepancies in the *Ne* absolute value calibration measured with another satellite in a similar orbit, the China Seismo-Electromagnetic Satellite (CSES). For instance, it is known that the absolute value of electron density measured by CSES-01 (i.e., the first satellite of the CSES series) is lower than the ones estimated by *Swarm* [33] and by the IRI-2016 ionospheric model (e.g., comparison in the Indonesia area in [34]). However, such an absolute difference (bias) does not affect our results because our algorithms work on residuals, i.e., they are based on variations measured along the track, which are well consistent between the two satellite missions [33,34].

This paper analyzes the nighttime and daytime tracks for both the magnetic field (B) and electron density (*Ne*) datasets. In fact, as the investigation approach is track-by-track, we do not expect that the solar-driven daytime activity can affect our result in terms of "false anomalies", even if some potential pre-earthquake anomalies could be lost inside the higher day noise (especially for *Ne*).

2.1.2. Earthquake Catalog Acquisition and Pre-Processing

We retrieved the M5+ global earthquakes (EQs) that occurred in the same analyzed time interval from the USGS global catalog, which provides an estimation of the hypocenter, origin time, and magnitude (generally, the moment magnitude Mw calculated from the centroid moment tensor inversion, or, in alternative, the short-period body wave magnitude mb or other values). We then applied a declustering technique using Reasenberg's [35] method to remove aftershocks and foreshocks. With respect to the previous work of De Santis et al. [29], here, a slightly larger time window is utilized—10 days before and 20 days after the events—to search for other possible related events of the same cluster, while the other parameters were unchanged (i.e., a radius of 10 km, magnitude cut-off of 5.0, confidence level of 95%, hypocenter uncertainty according to the catalog). For further details on the declustering of the catalog and how this impacts this work, the reader is invited to see Appendix A. In the declustered catalog, only the shallow (depth $\leq$ 50 km) EQs with a magnitude greater than or equal to 5.5 are selected as the seismic dataset for further processing. In the first 8 years of *Swarm*, slightly fewer than 2200 shallow M5.5+ earthquakes were finally selected.

2.2. Methods of Analysis

Firstly, the *Swarm* satellite data are pre-processed by MAgnetic *Swarm* anomaly detection by Spline analysis (MASS) or electron density (*Ne*) Anomaly Detection (NeAD) algorithms to search for anomalies. The approach applied to the magnetic field and electron density data is basically the same, and only some parameters are settled differently to consider the different sample frequencies and signal-to-noise ratios properly. MASS and NeAD perform the following operations while taking into account the data along the track between $-50°$ and $+50°$ geomagnetic latitude (see Figure S1 in the Supplementary Materials as an example of data processing):

1. First differences of the data: the difference between two consecutive data divided by the time interval between them (as a first order approximation of the temporal derivative).
2. Removal of the residual trend: a cubic spline coming from the fitting of the derivative of the data (as obtained from Step 1) is removed.
3. Analysis of the residuals obtained from Step 1 and Step 2 along the track: moving windows of 7° latitude are used, with an incremental shift of 1/5 of the window length (i.e., ~1.4° of latitude).
4. The root mean square (rms) is calculated inside the window and compared with the whole track's Root Mean Square (RMS). The anomaly is defined if $rms > k_t \cdot RMS$ (k_t is chosen as 2.5 or 3 if no-frequency investigation is performed; more details are given later on).

In addition, a new frequency analysis is carried out, and the Fast Fourier Transform (FFT) is calculated for each moving window and for the whole spectrum. Three sub frequency bands are divided according to their period ranges: (2–10) s, (10–25) s, and (25–50) s. In each frequency band, the anomaly is defined if the mean intensity overpasses the mean *FFT* intensity plus K_{FFT} times its standard deviation σ (for example, for the 2–10 s band, the anomaly is defined if $FFT(2s \leq p \leq 10s) > FFT + K_{FFT} \cdot \sigma(FFT)$. This criterion on K_{FFT} is in addition to Step 4 about k_t, but it is not applied for the basic anomaly extraction (i.e., when no frequency analysis is made). We recognize that the residuals with respect to the spectral mean are not Gaussian, but it is feasible to introduce the standard deviation simply as an operational measure of the data dispersion around the mean of the spectrum. Even if other quantities (e.g., the absolute mean deviation or the largest deviation) could be used, we found that the results are practically the same, so we preferred to use the standard deviation that can be easily and quickly estimated.

The two datasets (satellite anomalies and earthquakes) are investigated by the World wide Statistical Correlation (WSC) algorithm fully described in [29] that is only briefly summarized here.

The WSC algorithm first extracts the subset of anomalies and earthquakes to be correlated by a Superposed Epoch and Space Approach (SESA). The anomalies are selected by applying several thresholds. In particular, we only select tracks in geomagnetic quiet time (i.e., when the geomagnetic indices satisfy the following conditions: $|Dst| \leq 20$ nT and $a_p \leq 10$ nT), with all samples acquired in nominal satellite conditions (checked by the "Quality Flags" of the *Swarm* satellite, according to the respective operating manuals) and put conditions over the rms of the window and, for some analyses, an additional threshold on the frequency. For each earthquake of magnitude *M*, the anomalies inside its preparation area, as defined by a circular area with a Dobrovolsky's radius (in km $10^{(0.43 \cdot M)}$ [36]) that occurred in the analyzed window (for example, from 500 days before the earthquake until 500 days after it), are extracted. If an anomaly is univocally associated with one earthquake, this couple (anomaly–earthquake) is defined. In case the anomaly is compatible with more than one seismic event, one of the following four methods is used to select which earthquake is associated with the anomaly:

1. **Method 1 "All anomalies–EQs"**. This method selects all earthquakes compatible with the investigated anomaly. The advantage is that we do not apply any assumption for the analysis, and we can suppose that, in case the statistics are sufficient, the "wrong" anomaly–earthquake couples increase just the background without creating additional artificial concentrations of anomalies. On the other side, one disturbance can be produced only by one earthquake, and so, unless the signal is superposed (or under the satellite resolution), the anomaly could not be associated with more than one event. For this reason, we also introduce the following methods.
2. **Method 2 "Min [log(ΔT R)]"**. This method selects the closest earthquake in space ("R") and time (anticipation time $\Delta T = time_{EQ} - time_{anomaly}$). The selection is made by searching for the minimum of the following equation: $\log_{10}(|\Delta T \cdot R|)$. The criterion is based on the assumption that an anomaly is more likely to be produced by the closest earthquake in space and time.
3. **Method 3 "Max (magnitude)"**. This method selects the earthquake with the highest magnitude in the space and time domains of interest. The assumption is that a larger earthquake produces more anomalies before its occurrence that can be detectable in the ionosphere.
4. **Method 4 "Closer (Rikitake)"**. This method takes into account that a larger earthquake with a magnitude *M* is expected to have a longer anticipation time ΔT of its possible precursors, according to the Rikitake law [37], expressed as:

$$\Delta T = a + b \cdot M \tag{1}$$

With this method, the anticipation time ΔT is calculated using the a and b coefficients found for the *Swarm* satellite data (magnetic field or *Ne*) by De Santis et al. [29] and inserted in the Rikitake law expression. On the other side, the decimal logarithm of the real anticipation time is calculated, and the earthquake that presents the minimum difference with respect to the Rikitake law is selected.

Please note that the first three methods are the same as those used by De Santis et al. [29], while here, a new fourth method is also proposed. Hereafter, if no indication is given about the method, the analysis will follow Method 1, i.e., all anomalies are associated with all the compatible earthquakes. Finally, it is worth noticing that any method presents advantages and disadvantages, as we do not know the exact mechanism behind the generation of ionospheric disturbances (even if several theories have been proposed). We can rely on the concentrations of anomalies confirmed by several of these methods and expect that a future better understanding of the LAIC mechanism can improve the selection of the best of the above methods or help formulate a new one. Of course, all methods take advantage of the fact that all analyses are retrospective. It would be more difficult to adapt the methods for a real time analysis rather than establish a significant correlation between ionospheric anomalies and earthquakes, but this is not within the present scope of the work.

The results are illustrated in a "standard" representation graph (see, for example, Figure 1), where the horizontal axis represents the time of the anomaly with respect to the earthquake occurrence (i.e., negative days before the earthquake occurrence and positive days after), and the vertical axis is the distance of the detected anomaly in terms of degrees from the epicenter. Therefore, each anomaly is placed on this graph in terms of distance in time and space from its associated earthquake (according to one of the four previously listed methods). All the anomalies and earthquakes are superposed on the same graph, and, in the end, the results are binned in 50 horizontal and 10 vertical cells (or bins). For each bin, the number of anomalies falling in the bin is divided by its geographic area (circle or annulus), obtaining the anomalies' density represented by a color bar scale. For the first row (which represents the 50 temporal bins closer to the epicenters), the maximum concentration of such density is searched, and its location (pxmax), the number of anomalies (Nmax), the number of EQs associated with these anomalies (NEQmax), and the total sum of the released energy (E_{EQ}) is reported in the second line of the heading. Furthermore, for such a maximum density/concentration bin, other statistical quantities are evaluated and described in the following paragraph.

To evaluate if the results obtained from the WSC of real earthquake and anomaly data are statistically significant, 30 random space-time homogeneous distributions of anomalies, with the same number of anomalies as each real data analysis, have been generated. Then, the WSC algorithm has been applied with the same parameters used for the real satellite anomalies. In particular, for the maximum concentration, we estimate the quantity, D_{MAX}/D_0, that represents the density of anomalies in the maximum (D_{MAX}) with respect to a theoretical homogeneous distribution of anomalies (D_0):

$$D_{MAX}/D_0 = \frac{N_{anom}^{MAX} \cdot \Delta T^{Tot} \cdot Area^{Tot}}{N_{anom}^{Tot} \cdot \Delta T^{pixel} \cdot Area^{pixel} \cdot n_{EQ}} \tag{2}$$

where N_{anom}^{MAX} is the number of anomalies in the maximum pixel (which, in the title of the graphs, is indicated as "N_{MAX}") that has a duration ΔT^{pixel} and covers an area $Area^{pixel}$. The total number of identified anomalies (independently from the earthquakes) is N_{anom}^{Tot}, and it is also indicated in the first line of the heading of the graph as the second number (63,178 in the case of Figure 1). The whole analyzed area (i.e., the spherical zone between 50°S and 50°N of geomagnetic latitude) is $Area^{Tot}$, and the total analyzed time is ΔT^{Tot} (in this work, it is 8 years). The quantity D_{MAX}/D_0 has been calculated for real analysis, i.e., $[D_{MAX}/D_0]_{real}$, and its mean random value $\overline{[D_{MAX}/D_0]_{random}}$ and its standard deviation (σ) have been estimated over a set of 30 random simulations. By these quantities, we defined the two statistical parameters, d and n:

$$d = \frac{[D_{MAX}/D_0]_{real}}{\overline{[D_{MAX}/D_0]_{random}}}; n = \frac{[D_{MAX}/D_0]_{real} - \overline{[D_{MAX}/D_0]_{random}}}{\sigma([D_{MAX}/D_0]_{random})} \tag{3}$$

According to the above definitions, d represents how many times the real maximum concentration is higher than the one expected from random simulations, and n estimates how many standard deviations (of random simulations) the real concentration is higher than the random one. The number of random simulations was chosen as a function of the results in terms of the reached stability of the obtained values of d and n. We found that, generally, even with one or two random simulations, the estimation of d is accurate up to one decimal digit (if the number of anomalies is greater than 30,000), while n requires more random runs, i.e., at least 15 or 30 simulations for about 30,000 anomalies.

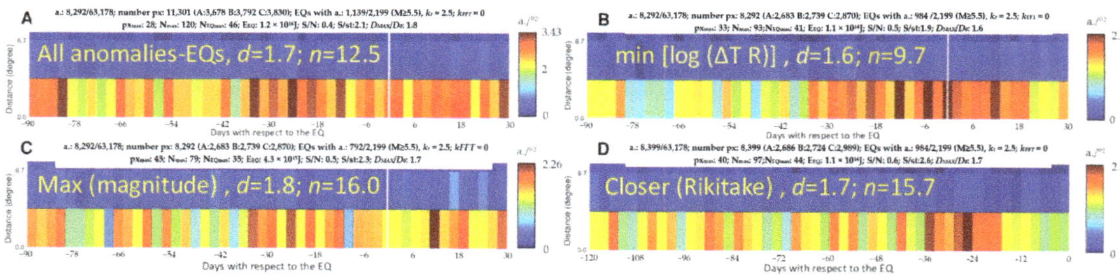

Figure 1. Correlation of *Swarm* magnetic field anomalies with M5.5+ earthquakes using four different methods: (**A**) all anomalies are associated with all compatible earthquakes; (**B**) each anomaly is associated with the closest earthquake in space and time, i.e., the minimum of log(ΔT R); (**C**) the anomaly is associated with the greatest magnitude earthquake; (**D**) the anomaly is associated with the earthquake closer to the Rikitake law, with coefficients from De Santis et al. [29]. The graphs are cut for the vertical section without significant information, showing only the closer band to the earthquake epicenter. The color bar represents the density of anomalies in terms of the number of anomalies per square degree.

When the concentration of the real anomalies is compatible with a homogeneous random simulation of anomalies, we expect $d \cong 1$ and $n \cong 0$. At the same time, we define a statistically significant concentration, i.e., a real distribution of anomalies with the largest concentration significantly differing from a random distribution, if $d \geq 1.5$ and $n \geq 4.0$, as suggested by De Santis et al. [29].

3. Results

We consider the three satellites of the *Swarm* mission and analyze two parameters: the Y-East component of the magnetic field and the electron density, *Ne*. All these results are shown in the following four subsections. Here, we present the figures with the most interesting results, while we provide other figures in the Supplementary Materials.

3.1. *Swarm* Magnetic Field Results

In this section, we applied the WSC algorithm to the *Swarm* magnetic field data recorded by the three main satellites of the constellation from 26 November 2013 until 25 November 2021, i.e., the first 8 years of available data. Figure 1 represents the WSC applied to the *Swarm* Y-East component magnetic field anomalies with M5.5+ earthquakes in a time window of 120 days (i.e., 90 days before each earthquake and 30 after for methods 1, 2, and 3, and 120 days before each earthquake for method 4) using the four methods to select which earthquake is associated with the anomaly in case of ambiguity. We note that all the methods identify the same group of anomalies from around −40 to −10 days, and, in some of them, it is also the absolute maximum concentration. It is very interesting that, even with Method 2, which minimizes the log (ΔT R), the absolute maximum concentration is 12 days prior to the earthquake and not the co-seismic, as the method would forcefully enhance the anomalies closer to the event (actually, it is only this method that presents a significant concentration at the earthquake occurrence, although it is not the largest one).

Figure S2 (in Supplementary Materials) reports the statistical correlation analysis for the *Swarm* Y-East magnetic field component anomalies (extracted with a threshold $k_t = 2.5$) with M5.5+ earthquakes from 500 days before until 500 days after each event using Method 1. With these criteria, we can obtain symmetric analyzed time and see that the maximum concentration of anomalies preceded the earthquake (of about 30 ± 10 days). Furthermore, the concentration of anomalies before the earthquake is slightly higher than that after the event. The total number of anomalies in the first row (i.e., the closest to the epicenter) before the earthquake is 16,630, while the total number after the earthquake is 16,436, showing that, before the earthquake, there are 194 more anomalies than the ones after it. Even if this number is just 1.2% more, it could support the idea that, inside the identified anomalies, there could be a part likely produced by ionospheric seismo-induced phenomena. A specific discussion that took into account the geomagnetic quiet and disturbed amount of time before and after each event is reported in Section 4.5, and the specific difference seems to be influenced by this aspect.

To search for a possible relationship between the anticipation time and magnitude of the earthquakes, the events have been divided into five magnitude bands (the same used in Figure 5 of De Santis et al. [30]): $5.5 \leq M \leq 5.9$, $6.0 \leq M \leq 6.4$, $6.5 \leq M \leq 6.9$, $7.0 \leq M \leq 7.4$, and $M \geq 7.5$ using Method 1. Figure 2 represents the WSC applied to 8 years of *Swarm* magnetic field anomalies and earthquakes divided into magnitude bands. Here, an extended period before the earthquake (1000 days instead of 500, as in [29]) is used, as now the statistics are better supported by a larger amount of data. It is worth noting that, in these analyses, a part of the events is not fully analyzed. In particular, the earthquakes that occur in the first 1000 days of the analyzed years have an incomplete time window before them.

In Figure 2B, we can see that, in the four larger magnitude bands (i.e., for M6+), the absolute maximum concentration increases perfectly with the magnitude. The anticipation time in the function of the magnitude is presented with a fit (as determined by the Rikitake law, i.e., $\log\Delta T = a + bM$) in Figure 2A, where $a = -0.96 \pm 0.73$, and $b = 0.51 \pm 0.10$. For the lower magnitude earthquake bands (i.e., $5.5 \leq M \leq 5.9$), the maximum concentrations are at −210, −150, and −90 days, but at the beginning of the graph (where we expect the maximum concentration corresponding to this band), there is also a high concentration of anomalies. We can explain such results in different ways, either independently or occurring simultaneously. One explanation is based on the fact that the lower magnitude earthquakes, i.e., those below M6.0, are less likely to produce distinguishable LAIC effects, while only a part of them can still show some possible ionospheric earthquake precursors. Furthermore, eventual ionospheric electromagnetic perturbations from lower magnitude earthquakes could be very much weaker, and the method presented in this paper could be insufficiently sensitive to detect all of them. An alternative explanation is that these lower magnitude earthquakes are so frequent that an anomaly can be associated with many of them, including those with a longer anticipation time. A third possible reason is that this lower magnitude band could have a typical anticipation time of less than 20 days, which is the most affected by the removal of earthquakes at these distances from larger earthquakes due to the applied declustering method. Despite these considerations, we noted that, in the Rikitake law of Figure 2A, the "not aligned" magnitude band with the fit is the one with earthquakes in the range of $6.0 \leq M \leq 6.4$. Future investigations are necessary to clarify such ambiguity.

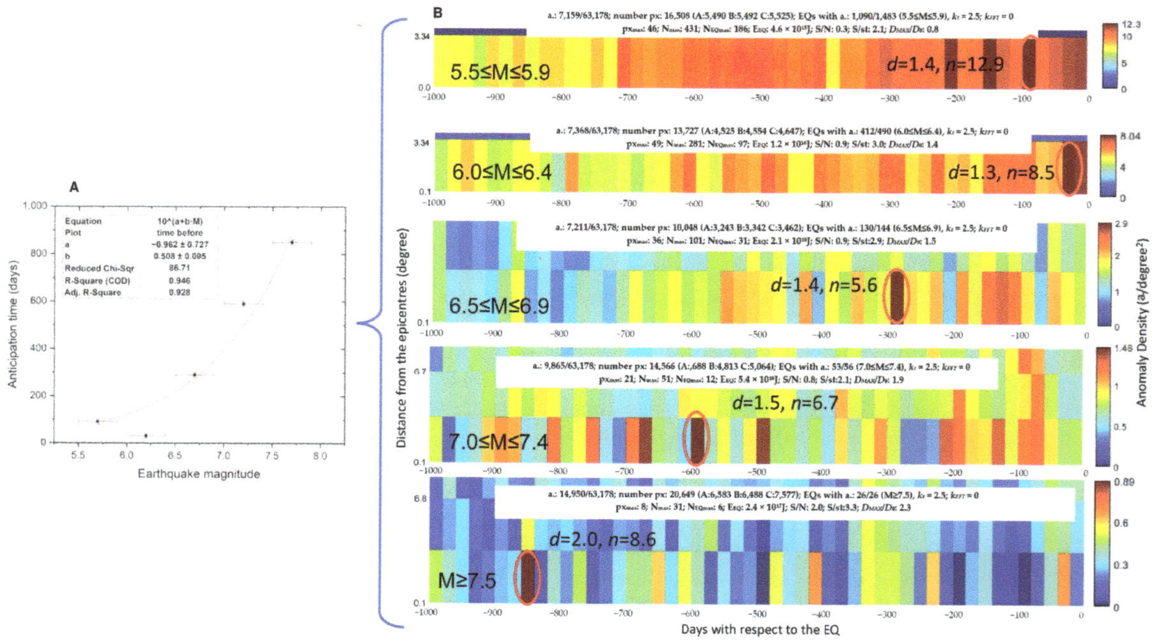

Figure 2. Correlation of *Swarm* magnetic field anomalies with earthquakes divided into five magnitude bands. (**A**) Fit of the anticipation time of maximum concentration (underlined by a red circle in subfigure (**B**) as a function of earthquake magnitude using the equation $\Delta T = 10^{a+b\cdot M}$. The error bar takes into account the width of the pixel. (**B**) Details of the analysis in the earthquake magnitude bands of $5.5 \leq M \leq 5.9$, $6.0 \leq M \leq 6.4$, $6.5 \leq M \leq 6.9$, $7.0 \leq M \leq 7.4$, and $M \geq 7.5$. Only the significant part of the superposed space-time graph is represented. The statistical indications of *d* (how many times the maximum real concentration is higher with respect to the random one) and *n* (how many standard deviations there are in which the real largest concentration is greater than the random one) are also shown. Such maximum concentrations (as objectively identified by the WSC algorithm) have been highlighted by a red or orange circle (the last one for a less significant concentration in terms of the Rikitake law). The color bar represents the density of the anomalies in terms of the number of anomalies per square degree.

A summary view of the WSC analysis of the *Swarm* magnetic field data applied to all the investigated M5.5+ earthquakes is provided in the Supplementary Materials in Figure S3, showing higher concentrations toward the earthquake occurrence.

To validate such results, an earthquake catalog with a shuffled magnitude has been produced, and we showed the results in Figure S4 in the Supplementary Materials. In the subplot on the left, it is possible to see that a correlation between the shuffled earthquake magnitudes and the real set is practically absent (R = 0.0004), while, on the right side of Figure S3, it is shown that the relationship between the anticipation time and the earthquake magnitude is lost, as expected after shuffling the earthquake catalog. Some minor correlations can still be found, but this could simply be due to the fact that, since we shuffled only magnitudes, the epicenter and the origin time of the events are still real, so the anomalies really preceded an earthquake.

Figure 3 shows the different analyses of the *Swarm* Y-East component magnetic field data, taking into account the signal's frequency correlated with M5.5+ earthquakes. For these analyses, we show the classical WSC with Method 1 (all possible combinations of anomalies and earthquakes) from 500 days before until 500 days after the earthquake. The periods are from the upward shorter period (2–10 s) (subpanels A) to the downward longer period (25–50 s) (subpanel C). We can note that it seems that the anticipation time increases with the period of the signal, this being the time between the maximum concentration of the anomalies and the occurrence of the earthquake. This observation could be crucial for a future system to predict earthquakes, as the frequency of the signal could give important information of the time-lapse to the earthquake. We tried to provide a rough linear fit (see Figure S5 in the Supplementary Materials) of such an anticipation time, obtaining:

$$\Delta T = -24\ (\pm 19) + 4.01\ (\pm 0.82)\cdot P \tag{4}$$

where "ΔT" is the anticipation time in days, and "P" is the period of the signal in seconds.

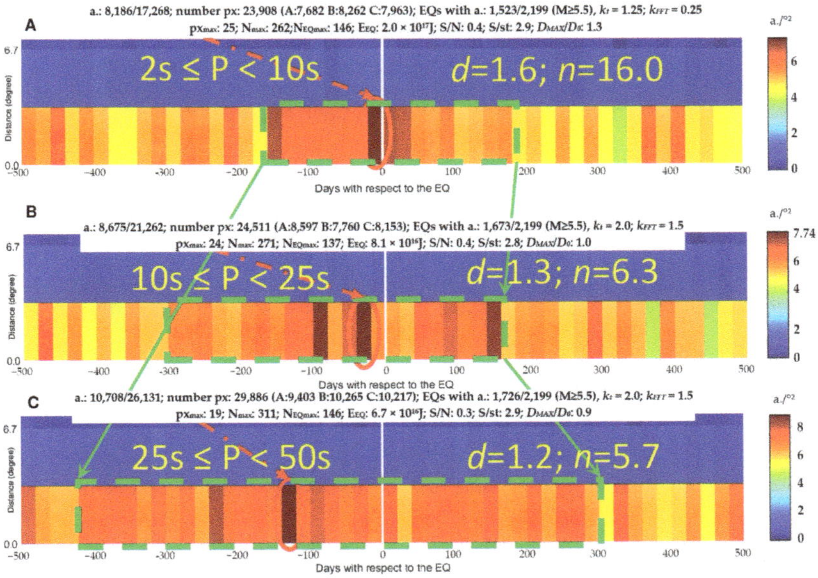

Figure 3. 8-year *Swarm* Y-mag data separated for frequency content in the anomalies correlated with M5.5+ earthquakes from 500 days before until 500 after their occurrence. The color bar represents the density of the anomalies in terms of the number of anomalies per square degree. The anomalies have been selected with higher signal content in three different period bands: (**A**) from 2 s to 10 s, (**B**) from 10 s to 25 s and (**C**) from 25 s to 50 s.

The fit provides a positive slope of about 4.0 days/s. The fact that the fit provides a negative intercept and then a possible negative anticipation time (and thus an effect after the earthquake) could be explained by the fact that the period stops at about 6 s before the earthquake's occurrence (i.e., $\Delta T = 0$, then $P = 6$ s). We provide a specific discussion on these results in Section 4.4.

We finally noted that there seems to be a very long period underlined by dashed green boxes in Figure 3, with higher anomaly concentrations with respect to the rest (the identification of the green boxes is taken by sight considering the "more red or brown" bins, subjectively). We speculate that the density of the anomaly in the tails (i.e., before or after the green dashed box) is not due to the earthquake and that it is in the background due to other sources, while the increase in the anomalies could be related to a sort of seismic activation and to the consequences of the seismic event after their occurrence. Even these higher anomaly regions increase their duration (before the earthquake) with the period of the investigated signal. Future studies are fundamental to investigate this aspect better.

Table S1 (in the Supplementary Materials) summarizes the main statistical results for all the investigated combinations of the 8-year *Swarm* magnetic field data correlated with M5.5+ shallow earthquakes. In particular, the two statistical quantities d and n are reported, along with the indication of when the maximum concentration of anomalies occurs in the bin closest to the epicenter (negative if it occurs before the earthquake and positive if it happens after) and how many earthquakes are associated with anomalies in the pixel with the maximum concentration of anomalies. When a concentration passed the statistical evaluation criteria defined in [29] (i.e., $d \geq 1.5$ and $n \geq 4.0$) and occurred before the earthquake, we underline the row in green.

3.2. *Swarm* Electron Density Results

In this section, we applied the WSC algorithm to the electron density (*Ne*) data recorded by the three main satellites of the *Swarm* constellation from 2 December 2013 until 1 December 2021, i.e., the first 8 years of available data. Most of the results of *Ne* are provided in the Supplementary Materials, as they have not been found to be statistically significant. Figure S6 (in the Supplementary Materials) provides the results of WSC applied to 8 years of *Swarm Ne* data correlated with M5.5+ shallow earthquakes using the above four methods to associate the anomalies and earthquakes. The results do not pass the statistical criteria defined in [29]: $d \geq 1.5$ and $n \geq 4.0$. Furthermore, the analyses with the minimization of log (ΔT R) and the maximum earthquake magnitude (methods 2 and 3) find a post-seismic bin as the maximum concentration. In fact, the analysis in the long symmetrical time from 500 days before until 500 days after the earthquakes with the same data (Figure S7 in the Supplementary Materials) shows low or null significance, and, overall, the number of anomalies after the earthquake (15,618) is greater than those before the earthquake (15,504); their difference being equal to 114 anomalies, which is lower than the square root of the anomalies, which is equal to 125 and could be considered as an approximation of their uncertainty, as further discussed in Section 4.4.

In the same way that we detected an increase in the anticipation time of the anomalies with the earthquake magnitude in the magnetic field data (Figure 2), we now apply the same approach to 8 years of the *Swarm Ne* data, as shown in Figure S8 in the Supplementary Materials. In this case, we tried to make a fit through the maximum concentrations of anomalies detected by the WSC algorithm in the first row (i.e., closer to the earthquake epicenter) of different magnitude ranges following the empirical Rikitake law. We note that the magnitude 7.5+ earthquakes analysis shows the absolute maximum concentration on the second row, i.e., between 380 km and 750 km from the epicenter. As such, a magnitude range is the one with fewer earthquakes (26 in total); this effect could be due to an insufficient number of cases to construct good statistics. Furthermore, it is important to note that these analyses did not pass the statistical evaluation criteria, and some of them even show a concentration of the same intensity as the one expected for a random simulation ($d \cong 1$) and a negative n, indicating a real largest concentration lower than the random one. Even if the concentration of anomalies is not significant, the maximum concentration in the row closer to the epicenter for magnitudes $\geq$6 seems to follow the Rikitake law, as can be seen in the fit shown in Figure S8A, which provides the following coefficients: $a = -3.5 \pm 2.1$ and $b = 0.83 \pm 0.27$. The maximum concentration in the magnitude band $5.5 \leq \mathrm{M} \leq 5.9$ is found 250 days before the earthquake, but the concentration 10 days before the earthquake is also high but lower than the maximum and, in the fit, has been used as the objective maximum extracted by the code. So, we can suppose that there may be some anomalies related to the earthquake, but the "anomaly definition" employed is too weak to extract all possible seismo-induced phenomena, consequently including many anomalies not related to earthquakes.

To improve the anomaly definition, we introduce the analysis of frequency content in the investigated window with the same approach as that one used for the magnetic field data. As can be seen in Figure 4, the significance of the results is improved for most of the analyses, passing the statistical criteria ($d \geq 1.5$ and $n \geq 4.0$). We note that, for the shorter period analysis in the symmetrical period from −500 days to 500 days with respect to the earthquake, the maximum concentration of the anomalies is post-seismic, and for the analysis in the signal period range between 10 s and 25 s from −90 days to 30 days, the maximum concentration is a "co-seismic" effect. Even if statistically significant, these results do not contribute to the main purpose of this paper, i.e., a better understanding of the preparatory phase of medium-large earthquakes. On the other hand, the investigation in the lower frequency band in which the longest period is equal to 25–50 s shows not only highly significant results but also pre-seismic maximum concentrations of anomalies both for the investigation in the time window from −90 days to 30 days and for the one from −500 days to 500 days at about −59 days and −290 days, respectively (see Table S2). Such a result seems to indicate that taking the frequency content of the anomalies into account could help to discriminate the ones possibly caused by the preparation phase of earthquakes from the other ones caused by other phenomena.

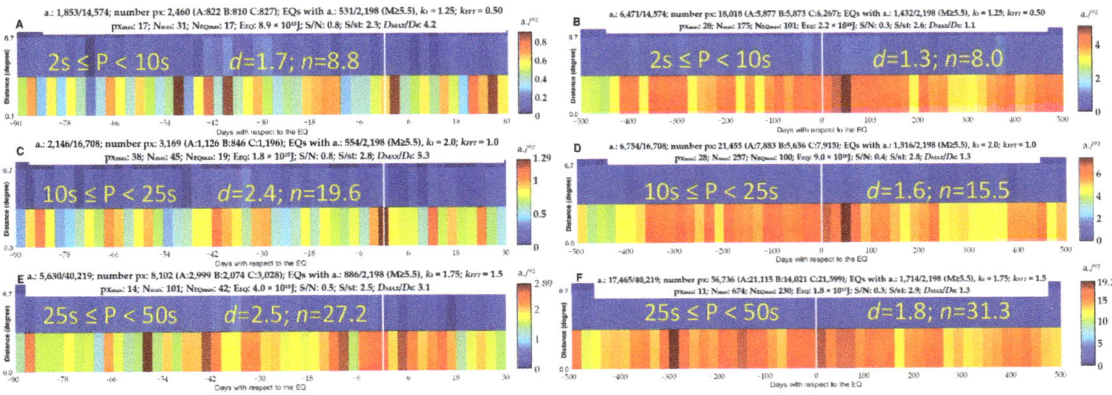

Figure 4. 8-year *Swarm* Ne data WSC analyses separated for frequency content in the anomalies correlated with M5.5+ earthquakes for 120 days (left) and 1000 days (right) around the EQ occurrence, in particular: (**A**) signal period from 2 s to 10 s and analysis from −90 days to +30 days with respect to the earthquake, (**B**) signal period from 2 s to 10 s and analysis from −500 days to +500 days with respect to the earthquake, (**C**) signal period from 10 s to 25 s and analysis from −90 days to +30 days with respect to the earthquake, (**D**) signal period from 10 s to 25 s and analysis from −500 days to +500 days with respect to the earthquake, (**E**) signal period from 25 s to 50 s and analysis from −90 days to +30 days with respect to the earthquake, and (**F**) signal period from 25 s to 50 s and analysis from −500 days to +500 days with respect to the earthquake. The color bar represents the density of the anomalies in terms of the number of anomalies per square degree.

Finally, Table S2 (in the Supplementary Materials) reports the main statistical parameters for all the performed analyses for 8 years of *Swarm Ne* data. The statistically significant results (i.e., the ones with $d \geq 1.5$ and $n \geq 4.0$) that precede the earthquake are underlined in green, as in Table S1. It is possible to note that just three out of sixteen results for *Ne* passed these criteria. Despite this, introducing the analysis of the signal frequency provides more significant results in the lowest frequency band (i.e., the period between 25 s and 50 s).

4. Discussion

4.1. General Comparison for Magnetic Field and Ne Results

Comparing the results of the electron density and magnetic field measured aboard the satellite, it is possible to note that the *Swarm Ne* results are not statistically significant. One reason for this could be the large corresponding number of detected anomalies, which is about 80,000. It will be interesting in future works to apply the same technique to CSES data, as, due to the satellite's Sun-synchronous orbit, it would permit to have *Ne* deep local nighttime (02:00 LT) data, always avoiding the effects of the solar irradiation (for example after sunset, when the ionosphere can show perturbation even until midnight or little later [38]). Therefore, the fixed local time reduces the signal variation in *Swarm* due to its gradual orbital precession. In addition, the crossing of critical local times close to the terminator can create perturbations or even false anomalies due to the shadow/ligh passage of the *Swarm* satellite along the same orbit.

Swarm satellites show better results for the magnetic field. This could be due to the intrinsically higher quality of magnetic instrumentation in comparison with that o Electric Field Instruments and to the satellite design (in particular, the star cameras or the same optical bench as the fluxgate magnetometers). Physically, this could also be explained as the magnetic field is usually less affected by the local time than *Ne*. Hence, it precession in *Swarm* satellites (see the discussion above) seems not to affect the usability o its measurements to search for possible pre-earthquake ionospheric disturbances.

In addition, a specific comparison of the coefficients found by the empirical Rikitake law in this work for the *Swarm* magnetic field and electron density data and the ones found by De Santis et al. [29] is provided in Figure S9 of the Supplementary Materials. We confirm the previous results and find that almost all the results are within the trend proposed in 1987 by Rikitake [37] for earthquake precursors of the quasi first kind (i.e., earth-currents resistivity, radon, underground water, and other geochemical elements) or only the ge omagnetic precursor (from the ground). This result is promising because it means that with the availability of new satellite data, we are confirming an empirical trend obtained decades ago for ground precursors and extending its validity to ionospheric precursors, as recently found in De Santis et al. [29].

4.2. Validation of the Results by Confusion Matrix Performance Evaluation and ROC Curves

The results have been evaluated by the so-called "confusion matrix", which is a table of two rows × two columns used to assess the performances of a prediction model, crossing the presence (or not) of an ionospheric anomaly followed (or not) by an earthquake in the given time within the pre-chosen distance and time.

Here, we considered the same anomalies presented in the results, but we evaluated a cell that has 6° geomagnetic latitude for 6° geographic longitude and 2.4 days (for investigations previously conducted 90 days before the earthquake) or 20 days (for investigations conducted 500 days before the earthquake). The evaluation of the cell has been selected among the classic four cases:

1. True positive (*TP*): in the cell, there is at least an anomaly, and an earthquake follows within the next 90 days
2. True negative (*TN*): in the cell, there are no anomalies, and no earthquakes follow in the next 90 days in the same cell.
3. False positive (*FP*): in the cell, there is one or more anomalies, and no earthquakes follow in the next 90 days in the same cell
4. False negative (*FN*): in the cell, there are no anomalies, and an earthquake follows in the next 90 days.

From the evaluation of the number of true and false positives and negatives, the accuracy "*Acc*" has been computed as the total number of good cases over the total cases:

$$Acc = \frac{TP + TN}{TP + TN + FP + FN} \tag{5}$$

Better performance with respect to a random prediction is achieved when *Acc* > 50%. We also calculated the Hit Rate (*HR*) and False Positive Rate (*FR*); the former expresses the percentage of predicted events, while the latter is the rate of false alarms:

$$HR = \frac{TP}{TP + FN} \quad FR = \frac{FP}{FP + TN} \tag{6}$$

An ideal prediction system would give the limit case with *HR* = 100% (all events are predicted) and *FR* = 0% (no false alarms), while a real system is better as close as possible to such limit case, and, typically, the earthquake prediction systems are not suitable for use with the population when the false positive rate is too high. Finally, the Alarmed Time-Space "AT" (also called the alarm rate in the literature [39]) has been estimated:

$$AT = \frac{TP + FP}{TP + TN + FP + FN} \tag{7}$$

This quantity is important to check that a good prediction is not just achieved regarding an analyzed space-time that is too large or even an entire space-time. For example, the affirmation that "*in the future, in the World, an M5.5+ earthquake will occur*" is surely true, but it is not a "prediction", because it does not predict the exact time, and the space domain is too large (the whole world). Given the same accuracy, a prediction is preferable with a lower AT. The alarmed time can also be compared with the number of positive cases ($TP + FN$) with respect to the total cases, as, even an ideal (so, perfect) prediction needs to alarm at least all of the space-time that any event occurred.

Table 1 reports the evaluation of performances in terms of confusion matrices for the *Swarm* magnetic field and electron density data considering 90 days and 500 days before the M5.5+ earthquakes. These tables reflect the analyses shown in Figures 1, 3, 4, S2, S4 and S5, even if we consider only the time before the earthquakes, as the confusion matrix is a tool to assess the prediction capability of a system, so only the ionospheric anomalies that preceded the earthquake can be "positively" evaluated. It is possible to note that all the analyses show a high accuracy, from 72% until 95%. Contrariwise, the hit rates tend to be rather low (i.e., from ~1% to ~19%), as several earthquakes are not preceded by anomalies in the considered interval of time, and, overall, several earthquakes did not show anomalies continuously in the 90 (or 500) days before their occurrence. The alarmed space-time windows are very low (i.e., around one or a few percent) for all the investigations with an alarm-window time of 90 days, and they are between 5.1% and 19.1% for the analyses performed in a time window of 500 days. Both the investigated time windows show alarmed times below the 50% threshold—in some cases, lower than 1%—which provide further evidence of the non-chance nature of the detected anomalies in relation to the following earthquakes. The accuracy is similar to that of a previous work performed on *Swarm* electromagnetic data and earthquakes but using a machine learning approach [40]. Furthermore, the overall performances reflected the previous considerations, with the scores for the magnetic field being generally better than those for the electron density. Further, better scores were always yielded when the frequency content of the signal was investigated, confirming that it is fundamental to take into account the signal frequency of the possible electromagnetic ionospheric precursor in future studies.

In order to further statically check the predicting capability of the magnetic field and electron density anomalies extracted in this work as the best candidates for earthquake precursors by WSC, the Receiving Operating Characteristic (ROC) graph has been calculated. ROC is a well-established method to test a precursor candidate, e.g., [41], and it is a graph where the false positive rate is conventionally reported on the horizontal axis and the hit rate is reported on the vertical one. The diagonal from (0%, 0%) to (100%, 100%) represents the random case, while the upper-left corner (0%, 100%) is an ideal case that predicts all the events without any false alarm. A good prediction is required to be as far from the diagonal toward the ideal case as possible.

In our case, we extract all the anomalies identified as statistically significant in this work, i.e., the pre-earthquake concentrations that show a statistical coefficient $d \geq 1.5$ and $n \geq 4.0$, as previously used in this work and first defined by De Santis et al. [29]. In this section, other concentrations that are different from the absolute maximum one that passed this statistical criterion have been also taken into account. Collecting all the anomalies from the magnetic field and electron density *Swarm* data and all the techniques, we produced a new dataset of the best anomalies that contains 4831 *Swarm* anomalous signals after cleaning the overcounted ones.

With the new dataset of the best WSC anomalies, the "Confusion matrix" and ROC diagram have been evaluated and reported in Table 2 and Figure 5, respectively. The step time of calculus has been set to be equal to the alerted time, and the window following the one with any *Swarm* electromagnetic anomaly is alerted if it contains any anomaly; otherwise, it is not alerted. Comparing Table 2 with Table 1, it is possible to fully confirm that the selections made in this paper extracted the statistically best pre-earthquake candidates permitted to reduce the false positive rate and increased the hit rate.

Table 1. Confusion matrices of the *Swarm* magnetic field and electron density anomalies identified in the 90 days or 500 days before the M5.5+ earthquakes.

			Magnetic Data		**Magnetic Data 2–10 s**		**Magnetic Data, 10–25 s**		**Magnetic Data, 25–50 s**	
90 days before the earthquake			Earthquake		Earthquake		Earthquake		Earthquake	
			Yes	No	Yes	No	Yes	No	Yes	No
	Anomalies	Yes	1396	25,654	573	9477	555	10,453	655	13,149
		No	50,454	1,091,776	51,277	1,107,953	51,295	1,106,977	51,195	1,104,281
	Hit and false positive rates		*HR* = 2.7%	*FR* = 0.23%	*HR*=1.11%	*FR*=0.85%	*HR* = 1.07%	*FR* = 0.94%	*HR* = 1.26%	*FR* = 1.18%
	Accuracy and Alarmed Time		*Acc* = 93.5%	*AT* = 2.31%	*Acc* = 94.8%	*AT* = 0.86%	*Acc* = 94.7%	*AT* = 0.94%	*Acc* = 94.5%	*AT* = 1.18%
500 days before the earthquake			Earthquake		Earthquake		Earthquake		Earthquake	
			Yes	No	Yes	No	Yes	No	Yes	No
	Anomalies	Yes	3255	19,575	1378	7826	1522	8800	1805	10,882
		No	15,325	102,965	17,202	114,714	17,058	113,740	16,775	111,658
	Hit and false positive rates		*HR* = 17.5%	*FR* = 16.0	*HR* = 7.42%	*FR* = 6.39%	*HR* = 8.19%	*FR* = 7.18%	*HR* = 9.71%	*FR* = 8.88%
	Accuracy and Alarmed Time		*Acc* = 75.3%	*AT* = 16.2%	*Acc* = 82.2%	*AT* = 6.52%	*Acc* = 81.7%	*AT* = 7.31%	*Acc* = 80.40%	*AT* = 8.99%
			***Ne* data**		***Ne* data 2–10 s**		***Ne* data, 10–25 s**		***Ne* data, 25–50 s**	
90 days before the earthquake			Earthquake		Earthquake		Earthquake		Earthquake	
			Yes	No	Yes	No	Yes	No	Yes	No
	Anomalies	Yes	1380	32,605	345	7984	434	7460	1100	16,156
		No	50,502	1,084,793	51,537	1,109,414	51,448	1,109,938	50,782	1,101,242
	Hit and false positive rates		*HR* = 2.66%	*FR* = 2.92%	*HR* = 0.66%	*FR* = 0.71%	*HR* = 0.84%	*FR* = 0.67%	*HR* = 2.12%	*FR* = 1.44%
	Accuracy and Alarmed Time		*Acc* = 92.9%	*AT* = 2.91%	*Acc* = 94.9%	*AT* = 0.71%	*Acc* = 95.0%	*AT* = 0.68%	*Acc* = 94.3%	*AT* = 1.48%
500 days before the earthquake			Earthquake		Earthquake		Earthquake		Earthquake	
			Yes	No	Yes	No	Yes	No	Yes	No
	Anomalies	Yes	3436	23,490	949	6744	1039	6101	2311	11,662
		No	15,164	99,030	17,651	115,776	17,561	116,419	16,289	110,858
	Hit and false positive rates		*HR* = 18.5%	*FR* = 19.2%	*HR* = 5.10%	*FR* = 5.50%	*HR* = 5.59%	*FR* = 4.98%	*HR* = 12.4%	*FR* = 9.52%
	Accuracy and Alarmed Time		*Acc* = 72.6%	*AT* = 19.1%	*Acc* = 82.7%	*AT* = 5.45%	*Acc* = 83.2%	*AT* = 5.06%	*Acc* = 80.2%	*AT* = 9.90%

Table 2. Confusion matrices of the best *Swarm* magnetic field and electron density anomalies identified in the WSC concentrations that passed the statistical significance test (i.e., the ones with $d \geq 1.5$ and $n \geq 4.0$).

		Alerted Time of 90 Days		**Alerted Time of 500 Days**	
		Earthquake		Earthquake	
		Yes	No	Yes	No
Anomalies	Yes	564	1112	406	478
	No	800	29204	258	4618
Hit and false positive rates		*HR* = 41.35%	*FR* = 3.67%	*HR* = 61.15%	*FR* = 9.38%
Accuracy and Alarmed Time		*Acc* = 94.0%	*AT* = 5.29%	*Acc* = 87.2%	*AT* = 15.35%

The ROC curve in bold blue in Figure 5 is obtained after testing several alerted times from 10 days (lower left side) to 2500 days (upper side). Two black arrows have underlined the two alerted times of 90 and 500 days, and they correspond to the one used in Table 2 and are mostly selected in this paper. In the same figures, the Area Under the Curve (AUC) has also been represented as a black inclined hatch for the random case (i.e., the one under the diagonal equal to 50% for the definition) and as a red inclined hatch for the real best *Swarm* anomalies, estimated to be equal to 83.6%. So, the AUC of the real data is 1.67 times better than that of a random predictor, confirming that WSC applied to *Swarm* data is able to extract potential pre-earthquake ionospheric disturbances.

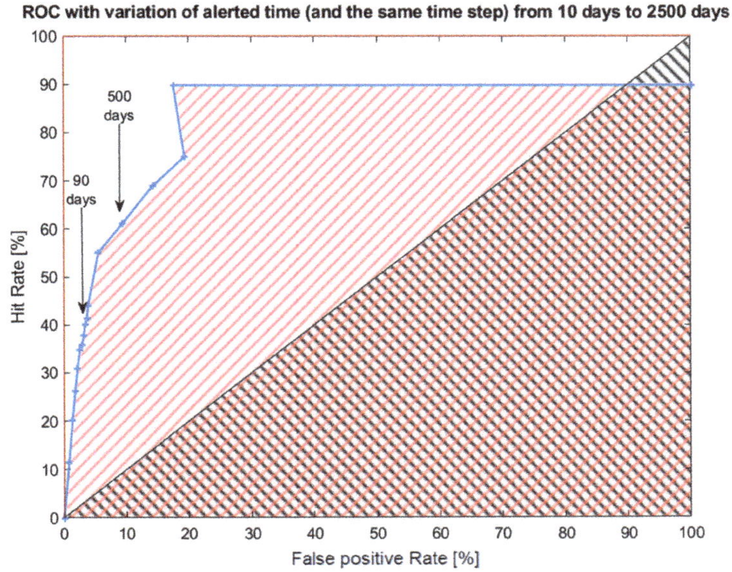

Figure 5. Receiving Operating Characteristic (ROC) graph for the best *Swarm* magnetic field and electron density anomalies (i.e., the ones in WSC concentrations with significant statistical parameters $d \geq 1.5$ and $n \geq 4.0$) calculated, varying the alerted time from 10 days to 2500 days. The Area Under the Curve (AUC) is represented as a black dashed pattern for a random case and as a red dashed pattern for the real *Swarm* case.

4.3. Improving the Superposed Epoch Approach by Taking into Account the Rikitake Law

The results of this paper, especially for the magnetic field anomalies, strongly confirm the empirical Rikitake law [37], which predicts that the logarithm of anticipation time linearly increases with earthquake magnitude. If such a law is true, then the superposed epoch approach would not work due to the mixing of several earthquakes with different magnitudes. It seems to work in the real case, but we think that this is due to the natural distribution of the earthquake magnitudes that follows the Gutenberg–Richter [42] earthquake distribution, privileging the smaller magnitude events. In this paper, we propose a new approach to addressing this issue when the epochs are superposed by "normalizing" the time scale using the Rikitike law, with the coefficients (a and b) being those proposed by De Santis et al. [29] for the *Swarm* magnetic field and electron density satellite data and referring to an M6.5 earthquake, following the next equation:

$$\Delta T' = \Delta T \cdot \frac{10^{(a+b\cdot 6.5)}}{10^{a+b\cdot M}} \quad (8)$$

where $\Delta T'$ is the normalized anticipation time, ΔT is the original time with respect to the earthquake, and M is the magnitude of the earthquake under analysis. Even if the formula can be, in principle, also applied to post-earthquake anomalies, we think that it is not proper to use it in such a way, as the formula is constructed on possible precursors. The reference magnitude (here, chosen as M6.5) can be, in principle, any magnitude, and the results are not affected by this choice. Its scope is to report the time in a different scale chosen as the number of days that preceded an M6.5 earthquake. M6.5 is an intermediate magnitude with a high number of events in the investigated period to be able to clearly show possible precursors, according to the results illustrated in this paper.

To select the window of analysis, we look at the results from [29] and plotted the expected anticipation time as a function of the earthquake magnitude (in Figure 6, as the orange curve). Based on these data, we decided to analyze 150 days before the earthquake for the magnetic field data and 85 days for the *Ne* measurements (when $M = 6.5$, i.e., our reference point). We would underline that, using Equation (4), the analyzed time before the earthquake is not fixed (as in the previous analyses), but it is different for any earthquake as a function of its magnitude, and it is represented as the blue curve in Figure 6, i.e., the result of applying Equation (4)—that is, how many days the normalized time corresponds for other magnitudes. For example, the analyzed time before an M8.0 earthquake is 4037 days for the magnetic field (or 1576 days for *Ne*) according to Equation (4), and the time interval from 4037 days before and until the event will be compressed in the horizontal axis (of Figures 7 and 8) and shown to be equivalent to 150 days to 0 days of an M6.5 earthquake. We note that, for the magnetic field, the larger magnitude earthquakes (M > 8) could show an anticipation time that is greater than that of the 8 years of *Swarm* data available at the moment (the red dashed horizontal line in Figure 6).

(A)

(B)

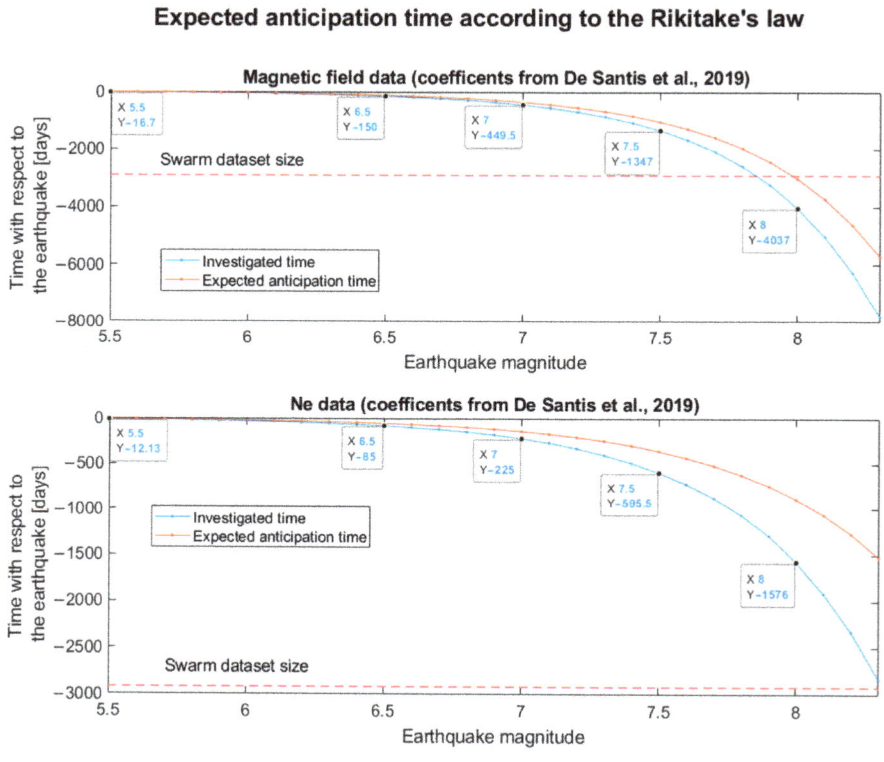

Figure 6. Expected anticipation time of magnetic field (**A**) and *Ne* (**B**) anomalies with respect to the earthquake, using the Rikitake law [37] and considering the coefficients found by De Santis et al. [29] for 4.7 years of *Swarm* data (orange line). The investigated time window is shown as a blue line, and the 8-year *Swarm* dataset maximum size is shown as a dashed red line. The data tips provide the time that has been analyzed before a specific magnitude earthquake, applying Equation (4).

We test and apply this approach to the most significant results of the *Swarm* data, which are the investigations of the magnetic field and *Ne* data with Method 1 and the *Swarm* magnetic field anomalies in the band of 2–10 s and the *Ne* anomalies in the band of 25–50 s, which are those with the best statistical indications from the previous analyses.

Figure 7 shows the results for the *Swarm* data without taking into account the frequency of the signal. The maximum concentration has a higher statistical significance with respect to the previous analyses: in fact, for the magnetic field data, d passes from 1.7 to 2.1, and for the *Ne* data, d passes from 1.3 to 1.5, which also means that, from a non-statistically significant result, we obtain a higher significance with this different superposed time approach, which implies that we could better investigate the identified anomalies to search for seismo-ionospheric disturbances.

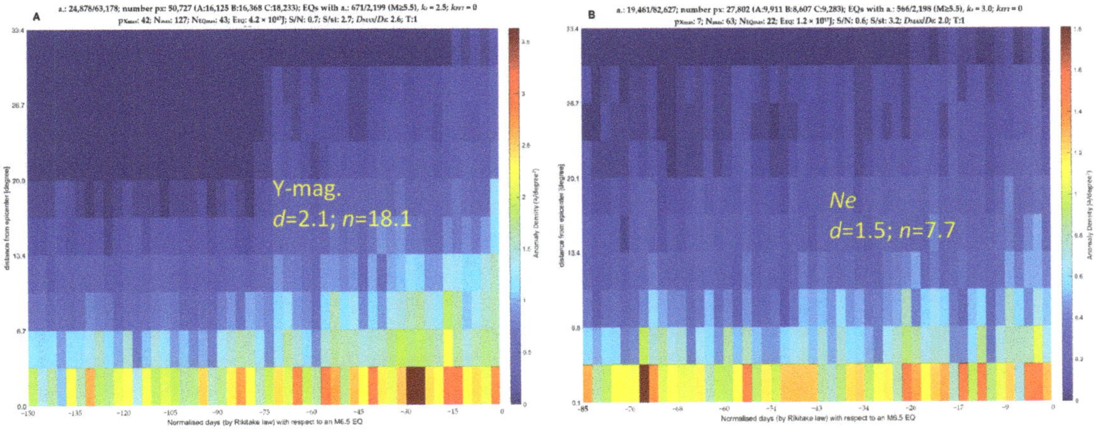

Figure 7. *Swarm* magnetic field data (Y-mag in (**A**)) and electron density (*Ne* in (**B**)) correlated with M5.5+ earthquakes using modified time according to the empirical Rikitake law.

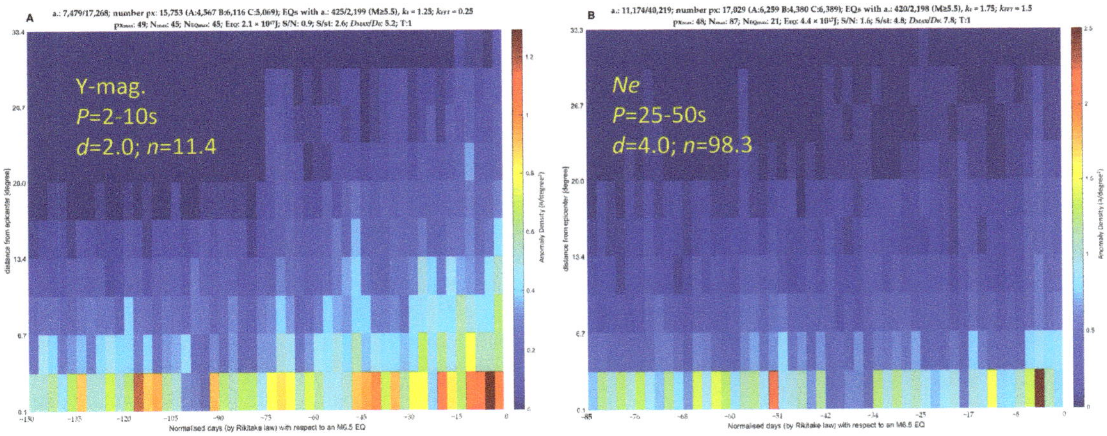

Figure 8. *Swarm* magnetic field data in the shorter period of 2–10 s (Y-mag in (**A**)) and electron density in the longer period of 25–50 s (*Ne* in (**B**)) correlated with M5.5+ earthquakes using modified time according to the empirical Rikitake law.

In the same way, in Figure 8, we apply this approach to the anomalies extracted by analyzing the frequency content of the magnetic field and *Ne* data. The frequency selected is the best from the previous analysis, so we obtained the magnetic field data in the quicker signal (2–10 s) and the *Ne* data in the slower signals (25–50 s). In this case, there is a general gain in the statistical improvements of both the *d* and *n* parameters (for the magnetic field data: *d* passes from 1.8 to 2.0, and *n* passes from 9.5 to 11.4; for the *Ne* data, *d* passes from 2.5 to 4.0, and *n* passes from 27.2 to 98.3, recording a very strong increase in statistical significance).

Finally, we underline that this approach seems very promising but requires a very large dataset. In fact, for the larger magnitude earthquakes, the whole time is not covered, even with 8 years of satellite data. On the other hand, for the visual inspection of the results, the immediate reading is lost with this approach: for example, the magnetic field data of

Figure 7A show several concentrations of anomalies between 75 days and 1 month before the M6.5 earthquakes, but such time corresponds to a different interval for any earthquake magnitude that can be estimated by Equation (4) and is represented in Figure 6A. However, as the empirical Rikitake law seems to describe the possible precursor well, it is fundamental to consider it and include it in a future prediction platform. In practice, detecting an anomaly can indicate a lower magnitude earthquake or a larger magnitude earthquake at a later time, and, if a lower magnitude earthquake does not happen immediately, after a number of days/weeks, the prediction can be adjusted, increasing the magnitude of the expected earthquake. Unfortunately, the larger number of false alarms (the anomalies not associated with any seismic events) makes it still impossible to implement such a prediction system.

4.4. Some Features of the Largest Concentrations

In the results shown in Figure 3, it seems to be observed that the content of the high frequency of the magnetic field signal of the possible earthquake precursor becomes more intense as the earthquake is approaching. There is not a simple explanation for this, and the present theories would suggest the contrary direction. From a seismological point of view, the crack dimension is supposed to increase toward the earthquake (e.g., [43,44]), the duration of the electromagnetic pulses eventually generated by the microfracture is supposed to be proportional to the crack dimension, and the emitted frequency is supposed to be proportional to the inverse of the pulse duration [6,7]. Even supposing that the fault is simplified as a two planar faces capacitor forming an oscillating circuit with the resistance and inductance of the ground (it can be the fluid patches in the rocks that act as a "transmission line" for the higher conductivity of fluids), the frequency "ω" is supposed to decrease with the crack size "l", which increases (a similar calculus that found that $\omega \propto \frac{1}{l}$ is provided by Kachakhidze et al. [45] and Liperovsky et al. [21]). These calculations and theories would not support the result presented in this paper, as they point to the opposite conclusion. However, first of all, we think that the problem is really open. Furthermore, we are not sure that the electromagnetic disturbances observed in the ionosphere are directly produced on the fault, as other phenomena can act at the ground-lower atmosphere interface, as suggested by Kuo et al. [10,11], who simulated the formation of ionospheric disturbances with accumulation on the Earth's surface of positive-holes released from the crack in the fault or a very different and complex chain of processes that influence the ionosphere as a consequence of the global atmospheric electric circuit alteration induced by air ionization from radon release, as theorized by Pulinets and Ouzounov [16]. In the last two mechanisms, it would be very difficult for the frequency of the supposed generated signal to be related to any crack evolution model, so it would be complicated to predict if the frequency of the magnetic field signal can increase or decrease toward the earthquake occurrence. Further studies will be necessary both in theory and in the observations, as shown here.

We finally investigated if the earthquakes involved in the absolute maximum concentrations of Figure 3A–C share some common features, such as focal mechanism or the location of the hypocenter beneath the sea or the land. A similar investigation on the *Swarm Ne* data and on earthquake occurrences over sea or land has been reported by He et al. [46]. Even if the authors did not find a particular "preference" in such classification, they found a better correlation for higher magnitude events, as was also detected in the present work.

First of all, we found that the earthquakes that are associated with the three maximum concentrations in the different analyzed signal bands (146 EQs, 137 EQs, and 146 EQs, respectively) are actually different events (except for two that are shared in all three concentrations), so the increase in frequency toward the earthquake occurrence could be apparent, and it is possible that the frequency of the generated signal depends on other parameters such as those investigated in the following. Table 3 reports the number of earthquakes and their percentage in the specific class. The focal mechanism has been checked, searching for the same earthquake in the global Centroid Moment Tensor (CMT) catalog [47,48] with a maximum tolerance of 2 min on the origin time, 2.5 on the magnitude estimation, and 2 degrees in the epicenter localization. Using one of the rake angles of the moment tensors solution, the focal mechanism has been classified as "reverse" (or "thrust") if +20° < rake < +160°, "normal" if −160° < rake < −20°, or "strike-slip" if −20° ≤ rake ≤ +20° or if | rake | ≥ 160°, following Cronin [49].

Table 3. Investigation of a possible relation with the focal mechanism with the recorded frequency of the magnetic field anomalous signals from the *Swarm* and sea or land epicenter location. The significant percentage of deviation from the standard earthquake distribution (i.e., the ones ≥ 5% in their absolute value) are marked by bold numbers. Green (red) numbers represent positive (negative) deviation percentages with respect to the reference.

Focal Mechanism	Earthquakes with Anomalies in the Band 2–10 s	Earthquakes with Anomalies in the Band 10–25 s	Earthquakes with Anomalies in the Band 25–50 s	All the Earthquakes
strike-slip	**50 (35.2%) −10.1%**	54 (39.4%) +0.6%	**54 (37.2%) −5.0%**	846 (39.2%)
reverse	**67 (47.2%) +10.5%**	**62 (45.3%) +6.0%**	64 (44.1%) +3.4%	922 (42.7%)
normal	25 (17.6%) −2.8%	**21 (15.3%) −15.4%**	27 (18.6%) +2.8%	391 (18.1%)
Sea or land				
Land	**16 (11.0%) −35.2%**	24 (17.5%) +3.6%	**30 (20.6%) +21.5%**	372 (16.9%)
Sea	**130 (89.0%) +7.2%**	113 (82.5%) −0.7%	116 (79.5%) −4.4%	1827 (83.1%)

The classification for the focal mechanism and the sea/land localization of the epicenters has been reported in the last column of Table 3 and for all the earthquakes investigated in this study. Further, it needs to be taken as a reference to search for any deviations from these percentages. We report in green (positive) or red (negative) the deviation percentage with respect to the reference, and in bold, we underline the significant difference for the focal mechanisms. Faster signals (2–10 s) seem more likely (+10.5% with respect to their natural distribution) to be produced by reverse focal mechanism earthquakes at the expense of strike-slip ones (−10.1%). The signals with a period in the range 10–25 s seem less likely (−15.4% with respect to their standard distribution) to be produced by normal fault earthquakes, and for the slower signals, the deviation percentages are not greater than 5%, so they are not really significant. Unfortunately, the detected deviations are not statistically significant, and deeper research in the future will be necessary to understand whether the focal mechanism has no influence on the possible ionospheric electromagnetic earthquake disturbances or if our study is too limited to have a clear detection of such dependence. On the other side, land earthquakes are more likely to show lower frequency signals (from faster to slower frequencies, they are −32.2%, +3.6%, and +21.5%), and, complementarily, the sea earthquakes seem likely to produce faster signals (from faster to slower frequencies, they are +7.2%, −0.7%, and −4.4%), as the percentage shows a clear trend with significant deviations for the highest and lowest frequency signals.

Finally, we find that only two earthquakes (above 146 with anomalies in 2–10 s, 137 in 10–25 s, and 146 in 25–50 s) are common events to all of these concentrations, so the results of this analysis are related to different earthquakes, and it is not clear if the frequency of the eventual electromagnetic wave produced by a single earthquake changes with the time, while the focal mechanism seems to have a slight or null influence on the signal frequency, and the localization in sea or land seems to play an important role for the frequency of the anomalies.

4.5. General Comparison of the Number of "Pre-" and "Post-" Earthquake Anomalies

Finally, we provide an objective comparison of the number of anomalies detected before and after the earthquake with a symmetrical analyzed time in Table 4. In this table, we provide the difference between the anomalies detected before and the ones detected after the earthquake. The percentage of such difference refers to the anomalies after the earthquake, and the uncertainty is estimated as the square root of the maximum number of anomalies. We consider the result significant if the difference is greater than the estimated uncertainty. A negative difference indicates that the number of anomalies detected after the earthquake is greater than the number of those detected before the earthquake. Normalization with respect to the total hours of geomagnetical quiet time is provided to take into account the possible imbalance in the investigated events due to our cut of the geomagnetic perturbed times (i.e., the ones when |Dst| > 20 nT and/or ap > 10 nT). The normalized quantities are provided in the second line of each cell with the label "Norm". Such imbalance due to geomagnetic activity can be due not only to sporadically geomagnetic storms but also to the different solar activity that was at a "minimum" at the beginning of the *Swarm* mission and is now in the increasing stage toward the next "maximum". The incomplete dataset in the first and last 500 days of analysis could enhance such a difference in the solar cycle. Most of the investigations have a significant differential number of anomalies that preceded the earthquake, supporting the hypothesis that some of these anomalies are induced by the preparation phase of earthquakes. The analysis with the higher percentage of pre-earthquake anomalies is the Y East magnetic field in the period band of 2–10 s. Concerning the *Swarm Ne* anomalies in the period band of 25–50 s, we note a relatively high percentage considering the high number of total anomalies (40,219) compared with the result of the "Y, no-band". Furthermore if we look at the general pattern, it is possible to suppose that the pre-seismic anomalies start about 400 days before the earthquake as the color of the concentration becomes darker (about orange, i.e., with a concentration of anomalies equal to or greater than seven anomalies/degree^2) until the earthquake and persists until about 250 days after the earthquake, when it seems to go back at a background level (which shows a yellow color corresponding to about 5.5 anomalies/degree^2). We recognize that such investigations are biased by the fact that higher magnitude earthquakes tend to show anomalies ahead of time, but such earthquakes are fewer in number (according to the Gutenberg–Richter magnitude distribution [42]), so most of the weight on the anomalies is posed by lower magnitude events. Even if 8 years of data seems like quite a long time, only a longer satellite dataset could help to better investigate this aspect. For this reason, we would like to underline the importance of maintaining the orbiting of the *Swarm* constellation (and other missions equipped with magnetometers, such as CSES) for as long as possible.

Table 4. Comparison of the number of anomalies in the closer bin to the epicenter before and after the earthquake for the symmetrical investigations (from −500 days to +500 days) of 8 years of *Swarm* data correlated with M5.5+ earthquakes by the WSC algorithm. A normalization ("Norm") is provided to take into account possible different geomagnetic activities occurring before and after each investigated event.

Parameter, Period Band	Anomalies before the Earthquake	Anomalies after the Earthquake	Difference of the Anomalies	Estimated Uncertainty	Is the Result Significant?
Y, no-band	16,630 Norm: 16,584	16,436 Norm: 16,481	194 (1.2%) Norm: 103 (0.6%)	129	No [1]
Y, 2–10 s	4959 Norm: 4953	4563 Norm: 4568	396 (8.7%) Norm: 385 (8.4%)	70	Yes
Y, 10–25 s	5078 Norm: 5066	4917 Norm: 4928	161 (3.3%) Norm: 137 (2.8%)	71	Yes
Y, 25–50 s	6297 Norm: 6273	5858 Norm: 5880	439 (7.5%) Norm: 392 (6.7%)	79	Yes
Ne, no-band	15,504 Norm: 15,469	15,618 Norm: 15,652	−114 (−0.7%) −182 (−1.2%)	125	Yes, but post-seismic

Table 4. *Cont.*

Parameter, Period Band	Anomalies before the Earthquake	Anomalies after the Earthquake	Difference of the Anomalies	Estimated Uncertainty	Is the Result Significant?
Ne, 2–10 s	3157 Norm: 3136	3337 Norm: 3360	−180 (−5.4%) Norm: −224 (−6.7%)	58	Yes, but post-seismic
Ne, 10–25 s	4661 Norm: 4669	4562 Norm: 4554	99 (2.2%) 114 (2.5%)	68	Yes
Ne, 25–50 s	13,033 Norm: 12,991	12,538 Norm: 12,579	495 (3.9%) Norm: 413 (3.3%)	114	Yes

[1] Such difference is not significant if we take into account the normalization, i.e., it could be due to unbalanced geomagnetic activity in the investigated period.

5. Conclusions

In this paper, we extended the work published by De Santis et al. [29] to a longer period (8 years) than that originally analyzed (4.7 years) of *Swarm* data correlated with M5.5+ shallow earthquakes. Here, we not only provide an analysis extended to the first 8 years of *Swarm* data but also introduce some new methodologies to improve the definition of the anomalies (by frequency analysis) and in terms of better investigation of the results—in particular, searching for possible influences of the seismo-tectonic context on the possible pre-earthquake electromagnetic signals. New methodologies and investigations are proposed to take into account the frequency of the signal and the different anticipation times of anomalies expected for several magnitude earthquakes. We found that several concentrations of ionospheric anomalies are statistically significant, well overpassing the given thresholds of $d \geq 1.5$ and $n \geq 4.0$—i.e., they are very much greater than the concentrations obtained from the space-time-homogenously-distributed anomalies, and we can infer that the earthquakes associated with such significant concentrations could be more likely "predicted", even if this is not the goal of the present work. By the ROC curve, we confirm that the anomalies selected by WSC and the aforementioned statistical criteria over "d" and "n" have predicting capabilities, showing an AUC that is 1.67 times better than that of a random predictor. Future studies could be oriented to filter the signals that are most prone to earthquake-induced anomalies (for example, Zhu et al. [50]) in order to better depict the signal "features" of possible seismo-induced phenomena in the ionosphere. Nevertheless, although such ionospheric pre-earthquake disturbances are still widely debated, there is some evidence, in the literature and in this paper, that they statistically exist.

Our main conclusions are:

1. The anticipation time "ΔT" of the anomaly increases with the magnitude of the incoming earthquakes following the Rikitake laws [37], with these specific coefficients being $\Delta T_{mag} = 10^{-0.96+0.51\cdot M}$ and $\Delta T_{Ne} = 10^{-3.46+0.83\cdot M}$ for the magnetic field "mag" and electron density data, respectively. The anticipation time of large earthquakes (M7.5+) seems to be some years before the event and has been detected.
2. The focal mechanism seems to have a small or null influence on the generated frequency of the possible pre-earthquake anomalies.
3. Earthquakes localized in the land areas tend to be preceded by lower frequency anomalous signals, while sea earthquakes are more likely to be preceded by faster signal anomalies.
4. The *Swarm* magnetic field signal anomalies generally show a better correlation with earthquakes than the electron density ones do.
5. A more selective set of parameters, achieved here by the investigation of the signal frequency, reduces the size of the anomaly dataset, and it is shown that the possible correlation with the seismic event has a higher statistical significance for both the magnetic field and *Ne* observations.
6. Frequency analysis seems to be fundamental in some cases: for electron density, we find a higher correlation with anomalies, with a signal period in the range of 25–50 s.

7. All the results in this paper have been tested with the "confusion matrix" approach, reaching an accuracy from 75% to 95% and an alarmed time-space from 0.7% to 19.1%. The real results show a predicting capability that is 1.67 times better than that of a random predictor, according to the AUC of the ROC curve, which further proves a prediction capability of the best detected ionospheric anomalies by WSC.

The future perspective of this work can include, but is not limited to, an improvement of the anomaly definition criteria and the extension to a larger dataset and to other satellites such as CSES-01. By the way, CNSA, together with the Italian Space Agency (ASI), is preparing a second CSES-02 satellite that is planned to be launched next year, forming the first-ever pair of satellites fully dedicated to searching for the possible ionospheric precursors of strong earthquakes.

Finally, we think that this study is fundamental to assessing the Lithosphere-Atmosphere-Ionosphere Coupling mechanism and for a future prediction system. Further, it integrates such kinds of studies with seismological investigations of the Earth's surface and chemical–physical analyses of the atmosphere (e.g., a preliminary comparison between magnetic anomalies and atmospheric surface temperature has been performed by Ghramy et al. [51]). So, as the earthquake is a complex phenomenon, by looking at all of the geo-layers together (as suggested in the "Geosystemics" approach by De Santis et al. [52]), it will be possible to better understand its physical features in its preparation phase and, hopefully, predict it one day.

Supplementary Materials: The following are available online at https://www.mdpi.com/article/10.3390/rs14112649/s1, Figure S1: Example of data processing to extract the anomaly, Figure S2: WSC applied to *Swarm* mag. from −500 to 500 days, Figure S3: WSC applied to *Swarm* mag. from −1000 days until the earthquake origin time, Figure S4: WSC applied to the shuffled magnitude earthquake catalog, Figure S5: Fit of anticipation time versus signal period, Table S1: Statistical evaluation of the results for the *Swarm* magnetic field, Figure S6: WSC applied to *Swarm Ne* with four different methods, Figure S7: WSC applied to *Swarm Ne* from −500 to 500 days, Figure S8: WSC applied to *Swarm Ne* in five earthquake magnitude bands, Table S2: Statistical evaluation of the results for the *Swarm Ne*, Figure S9: Comparison of the Rikitake law coefficients obtained in this work with the previous ones.

Author Contributions: Conceptualization, D.M. and A.D.S.; methodology, D.M., A.D.S. and S.A.C.; software, D.M., S.A.C. and M.S.; validation, S.A.C. and M.S.; formal analysis, D.M.; investigation, D.M.; resources, K.Z. and D.M.; data curation, D.M.; interpretation, D.M. and A.D.S.; writing—original draft preparation, D.M.; writing—review and editing, all authors; visualization, D.M.; supervision, A.D.S. and K.Z.; project administration, K.Z., D.M. and A.D.S.; funding acquisition, K.Z., D.M. and A.D.S. All authors have read and agreed to the published version of the manuscript.

Funding: This research was funded by the National Natural Science Foundation of China, grant number 41974084; the China Postdoctoral Science Foundation, grant number 2021M691190; the Italian Space Agency, grant number 2020-32-HH.0 (Limadou-Science+); the International Cooperation Project of the Department of Science and Technology of Jilin Province, grant number 20200801036GH; and the Italian Ministry of University and Research, grant number D53J19000170001 (Pianeta Dinamico—Working Earth).

Data Availability Statement: The *Swarm* satellite data are freely available from ftp and http servers (swarm-diss.eo.esa.int, last accessed on 3 January 2022). The USGS Earthquake catalog is freely accessible at https://earthquake.usgs.gov/earthquakes/map (accessed on 1 February 2022).

Acknowledgments: We warmly thank Francisco Javier Pavón-Carrasco for the software developed to analyze the *Swarm* data, and we would like to acknowledge Guido Ventura, Rita di Giovambattista, Loredana Perrone, Alessandro Piscini, Roger Haagmans, Giorgiana De Franceschi, Luca Spogli, Claudio Cesaroni, Ilaria Spassiani, and Matteo Taroni, who participated in some of the discussions about this work and its previous investigations carried out in the framework of the SAFE (SwArm For Earthquake study) ESA-funded project and in the following activities and projects.

Conflicts of Interest: The authors declare no conflict of interest. The funders had no role in the design of the study; in the collection, analyses, or interpretation of data; in the writing of the manuscript; or in the decision to publish the results.

Appendix A. Evaluation of the Declustering of the Earthquake Catalog on the Worldwide Statistical Correlation Results

This appendix is dedicated to evaluating how the decluster technique that was used to pre-process the earthquake catalog could have affected the results.

Determining if an earthquake is an isolated event or part of a seismic swarm or sequence—and, in this last case, discriminating between foreshocks, mainshocks, and aftershocks—has been always a very challenging topic for seismologists [53]. Very recent frontiers of seismology are even trying to evaluate in real-time if an earthquake is a foreshock or a mainshock by a traffic-light scheme based on the evaluation of the *b*-value of the Gutenberg–Richter distribution [54].

Declustering an earthquake catalog means the removal of the foreshocks and aftershocks in order to identify the "clusters" of seismicity led by the mainshock. The detection of the cluster is a difficult problem, and several approaches are available, such as that of Reasenberg [35] that we applied in this paper, which uses a radius around the event (a radius of 10 km for the present paper) and a time window to search for foreshocks (with 10 days in advance, for this paper) and aftershocks (in this paper: 20 days). If another earthquake is detected inside the space-time researched area, the same approach is applied until no other events are detected. The algorithm also takes into account the uncertainty of the earthquake localization that we selected from the values directly available in the USGS catalog. A different choice of such free parameters of the algorithm yields obviously different results. For example, in De Santis et al. [29], a shorter time window was used (1 day before and 10 days after), but we found that some events that occurred in the same tectonic context in a short time were not detected as the same cluster for such strict time windows; therefore, in the present work, we decided to use a slightly longer time window. As already presented in the discussion section, the statistical results are very consistent among the two works, demonstrating that, even with different decluster parameters in the Reasenberg approach and the slightly different earthquake catalog, the main results of our Worldwide Statistical Correlation (WSC) algorithm are not affected significantly. The reason for this is probably that, in the final WSC results, the bins are generally large in space (3.34°) and time (from 2.4 days to 20 days), so the low "space and time" resolution homogenizes the differences among the two declustered earthquake catalogs. Here, we want to explore how the results are affected by using another declustering technique—in particular, that provided by Urhammer [55]—or by using the original undeclustered catalog. More recent seismological techniques have been developed, such as the Epidemic-Type Aftershock Sequence (ETAS) model that was also applied to decluster earthquake catalogs [56]. The ETAS technique permits the obtention of the probability that an earthquake can trigger another one (or be triggered by another one) and also has some forecast capabilities [57]. ETAS declustering has been applied to global seismicity by Nandan et al. [58], but this is beyond of the scope of this paper (to test it by WSC). We propose, as a future improvement, the study of the ionospheric electromagnetic disturbances combined with the ETAS investigation of seismicity. Different decluster techniques have been compared by Mizrahi et al. [59] in terms of the Gutenberg–Richter parameters obtained after declustering the catalog and were also compared with the original catalog. From [59], it is possible to note that Reasenberg [35] and Urhammer [55] provide a different, non-superposed set of *a*- and *b*-values of the Gutenberg–Richter earthquake magnitude distribution, so we consider them as independent for the next test. We reanalyzed the *Swarm* magnetic field data presented in Figure 1 with the Urhammer declustered catalog and the earthquakes not declustered by associating all the anomalies to all the earthquakes (Method 1). In addition, we also propose Method 3 (maximum magnitude) to analyze the non-declustered catalog, as it could, in principle, provide similar results to a declustered catalog. In fact, the declustering means extracting the mainshocks that have a higher magnitude than the foreshocks and aftershocks. So, associating the anomaly with the maximum magnitude earthquake could mean linking it with the mainshock. Contrariwise, we need to note that the seismic radius used by the decluster techniques tends to be very much smaller (e.g., 10 km) than the Dobrovolsky's radius used by our searching algorithm (e.g., ~600 km for M6.5), so it is not the same approach as that using the declustered catalog.

Figure A1 reports the results of the WSC applied to 8 years of *Swarm* magnetic field data and M5.5+ shallow earthquakes declustered by the above-mentioned techniques (on the left side: A and C) or not declustered (on the right side: B and D). In subfigure D, the method of "Max (magnitude)" has been applied to the WSC using the non-declustered earthquake catalog (i.e., the original one). It is possible to note that, in any case, some concentrations of anomalies at about 30 days and 80 days before the earthquake are underlined. Such evidence provides more proof that these concentrations are real and, furthermore, that they are not dependent on the way in which the earthquake catalog is treated. Comparing the two results with the declustered catalogs (subfigures A and C), it is possible to note that the results are very similar, even if the decluster technique is different. The pattern of concentrations is almost equal, with slightly differences, and the statistical significance is comparable. Even without declustering the earthquakes, some high-concentration bins are preserved, but such a choice leads to the overcounting of the anomalies with a value of 1.68 anomalies–EQ links over the number of anomalies associated with at least one earthquake. The overcounting is lower with the declustering by Reasenberg (1.36) [35] compared to that obtained by applying Urhammer's technique (1.41) [55], so even if the statistical significance in terms of d and n is a bit lower with Reasenberg [35], we think that such a technique is a better choice than that of Urhammer [55], as it has a lower overcounting of anomalies. Finally, the choice of the maximum magnitude with the non-declustered earthquake catalog confirms the other results; however, it also loses some seismic information that the declustering algorithm takes into account. In fact, this technique operates on a distance that, in several cases, goes behind the typical stress-interaction region between the earthquakes, as already mentioned above.

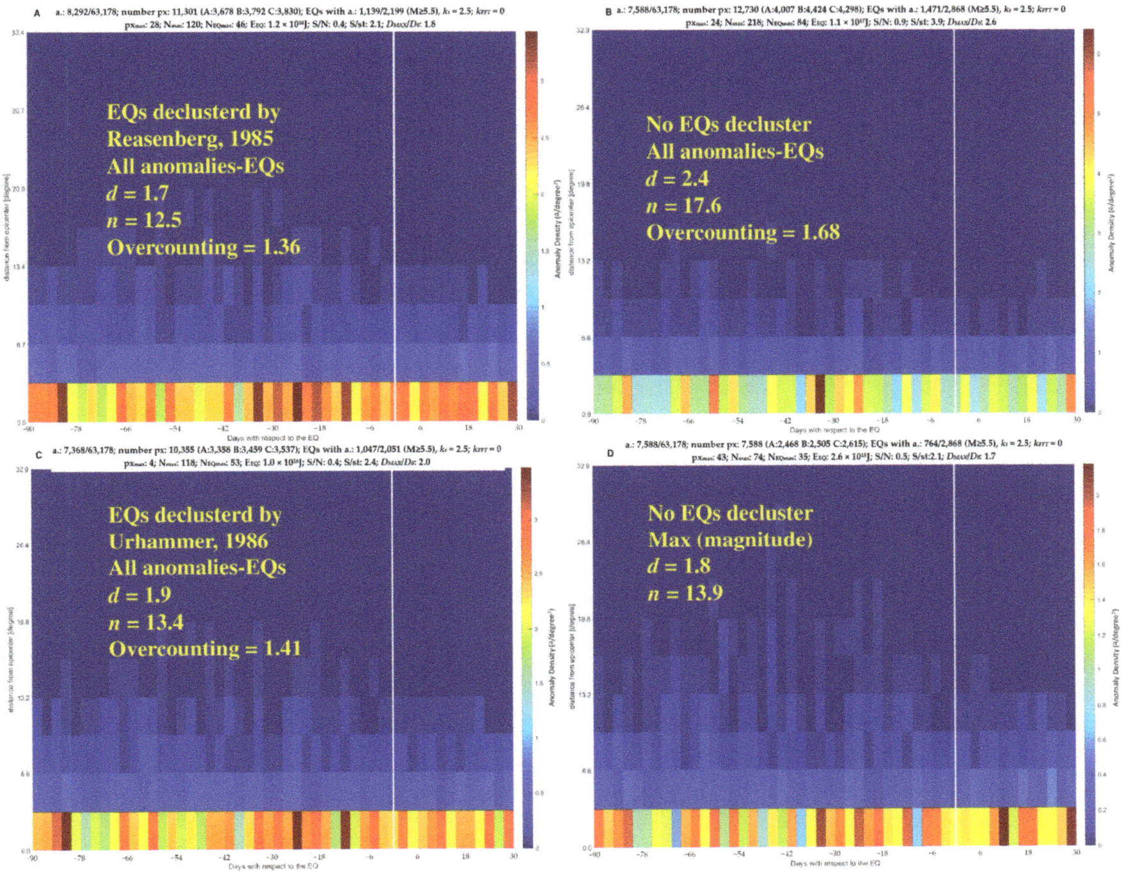

Figure A1. Comparison of the results of the Worldwide Statistical Correlation using two different declustering methods or by not declustering the catalog. The *Swarm* magnetic field data investigated from 90 days before the earthquake until 30 days after it are used for this example. The earthquake catalog has been declustered with: (**A**) Reasenberg, 1985 [35]; (**C**) Urhammer, 1986 [55]. The earthquake catalog used for the WSC of (**B**,**D**) has not been declustered. In (**A**–**C**), all anomalies have been assigned to any compatible earthquake, and the overcounting factor (the number of links between the anomalies and EQs divided by the number of anomalies with EQs) has been reported, while, in (**D**), each anomaly has been connected only to the higher magnitude earthquake among the possible ones (thus, there is no overcounting).

References

1. Rikitake, T. Classification of earthquake precursors. *Tectonophysics* **1979**, *54*, 293–309. [CrossRef]
2. Conti, L.; Picozza, P.; Sotgiu, A. A Critical Review of Ground Based Observations of Earthquake Precursors. *Front. Earth Sci.* **2021**, *9*, 676766. [CrossRef]
3. Picozza, P.; Conti, L.; Sotgiu, A. Looking for Earthquake Precursors From Space: A Critical Review. *Front. Earth Sci.* **2021**, *9*. [CrossRef]
4. Fraser-Smith, A.C.; Bernardi, A.; McGill, P.R.; Ladd, M.E.; Helliwell, R.A.; Villard, O.G., Jr. Low-frequency magnetic field measurements near the epicenter of the M_s7.1 Loma Prieta Earthquake. *Geophys. Res. Lett.* **1990**, *17*, 1465–1468. [CrossRef]
5. Molchanov, O.A.; Kopytenko, Y.A.; Voronov, P.M.; Matiashvili, T.G.; Fraser-Smith, A.C.; Bernardi, A. Results of ULF magnetic field measurements near the epicenters of the Spitak (M_s = 6.9) and Loma Prieta (M_s = 7.1) earthquakes: Comparative analysis. *Geophys. Res. Lett.* **1992**, *19*, 1495–1498. [CrossRef]
6. Molchanov, O.A.; Hayakawa, M. Generation of ULF electromagnetic emissions by microfracturing. *Geophys. Res. Lett.* **1995**, *22*, 3091–3094. [CrossRef]

7. Molchanov, O.; Hayakawa, M. On the generation mechanism of ULF seismogenic electromagnetic emissions. *Phys. Earth Planet. Inter.* **1998**, *105*, 201–210. [CrossRef]
8. Freund, F.T. Pre-earthquake signals: Underlying physical processes. *J. Southeast Asian Earth Sci.* **2011**, *41*, 383–400. [CrossRef]
9. Freund, F.; Ouillon, G.; Scoville, J.; Sornette, D. Earthquake precursors in the light of peroxy defects theory: Critical review of systematic observations. *Eur. Phys. J. Spec. Top.* **2021**, *230*, 7–46. [CrossRef]
10. Kuo, C.L.; Lee, L.C.; Huba, J.D. An improved coupling model for the lithosphere-atmosphere-ionosphere system. *J. Geophys. Res. Space Phys.* **2014**, *119*, 3189–3205. [CrossRef]
11. Kuo, C.-L.; Ho, Y.-Y.; Lee, L.-C. Electrical Coupling Between the Ionosphere and Surface Charges in the Earthquake Fault Zone. In *Pre-Earthquake Processes: A Multidisciplinary Approach to Earthquake Prediction Studies*; John Wiley & Sons, Inc.: Hoboken, NJ, USA, 2018; pp. 99–124. [CrossRef]
12. Prokhorov, B.E.; Zolotov, O.V. Comment on "An improved coupling model for the lithosphere-atmosphere-ionosphere system" by Kuo et al. [2014]. *J. Geophys. Res. Space Phys.* **2017**, *122*, 4865–4868. [CrossRef]
13. Kuo, C.-L.; Lee, L.-C. Reply to comment by B. E. Prokhorov and O. V. Zolotov on "An improved coupling model for the lithosphere-atmosphere-ionosphere system". *J. Geophys. Res. Space Phys.* **2017**, *122*, 4869–4874. [CrossRef]
14. Denisenko, V.V.; Boudjada, M.Y.; Lammer, H. Propagation of Seismogenic Electric Currents Through the Earth's Atmosphere. *J. Geophys. Res. Space Phys.* **2018**, *123*, 4290–4297. [CrossRef]
15. Denisenko, V.V.; Nesterov, S.A.; Boudjada, M.Y.; Lammer, H. A mathematical model of quasistationary electric field penetration from ground to the ionosphere with inclined magnetic field. *J. Atmospheric Sol.-Terr. Phys.* **2018**, *179*, 527–537. [CrossRef]
16. Pulinets, S.A.; Ouzounov, D. Lithosphere–Atmosphere–Ionosphere Coupling (LAIC) model—An unified concept for earthquake precursors validation. *J. Asian Earth Sci.* **2011**, *41*, 371–382. [CrossRef]
17. Scholz, C.H.; Sykes, L.R.; Aggarwal, Y.P. Earthquake Prediction: A Physical Basis. *Science* **1973**, *181*, 803–810. [CrossRef]
18. Ventura, G.; Di Giovambattista, R. Fluid pressure, stress field and propagation style of coalescing thrusts from the analysis of the 20 May 2012 M_L5.9 Emilia earthquake (Northern Apennines, Italy). *Terra Nova* **2013**, *25*, 72–78. [CrossRef]
19. Fidani, C.; Orsini, M.; Iezzi, G.; Vicentini, N.; Stoppa, F. Electric and Magnetic Recordings by Chieti CIEN Station During the Intense 2016–2017 Seismic Swarms in Central Italy. *Front. Earth Sci.* **2020**, *8*, 536332. [CrossRef]
20. Marchetti, D.; De Santis, A.; D'Arcangelo, S.; Poggio, F.; Jin, S.; Piscini, A.; Campuzano, S.A. Magnetic Field and Electron Density Anomalies from Swarm Satellites Preceding the Major Earthquakes of the 2016–2017 Amatrice-Norcia (Central Italy) Seismic Sequence. *Pure Appl. Geophys.* **2019**, *177*, 305–319. [CrossRef]
21. Liperovsky, V.A.; Pokhotelov, O.A.; Meister, C.-V.; Liperovskaya, E.V. Physical models of coupling in the lithosphere-atmosphere-ionosphere system before earthquakes. *Geomagn. Aeron.* **2008**, *48*, 795–806. [CrossRef]
22. Parrot, M.; Li, M. Demeter results related to seismic activity. *URSI Radio Sci. Bull.* **2015**, *2015*, 18–25. [CrossRef]
23. Yan, R.; Parrot, M.; Pinçon, J.-L. Statistical Study on Variations of the Ionospheric Ion Density Observed by DEMETER and Related to Seismic Activities. *J. Geophys. Res. Space Phys.* **2017**, *122*, 12421–12429. [CrossRef]
24. Ouyang, X.Y.; Parrot, M.; Bortnik, J. ULF Wave Activity Observed in the Nighttime Ionosphere Above and Some Hours Before Strong Earthquakes. *J. Geophys. Res. Space Phys.* **2020**, *125*, e2020JA028396. [CrossRef]
25. Zolotov, O.V. Ionosphere quasistatic electric fields disturbances over seismically active regions as inferred from satellite-based observations: A review. *Russ. J. Phys. Chem. B* **2015**, *9*, 785–788. [CrossRef]
26. Akhoondzadeh, M.; De Santis, A.; Marchetti, D.; Piscini, A.; Jin, S. Anomalous seismo-LAI variations potentially associated with the 2017 Mw = 7.3 Sarpol-e Zahab (Iran) earthquake from Swarm satellites, GPS-TEC and climatological data. *Adv. Space Res.* **2019**, *64*, 143–158. [CrossRef]
27. De Santis, A.; Marchetti, D.; Spogli, L.; Cianchini, G.; Pavón-Carrasco, F.J.; De Franceschi, G.; Di Giovambattista, R.; Perrone, L.; Qamili, E.; Cesaroni, C.; et al. Magnetic Field and Electron Density Data Analysis from Swarm Satellites Searching for Ionospheric Effects by Great Earthquakes: 12 Case Studies from 2014 to 2016. *Atmosphere* **2019**, *10*, 371. [CrossRef]
28. Marchetti, D.; De Santis, A.; Campuzano, S.A.; Soldani, M.; Piscini, A.; Sabbagh, D.; Cianchini, G.; Perrone, L.; Orlando, M. *Swarm* Satellite Magnetic Field Data Analysis Prior to 2019 Mw = 7.1 Ridgecrest (California, USA) Earthquake. *Geosciences* **2020**, *10*, 502. [CrossRef]
29. De Santis, A.; Marchetti, D.; Pavón-Carrasco, F.J.; Cianchini, G.; Perrone, L.; Abbattista, C.; Alfonsi, L.; Amoruso, L.; Campuzano, S.A.; Carbone, M.; et al. Precursory worldwide signatures of earthquake occurrences on Swarm satellite data. *Sci. Rep.* **2019**, *9*, 20287. [CrossRef]
30. De Santis, A.; Marchetti, D.; Perrone, L.; Campuzano, S.A.; Cianchini, G.; Cesaroni, C.; Di Mauro, D.; Orlando, M.; Piscini, A.; Sabbagh, D.; et al. Statistical correlation analysis of strong earthquakes and ionospheric electron density anomalies as observed by CSES-01. *Nuovo Cim. C* **2021**, *44*, 1–4. [CrossRef]
31. Friis-Christensen, E.; Lühr, H.; Hulot, G. *Swarm*: A constellation to study the Earth's magnetic field. *Earth Planets Space* **2006**, *58*, 351–358. [CrossRef]
32. Pinheiro, K.J.; Jackson, A.; Finlay, C.C. Measurements and uncertainties of the occurrence time of the 1969, 1978, 1991, and 1999 geomagnetic jerks. *Geochem. Geophys. Geosyst.* **2011**, *12*, Q10015. [CrossRef]
33. Yan, R.; Zhima, Z.; Xiong, C.; Shen, X.; Huang, J.; Guan, Y.; Zhu, X.; Liu, C. Comparison of Electron Density and Temperature From the CSES Satellite With Other Space-Borne and Ground-Based Observations. *J. Geophys. Res. Space Phys.* **2020**, *125*, e2019ja027747. [CrossRef]

34. Marchetti, D.; De Santis, A.; Shen, X.; Campuzano, S.A.; Perrone, L.; Piscini, A.; Di Giovambattista, R.; Jin, S.; Ippolito, A.; Cianchini, G.; et al. Possible Lithosphere-Atmosphere-Ionosphere Coupling effects prior to the 2018 Mw = 7.5 Indonesia earthquake from seismic, atmospheric and ionospheric data. *J. Southeast Asian Earth Sci.* **2019**, *188*, 104097. [CrossRef]
35. Reasenberg, P. Second-order moment of central California seismicity, 1969–1982. *J. Geophys. Res. Earth Surf.* **1985**, *90*, 5479–5495. [CrossRef]
36. Dobrovolsky, I.P.; Zubkov, S.I.; Miachkin, V.I. Estimation of the size of earthquake preparation zones. *Pure Appl. Geophys.* **1979**, *117*, 1025–1044. [CrossRef]
37. Rikitake, T. Earthquake precursors in Japan: Precursor time and detectability. *Tectonophysics* **1987**, *136*, 265–282. [CrossRef]
38. Song, R.; Hattori, K.; Zhang, X.; Liu, J.; Yoshino, C. Detecting the Ionospheric Disturbances in Japan Using the Three-Dimensional Computerized Tomography. *J. Geophys. Res. Space Phys.* **2021**, *126*, e2020ja028561. [CrossRef]
39. Shcherbakov, R.; Turcotte, D.L.; Rundle, J.B.; Tiampo, K.F.; Holliday, J.R. Forecasting the Locations of Future Large Earthquakes: An Analysis and Verification. *Pure Appl. Geophys.* **2010**, *167*, 743–749. [CrossRef]
40. Xiong, P.; Marchetti, D.; De Santis, A.; Zhang, X.; Shen, X. SafeNet: SwArm for Earthquake Perturbations Identification Using Deep Learning Networks. *Remote Sens.* **2021**, *13*, 5033. [CrossRef]
41. Fawcett, T. An Introduction to ROC analysis. *Pattern Recogn. Lett.* **2006**, *27*, 861–874. [CrossRef]
42. Gutenberg, B.; Richter, C.F. *Seismicity of the Earth and Associated Phenomena*, 2nd ed.; Princeton University Press: Princeton, NJ, USA, 1954.
43. McLaskey, G.C. Earthquake Initiation From Laboratory Observations and Implications for Foreshocks. *J. Geophys. Res. Solid Earth* **2019**, *124*, 12882–12904. [CrossRef]
44. Scholz, C. *The Mechanics of Earthquakes and Faulting*, 3rd ed.; Cambridge University Press: Cambridge, UK, 2019. [CrossRef]
45. Kachakhidze, M.K.; Kachakhidze, N.K.; Kaladze, T.D. A model of the generation of electromagnetic emissions detected prior to earthquakes. *Phys. Chem. Earth Parts A/B/C* **2015**, *85–86*, 78–81. [CrossRef]
46. He, Y.; Zhao, X.; Yang, D.; Wu, Y.; Li, Q. A study to investigate the relationship between ionospheric disturbance and seismic activity based on Swarm satellite data. *Phys. Earth Planet. Inter.* **2021**, *323*, 106826. [CrossRef]
47. Dziewonski, A.M.; Chou, T.-A.; Woodhouse, J.H. Determination of earthquake source parameters from waveform data for studies of global and regional seismicity. *J. Geophys. Res. Earth Surf.* **1981**, *86*, 2825–2852. [CrossRef]
48. Ekström, G.; Nettles, M.; Dziewoński, A. The global CMT project 2004–2010: Centroid-moment tensors for 13,017 earthquakes. *Phys. Earth Planet. Inter.* **2012**, *200*, 1–9. [CrossRef]
49. Cronin Vincent, S. A Primer on Focal Mechanism Solutions for Geologists. Lecture 11 of "OCEAN/ESS 410, Fall 2014 Marine Geology and Geophysics". 2010. Available online: https://www.ocean.washington.edu/courses/oc410/backgroundreading.html (accessed on 8 January 2022).
50. Zhu, K.; Fan, M.; He, X.; Marchetti, D.; Li, K.; Yu, Z.; Chi, C.; Sun, H.; Cheng, Y. Analysis of Swarm Satellite Magnetic Field Data Before the 2016 Ecuador (Mw = 7.8) Earthquake Based on Non-negative Matrix Factorization. *Front. Earth Sci.* **2021**, *9*, 621976. [CrossRef]
51. Ghamry, E.; Mohamed, E.K.; Abdalzaher, M.S.; Elwekeil, M.; Marchetti, D.; De Santis, A.; Hegy, M.; Yoshikawa, A.; Fathy, A. Integrating Pre-Earthquake Signatures From Different Precursor Tools. *IEEE Access* **2021**, *9*, 33268–33283. [CrossRef]
52. De Santis, A.; Abbattista, C.; Alfonsi, L.; Amoruso, L.; Campuzano, S.A.; Carbone, M.; Cesaroni, C.; Cianchini, G.; De Franceschi, G.; De Santis, A.; et al. Geosystemics View of Earthquakes. *Entropy* **2019**, *21*, 412. [CrossRef] [PubMed]
53. Jordan, T.H.; Chen, Y.-T.; Gasparini, P.; Madariaga, R.; Main, I.; Marzocchi, W.; Papadopoulos, G.; Sobolev, G.; Yamaoka, K.; Zschau, J. Operational Earthquake Forecasting—State of Knowledge and Guidelines for Utilization. *Ann. Geophys.* **2011**, *54*, 316–391. [CrossRef]
54. Gulia, L.; Wiemer, S. Real-time discrimination of earthquake foreshocks and aftershocks. *Nature* **2019**, *574*, 193–199. [CrossRef]
55. Uhrhammer, R. Characteristics of Northern and Central California Seismicity. *Earthq. Notes* **1986**, *57*, 21.
56. Console, R.; Jackson, D.D.; Kagan, Y.Y. Using the ETAS Model for Catalog Declustering and Seismic Background Assessment. *Pure Appl. Geophys.* **2010**, *167*, 819–830. [CrossRef]
57. Nandan, S.; Ouillon, G.; Sornette, D.; Wiemer, S. Forecasting the Rates of Future Aftershocks of All Generations Is Essential to Develop Better Earthquake Forecast Models. *J. Geophys. Res. Solid Earth* **2019**, *124*, 8404–8425. [CrossRef]
58. Nandan, S.; Kamer, Y.; Ouillon, G.; Hiemer, S.; Sornette, D. Global models for short-term earthquake forecasting and predictive skill assessment. *Eur. Phys. J. Spec. Top.* **2021**, *230*, 425–449. [CrossRef]
59. Mizrahi, L.; Nandan, S.; Wiemer, S. The Effect of Declustering on the Size Distribution of Mainshocks. *Seism. Res. Lett.* **2021**, *92*, 2333–2342. [CrossRef]

remote sensing

Article

Study on Electron Density Anomalies Possibly Related to Earthquakes Based on CSES Observations

Chengcheng Han [1,2], Rui Yan [1,2,*], Dedalo Marchetti [3], Weixing Pu [1], Zeren Zhima [1], Dapeng Liu [1], Song Xu [1], Hengxin Lu [1] and Na Zhou [1]

1 National Institute of Natural Hazards, Ministry of Emergency Management of China, Beijing 100085, China; 21661317@st.cidp.edu.cn (C.H.); puweixing22@mails.ucas.ac.cn (W.P.); zerenzhima@ninhm.ac.cn (Z.Z.); dapengliu@ninhm.ac.cn (D.L.); songxu@ninhm.ac.cn (S.X.); hengxinlu@ninhm.ac.cn (H.L.); nazhou@ninhm.ac.cn (N.Z.)
2 School of Information Engineering, Institute of Disaster Prevention, Langfang 065201, China
3 College of Instrumentation and Electrical Engineering, Jilin University, Changchun 130061, China; dedalomarchetti@jlu.edu.cn
* Correspondence: ruiyan@ninhm.ac.cn

Abstract: This research examines the correlation between seismic activity and variations in ionospheric electron density (Ne) using the data from the Langmuir probe (LAP) onboard the China Seismo-Electromagnetic Satellite (CSES) during nighttime. Statistical analysis of Ms $\geq$ 6.8 earthquakes that occurred globally between August 2018 and March 2023 is conducted, as well as Ms $\geq$ 6.0 earthquakes in China during the same period, using the quartile analysis method for fixed revisiting orbits. The main conclusions are that: (1) the larger the magnitude of the earthquake, the more anomalous the phenomena that appear; (2) the anomalies on the east side of the epicenter are significantly higher than those on the west side, and the anomalies in the Northern Hemisphere are mostly distributed southward from the epicenter, while those in the Southern Hemisphere are mostly distributed northward from the epicenter; (3) anomalies appear with a higher frequency on several specific time intervals, including the day of the earthquake (likely co-seismic effect) and 2, 7, and 11 days before the earthquake (possible precursor candidates); and (4) for the 15 earthquakes of Ms $\geq$ 6.0 in China over the past five years, anomalous Ne mainly occurred southwest of the epicenter, with the highest frequency observed 5 days before the earthquake, and there were continuous anomalous phenomena between 9 days and 5 days before the earthquake. This study concludes that Ne, measured by CSES, can play a fundamental role in studying earthquake-related ionospheric disturbances.

Keywords: CSES; electron density; earthquake; seismic anomaly

Citation: Han, C.; Yan, R.; Marchetti, D.; Pu, W.; Zhima, Z.; Liu, D.; Xu, S.; Lu, H.; Zhou, N. Study on Electron Density Anomalies Possibly Related to Earthquakes Based on CSES Observations. *Remote Sens.* **2023**, *15*, 3354. https://doi.org/10.3390/rs15133354

Academic Editor: Roberto Orosei

Received: 14 June 2023
Revised: 27 June 2023
Accepted: 27 June 2023
Published: 30 June 2023

1. Introduction

Earthquakes are sudden natural disasters that can cause extensive destruction and pose a significant threat to human life and property. Earthquake is one of the most serious natural disasters faced by humanity, causing more casualties than any other natural disaster. Globally, over 5 million earthquakes are estimated every year, including around 100 strong earthquakes of Ms $\geq$ 6.0, 18 large earthquakes of Ms $\geq$ 7.0, and one or two mega-earthquakes of Ms $\geq$ 8.0. These events may significantly affect people's safety and the economic development of society, making it crucially important to conduct sufficient research on earthquakes and attempt to build proper systems for early warning and forecasting [1].

With the development of space exploration technology, the observation of ionospheric anomalies by satellite has become a research hotspot. A large number of observational studies have shown that there is indeed a correlation between earthquakes and ionospheric anomalies (e.g., [2]), and these ionospheric disturbances possibly related to "earthquake precursors" have obtained significant interest. Utilizing satellite earth observation, researchers can access all-weather, short-term, global coverage, and rich dynamic information that displays enormous potential for application in earthquake mechanism research, monitoring and forecasting, disaster prevention, and emergency relief efforts. It also provides favorable support for short-term forecasting of earthquakes and has emerged as an increasingly important support tool for earthquake monitoring and forecasting and scientific research. The study of seismic ionospheric effects has also become one of the most challenging topics in the fields of earthquake science and ionospheric science [3].

Previous research has demonstrated the ionosphere's response to seismic activity. Specifically, ionospheric plasma parameter changes were reported as early as 1965, when ionospheric anomalies were observed before the 28 March 1964 Alaska earthquake, based on ionosonde data [4]. Subsequent studies using data from the French DEMETER satellite since its launch in 2004 have further expanded our understanding of this phenomenon, resulting in numerous reports of variations in ionospheric parameters preceding earthquakes. Scholars have studied seismic ionospheric anomalies using various methods. Zhang et al. [5] found that multiple parameters, including ion density, temperature, and electromagnetic field, exhibited synchronous perturbation anomalies before the M7.9 earthquake in Chile on 14 November 2007. Further analysis of the Ms7.1 Yushu earthquake on 14 April 2010 revealed that multiple parameters showed increased anomalous change one day before the earthquake [6]. By analyzing DEMETER data ten days before and one day after the M8.8 earthquake in Chile on 27 February 2010, Liu et al. [3] observed significant perturbations in various plasma parameters near the epicenter. Case studies indicate that strong anomalous perturbations typically occur in the ionosphere before and after moderate to strong earthquakes worldwide, and the larger the magnitude of the earthquake, the more obvious the anomaly [7].

In order to identify common features in earthquake ionospheric phenomena, statistical studies were conducted alongside seismic case studies. Several analyses investigated the correlation between ionospheric anomalies and seismic events, ultimately confirming a statistically significant relationship between them. Kon et al. [8] employed the superposed epoch method to scrutinize earthquakes in Japan from 1998 to 2010 with a focal depth less than 40 km and $M \geq 6.0$. Significant positive TEC anomalies were found within a range of 1000 km from the epicenter 1–5 days prior to earthquakes. Liu et al. [9] examined plasma parameter disturbances from 82 earthquakes with $Ms \geq 7.0$ occurring during 2005–2010, from 10 days before the earthquake until 2 days after it. They found that shallow earthquakes (depth < 70 km) exhibited more perturbations before their occurrence than deep earthquakes (depth $\geq$ 70 km).

Using ion density data detected by the DEMETER satellite from 2004 to 2010, Yan et al. [2] scrutinized global earthquakes with $M \geq 4.8$ within the same timeframe using the superposed epoch method. They discovered that anomalies mainly occurred within a range of 200 km from the epicenter and 5 days before the earthquake. Additionally, He et al. [10] categorized seismic events based on different depth, magnitude, and location, ultimately discovering anomalies in ionospheric electron density on a spatial scale of around 350 km before earthquakes. De Santis et al. [11] and Marchetti et al. [12] investigated several years of Swarm magnetic and electron density data and correlated the extracted anomalies with shallow (depth $\leq$ 50 km) $M \geq 5.5$ worldwide earthquakes by superposed epoch and space approaches. In addition, in the latter study, the eventual influence of focal mechanism and sea or land location was explored. Both papers identified that the anticipation time of the ionospheric anomalies increases with earthquake magnitude according and extending to in situ space data following the Rikitake [13] law. Similar preliminary results have been obtained for the China Seismo Electromagnetic Satellite (CSES-01), identifying that statistically, the number of anomalies preceding M5.5+ earthquakes is higher than those following the same events [14].

Previous research findings suggest that the ionosphere undergoes significant disturbances prior to earthquakes, as evidenced by both seismic case studies and statistical analyses. The ionospheric anomalies induced by the earthquake preparation process observed by satellite have garnered significant interest in the seismological community, and their monitoring and investigation are considered one of the effective approaches to studying earthquake ionospheric precursors [15].

The DEMETER satellite was the first in the world to be dedicated to detecting and researching ionospheric anomalies related to earthquakes. Following this achievement, China developed its own electromagnetic monitoring satellite, CSES-01, which was successfully launched on 2 February 2018, after ten years of scientific demonstration and five years of research and development. At present, CSES is the only satellite in flight explicitly built and dedicated to the investigation of ionospheric phenomena induced by earthquakes. A second CSES satellite is in its final building stage, and China plans to launch it within a year. This satellite monitors a wide range of parameters, including plasma properties, detection of high-energy particles, and electric and magnetic field measurements. It is the second satellite primarily dedicated to exploring earthquake-related ionospheric disturbance phenomena after DEMETER, thus enriching information in this field [16]. Real-time dynamic ionospheric monitoring and seismic precursor tracking may open new ways of earthquake monitoring and forecasting. Several studies on ionospheric anomalies have already been conducted based on data acquired from the CSES satellite, including electromagnetic fields [17] and plasma in situ [18]. Some multiparametric investigations of the preparation phase of the earthquakes integrated the CSES Ne and magnetic field data inside a variety of observations, showing the great potential of CSES to better understand the anomalous behavior of the ionosphere before large earthquakes, such as the Mw7.5 Indonesia 2018, Mw7.2 Kermadec Islands 2019, and Mw7.6 Papua New Guinea 2019 earthquakes [19–21]. However, studies related to seismic ionosphere characteristics using CSES data are still being accumulated and are not yet sufficient. Therefore, it is crucial to explore seismic ionospheric studies by utilizing a large amount of ionospheric observation data to investigate seismic ionospheric anomalous phenomena characteristics and their coupling and then explore the process and mechanism of seismic nucleation and development.

This paper analyzes global earthquakes of Ms $\geq$ 6.8 and earthquakes of Ms $\geq$ 6.0 in China using electron density (Ne) data from the Langmuir probe (LAP) onboard the CSES-01 (also known as Zhangheng-1 Electromagnetic Satellite) on a case-by-case basis, built on previous research experience. The earthquake cases are explored thoroughly, classified and summarized to investigate the underlying relationship between seismic events and ionospheric phenomena in terms of the seismic parameters.

2. Data and Processing Methods

2.1. Data Introduction

The China Seismic Electromagnetic Monitoring Experiment Satellite (CSES), also known as the Zhangheng-1 satellite, was launched on 2 February 2018. It is the first space-based platform for seismic observation and geophysical field measurements in China, orbiting at an altitude of 507 km with a fixed local time for ascending node at 2:00 and descending node at 14:00 respectively, and a revisiting period of five days [16]. The unique altitude and local time provide stable and long-term observations to explore the physical characteristics of the ionosphere, as well as the ionospheric phenomena and possibly understanding the mechanisms of earthquake-related signals. The satellite carries eight instruments onboard, including a search coil magnetometer (SCM), electric field detector (EFD), high-precision magnetometer (HPM), GNSS occultation receiver (GOR), plasma analyzer package (PAP), Langmuir probe (LAP), high-energy particle package/detector (HEPP/HEPD), and tri-band beacon (TBB), enabling continuous investigations of ionospheric plasma, electromagnetic fields and waves, and energetic particles [16].

This paper analyzes electron density (Ne) detected by the Langmuir probe (LAP) onboard the CSES, which is a key parameter that characterizes the ionospheric state. The LAP can measure Ne in the range of 5×10^2–1×10^7 e^-/cm^{-3}, with a relative measurement accuracy of 10%. The satellite's operating modes include survey mode and burst mode. The survey mode is used to detect global electron density and electron temperature, with a time resolution of one observation data every 3 s, while the burst mode is used to detect possible earthquake-related signals above the global main seismic belts, with a time resolution of 1.5 s [22]. Typically, all payloads work continuously in the latitude range of [−65°, 65°] [16]. After in-orbit testing and a series of cross-validation studies [18,23], it can be confirmed that although the absolute value of Ne may differ, its relative variations are well in agreement with other ground and space missions (e.g., Swarm) and consequently can be used in ionospheric research, such as space weather, plasma bubbles, and earthquake ionospheric anomalies.

2.2. Selection of Earthquakes and Satellite Data

This paper focuses on 57 global shallow earthquakes (with a focal depth less than 100 km) with Ms $\geq$ 6.8 that occurred from August 2018 to March 2023, and 15 earthquakes with Ms $\geq$ 6.0 in China in the past five years. We retrieved the earthquake cases from website of the China Seismic Earthquake Network Center: http://www.ceic.ac.cn/speedsearch (accessed on 1 May 2023). The distribution of the epicenters is shown in Figure 1. The selection of earthquakes is based on several considerations, as follows. (1) Earthquakes with a focal depth of less than 100 km are selected because deep earthquakes are unlikely to produce ionospheric effects [9]. (2) The magnitude threshold is a compromise, as only larger earthquakes are more prone to show some pre-earthquake effects impacting the ionosphere [7], and on the other hand, they are rarer, so the limit needs to be lower to achieve a significant statistic. (3) Only earthquakes with epicenter latitudes within ±55° are selected, considering the observation latitude range of CSES LAP and the large amount of auroral activity from polar regions in high-latitude areas. (4) The number of selected earthquakes needs to be sufficient to ensure the research results are universal.

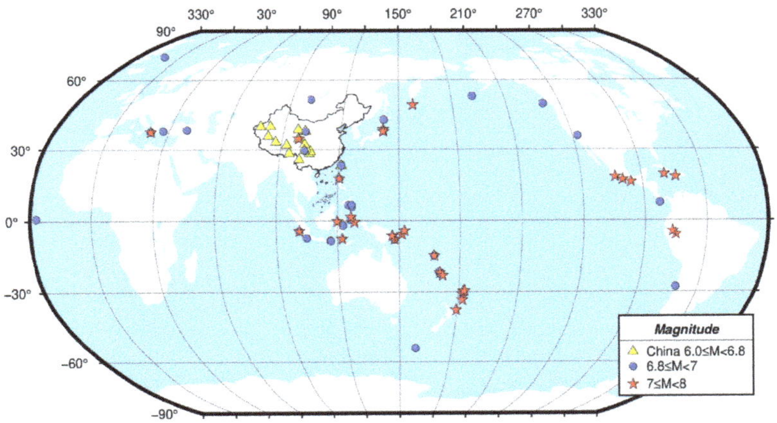

Figure 1. Distribution of global Ms $\geq$ 6.8 and Ms $\geq$ 6.0 EQs in China.

To ensure the best accuracy in extracting possible earthquake-related information, we selected only nighttime (ascending orbit) data for analysis, as solar radiation in the ionosphere during the day may have a significant impact [24]. To eliminate data interference caused by changes in geomagnetic activity, data with $|Dst| \geq 40$ nT or Kp $\geq$ 3 are excluded to remove geomagnetic environment disturbances, especially geomagnetic storms [9]. Additionally, intrinsic data problems are eliminated as much as possible, such as the data at the terminator transition point in Yan's study [23].

Previous research has shown that ionospheric disturbances have been observed a few days prior to certain earthquake activities, particularly within a 25-day timeframe before the earthquake [10]. Additionally, abnormal disturbances have been recorded on the day of the earthquake and within 5 days after the earthquake [25]. Furthermore, ionospheric disturbances within a 1000 km radius from the epicenter have been identified as particularly significant [8]. Therefore, this study focuses on the period from 25 days before the earthquake to 5 days after the earthquake, searching for ionospheric disturbances within a range of 1000 km from the epicenter.

2.3. Data Processing Methods

As outlined in Section 2.2, data in nighttime orbits within a ±10° (i.e., 1000 km) range around the epicenter from 25 days before to 5 days after the earthquake were chosen. Figure 2 provides an example of orbit selection, displaying the distribution map of orbit position of the Ms7.0 earthquake in the southern waters of Sumatra, Indonesia (4.31°S, 101.15°E). The red dot indicate the epicenter's location and the blue lines represent the ground projection of the chosen orbits for studying the earthquake.

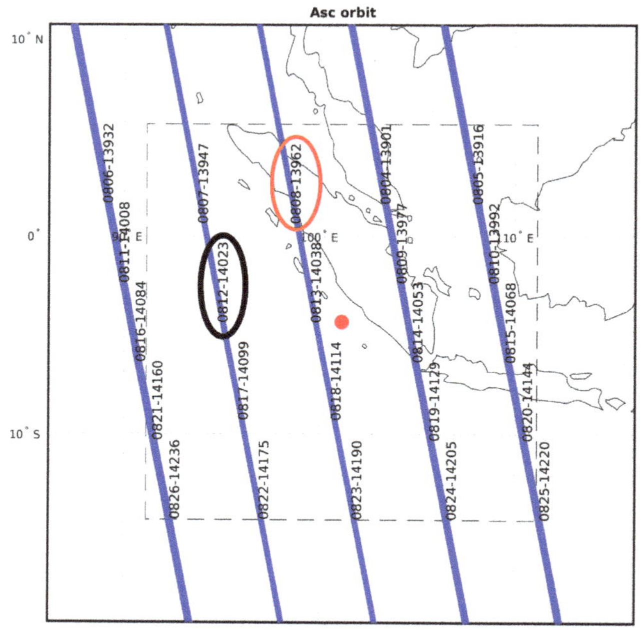

Figure 2. Distribution of the orbits selected and epicenter of the Ms7.0 earthquake in the southern waters of Sumatra, Indonesia on 19 August 2020. Only nighttime (Ascending = Asc) orbits were selected. Red and black circles underline the position of two anomalies shown in the time series of Ne in the following picture.

In order to extract eventual anomalies, the quartile analysis method was implemented, which involves several steps.

Firstly, observations within a ±10° radius around the epicenter were obtained for the period ranging from 25 days before to 5 days after the earthquake.

Secondly, for each current orbit (C_0) obtained in the first step, the corresponding orbits acquired during its revisiting orbits in the past 30 days were collected. Each current orbit (C_0) corresponds to a set of 6 revisiting orbits.

The third step involves resampling each group of 6 revisited orbits with a time resolution of 0.3° (survey mode)/0.15° (burst mode) and calculating the median and upper and lower quartile difference (IQR) of the 6 revisited orbits in each sampling interval. The median is represented as C_b, while $C_b \pm 1.5$ IQR is represented as the upper and lower bounds C_u and C_1.

In the fourth step, the current observation data orbit (C_0) is also resampled by 0.3° (survey mode)/0.15° (burst mode), and then compared with the upper and lower boundaries obtained from the 30-day background calculation for analysis. The variation in current observation data relative to the background one is then computed using the following formula:

$$\begin{aligned} \%\mathrm{Dev(increase)} &= \frac{C_0 - C_u}{C_u} \times 100\% \\ \%\mathrm{Dev(decrease)} &= \frac{C_0 - C_1}{C_1} \times 100\% \end{aligned} \quad (1)$$

where "Dev" is the abbreviation for deviation (difference) and the "%" sign indicates the percentage change. The "increase" and "decrease" in parentheses represent upward and downward trends, respectively. C_0 represents the current observation data, C_u represents the upper limit value, and C_1 represents the lower limit value. Based on previous research experience with a large number of earthquake cases, this study considers Ne abnormal changes with a magnitude of $\geq$50% as abnormal. A similar technique based on the interquartile range and revisiting orbits has already been applied to CSES Ne data before La Palma 2021 volcano eruption, obtaining interesting results in agreement with the other geophysical investigated observables [26].

3. Analysis of Typical Earthquake Cases

Based on carefully selected and processed data, we conducted a comprehensive analysis of anomalies before and after 57 worldwide earthquakes with Ms $\geq$ 6.8. This chapter provides a detailed account of the anomaly extraction results and the criteria used to determine anomalies for two typical earthquake cases, one with significant anomalies and one without.

3.1. *An Example of an Earthquake with Clear Anomalies*

According to the China Earthquake Network Center, an Ms7.0 earthquake occurred in the south of Sumatra, Indonesia (4.31°S, 101.15°E), with a depth of 10 km, on 19 August 2020. The epicenter, together with the orbit distribution map, is shown in Figure 2, while Figure 3 displays the changes in Ne around the epicenter on the southern side of Sumatra from 25 days before the earthquake to 5 days after the earthquake (from 25 July to 25 August). The panels from top to bottom show the geomagnetic indices (planetary Kp and equatorial disturbance storm index Dst), the Ne time-series variation, and the Ne abnormal disturbance amplitude. Based on our experience of abnormal observations from more than five years of CSES observations, further analysis is required when the Ne abnormal amplitude exceeds 50%, in order to exclude or support the possible seismic source for such abnormalities. Despite this, it is possible to note that there is a slight increase above the threshold of Ne on 2 August 2019, in correspondence to geomagnetic activity, as underlined by high values of kp and drop of Dst, which underline the impact of a moderate geomagnetic storm in the Earth's environment. On the other hand, Figure 3 shows that on 8 August (11 days before the earthquake), the Ne abnormal amplitude was $\geq$100%, and it was a continuous abnormal disturbance change. On that day, Dst was <40 nT and Kp < 3, which can exclude the influence of space weather. Therefore, it is judged that the abnormal increase is still likely to be related to this earthquake. Similarly, on 12 August (7 days before the earthquake), Ne slightly increased (about 80%) again and in very quiet geomagnetic conditions, indicating that it may have some correlation with the earthquake.

Based on the positions of each orbit passing around the epicenter, as shown in Figure 2, the red and black circles represent the position of anomalous orbit data. The abnormal phenomenon on 8 August occurred in the closest orbit to the epicenter (orbit number 13962), with an abnormal orientation in the northwest direction of the epicenter, about 2° away from the epicenter (red ellipse in Figure 2); the abnormality on 12 August was located in the northwest direction of the epicenter, about 7° away from the epicenter (black ellipse in Figure 2). The highest longitudinal distance from the epicenter of the second anomaly may explain the lower intensity of the anomaly and further support its link to the seismic event, as it is expected that the anomalies are higher as the earthquake is closer and vanish after a certain distance.

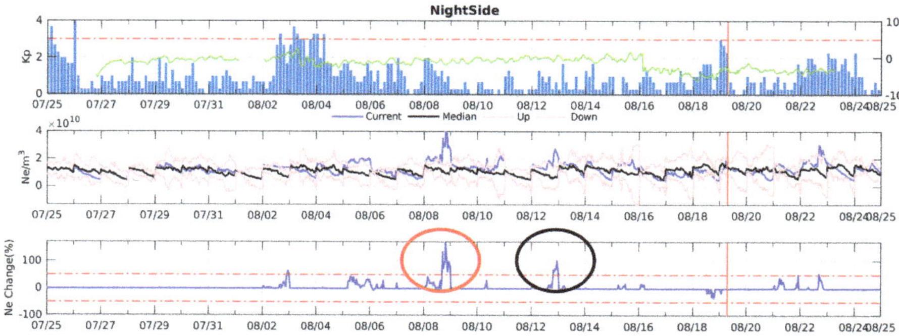

Figure 3. The electron density disturbances over Sumatra EQ influential area occurred on the 19 August 2020. The space weather index Kp is represented by blue bars, and the Dst index is represented by the cyan curve. The red horizontal dashed line in the first panel represents the values of Kp = 3, while the red horizontal dashed lines in the third panel represent the upper and lower boundaries with 50%. The red vertical lines mark the time of the earthquake occurring.

3.2. *An Example of an Earthquake without Anomalies*

On 16 October 2018, an Ms6.9 earthquake occurred in the southeast of the Loyalty Islands (21.65°S, 169.61°E), with a depth of 10 km. Figure 4 shows the geomagnetic index trends and changes in Ne abnormal amplitude obtained within a range of 1000 km around the epicenter from 25 days before the earthquake to 5 days after the earthquake (from 21 September to 22 October). From Figure 4, it can be seen that the Ne amplitude changes were weak, i.e., always inside the negative and positive thresholds (dashed red lines), indicating no anomaly during the investigated period.

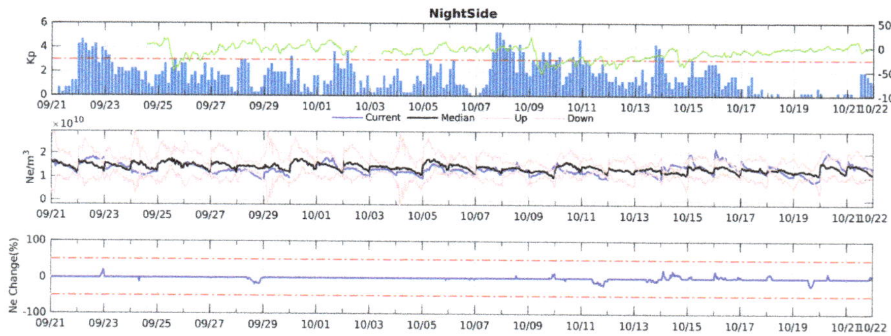

Figure 4. Electron density disturbances over the Ms6.9 Loyalty Islands EQ on 16 October 2018.

4. Statistical Analysis of Global Earthquakes with Ms $\geq$ 6.8

Considering the analysis preprocessing and the discrimination criteria in the previous section, a statistical investigation on the selected 57 earthquakes is conducted in this chapter, searching for significance of the possible relationships between seismic events and ionospheric effects.

4.1. *Statistics on the Number of Anomalies*

The anomalies observed before and after each earthquake were derived from analyzing exactly the same period for each case study. To explore the possible correlation between earthquake magnitude and the number of anomalies, the earthquake cases were categorized into three groups based on the magnitude Ms (refer to Table 1). As shown in Table 1, there were 7 earthquakes with Ms $\geq$ 7.5, and anomalies were observed in 71.4% of this group, with most cases having only one anomaly. Among the 24 earthquakes with 7 $\leq$ Ms < 7.5, two of them had more than three anomalies, six had two anomalies, and nine had one anomaly (accounting for 37.5% of this group), while seven earthquakes had no anomalies, accounting for 29.2%. The proportion of earthquakes with at least one anomaly is 70.8% in this group. For the 26 earthquakes with $6.8 \leq Ms < 7$, two of them had more than three anomalies, two had two anomalies, ten had one anomaly (accounting for 38.5% of this group), and no anomalies were observed in 12 earthquakes (46.2% of this group). It was calculated that earthquakes with anomalies accounted for 53.8% of this group. Overall, the data suggest that the larger the magnitude, the higher the likelihood of observing anomalies, supporting at least some of the anomalies being likely induced by the earthquakes.

Table 1. Number of earthquakes (EQs) with anomalies of different magnitudes.

Ms	Count of EQs with ≥3 Anomalies	Count of EQs with ≥2 Anomalies	Count of EQs with One Anomaly	Count of EQs with Anomalies	Count of EQs with No Anomalies
$Ms \geq 7.5$	0	0	5 (71.4%)	5 (71.4%)	2 (28.6%)
$7 \leq Ms < 7.5$	2 (8.3%)	6 (25%)	9 (37.5%)	17 (70.8%)	7 (29.2%)
$6.8 \leq Ms < 7$	2 (7.7%)	2 (7.7%)	10 (38.5%)	14 (53.8%)	12 (46.2%)

In the following studies about the location and time of abnormal phenomena, the statistics are based on the number of abnormal phenomena rather than the number of earthquakes.

4.2. Typical Regional Distribution of Anomalies

To investigate the spatial distribution of abnormal phenomena in different directions, we searched for a predominant direction for such phenomena. In analogy with the analysis of the location of earthquake anomalies in Section 3, we gathered statistics of anomalies, dividing them in eight directions: north, northeast, east, southeast, south, southwest, west, and northwest. Figure 5 displays the distribution of abnormal phenomena relative to the epicenter. From Figure 5, it can be observed that the northeast and southeast directions from the epicenter have the highest relative number of anomalies, both accounting for 19%. Abnormal phenomena in the northwest and southwest directions with respect to the epicenter accounted for 17% each, while abnormal phenomena in the east direction accounted for 16%. Abnormal phenomena in the west direction accounted for 7%, while the least abnormal phenomena were found in the north and south directions, accounting for only 4% and 1%, respectively. When looking at this pie chart with left and right sides, respectively, representing the west and east sides of the epicenter, it can be seen that abnormal phenomena on the east side of the epicenter were significantly higher than those on the west side.

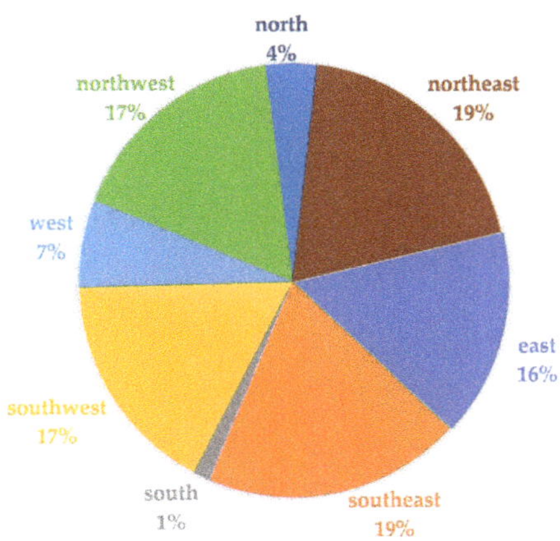

Figure 5. Percentage of spatial distribution directions of earthquake ionospheric disturbances.

4.3. The Time when the Anomalies Mostly Occur

Anomalies observed on the day of the earthquake were labeled as "0 days," while those observed before the earthquake are indicated with negative number (e.g., an anomaly one day before the event was labeled as "−1 day") and those observed after the earthquake present positive numbers (e.g., an anomaly one day after the earthquake was labeled as "1 day"). A summary of results is presented in Figure 6. Among the 57 cases analyzed anomalies were observed on every day except for −20, −18, −9, −8, and −5 days with respect to the earthquake. However, there were more anomalies observed on −11, −7 and −2 days with respect to the earthquake, as well as on the day of the earthquake. The concentration of anomalies was higher 11 days before the earthquake and near the day of the earthquake, with two smaller peaks. Finally, it seems that from about 2 weeks before the earthquake up to its occurrence, the number of recorded anomalies increased. This suggests that the incoming earthquakes induced some ionospheric disturbances, which in agreement with previous statistical preliminary investigations made on CSES Ne that reported more anomalies before the seismic events than after [27].

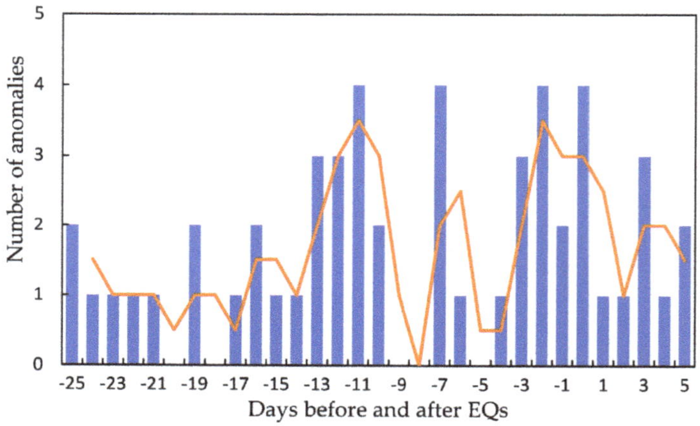

Figure 6. Statistical analysis of the time of multiple occurrences of anomalies for global earthquake with Ms $\geq$ 6.8.

4.4. Possible Influence of Land or Marine Earthquake Location on Ionospheric Anomalies

The 57 selected earthquakes were categorized into two groups based on their epicenter location: land or marine earthquakes. Statistical analyses were conducted on both groups, and Table 2 displays the total number of earthquakes, the number of earthquakes with anomalies, and the ratio of earthquakes with anomalies to the total amount of earthquake for each type of location. The table shows that there were 18 land earthquakes and 39 ocean earthquakes, so most earthquakes occurred in the ocean. Despite the fact that the total number of land earthquakes was lower than that of ocean earthquakes, the probability of occurrences of anomalies before land earthquakes was higher than that one in ocean earthquakes. The role of the ocean in ionospheric seismo-induced electromagnetic disturbances is debated, as well as the LAIC modeling. In fact, even though some models would theoretically fail in the sea context, the empirical results (including those of this paper) show that marine earthquakes seem to be at least as productive as land ones in terms of generation of ionospheric electromagnetic anomalies [28]. In addition, Marchetti et al. [12] found that the frequency of magnetic signals before M5.5+ earthquakes seems higher for marine epicenters compared with those above sea level.

Table 2. Statistical study on land and marine earthquakes.

Counts \ Earthquake Types	Land Earthquakes	Marine Earthquakes
Total number of EQs	18	39
Number of EQs with anomalies	13	23
Probability of anomalies	72.2%	59%

After classifying earthquakes into land and marine categories, the number of examples for both categories decreased. Therefore, only four directions—northwest, southwest, southeast, and northeast—are used to analyze the direction in which anomalies tend to appear. Figure 7 compares the proportion of anomalies in four directions for land and ocean earthquakes. For land earthquakes, the proportion in the northwest direction was highest and the one in the southwest direction lowest. Conversely, the proportion for marine earthquakes was highest in the northeast direction and lowest in the southeast direction. The abnormal distribution of land earthquakes in the northwest direction was significantly higher than that of marine earthquakes, while it was significantly lower in the southwest direction than that of marine earthquakes.

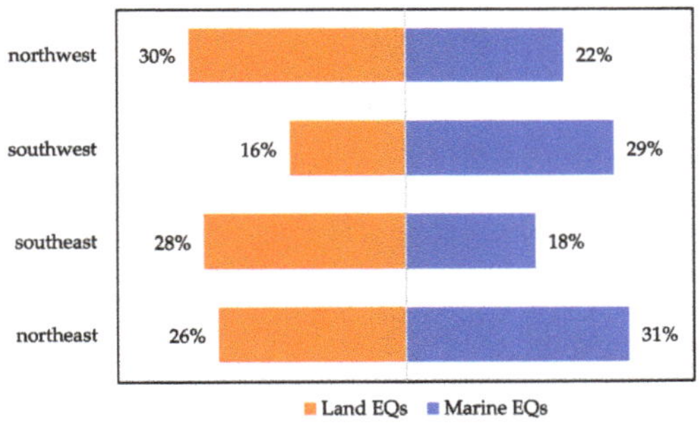

Figure 7. Bar chart comparing the orientation of land and marine EQ anomalies.

In Figure 8, some variations can be observed, such as an increase in anomalies 13, 11 days before and 3 days after the earthquake for land events and 7 and 2 days before the earthquake for marine events, an extremely low number of anomalies and not statistically significant. Despite this, it is possible to note a trend of more anomalies from 12 days before the earthquake up to 1 day after the earthquake for marine events. This is the same result depicted in Figure 6, but here it is possible to note that it seems to be related to marine earthquakes.

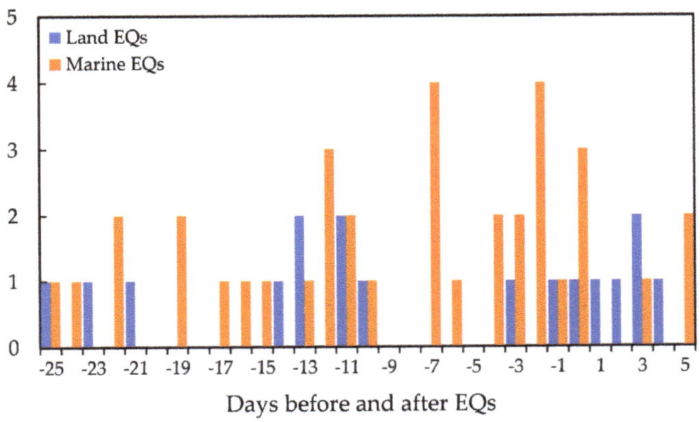

Figure 8. Comparison of the time of occurrence of most land and marine EQ anomalies.

4.5. Characteristics of Earthquakes in the Northern and Southern Hemispheres

The ionospheric disturbances caused by earthquakes are shifted along the direction of magnetic field lines of force due to electromagnetism, i.e., basically Maxwell equations. The 57 earthquake cases are divided into Southern and Northern Hemisphere groups, and the locations of their anomalies appearing in the Northern and Southern Hemisphere earthquakes are analyzed separately, as well as the comparative analysis of their temporal distribution.

As the number of earthquake cases significantly decreases after the division of the Northern and Southern Hemispheres, the anomalous orientation is divided into only four orientations—northeast, southeast, southwest, and northwest—and the characteristics of the distribution of anomalous orientations in the Northern and Southern Hemispheres are analyzed separately. Figure 9 displays the pie charts of the percentage distribution of earthquake anomaly orientation in the Northern Hemisphere (**a**) and Southern Hemisphere (**b**), respectively, while Figure 10 shows the radar charts of the percentage distribution of earthquake anomaly orientation in the Northern and Southern Hemispheres. By classifying northwest and northeast as directions north of the epicenter and southwest and southeast as directions south of the epicenter, it is evident that the Northern Hemisphere anomalies are predominantly distributed in the direction south of the epicenter, while the southern hemisphere anomalies are mostly distributed in the direction north of the epicenter. This may indicate that the anomalies point toward the equator, in agreement with Parrot's theory that the preparation phase of an earthquake not only increases the level of electron density but also shift (EIA crests) [28]. Although this study is based on nighttime data, and as such without the classic EIA, we still think that this result is related to the same phenomena. In particular, we can consider that previous investigations were based on the DEMETER satellite and this one on the new mission CSES, and the comparison is not only crucial but seems to support the reliability of the results from both missions. Further studies are necessary to strengthen these pieces of evidence and compare them with other satellite missions and ground data.

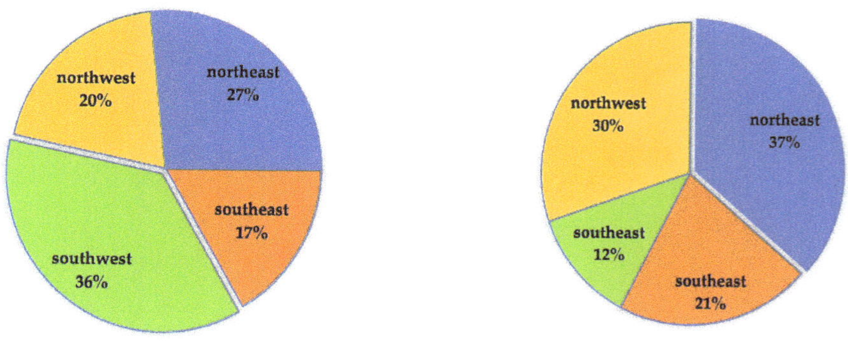

(**a**) (**b**)

Figure 9. (**a**) Percentage distribution of abnormal directions for EQs in the Northern Hemisphere; (**b**) Percentage distribution of abnormal directions for EQs in the Southern Hemisphere.

Similar to the investigation presented in Figure 8 about land/marine time distribution, here we divided the anomalies in the Northern and Southern Hemispheres, as depicted in Figure 11. Although the number of anomalies is not so high, some observations can be made. In fact, some days present more anomalies in a particular group of earthquakes. For example, the number of anomalies occurring 12 days prior to and on the day of the earthquake was significantly higher in the Northern Hemisphere. In addition, the number of anomalies occurring 11 days prior to, 2 days prior to, and 3 days following the earthquake was found to be significantly lower in the Northern Hemisphere. On the other hand, the number of anomalies 11, 7 and 2 days before the earthquakes in the Southern Hemisphere is significantly higher. Finally, the number of anomalies occurring at 25, 19, and 10 days prior to the earthquake, as well as 3 days prior to and 5 days following the earthquake, were found to be comparable for both Northern and Southern Hemisphere earthquakes. The different time patterns in the two hemispheres are hard to explain, but they may be related to the different orientation of the magnetic field and/or the presence of the South Atlantic magnetic anomaly, which can induce a delay in the LAIC mechanism, and further theoretical studies are necessary to understand better such mechanisms.

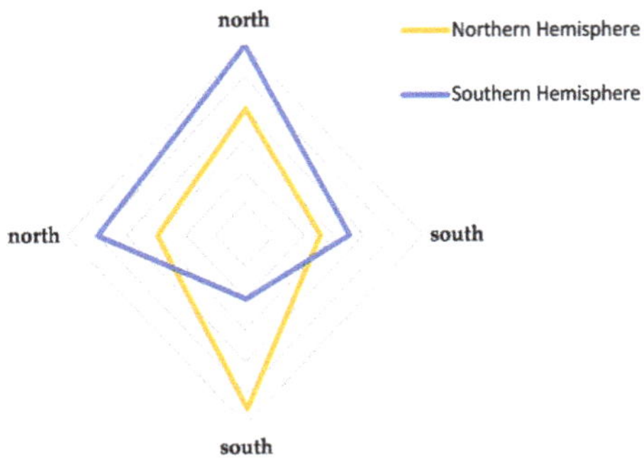

Figure 10. Radar chart of percentage distribution of abnormal directions for EQs in the Northern and Southern Hemispheres.

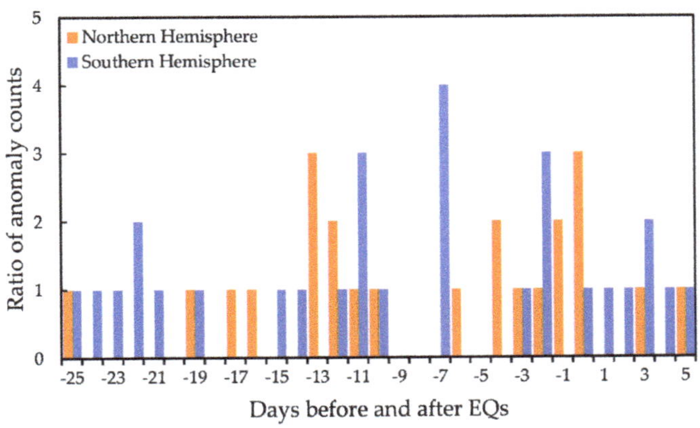

Figure 11. Comparison of abnormal occurrence time for earthquakes in the Northern and Southern Hemispheres.

4.6. Analysis of Earthquakes of Ms ≥ 6.0 in China

Based on the analysis of earthquakes with Ms ≥ 6.8 worldwide, this paper also conducted statistical analysis on 15 earthquakes with Ms ≥ 6.0 in China during the same time range, using the method described in Section 2.3. In this paper, we only focus on the land earthquakes in China, and the ocean earthquakes and island earthquakes are not considered. For each earthquake event, the abnormal parameters, its occurrence time, amplitude, and direction were then statistically analyzed, in analogy with global investigations. The specific results are shown in Table 3. Among the 15 earthquakes with Ms ≥ 6.0, pre-earthquake anomalies were observed before 10 earthquakes, with a total of 19 abnormal phenomena, including 2 post-earthquake anomalies and 1 anomaly on the day of the earthquake (i.e., likely a co-seismic effect).

Table 3. Results of the analysis of EQs with Ms ≥ 6.0 in China.

No.	Position of Epicenter	CST	Ms	Parameters	Time of Anomalies Occurrence	Location of Anomalies
1	**Motuo, Tibet**	2019-04-24 04:15:48	6.3	None		
2	**Changning, Sichuan**	2019-06-17 22:55:43	6.0	Ne	6.1 (−16 days)	Southwest of epicenter
3	**Jiashi, Xinjiang**	2020-01-19 21:27:55	6.4	Ne	1.14 (−5 days)	Southeast–northeast of epicenter
					12.26 (−24 days)	Northeast of epicenter
4	**Yutian, Xinjiang**	2020-06-26 05:05:20	6.4	Ne	6.2 (−24 days)	Southwest of epicenter
5	**Nima, Tibet**	2020-07-23 04:07:20	6.6	Ne	7.14 (−9 days)	Southeast–northeast of epicenter
					7.6 (−17 days)	Southwest of epicenter
6	**Biru, Tibet**	2021-03-19 14:11:26	6.1	None		
7	**Yangbi, Yunnan**	2021-05-21 21:48:34	6.4	Ne	5.16 (−5 days)	Southwest–northwest of epicenter
					5.15 (−6 days)	Southwest of epicenter
					5.14 (−7 days)	Southwest of epicenter
8	**Maduo, Qinghai**	2021-05-22 02:04:11	7.4	Ne	5.21 (−1 days)	Southwest of epicenter
					5.16 (−6 days)	Southwest of epicenter
9	**Luxian, Sichuan**	2021-9-16 04:33:31	6.0	None		
10	**Menyuan, Qinghai**	2022-01-08 01:45:27	6.9	None		
11	**Delingha, Qinghai**	2022-03-26 00:21:02	6.0	Ne	3.26 (0 day)	Northwest-southwest of epicenter
					3.29 (3 days)	Southeast of epicenter
					3.30 (4 days)	Southwest of epicenter
12	**Lushan, Sichuan**	2022-06-01 17:00:08	6.1	Ne	5.19 (−13 days)	Northeast of epicenter
					5.25 (−7 days)	Northwest of epicenter
13	**Barkam, Sichuan**	2022-06-10 01:28:34	6.0	Ne	5.19 (−22 days)	Northeast of epicenter
					5.25 (−16 days)	Southwest of epicenter
14	**Luding, Sichuan**	2022-09-05 12:52:18	6.8	None		
15	**Shayar, Xinjiang**	2023-01-30 07:49:39	6.1	None		

Upon analysis of the anomalous concentration area of 15 earthquake cases, as shown in Figure 12, this is mainly located in the southwest direction of the epicenter (accounting for 50%), while it appears less frequently in the southeast direction of the epicenter (only accounting for 13%). This result further confirms our previous global investigation, considering China is located in the Northern Hemisphere.

The statistical analysis of the anomalous occurrence time is shown in Figure 13, using the marking method in Section 4.3. Anomalies appeared continuously for 5 days from 9 days to 5 days before the earthquake, with the highest frequency of anomalies occurring 5 days before the earthquake. This result further confirms the DEMETER statistical investigations carried out by Li and Parrot [29], with a very similar anticipation time. It is noteworthy that due to the revisiting time of the satellite, there cannot be high precision in the determination of the timing of anomalous, with an uncertainty of a few days at present. Future constellation of CSES satellites will help to reduce the revisiting time, and consequently, the uncertainty in the time of occurrence of the anomalies.

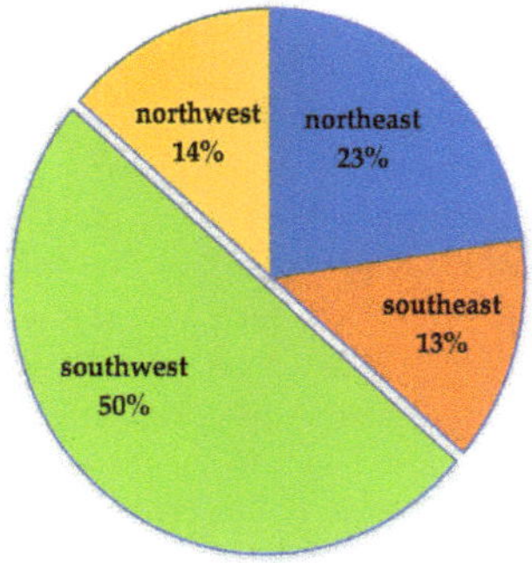

Figure 12. Percentage distribution of spatial distribution directions of earthquake ionospheric disturbances.

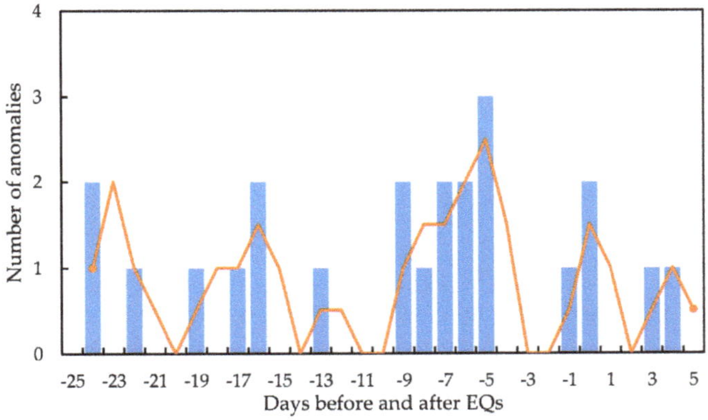

Figure 13. Statistical analysis of the time of multiple occurrences of anomalies for earthquake with Ms $\geq$ 6.0 in China.

5. Discussions and Conclusions

In this paper, the Ne data from the LAP onboard CSES were used to analyze earthquakes with Ms $\geq$ 6.8 worldwide and Ms $\geq$ 6.0 in China. The analysis of single earthquake cases suggests the existence of earthquake ionospheric anomalies. On the other hand, through statistical analysis of earthquake cases, it was possible to prove statistically previous research, and they were generally consistent.

In terms of occurrence time of anomalies, Zhu et al. [30] used the stacking epoch method, and found that the abnormal changes in ionospheric Ne during nighttime mainly occurred within two time periods: 1–7 days and 13–15 days before the earthquake. He et al. [31] found that the statistical samples of anomalies from 10 days before the earthquake to around 3 days after the earthquake were relatively rich, and the range of abnormal occurrence time for earthquakes with a magnitude of 6+ was between 0 and 6.5 days before the earthquake. Li and Parrot [29] found that the number of disturbances on the day of the earthquake was higher and then steadily decreased in the days leading up to the earthquake. Li et al. [32] found that the abnormal ionospheric disturbances related to earthquakes were most significant one week before the earthquake, and the frequency of occurrence increased as the earthquake approached. These studies are generally consistent with the statistical results in Figure 6 in this paper.

Regarding research on anomalous occurrence direction, He et al. [31] conducted statistical analysis on multiple parameters of ionospheric disturbances, presenting the spatial distribution and frequency of anomalous occurrence in different directions. They discovered that over 50% of anomalies were located in the north and south directions of the epicenter, which could be attributed to the influence of the geomagnetic field, making it easier to detect abnormal changes on the south and north sides of the epicenter. However, this paper analyzed only one parameter—Ne, as depicted in Figure 5—and anomalies were more concentrated on the east side of the epicenter. This could be due to the limited number of parameters analyzed and the small number of earthquake cases. Further research is necessary to explore the characteristics of anomaly direction. He et al. [10] utilized data from the DEMETER satellite to demonstrate that Ne anomalies near the epicenter can be found in both the Northern and Southern Hemispheres. However, the location of anomalies shifts slightly northward in the Northern Hemisphere and slightly southward in the Southern Hemisphere. The results presented in Figures 9 and 10 of this paper are consistent with these findings.

He et al. [10] found that Ne related to both land and ocean earthquakes showed anomalies close to the epicenter, but the anomalies of ocean earthquakes were more significant. Table 2 also shows that although the total number of land earthquakes is smaller than that of marine earthquakes, the probability of anomaly occurrence is higher than that of marine earthquakes. Due to the low total number of events, further data will be necessary to better investigate this, which seems to be different from previous investigations. Another possibility could be related to the sensibility of the observable, for example, for the Swarm mission, it was suggested that the magnetic field might be more sensible to pre-earthquake phenomena than Ne [12].

Despite numerous earthquake cases, scholars have proposed various earthquake–ionosphere coupling models, without a unified understanding of the coupling mechanism [33]. Therefore, to uncover the inherent laws of earthquakes and ionospheric phenomena, it is necessary to conduct a more comprehensive analysis of earthquake–ionospheric anomalies. This includes increasing the number of earthquake case studies and carrying out a series of multi-parameter earthquake case studies to accumulate more research experience. By doing so, we can try to explore the universal characteristics of earthquake–ionosphere coupling mechanisms with different seismic source mechanisms and reveal the underlying laws of earthquakes and ionospheric phenomena. Some investigations even suggest that more mechanisms are possible before the same earthquake, as proposed by Zhang et al. [34] investigating the Mw6.7 Lushan (China) 2013 earthquake.

Although short-term earthquake forecasting remains one of the most debated and challenging goals in earth science, it is indisputable that some moderate earthquakes are accompanied by ionospheric disturbances before and after the event [9,35]. Li et al. [35] analyzed CSES data recorded in China for the 21 May 2021 Maduo Ms7.4 earthquake and the 8 January 2022 Menyuan Ms6.9 earthquake, discovering effective ionospheric disturbances two weeks and one week before these two events, respectively. This confirms that ionospheric information can be utilized for short-term earthquake forecasting.

This paper analyzes Ne anomalies in 57 earthquakes with Ms $\geq$ 6.8 worldwide, as well as 15 earthquakes with Ms $\geq$ 6.0 in China. The study primarily yields the following conclusions.

(1) In terms of the number of anomalies, the proportion of anomalies occurring before and after an earthquake tends to increase as the magnitude increases. Additionally, although the total number of land earthquakes is smaller than that of ocean earthquakes, the probability of anomalies seems higher for land earthquakes than for marine earthquakes.

(2) In terms of areas of concentration of anomalies, it is generally observed that the east side of the epicenter exhibits significantly higher anomalies than the west side. The highest concentration of anomalies is found in the northwest direction of land earthquakes and in the northeast direction of marine earthquakes. Anomalies in the Northern Hemisphere are mostly distributed towards the south of the epicenter, while those in the Southern Hemisphere are mostly distributed towards the north of the epicenter.

(3) In terms of the time of anomaly occurrence, it was observed that the frequency of anomalies was higher 7 days before, 11 days before, and in the vicinity of the day of

the earthquake. This feature remains consistent across earthquakes classified as land and marine, as well as those in the Northern and Southern Hemispheres.

(4) Regarding earthquakes with Ms $\geq$ 6.0 in China over the past five years, it was observed that Ne responds significantly to anomalies southwest of the epicenter in agreement with global analysis. Furthermore, the Ne anomaly exhibits the highest frequency of anomalies 5 days before the earthquake and occurs continuously from 9 days before to 5 days before the earthquake.

These research findings suggest that the Ne of CSES can play a fundamental role in studying ionospheric disturbances related to earthquakes. While most detected ionospheric disturbances seem to be related to earthquakes under this set of conditions, there could still be some anomalies that have not been detected or may be due to other unknown sources. In the future, more detailed research will be conducted to explore the response characteristics of ionospheric anomalies in different types of earthquakes in order to identify common features. This will involve searching for ionospheric disturbances in the complete satellite ionospheric data set, eliminating disturbances caused by known ionospheric phenomena such as solar activity, eliminating disturbances that are not in the earthquake zone, and checking whether each selected disturbance corresponds to future earthquakes. To better understand the nature of these short-term earthquake precursors, future research will be based on CSES's long-term data and other multi-parameter observations from ground and space data.

Author Contributions: Conceptualization, R.Y.; Methodology, C.H.; Software, C.H. and W.P.; Validation, D.L.; Formal analysis, D.M. and Z.Z.; Investigation, Z.Z.; Resources, S.X. and N.Z.; Data curation, H.L. and N.Z.; Writing—original draft, C.H. and R.Y.; Writing—review & editing, R.Y. and D.M. All authors have read and agreed to the published version of the manuscript.

Funding: This work was supported by the Fundamental Research Funds for NINH (grant ZDJ2019-22), the APSCO Earthquake Project Phase II, the Dragon Five cooperation (grant 59236), the International Space Science Institute in Beijing (ISSI-BJ; grant 23-583), the Stable-Support Scientific Project of the CRIRP (grant A132001W07), and the National Natural Science Foundation of China (grant 41974084).

Data Availability Statement: CSES data are freely available upon registration and approval on the CSES web portal: www.leos.ac.cn (accessed on 9 June 2023).

Acknowledgments: The authors acknowledge the CSES mission (http://www.leos.ac.cn (accessed on 1 May 2023) for providing the electron density data, which is funded by the China National Space Administration (CNSA) and China Earthquake Administration (CEA). CSES data can be downloaded from http://www.leos.ac.cn. Thanks for the Kp index download from site https://wdc.kugi.kyotou.ac.jp/wdc/Sec3.html (accessed on 1 May 2023).

Conflicts of Interest: The authors declare no conflict of interest. The funders had no role in the design of the study, collection, analyses, or interpretation of data, writing of the manuscript, or decision to publish the results.

References

1. Wu, Z.H.; Zhao, G.M. The earthquake prediction status and related problems: A review. *Geol. Bull. China* **2013**, *32*, 1493–1512.
2. Yan, R.; Parrot, M.; Pinçon, J.-L. Statistical Study on Variations of the Ionospheric Ion Density Observed by DEMETER and Related to Seismic Activities: Ionospheric Density and Seismic Activity. *J. Geophys. Res. Space Phys.* **2017**, *122*, 12421–12429. [CrossRef]
3. Liu, J.; Wan, W.X.; Huang, J.P.; Zhang, X.; Shu-Fan, Z.; Ouyang, X.-Y.; Zhima, Z. Electron density perturbation before Chile M8.8 earthquake. *Chin. J. Geophys.* **2011**, *54*, 2717–2725. (In Chinese) [CrossRef]
4. Davies, K.; Baker, D.M. Ionospheric Effects Observed around the Time of the Alaskan Earthquake of March 28, 1964. *J. Geophys. Res.* **1965**, *70*, 2251–2253. [CrossRef]
5. Zhang, X.M.; Qian, J.D.; Ouyang, X.Y.; Shen, X.; Cai, J.; Zhao, S. Ionospheric electromagnetic disturbances observed on DEMETER satellite before an earthquake of M7.9 in Chile. *Prog. Geophys.* **2009**, *24*, 1196–1203. (In Chinese) [CrossRef]
6. Zhang, X.M.; Liu, J.; Zhao, B.Q.; Xu, T.; Shen, X.H.; Yao, L. Analysis on ionospheric perturbations before Yushuearthquake. *Chin. J. Space Sci.* **2014**, *34*, 822–829. [CrossRef]
7. Le, H.; Liu, J.Y.; Liu, L. A Statistical Analysis of Ionospheric Anomalies before 736 M6.0+ Earthquakes during 2002–2010. *J. Geophys. Res.* **2011**, *116*, A02303. [CrossRef]

8. Kon, S.; Nishihashi, M.; Hattori, K. Ionospheric Anomalies Possibly Associated with M $\geq$ 6.0 Earthquakes in the Japan Area during 1998–2010: Case Studies and Statistical Study. *J. Asian Earth Sci.* **2011**, *41*, 410–420. [CrossRef]
9. Liu, J.; Huang, J.; Zhang, X. Ionospheric Perturbations in Plasma Parameters before Global Strong Earthquakes. *Adv. Space Res* **2014**, *53*, 776–787. [CrossRef]
10. He, Y.; Yang, D.; Qian, J.; Parrot, M. Response of the Ionospheric Electron Density to Different Types of Seismic Events. *Nat Hazards Earth Syst. Sci.* **2011**, *11*, 2173–2180. [CrossRef]
11. De Santis, A.; Marchetti, D.; Pavón-Carrasco, F.J.; Cianchini, G.; Perrone, L.; Abbattista, C.; Alfonsi, L.; Amoruso, L.; Campuzano S.A.; Carbone, M.; et al. Precursory Worldwide Signatures of Earthquake Occurrences on Swarm Satellite Data. *Sci. Rep.* **2019** *9*, 20287. [CrossRef]
12. Marchetti, D.; De Santis, A.; Campuzano, S.A.; Zhu, K.; Soldani, M.; D'Arcangelo, S.; Orlando, M.; Wang, T.; Cianchini, G. Di Mauro, D.; et al. Worldwide Statistical Correlation of Eight Years of Swarm Satellite Data with M5.5+ Earthquakes: New Hints about the Preseismic Phenomena from Space. *Remote Sens.* **2022**, *14*, 2649. [CrossRef]
13. Rikitake, T. Earthquake Precursors in Japan: Precursor Time and Detectability. *Tectonophysics* **1987**, *136*, 265–282. [CrossRef]
14. De Santis, A.; Marchetti, D.; Perrone, L.; Campuzano, S.A.; Cianchini, G. Statistical Correlation Analysis of Strong Earthquakes and Ionospheric Electron Density Anomalies as Observed by CSES-01. *Il Nuovo Cim. C* **2021**, *44*, 1–4. [CrossRef]
15. Hayakawa, M.; Schekotov, A.; Potirakis, S.M.; Eftaxias, K.; Li, Q.; Asano, T. An Integrated Study of ULF Magnetic Field Variations in Association with the 2008 Sichuan Earthquake, on the Basis of Statistical and Critical Analyses. *Open J. Earthq. Res.* **2015**, *4*, 85–93. [CrossRef]
16. Shen, X.H.; Zhang, X.; Yuan, S.; Wang, L.; Cao, J.; Huang, J.; Zhu, X.; Piergiorgio, P.; Dai, J. The State-of-the-Art of the China Seismo-Electromagnetic Satellite Mission. *Sci. China Technol. Sci.* **2018**, *61*, 634–642. [CrossRef]
17. Zhima, Z.; Yan, R.; Lin, J.; Wang, Q.; Yang, Y.; Lv, F.; Huang, J.; Cui, J.; Liu, Q.; Zhao, S.; et al. The Possible Seismo-Ionospheric Perturbations Recorded by the China-Seismo-Electromagnetic Satellite. *Remote Sens.* **2022**, *14*, 905. [CrossRef]
18. Yan, R.; Zhima, Z.; Xiong, C.; Shen, X.; Huang, J.; Guan, Y.; Zhu, X.; Liu, C. Comparison of Electron Density and Temperature From the CSES Satellite With Other Space-Borne and Ground-Based Observations. *J. Geophys. Res. Space Phys.* **2020**, *125*, e2019JA027747. [CrossRef]
19. Marchetti, D.; De Santis, A.; Shen, X.; Campuzano, S.A.; Perrone, L.; Piscini, A.; Di Giovambattista, R.; Jin, S.; Ippolito, A.; Cianchini, G.; et al. Possible Lithosphere-Atmosphere-Ionosphere Coupling Effects Prior to the 2018 Mw = 7.5 Indonesia Earthquake from Seismic, Atmospheric and Ionospheric Data. *J. Asian Earth Sci.* **2020**, *188*, 104097. [CrossRef]
20. De Santis, A.; Perrone, L.; Calcara, M.; Campuzano, S.A.; Cianchini, G.; D'Arcangelo, S.; Di Mauro, D.; Marchetti, D.; Nardi, A.; Orlando, M.; et al. A Comprehensive Multiparametric and Multilayer Approach to Study the Preparation Phase of Large Earthquakes from Ground to Space: The Case Study of the June 15 2019, M7.2 Kermadec Islands (New Zealand) Earthquake. *Remote Sens. Environ.* **2022**, *283*, 113325. [CrossRef]
21. Akhoondzadeh, M.; De Santis, A.; Marchetti, D.; Shen, X. Swarm-TEC Satellite Measurements as a Potential Earthquake Precursor Together with Other Swarm and CSES Data: The Case of Mw7.6 2019 Papua New Guinea Seismic Event. *Front. Earth Sci.* **2022**, *10*, 820189. [CrossRef]
22. Yan, R.; Guan, Y.; Shen, X.; Huang, J.; Zhang, X.; Liu, C.; Liu, D. The Langmuir Probe Onboard CSES: Data Inversion Analysis Method and First Results. *Earth Planet. Phys.* **2018**, *2*, 479–488. [CrossRef]
23. Yan, R.; Xiong, C.; Zhima, Z.; Shen, X.; Liu, D.; Liu, C.; Guan, Y.; Zhu, K.; Zheng, L.; Lv, F. Correlation Between Ne and Te Around 14:00 LT in the Topside Ionosphere Observed by CSES, Swarm and CHAMP Satellites. *Front. Earth Sci.* **2022**, *10*, 14. [CrossRef]
24. Cao, J.B.; Yang, J.Y.; Yuan, S.G.; Shen, X.; Liu, Y.; Yan, C.; Li, W.; Chen, T. In-flight observations of electromagnetic interferences emitted by satellite. *Sci. China Ser. E Tech. Sci.* **2009**, *52*, 2112–2118. [CrossRef]
25. Sarkar, S.; Choudhary, S.; Sonakia, A.; Vishwakarma, A.; Gwal, A.K. Ionospheric Anomalies Associated with the Haiti Earthquake of 12 January 2010 Observed by DEMETER Satellite. *Nat. Hazards Earth Syst. Sci.* **2012**, *12*, 671–678. [CrossRef]
26. Marchetti, D.; Zhu, K.; Zhang, H.; Zhima, Z.; Yan, R.; Shen, X.; Chen, W.; Cheng, Y.; He, X.; Wang, T.; et al. Clues of Lithosphere, Atmosphere and Ionosphere Variations Possibly Related to the Preparation of La Palma 19 September 2021 Volcano Eruption. *Remote Sens.* **2022**, *14*, 5001. [CrossRef]
27. Marchetti, D.; Zhu, K.; Yan, R.; Zhima, Z.; Shen, X.; Chen, W.; Cheng, Y.; Fan, M.; Wang, T.; Wen, J.; et al. Ionospheric Effects of Natural Hazards in Geophysics: From Single Examples to Statistical Studies Applied to M5.5+ Earthquakes. *Proceedings* **2023**, *87*, 34. [CrossRef]
28. Parrot, M.; Tramutoli, V.; Liu, T.J.Y.; Pulinets, S.; Ouzounov, D.; Genzano, N.; Lisi, M.; Hattori, K.; Namgaladze, A. Atmospheric and ionospheric coupling phenomena associated with large earthquakes. *Eur. Phys. J. Spec. Top.* **2021**, *230*, 197–225. [CrossRef]
29. Li, M.; Parrot, M. Statistical analysis of an ionospheric parameter as a base for earthquake prediction. *J. Geophys. Res. Space Phys.* **2013**, *118*, 3731–3739. [CrossRef]
30. Zhu, K.; Zheng, L.; Yan, R.; Shen, X.; Zeren, Z.; Xu, S.; Chu, W.; Liu, D.; Zhou, N.; Guo, F. The Variations of Electron Density and Temperature Related to Seismic Activities Observed by CSES. *Nat. Hazards Res.* **2021**, *1*, 88–94. [CrossRef]
31. He, Y.F. Analysis and summary of results based on the study of seismic ionospheric phenomena. *Earthq. Res. China* **2020**, *36*, 244–257.
32. Li, M.; Wang, F.Q.; Zhang, X.D.; Tan, H.-D.; Kang, C.-L.; Xie, T. Time spatial statistical characteristics of seismic influence on ionosphere. *Progress. Geophys.* **2014**, *29*, 0498–0504. (In Chinese) [CrossRef]

33. Pulinets, S.; Ouzounov, D.; Karelin, A.; Davidenko, D. Lithosphere–Atmosphere–Ionosphere–Magnetosphere Coupling—A Concept for Pre-earthquake Signals Generation. In *Pre-Earthquake Processes: A Multidisciplinary Approach to Earthquake Prediction Studies*; John Wiley & Sons: Hoboken, NJ, USA, 2018; pp. 77–98.
34. Zhang, Y.; Wang, T.; Chen, W.; Zhu, K.; Marchetti, D.; Cheng, Y.; Fan, M.; Wang, S.; Wen, J.; Zhang, D.; et al. Are There One or More Geophysical Coupling Mechanisms before Earthquakes? The Case Study of Lushan (China) 2013. *Remote Sens.* **2023**, *15*, 1521. [CrossRef]
35. Li, M.; Wang, H.; Liu, J.; Shen, X. Two Large Earthquakes Registered by the CSES Satellite during Its Earthquake Prediction Practice in China. *Atmosphere* **2022**, *13*, 751. [CrossRef]

remote sensing

MDPI

Communication

A New Analysis Method for Magnetic Disturbances Possibly Related to Earthquakes Observed by Satellites

Xin-Yan Ouyang [1,*], Yong-Fu Wang [2], Xue-Min Zhang [1], Ya-Lu Wang [1] and Ying-Yan Wu [1]

[1] Institute of Earthquake Forecasting, China Earthquake Administration, Beijing 100036, China; zxm@ief.ac.cn (X.-M.Z.); wangyalu@ief.ac.cn (Y.-L.W.); wuyingyan@ief.ac.cn (Y.-Y.W.)
[2] School of Earth and Space Sciences, Peking University, Beijing 100871, China; wyffrank@pku.edu.cn
* Correspondence: oyxy@ief.ac.cn

Abstract: Studies on magnetic disturbances in ultralow frequency ranges related to earthquakes observed by satellites are still limited. Based on Swarm satellites, this paper proposes a new analysis method to investigate pre-earthquake magnetic disturbances by excluding some known non-earthquake magnetic effects that are not confined to those caused by intense geomagnetic activity. This method is demonstrated by two earthquake cases. One is an interplate earthquake, and the other is an intraplate earthquake. Magnetic disturbances around these two earthquakes are associated with solar wind and geomagnetic activity indices, electron density and field-aligned currents. Magnetic disturbances several days before earthquakes do not show clear relations with the already known magnetic effects. These nightside disturbances (LT~17/18, ~02), possibly related to earthquakes observed by Swarm satellites, oscillate in the transverse magnetic field below 2 Hz, propagate along the background magnetic field and are mostly linearly polarized.

Keywords: magnetic disturbances; earthquakes; plasma depletions; field-aligned currents; polarization parameters

Citation: Ouyang, X.-Y.; Wang, Y.-F.; Zhang, X.-M.; Wang, Y.-L.; Wu, Y.-Y. A New Analysis Method for Magnetic Disturbances Possibly Related to Earthquakes Observed by Satellites. *Remote Sens.* **2022**, *14*, 2709. https://doi.org/10.3390/rs14112709

Academic Editors: Kaiguang Zhu, Yunbin Yuan and Dedalo Marchetti

Received: 13 April 2022
Accepted: 31 May 2022
Published: 5 June 2022

1. Introduction

Electromagnetic variations associated with earthquakes have been reported in a wide range of frequencies from quasi-DC to very high frequency by ground or satellite observations. Electromagnetic disturbances in extremely low frequency (ELF, f < 3 kHz) or very low frequency (VLF, f < 30 kHz) ranges are extensively studied. Especially for VLF radio waves from ground-based transmitters, numerous studies have shown perturbations in the VLF phase and amplitude before earthquakes, e.g., [1,2]. Compared with ELF/VLF, electromagnetic observations in ultralow frequency (ULF, f < 10 Hz) are easier to obtain by both ground-based and space-based magnetometers. Furthermore, ground observations of electromagnetic disturbances in ULF are considered the most promising means for monitoring possible earthquake precursors because larger skin depths are comparable to the depths of hypocenters, e.g., [3]. A number of case and statistical studies based on ground observations have verified the existence of ULF electromagnetic disturbances, e.g., [4–8].

Theoretical calculations have shown that for an electromagnetic field of 0.01–100 Hz induced by seismic sources, only the fields from a magnetic type source can penetrate into the ionosphere and magnetosphere, and penetration is possible only in the ULF band (f < 10–20 Hz) [9]. Therefore, it is possible for ULF disturbances to be observed by satellites, which can offer widely covered and high-precision measurements. A few case studies have demonstrated anomalous electromagnetic signals in the ULF range before earthquakes observed by satellites, e.g., [10–13]. The most recent statistical study confirmed the association between earthquakes and ULF wave activity in the nighttime ionosphere [14]. It is common in pre-earthquake effect studies to use geomagnetic activity indices to exclude disturbances in the space environment. However, electromagnetic disturbances in the ionosphere are affected by many energy sources other than intense geomagnetic activity. In addition, most previous studies on ULF disturbances related to earthquakes are based on the electric field in the DC/ULF band recorded by the DEMETER satellite [10–12,14,15]. However, only using electric field data limits further analyses of wave properties such as wave mode and polarization parameters. As a result, it becomes more difficult to deepen our understanding of ULF electromagnetic disturbances.

This study aims to investigate pre-earthquake ULF magnetic disturbances by excluding already known non-earthquake factors demonstrated by two different earthquake cases, interplate and intraplate earthquakes. It is hoped that this research will contribute to the practice of earthquake precursor analysis by analyzing magnetic field disturbances observed by satellites. Section 2 briefly describes the Swarm constellation and its data, as well as the main methods used in this paper. Section 3 presents magnetic field disturbances and their associations with solar wind and geomagnetic activity, electron density and field-aligned currents. The wave properties of magnetic field disturbances are also analyzed. Section 4 discusses the relationship between these disturbances and plasma depletions and midlatitude magnetic fluctuations. Section 5 summarizes our findings.

2. Data and Methods

Swarm is a low-Earth-orbiting constellation made up of three identical satellites: Alpha (A), Bravo (B) and Charlie (C). The three satellites were launched on 22 November 2013 to polar orbits at an ~87° inclination angle. Swarm A and C fly side-by-side at an initial altitude of ~460 km, and Swarm B cruises at a higher altitude of ~510 km. The main objective of Swarm is to measure the Earth's magnetic field. The vector field magnetometer (VFM) can provide three components of the magnetic field with sampling rates of 1 and 50 Hz. The Langmuir probe (LP) provides measurements of electron density and electron temperature at rates of 1 and 2 Hz. The Swarm level 2 FAC-dual product includes dual ionospheric radial currents (IRCs) and field-aligned currents (FACs) with a cadence of 1 Hz, which are produced per day and derived from the lower satellite pair Swarm A and C.

Three components of the magnetic field in North-East-Center (NEC) coordinates are transformed to local field-aligned coordinates (see Figure A1); these coordinates have three components in the radial (Br, toward outer L-shells), azimuthal (Ba, eastward) and parallel (Bp, parallel to the background magnetic field) directions. To highlight the magnetic disturbances, we apply a Savitzky–Golay smoothing filter [16] to define the trend of the magnetic field with a polynomial order of 2 and a frame length of 25 s·fs (sampling frequency). The residual magnetic field is obtained by subtracting the trend from the original magnetic field (see Figure A1c). A similar detrending method is used in studies on electromagnetic ion cyclotron waves, e.g., [17]. Polarization parameters (e.g., wave normal angle, ellipticity) are calculated by the methodology of Means [18,19] to study the possible origin and propagation characteristics of magnetic disturbances.

First, we choose magnetic data according to the estimated earthquake preparation area. Second, residual magnetic data are obtained by subtracting the background magnetic field, which is defined by the Savitzky–Golay smoothing filter. This method of obtaining the residual magnetic field has a faster running speed than subtracting the background magnetic field deduced from CHAOS. This detrending method with a faster processing speed is more convenient for the daily monitoring of magnetic disturbances possibly related to earthquakes. Third, the influences of solar wind and geomagnetic disturbances are excluded. Fourth, we analyze whether the remaining magnetic disturbances are related to significant plasma variations and intense FACs. If these magnetic disturbances are not closely connected with already-known main factors, we further investigate their wave properties.

3. Results

3.1. Overview of Magnetic Disturbances Observed by Swarm B

We show the results of an M7.7 earthquake case with a depth of 10 km that occurred at UTC 13:19:55 on 10 February 2021. The epicenter is located at 23.05°S and 171.66°E. This earthquake occurred as the result of low angle thrust faulting on or near the plate boundary interface between the Australia and Pacific plates. The location of this earthquake is close to the New Hebrides trench, where the Australia lithosphere converges with and sinks beneath the Pacific plate, descending into the mantle and forming the New Hebrides/Vanuatu subduction zone. This earthquake occurred on the subduction zone interface between the Australia and Pacific plates. (https://earthquake.usgs.gov/earthquakes/eventpage/us6000dg77/executive, (accessed on 13 May 2022)). According to Dobrovolsky's formula [20], it is known that the radius of the estimated earthquake preparation zone is approximately 2000 km, corresponding to an approximately 19° latitudinal range. Figure 1 shows the residual magnetic field variations observed by Swarm B in the range of 20° latitude by 20° longitude centered on the epicenter 15 days before and 7 days after the earthquake The time range of our analysis is mainly based on previous studies on the time of pre-earthquake ionospheric effects, e.g., [14]. There are 39 orbits passing the area during this period and sometimes two orbits on the same day. The local time is approximately 16 or 17. Figure 1 presents the residual magnetic field recorded by each orbit, and it is known that clear magnetic disturbances appear in radial and azimuthal components on different days. As these magnetic disturbances may be affected by many factors, it is important to exclude some known causes. We will first analyze disturbances related to space weather events. Figure 2 provides the solar wind parameters and SYM-H index from 15 days before to 7 days after the earthquake, marked as day −15 to day 7. For the magnetic disturbances around day −15, SYM-H is less than −30 nT on day −15, with a recovery phase in the following days. Large amplitude and broad range disturbances, especially at higher latitudes, can be seen from orbits on day −15 and day −14, which suggests that these disturbances are affected by the weak magnetic storm on day −15 (SYM-H < −30 nT). In addition, there are two small jumps of dynamic pressure between day −9 and day −8 and between day −4 and day −3, followed by a small magnetic storm on day −3. Therefore, disturbances on day −9, day −8 and day −3 are eliminated from our study. Pulses on day −10 and day −2 shown in Figure 1a,b are not considered since they do not last long enough.

Figure 3a presents the remaining Swarm B orbits (LT~17) with magnetic disturbances in Br and Ba components after the above analysis, and Figure 3b–e show comparisons of these magnetic disturbances with the simultaneously measured electron density (Ne). There are no distinct perturbations in either the original or residual Ne.

3.2. Magnetic Disturbances Observed by Swarm A

Figure 4 shows observations from Swarm A at a similar local time (LT~18) and on the same day as that in Figure 3. We can see clear disturbances in the residual magnetic field in Br and Ba components in Figure 4b,c. Figure 4d,e present the original and residual Ne. In Figure 4e, residual Ne perturbations with a scale of 10^3 cm^{-3} can be seen from the orbit on day −12 at approximately −22° latitude and day −11 at approximately −15° latitude, which are simultaneous with magnetic disturbances in Br and Ba components. On other days, there are no clear perturbations in the residual Ne.

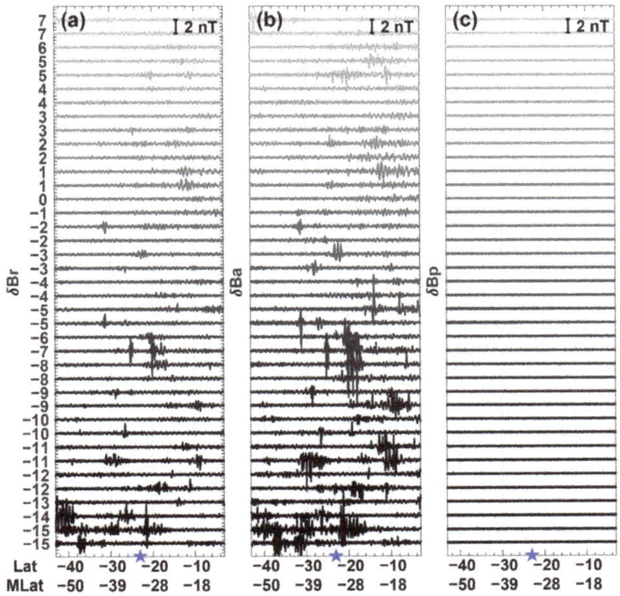

Figure 1. Magnetic disturbances observed by Swarm B 15 days before and 7 days after the M7.7 earthquake, and in the range of 20° latitude by 20° longitude centered on the M7.7 epicenter. (**a–c**) present the residual magnetic field in the radial, azimuthal and parallel directions. The stars on the horizontal axes indicate the epicenter's latitude, and the y-axis shows the time differences (ΔT in days) of each orbit relative to the day of the M7.7 earthquake.

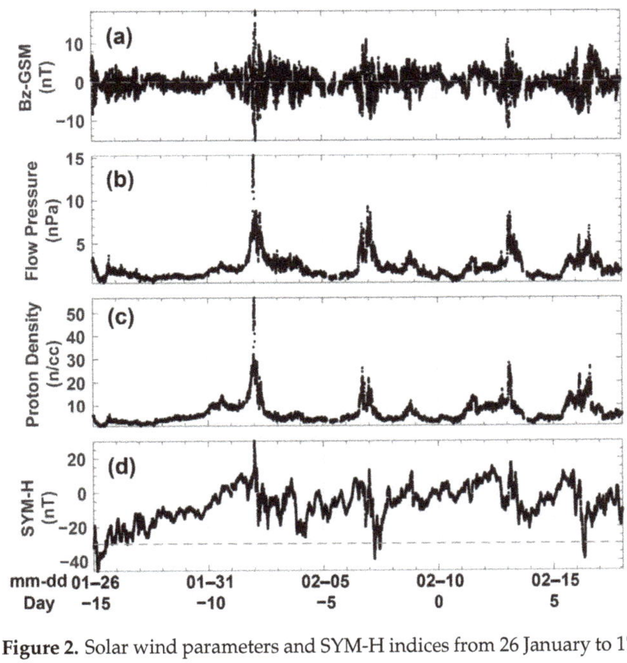

Figure 2. Solar wind parameters and SYM-H indices from 26 January to 17 February 2021, i.e., 15 days before to 7 days after the M7.7 earthquake. (**a**) IMF Bz in the GSM coordinate system, and the gray dashed line indicates Bz = 0; (**b**) the solar wind dynamic pressure; (**c**) the solar wind proton density; and (**d**) the SYM-H index. The gray dashed line indicates −30 nT.

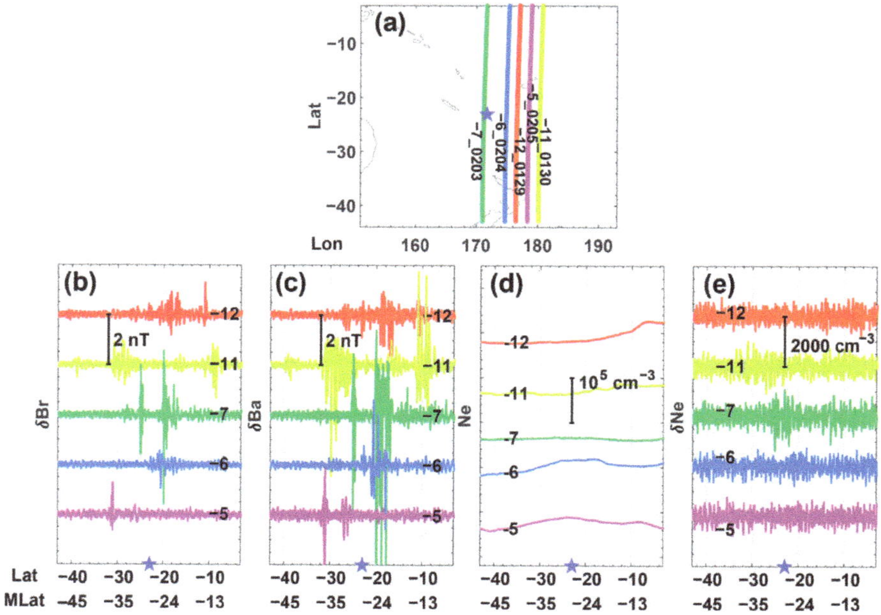

Figure 3. After removing the main influences of solar wind and SYM-H, a comparison between magnetic disturbances and electron density (Ne) was performed. (**a**) Swarm B orbits with magnetic disturbances (LT~17). The blue star indicates the epicenter. Numbers around orbits show time differences (ΔT in days) and the date of each orbit; (**b**,**c**) residual magnetic field in Br and Ba; (**d**) original Ne; (**e**) residual Ne.

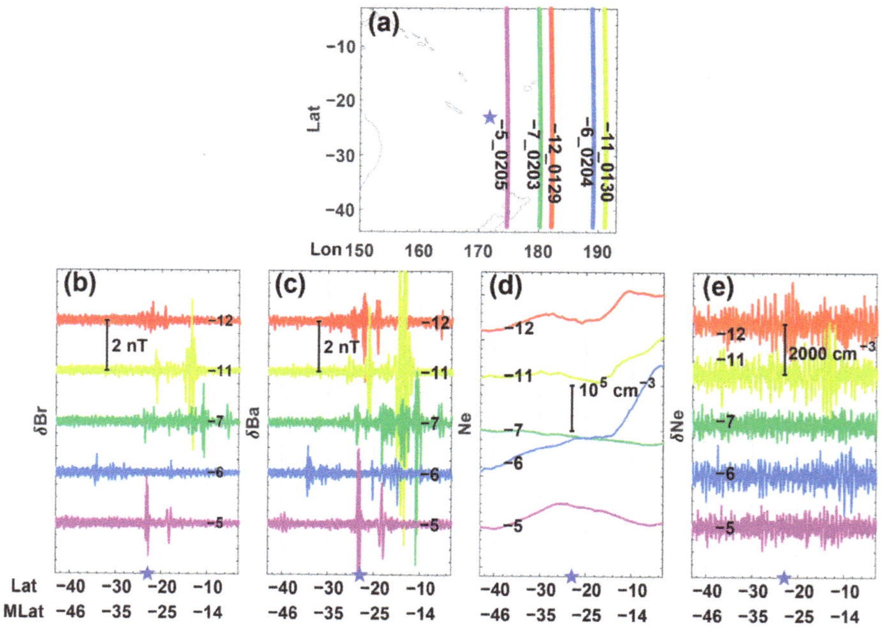

Figure 4. Observations from Swarm A at a similar local time (LT ~18) and on the same day as Swarm B orbits recorded magnetic disturbances. (**a**) Swarm A orbits; the blue star indicates the epicenter; (**b**,**c**) residual magnetic field in Br and Ba; (**d**) original Ne; (**e**) residual Ne.

3.3. Relationship with Field-Aligned Currents (FACs)

This section shows FACs in the area of 20° latitude by 20° longitude centered on the epicenter. Since the FAC deduced from single-satellite magnetic field data needs to make a number of assumptions, we use more realistic FACs computed from the lower satellite pair Swarm A and C (Swarm Level 2 FAC-dual product) [21,22]. Figure 5 presents tracks and FACs in this area on disturbed days −12, −11, −7, −6 and −5. Although the tracks in Figure 5a do not completely correspond to the orbits of Swarm A in Figure 4, we show FACs in the same area of interest that can be used as a reference. From Figures 3a and 4a, it is known that the disturbed orbits are mainly around the east side of the epicenter, and the right tracks in Figure 5a may give reference FACs for these disturbed orbits. Figure 5g–k show FACs along the right overlapped tracks, and variations in FACs are almost the same on different disturbed days. A comparison of Figure 5a with Figure 4a shows that the orbit on day −12 in Figure 4a is the closest to the right tracks in Figure 5a. We can see magnetic disturbances with larger amplitudes in δBa on day −12 above the epicenter in Figure 4c. Figure 5g shows FACs on day -12, and they have values of zero to median above the epicenter. From Figure 5, it is known that FACs are very weak, changing around the median FACs and mostly within 0.04 μA/m^2.

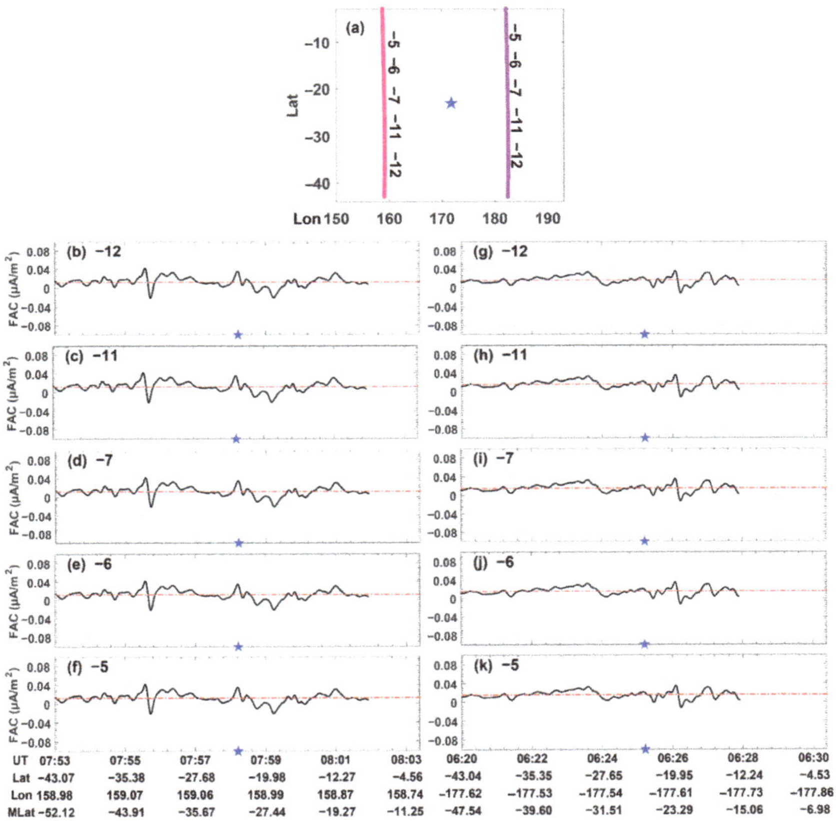

Figure 5. Field-aligned currents (FACs) in the range of 20° latitude by 20° longitude centered on the M7.7 epicenter on the disturbed days. (**a**) Tracks in this area; the blue star shows the epicenter; (**b–f**) FACs on days −12, −11, −7, −6 and −5 along the left tracks in (**a**); (**g–k**) FACs on days −12, −11, −7, −6 and −5 along the right tracks in (**a**). The blue star indicates the epicenter's latitude, and the red dashed lines represent the median FACs along the corresponding tracks in the area of interest.

3.4. Analysis of Polarization Parameters

Since these magnetic disturbances do not have a clear relationship with FACs, further analysis of wave properties is presented. Figures 6 and 7 show the power spectral density and polarization parameters of Swarm B and A orbits in Figures 3 and 4, respectively. These disturbances are located below 2 Hz, and they oscillate in the perpendicular magnetic field (Br and Ba components). Combined with polarization parameters, we know that these disturbances propagate along the background magnetic field and are mostly linearly polarized.

3.5. An Intraplate Earthquake Case Study

In this section, we show magnetic disturbances of an M7.0 earthquake case with a depth of 21 km that occurred at UTC 11:51:27 on 30 October 2020. This earthquake is located at 37.90°N, 26.78°E, and is considered an intraplate event as the result of normal faulting at a shallow crustal depth within the Eurasia tectonic plate (https://earthquake.usgs.gov/earthquakes/eventpage/us7000c7y0/executive, (accessed on 13 May 2022)). The analysis method described in the above sections is applied to this earthquake. The radius of the estimated preparation area is about 1000 km, around 10° latitudinal range. Figure 8 presents residual magnetic field 15 days before and 7 days after the earthquake. There are no orbits at 7 days after the earthquake when the range is limited to 10° latitude by 10° longitude centered on the epicenter. Magnetic disturbances in Br and Ba components occurred at $\Delta T = -8$, -3 and -1 days. Combined with solar wind parameters and SYM-H indices in Figure 9, we know that above magnetic disturbances are not influenced by space weather events. Similar to the analysis of the M7.7 earthquake, these magnetic disturbances are related to simultaneous electron density (see Figure A2) and referenced FACs at similar local times (see Figure A3). Their correlations are not found. At last, we further analyze power spectral density and polarization parameters for these disturbances (see Figure A4).

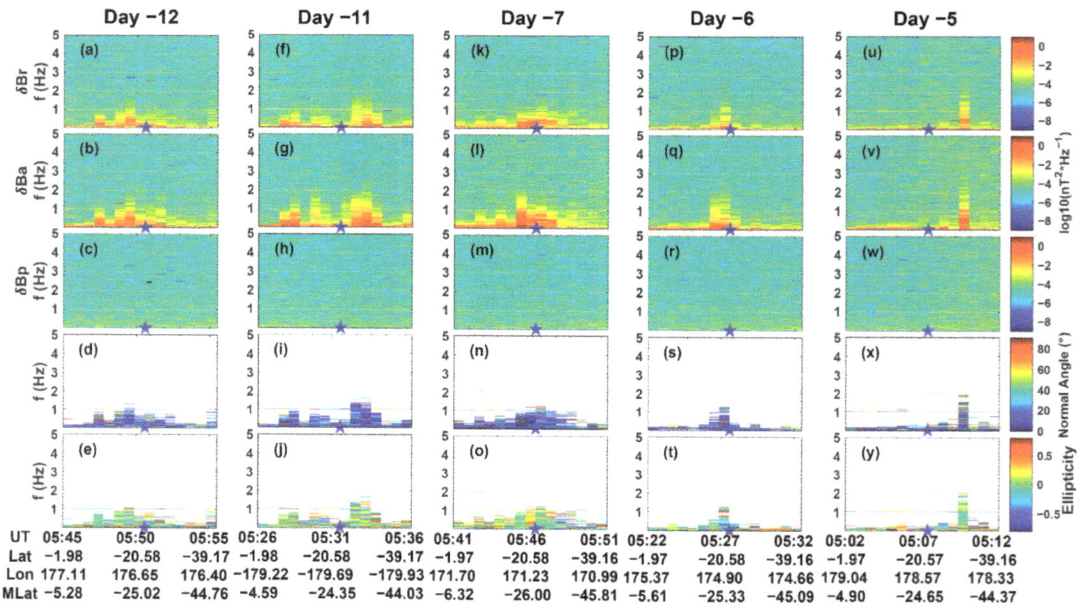

Figure 6. Power spectral density and polarization parameters corresponding to Swarm B orbits in Figure 3. Different columns represent different orbits. For example, the first column shows the power spectral density in panels (**a**–**c**), as well as the wave normal angle and ellipticity in panels (**d**,**e**) on day −12. The second to fifth columns (**f**–**y**) show the power spectral density and polarization parameters on days −11, −7, −6 and −5. The blue star indicates the epicenter.

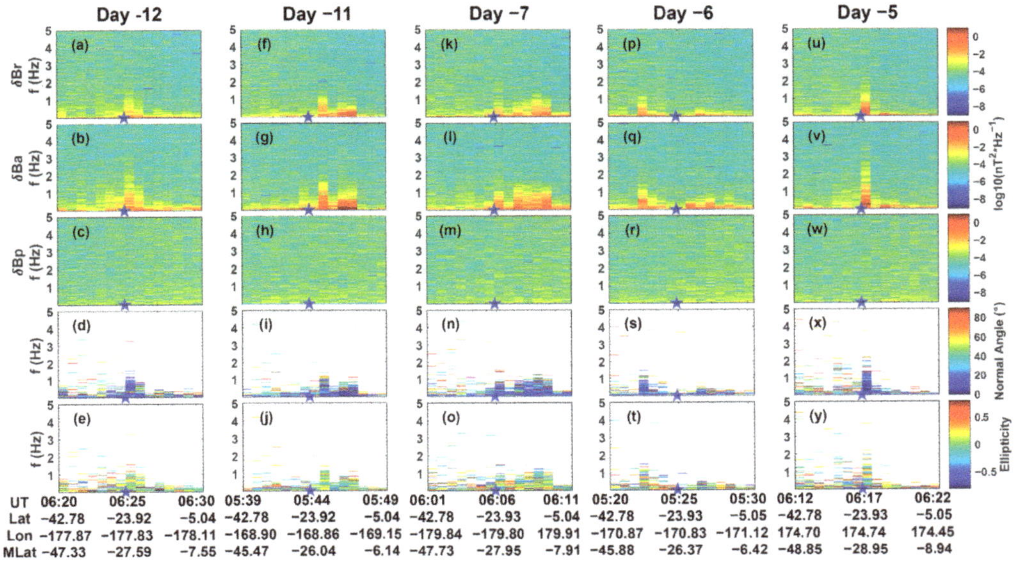

Figure 7. Power spectral density and polarization parameters corresponding to Swarm A orbits shown in Figure 4. The first to the fifth columns show the power spectral density (the first three rows (**a–c**,**f–h**,**k–m**,**p–r**,**u–w**)) and polarization parameters (the fourth and the fifth rows (**d**,**e**,**i**,**j**,**n**,**o**,**s**,**t**,**x**,**y**)) on days −12, −11, −7, −6 and −5. The blue star indicates the epicenter.

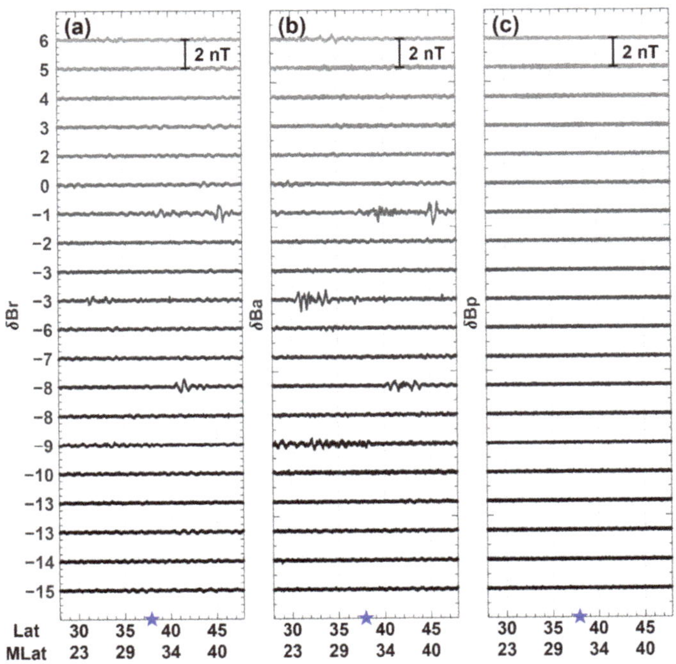

Figure 8. Magnetic disturbances observed by Swarm B 15 days before and 7 days after the M7.0 earthquake and in the range of 10° latitude by 10° longitude centered on an M7.0 epicenter. Panels (**a–c**) show the residual magnetic field in radial, azimuthal and parallel directions. Blue stars on x-axes represent the epicenter's latitude. y-axis indicates time-differences (ΔT in days) between Swarm observations and the earthquake time.

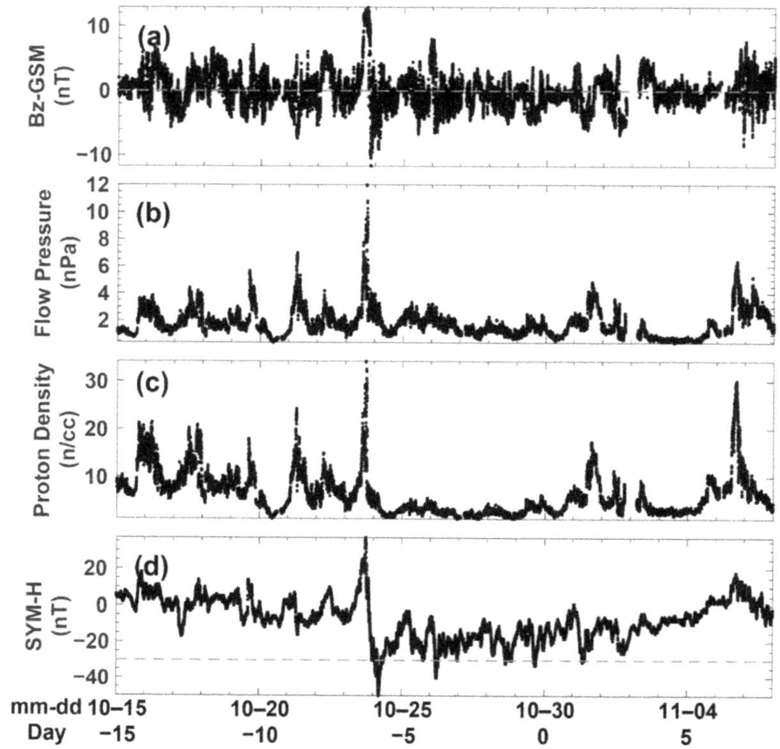

Figure 9. Solar wind parameters and SYM-H indices from 15 October to 6 November, i.e., 15 days before to 7 days after the M7.0 earthquake. (**a**) IMF Bz in GSM coordinates; (**b**) the solar wind dynamic pressure; (**c**) the solar wind proton density and (**d**) the SYM-H index. The dashed line indicates −30 nT.

4. Discussion

This work analyzes the magnetic disturbances related to interplate and intraplate earthquakes recorded by Swarm satellites. Magnetic disturbances associated with the interplate shallow earthquake (M7.7) have larger amplitudes than those related to the intraplate shallow earthquake (M7.0). In this paper, we propose a new method for analyzing magnetic disturbances possibly related to earthquakes. With our method, we try to eliminate already-known factors that can cause magnetic effects. The most common approach is to check solar wind parameters and geomagnetic indices to exclude the influences of space weather events. In addition, magnetic signatures of equatorial plasma depletions (EPDs) and midlatitude magnetic fluctuations are investigated by jointly analyzing electron density (Ne) and field-aligned currents (FACs).

Magnetic field variations in this work are observed in the nightside (LT~17/18, ~02). Previous studies have shown magnetic perturbations related to postsunset equatorial plasma depletions. Their occurrence rate is the lowest at 18 LT and very low at 02 LT. The highest is between 21 and 22 LT [23]. These magnetic field fluctuations are caused by two types of currents: diamagnetic currents flowing across the depletions, producing magnetic perturbations parallel to the ambient field and field-aligned currents (FACs) flowing within the edges of the depletions, generating transverse magnetic field perturbations [24,25]. From Figures 1 and 8, we know that the magnetic field fluctuates in perpendicular components. Figures 3 and 4 present transverse magnetic perturbations and simultaneous Ne. There are no Ne depletions shown in Figures 3d and 4d. Residual Ne disturbances with a scale of 10^3 cm^{-3} on day -11 can be seen in Figure 4e. It is much weaker than the density change of at least 10^5 cm^{-3} to generate a detectable magnetic signal [23,26]. Similar changes of Ne in Figure A2 are presented. These features in Ne suggest that there is no clear relationship between equatorial plasma depletions and magnetic disturbances.

Another influence comes from midlatitude magnetic fluctuations (MMFs) [26]. Nighttime medium-scale MMFs have been related to nighttime medium-scale traveling ionospheric disturbances (MSTIDs) [27]. Their work interprets MMFs in terms of FACs, and most estimated FACs have amplitudes of approximately 0.15 μA/m^2. Small-scale MMFs (<10 km) at midlatitudes and low latitudes are explained as narrow, intense interhemispheric FACs with several μA/m^2 [28]. FACs in the vicinity of the epicenter in Figures 5 and A3 are very weak, almost around the background FACs indicated by red dashed lines, which are in a similar amplitude as given by [21]. Although magnetic perturbations in this work are in perpendicular magnetic field components, they are not significantly affected by enhanced interhemispheric FACs.

5. Conclusions

This paper presents interplate and intraplate earthquake case studies to demonstrate the elimination of some known magnetic disturbances induced by non-earthquake sources. We present magnetic disturbances around earthquakes recorded by Swarm satellites. By correlating these magnetic disturbances with space weather events, equatorial plasma depletions and midlatitude magnetic fluctuations, we found that magnetic disturbances several days before earthquakes do not show clear associations with already known non-earthquake disturbances. These magnetic disturbances possibly related to earthquakes oscillate below 2 Hz in the transverse magnetic field, propagating along the background magnetic field and mostly linearly polarized.

The method proposed in this study can be used in the daily practice of analyzing earthquake precursors by using magnetic data observed by satellites. Further work could conduct to categorize different magnetic disturbances and deepen the understanding of their origins. For magnetic disturbances possibly related to earthquakes observed by satellites, combining geomagnetic data to discuss and study correlations between magnetic disturbances and earthquakes in different seismotectonic circumstances is also needed in future research.

Author Contributions: Conceptualization, X.-Y.O.; funding acquisition, X.-Y.O.; investigation, X.-Y.O.; methodology, X.-Y.O.; visualization, X.-Y.O. and Y.-F.W.; writing—original draft, X.-Y.O.; writing—review and editing, Y.-F.W., X.-M.Z., Y.-L.W. and Y.-Y.W. All authors have read and agreed to the published version of the manuscript.

Funding: This research was funded by the APSCO, dragon 5 cooperation proposal and the National Key R&D Program of China, grant number "the APSCO Earthquake Research Project Phase II, #59308 and 2018YFC1503506". Yong-Fu Wang is supported by the Innovation Fund from Joint Innovation Center of Space Science (Aerospace Shanghai).

Data Availability Statement: The Swarm data can be accessed at https://earth.esa.int/swarm (accessed on 20 December 2021). The solar wind and geomagnetic activity data can be acquired from the OMNIWeb service at https://omniweb.gsfc.nasa.gov (accessed on 22 December 2021). The earthquake information is available at https://earthquake.usgs.gov/earthquakes/search (accessed on 10 December 2021).

Acknowledgments: The authors would like to thank Hui Wang (Wuhan University), Yueren Xu, Lianqing Zhou and Qi Liu (IEF, CEA) for helpful discussions.

Conflicts of Interest: The authors declare no conflict of interest.

Appendix A

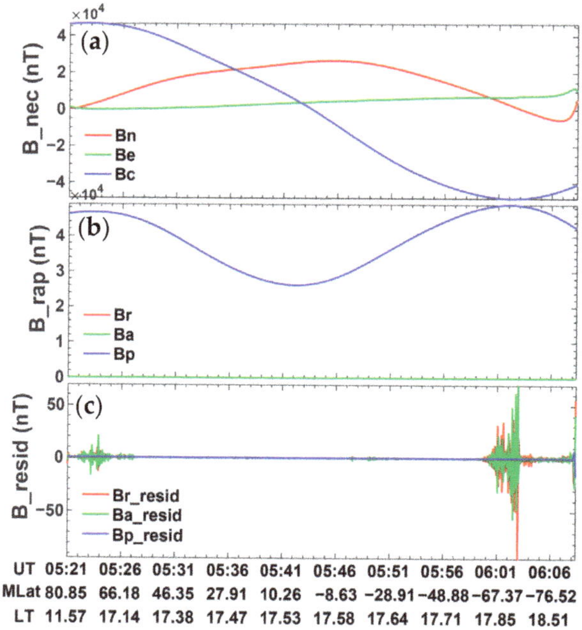

Figure A1. (**a**) Three magnetic field components in North-East-Center (NEC) coordinates; (**b**) three magnetic field components in local field-aligned coordinates; (**c**) residual magnetic field in local field-aligned coordinates.

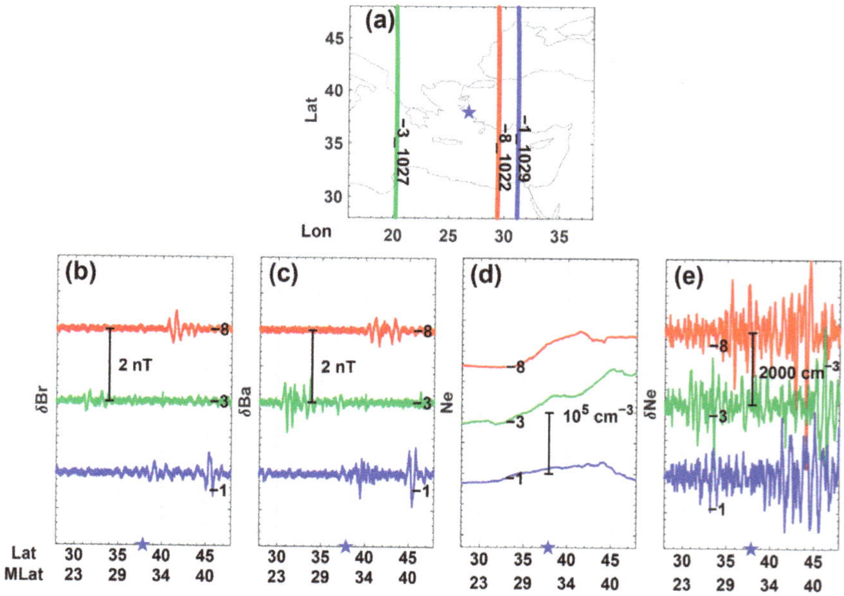

Figure A2. Remaining magnetic disturbances (LT~02) after removing the main influences of space weather events. (**a**) Swarm B orbits with magnetic disturbances in the range of 10° latitude by 10° longitude centered on the M7.0 earthquake. Annotations in panel (**a**) show time differences (ΔT in days) and dates (mmdd) of each orbit; (**b**,**c**) present the residual magnetic field in Br and Ba components; (**d**) original Ne; and (**e**) residual Ne. The blue star indicates the epicenter.

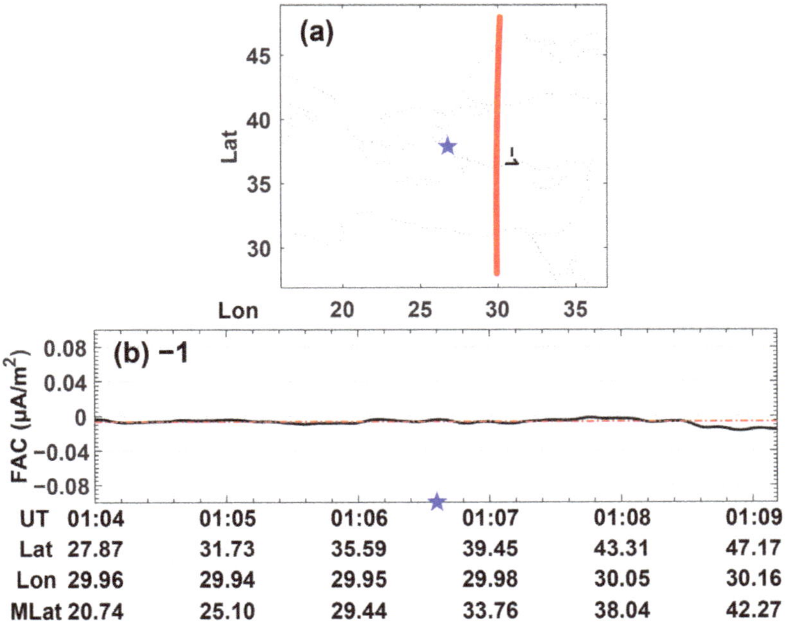

Figure A3. Field-aligned currents (LT~03) derived from the Swarm A and C pair in the range of 10° latitude by 10° longitude centered on the M7.0 earthquake. There is no FAC data at ΔT = −8 and −3 days. (**a**) Orbits on ΔT = −1 day, blue star indicates the epicenter; (**b**) FACs along orbits on day −1, and blue star shows the epicenter's latitude.

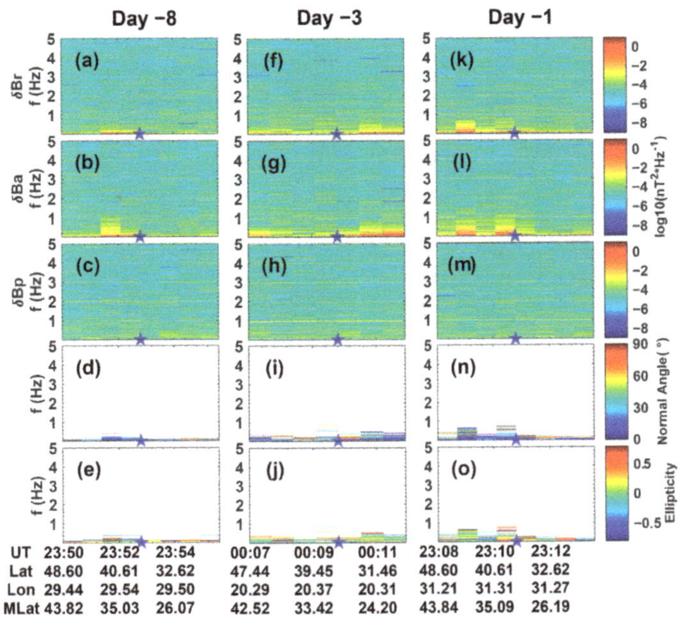

Figure A4. Power spectral density and polarization parameters of disturbed orbits at ΔT = −8, −3 and −1 days. For example, (**a–c**) show the power spectral density of Br, Ba and Bp components at ΔT = −8 days; (**d**,**e**) present wave normal angle and ellipticity at ΔT = −8 days. The second and third columns (**f–o**) display results at ΔT = −3 and −1 days. The blue star indicates the epicenter.

References

1. Zhao, S.; Shen, X.; Liao, L.; Zhima, Z.; Zhou, C.; Wang, Z.; Cui, J.; Lu, H. Investigation of Precursors in VLF Subionospheric Signals Related to Strong Earthquakes (M > 7) in Western China and Possible Explanations. *Remote Sens.* **2020**, *12*, 3563. [CrossRef]
2. Zhang, X.; Wang, Y.; Boudjada, M.Y.; Liu, J.; Magnes, W.; Zhou, Y.; Du, X. Multi-Experiment Observations of Ionospheric Disturbances as Precursory Effects of the Indonesian Ms6.9 Earthquake on 5 August 2018. *Remote Sens.* **2020**, *12*, 4050. [CrossRef]
3. Yumoto, K.; Ikemoto, S.; Cardinal, M.G.; Hayakawa, M.; Hattori, K.; Liu, J.Y.; Saroso, S.; Ruhimat, M.; Husni, M.; Widarto, D.; et al. A new ULF wave analysis for Seismo-Electromagnetics using CPMN/MAGDAS data. *Phys. Chem. Earth Parts A B C* **2009**, *34*, 360–366. [CrossRef]
4. Fraser-Smith, A.C.; Bernardi, A.; McGill, P.R.; Ladd, M.E.; Helliwell, R.A.; Villard, O.G. Low-frequency magnetic field measurements near the epicenter of the Ms 7.1 Loma Prieta earthquake. *Geophys. Res. Lett.* **1990**, *17*, 1465–1468. [CrossRef]
5. Hayakawa, M.; Kawate, R.; Molchanov, O.A.; Yumoto, K. Results of ultra-low-frequency magnetic field measurements during the Guam Earthquake of 8 August 1993. *Geophys. Res. Lett.* **1996**, *23*, 241–244. [CrossRef]
6. Bortnik, J.; Cutler, J.W.; Dunson, C.; Bleier, T.E. The possible statistical relation of Pc1 pulsations to Earthquake occurrence at low latitudes. *Ann. Geophys.* **2008**, *26*, 2825–2836. [CrossRef]
7. Han, P.; Hattori, K.; Hirokawa, M.; Zhuang, J.; Chen, C.-H.; Febriani, F.; Yamaguchi, H.; Yoshino, C.; Liu, J.-Y.; Yoshida, S. Statistical analysis of ULF seismomagnetic phenomena at Kakioka, Japan, during 2001–2010. *J. Geophys. Res. Space Phys.* **2014**, *119*, 4998–5011. [CrossRef]
8. Hattori, K.; Han, P.; Yoshino, C.; Febriani, F.; Yamaguchi, H.; Chen, C.-H. Investigation of ULF Seismo-Magnetic Phenomena in Kanto, Japan During 2000–2010: Case Studies and Statistical Studies. *Surv. Geophys.* **2013**, *34*, 293–316. [CrossRef]
9. Molchanov, O.; Hayakawa, M.; Rafalsky, V. Penetration characteristics of electromagnetic emissions from an underground seismic source into the atmosphere, ionosphere, and magnetosphere. *J. Geophys. Res. Space Phys.* **1995**, *100*, 1691–1712. [CrossRef]
10. Athanasiou, M.A.; Anagnostopoulos, G.C.; Iliopoulos, A.C.; Pavlos, G.P.; David, C.N. Enhanced ULF radiation observed by DEMETER two months around the strong 2010 Haiti earthquake. *Nat. Hazards Earth Syst. Sci.* **2011**, *11*, 1091–1098. [CrossRef]
11. Walker, S.N.; Kadirkamanathan, V.; Pokhotelov, O.A. Changes in the ultra-low frequency wave field during the precursor phase to the Sichuan earthquake: DEMETER observations. *Ann. Geophys.* **2013**, *31*, 1597–1603. [CrossRef]
12. Zhang, X.; Chen, H.; Liu, J.; Shen, X.; Miao, Y.; Du, X.; Qian, J. Ground-based and satellite DC-ULF electric field anomalies around Wenchuan M8.0 earthquake. *Adv. Space Res.* **2012**, *50*, 85–95. [CrossRef]
13. De Santis, A.; Balasis, G.; Pavón-Carrasco, F.J.; Cianchini, G.; Mandea, M. Potential earthquake precursory pattern from space: The 2015 Nepal event as seen by magnetic Swarm satellites. *Earth Planet. Sci. Lett.* **2017**, *461*, 119–126. [CrossRef]
14. Ouyang, X.Y.; Parrot, M.; Bortnik, J. ULF Wave Activity Observed in the Nighttime Ionosphere Above and Some Hours Before Strong Earthquakes. *J. Geophys. Res. Space Phys.* **2020**, *125*, e2020JA028396. [CrossRef]
15. Berthelier, J.J.; Godefroy, M.; Leblanc, F.; Malingre, M.; Menvielle, M.; Lagoutte, D.; Brochot, J.Y.; Colin, F.; Elie, F.; Legendre, C. ICE, the electric field experiment on DEMETER. *Planet. Space Sci.* **2006**, *54*, 456–471. [CrossRef]
16. Savitzky, A.; Golay, M.J.E. Smoothing and Differentiation of Data by Simplified Least Squares Procedures. *Anal. Chem.* **1964**, *36*, 1627–1639. [CrossRef]
17. Kim, H.; Hwang, J.; Park, J.; Miyashita, Y.; Shiokawa, K.; Mann, I.R.; Raita, T.; Lee, J. Large-Scale Ducting of Pc1 Pulsations Observed by Swarm Satellites and Multiple Ground Networks. *Geophys. Res. Lett.* **2018**, *45*, 12–703. [CrossRef]
18. Means, J.D. Use of the three-dimensional covariance matrix in analyzing the polarization properties of plane waves. *J. Geophys. Res.* **1972**, *77*, 5551–5559. [CrossRef]
19. Bortnik, J.; Cutler, J.W.; Dunson, C.; Bleier, T.E. An automatic wave detection algorithm applied to Pc1 pulsations. *J. Geophys. Res. Space Phys.* **2007**, *112*, A04204. [CrossRef]
20. Dobrovolsky, I.P.; Zubkov, S.I.; Miachkin, V.I. Estimation of the size of earthquake preparation zones. *Pure Appl. Geophys.* **1979**, *117*, 1025–1044. [CrossRef]
21. Lühr, H.; Kervalishvili, G.; Michaelis, I.; Rauberg, J.; Ritter, P.; Park, J.; Merayo, J.M.G.; Brauer, P. The interhemispheric and F region dynamo currents revisited with the Swarm constellation. *Geophys. Res. Lett.* **2015**, *42*, 3069–3075. [CrossRef]
22. Ritter, P.; Lühr, H.; Rauberg, J. Determining field-aligned currents with the Swarm constellation mission. *Earth Planets Space* **2013**, *65*, 9. [CrossRef]
23. Stolle, C.; Lühr, H.; Rother, M.; Balasis, G. Magnetic signatures of equatorial spread F as observed by the CHAMP satellite. *J. Geophys. Res. Space Phys.* **2006**, *111*, A02304. [CrossRef]
24. Rodríguez-Zuluaga, J.; Stolle, C.; Park, J. On the direction of the Poynting flux associated with equatorial plasma depletions as derived from Swarm. *Geophys. Res. Lett.* **2017**, *44*, 5884–5891. [CrossRef]
25. Rodríguez-Zuluaga, J.; Stolle, C.; Yamazaki, Y.; Lühr, H.; Park, J.; Scherliess, L.; Chau, J.L. On the Balance Between Plasma and Magnetic Pressure Across Equatorial Plasma Depletions. *J. Geophys. Res. Space Phys.* **2019**, *124*, 5936–5944. [CrossRef]
26. Yokoyama, T.; Stolle, C. Low and Midlatitude Ionospheric Plasma Density Irregularities and Their Effects on Geomagnetic Field. *Space Sci Rev.* **2017**, *206*, 495–519. [CrossRef]

27. Park, J.; Lühr, H.; Stolle, C.; Rother, M.; Min, K.W.; Chung, J.-K.; Kim, Y.H.; Michaelis, I.; Noja, M. Magnetic signatures of medium-scale traveling ionospheric disturbances as observed by CHAMP. *J. Geophys. Res. Space Phys.* **2009**, *114*, A03307. [CrossRef]
28. Yin, F.; Lühr, H.; Park, J.; Wang, L. Comprehensive Analysis of the Magnetic Signatures of Small-Scale Traveling Ionospheric Disturbances, as Observed by Swarm. *J. Geophys. Res. Space Phys.* **2019**, *124*, 10794–10815. [CrossRef]

remote sensing

Article

The Ionospheric Plasma Perturbations before a Sequence of Strong Earthquakes in Southeast Asia and Northern Oceania in 2018

Dapeng Liu [1,*], Zhima Zeren [1], He Huang [1], Dehe Yang [1], Rui Yan [1], Qiao Wang [1], Xuhui Shen [2], Chao Liu [2] and Yibing Guan [2]

[1] National Institute of Natural Hazards, Ministry of Emergency Management of China, Beijing 100085, China; zerenzhima@ninhm.ac.cn (Z.Z.); hehuang@ninhm.ac.cn (H.H.); deheyang@ninhm.ac.cn (D.Y.); ruiyan@ninhm.ac.cn (R.Y.); qiaowang@ninhm.ac.cn (Q.W.)
[2] National Space Science Center, Chinese Academy of Sciences, Beijing 100190, China; shenxuhui@nssc.ac.cn (X.S.); liuch@nssc.ac.cn (C.L.); gyb@nssc.ac.cn (Y.G.)
* Correspondence: dapengliu@ninhm.ac.cn

Abstract: From August to October 2018, a series of strong earthquake (EQ) events occurred in southeast Asia and northern Oceania (22°S to 0°N, 115°E to 170°E) within 50 days. In this paper, we analyze the features of ionospheric plasma perturbations, recorded by the Plasma Analyzer Package (PAP) and Langmuir probe (LAP) onboard the China Seismo-Electromagnetic Satellite (CSES-01), before four EQs with magnitudes of Ms 6.9 to Ms 7.4. The ion parameters such as the oxygen ion density (No^+), the ion drift velocity in the vertical direction (V_z) under the conditions of geomagnetic storms, and strong EQs are compared. The results show that within 1 to 15 days before the strong EQs, the No^+ and the electron density (*Ne*) increased while the electron temperature (*Te*) decreased synchronously. Meanwhile, the V_z significantly increased along the ground-to-space direction. The relative variation of No^+ and V_z before the strong EQs is more prominent, and the V_z is not easily influenced by the geomagnetic storm but is susceptible to the seismic activities. Our results suggest that the anomaly of ionospheric plasma perturbations occurring in this area is possibly related to the pre-EQ signatures.

Keywords: CSES; ionospheric disturbance; plasma; earthquake precursor; ion density; ion drift velocity

Citation: Liu, D.; Zeren, Z.; Huang, H.; Yang, D.; Yan, R.; Wang, Q.; Shen, X.; Liu, C.; Guan, Y. The Ionospheric Plasma Perturbations before a Sequence of Strong Earthquakes in Southeast Asia and Northern Oceania in 2018. *Remote Sens.* **2023**, *15*, 5735. https://doi.org/10.3390/rs15245735

Academic Editors: Dedalo Marchetti, Yunbin Yuan and Kaiguang Zhu

Received: 31 October 2023
Revised: 9 December 2023
Accepted: 11 December 2023
Published: 15 December 2023

1. Introduction

An earthquake (EQ) is one of the most destructive natural disasters, and discerning seismic precursors has been one of the most challenging topics for scientists in the last few decades. With the rapid development of aerospace technology, the Low Earth Orbit (LEO) satellite in the upper ionosphere provides unique opportunities to identify possible pre-EQ anomalies. Many researchers have confirmed the disturbances of the electromagnetic field and plasma parameters appearing before some large EQs [1–7].

The Detection of Electro-Magnetic Emissions Transmitted from Earthquake Regions (DEMETER), which was the first satellite in the world dedicated to studying seismic ionospheric effects, was launched by France in 2004 on a sun-synchronous orbit at ~710 km altitude. The DEMETER satellite was designed to detect and characterize the ionosphere's electromagnetic fields, plasma parameters, and energetic particles [8,9]. Later, the Swarm satellite constellation was launched by the European Space Agency in 2013 to accurately measure the different magnetic signals that arise from the Earth's core, crust, and the environment around the Earth [5,10].

Liu et al. [11] examined variations of ionospheric *fo*F2, recorded by the Chung-Li ionosonde before M $\geq$ 6.0 EQs in Taiwan from 1994 to 1999, to understand seismo-ionospheric precursor signatures. Ouyang et al. [12] analyzed the disturbance of oxygen ion density detected by the DEMETER satellite before the Ms 7.9 Tonga Islands EQ in 2006. Piersanti et al. [13] analyzed the Bayan Mw 6.9 EQ, which occurred in 2018, using ground and satellite data, and they built a new lithospheric–atmospheric–ionospheric–magnetospheric coupling model to analyze the anomalous observations before and at the moment of the EQ occurrence.

In recent years, the China Seismo-Electromagnetic Satellite (CSES-01), which is also called ZHANGHENG-1, was launched on 2 February 2018, on a sun-synchronous orbit at an altitude of ~507 km with a high orbital inclination of 97.4° and a recursive period of 5 days. The ascending/descending node local time (LT) is fixed at 2:00/14:00, respectively. The main scientific objectives of CSES-01 are to detect the electromagnetic field/waves, ionospheric plasma, and charged particles intended to survey the space environment and the ionospheric disturbances that may be associated with destructive EQs, volcano explosions, tsunamis, and space weather anomalies, and to analyze the Lithosphere-Atmosphere-Ionosphere (LAI) coupling mechanism [14]. In addition, China and Italy are jointly building the second one, i.e., CSES-02, which will be launched in 2024. CSES-02 will also provide almost the same physical field parameters as CSES-01 but with numerous specific optimizations based on CSES-01's in-orbit performance.

Using the CSES-01 data, the pre-EQ ionospheric anomalies have already been widely studied. Li et al. [15] statistically analyzed the seismo-ionospheric effects of electron density and oxygen ion density detected by the DEMETER and CSES-01 satellites. They suggested that the CSES-01 could effectively detect ionospheric disturbances during strong EQs, as the DEMETER satellite does. Liu et al. [14] reported the simultaneous disturbances of ion drift velocity and oxygen ion density recorded by the orbit closest to the epicenter five days before the Venezuelan Ms 7.3 EQ in 2018. Li et al. [16] analyzed possible perturbations in the electric field detected by CSES-01 before the Mw 7.7 Caribbean Sea EQ occurred in 2020. Song et al. [17] reported the possible plasma disturbance characteristic before four shallow-focus EQs in Indonesia were detected by CSES-01 in 2018.

More recently, Zeren et al. [18] put forward the routine data preprocessing and seismo-ionospheric information extraction methods by using CSES-01 data and other multi-source data and provided the results of the possible seismo-ionospheric anomalies that appeared during the Ms $\geq$ 7 shallow EQs recorded by CSES-01. Liu et al. [19] analyzed the seismo-ionospheric electron density disturbances observed by CSES-01 through a spatial analysis method.

These new scientific research results bring hope for a breakthrough in EQ forecasting. However, due to the insufficient amount of experimental evidence supporting those theories, there are also some debates on the possibility of seismo-ionospheric precursors; thus, the existence of ionospheric perturbations before EQs is controversial [9,13,20]. One of the main challenges for seeking seismic precursors from the ionosphere is that strong EQ events happen so occasionally around the globe, and there is a large time period and space among them, so it is difficult to compare and summarize the regularity and relevance of EQ precursors.

From August to October 2018, a series of strong EQ events occurred in southeast Asia and northern Oceania (22°S to 0°N, 115°E to 170°E), providing a rare opportunity for EQ precursor research. In this study, we analyze the characteristics of ionospheric plasma perturbations before four strong EQs with magnitudes equal to or larger than Ms 6.9 occurred during this period by using the observations from the Plasma Analyzer Package (PAP) and Langmuir Probe (LAP) onboard CSES-01. Our findings support the possibility of using in-situ plasma observation results from the LEO satellite to examine EQ precursor signatures in the ionosphere.

2. Data and Methods

2.1. Seismic Data

According to the China Earthquake Networks Center (CENC) and the United States Geological Survey (USGS), four strong EQ events with magnitudes Ms 6.9 to Ms 7.4

occurred in southeast Asia and northern Oceania from August to October 2018. The detailed information is described in Table 1 and Figure 1. These strong EQs happened in the adjacent areas—south of the circum-Pacific seismic belt—within 50 days, and the focal depths are all between 17 km and 39 km.

Table 1. Information on the four strong EQs that occurred in southeast Asia and northern Oceania in 2018.

No.	Region	Date and Time (UTC)	Magnitude (Ms/Mw) [1]	Latitude (°N)	Longitude (°E)	Focal Depth (km)
1	Loyalty Islands	29 August 2018 at 03:51:54	7.1/7.1	−21.95	170.10	20
2	Southeast of Loyalty Islands	16 August 2018 at 01:03:42	6.9/6.5	−21.65	169.61	17
3	Indonesia	28 September 2018 at 10:02:44	7.4/7.5	−0.25	119.90	20
4	Papua New Guinea	10 October 2018 at 20:48:18	7.1/7.0	−5.70	151.25	39

[1] The moment magnitude Mw and focal depth were retrieved from USGS, and the other information was retrieved from CENC.

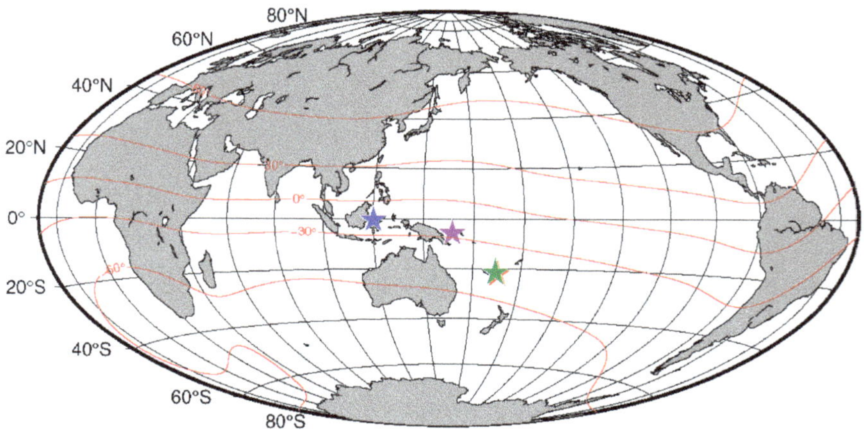

Figure 1. Location of the four investigated EQs that occurred in southeast Asia and northern Oceania in 2018. The stars represent the epicenters, and the red lines are the geomagnetic latitude lines. The green star represents the Loyalty Islands Ms 7.1 EQ, and the red star represents the Southeast of Loyalty Islands Ms 6.9 EQ, and the blue star represents the Indonesia Ms 7.4 EQ, and the purple star represents the Papua New Guinea Ms 7.1 EQ.

Among them, two EQs happened in the Loyalty Islands, and southeast of the Loyalty Islands are less than 100 km away. The Indonesia Ms 7.4 EQ and the Papua New Guinea Ms 7.1 EQ occurred at similar latitudes, with a distance of less than 5.5° in latitude. Similar occurrence characteristics provide an excellent research opportunity for studying EQ precursors.

2.2. CSES-01 Satellite Data

The CSES-01 satellite was specially developed to detect the abnormal ionospheric disturbances associated with strong seismic activities. It has eight types of scientific payloads that can observe the ionospheric background environment and perturbations in the electromagnetic waves and field, plasma, and high-energy particles. Among them, the PAP and LAP were designed to obtain plasma parameters (ion and electron properties) through in-situ detection. The PAP provides ion density (N_i, including hydrogen ion, helium ion, and oxygen ion), ion temperature, and ion drift velocity (V_i). LAP data provides electron density (*Ne*) and electron temperature (*Te*), respectively. They have a large measurement range that covers the plasma change at ~507 km altitude and high accuracy to discern weak abnormal information. The measurement range of *Ni* and *Ne* is 5×10^8–1×10^{13} m^{-3}, and *Te* is 500 K–10,000 K, with a relative accuracy of 10%. The measurement range of V_i is −3000–+3000 m/s, with a relative accuracy of 20 m/s [14,18].

The PAP and LAP onboard CSES-01 operate between latitudes of ±65 °, providing 30 or 31 sets of ascending and descending half-orbit data files in the Hierarchical Data Format (HDF) format every day. The PAP samples every 1 s in the whole observation area, while the LAP samples every 1.5 s above China and the main seismic belt around the world, and samples every 3 s in other regions [14,18]. In this research, we processed the Level 2 standard data of PAP and LAP, which are calibrated physical data with geographic coordinates and time information.

It should be noted that the PAP was contaminated after launch, leading to lower absolute values of N_i and larger V_i than the expected ones in the orbit position. After a lot of in-flight commissioning tests and simulation analysis, it was considered that this phenomenon was due to the decrease in the sensitivity of the sensors caused by the contamination layers on the surface of the PAP's sensors. Due to the problems with the absolute values, the PAP data have not been widely studied yet [14].

Nonetheless, the calibration works and some observation results showed that the state of the contaminated layers on the sensors did not change apparently in 2018, therefore, the relative variations of N_i and V_i were stable in a short time, which had the capability of describing the ionospheric anomaly information. Liu et al. [14] discussed the reliability and application of PAP data in 2018 based on the distribution characteristics of ion parameters around the magnetic equator and the observations of ionospheric disturbance phenomena induced by natural sources (e.g., strong EQ and geomagnetic storms) and artificial sources (e.g., ground-based powerful VLF transmitters), which indicated the reliability of the relative variation of PAP data. Yang et al. [21] reported significant ionospheric responses to the intense geomagnetic storm event on 25 August 2018, using observations from PAP and the other payloads onboard CSES-01. After the comprehensive analysis, we suggested that the oxygen ion density (No^+) and ion drift velocity in the vertical direction (V_z) of PAP in 2018 are reliable when studying the relative variations in a short time (not exceeding one month).

Meanwhile, the LAP data are considered to have good performance based on comparison with the Swarm constellation and ground-based observations [18]. Li et al. [15] made a preliminary statistical analysis of the seismic information on ionospheric plasma parameters detected by CSES-01 from August 2018 to November 2019. The results show that the CSES-01 can effectively record strong EQ precursors, and No^+ seems more sensitive to seismic activities than *Ne*. This viewpoint needs further investigation with more observation data.

In this study, we mainly used No^+, V_z, *Ne*, and *Te* to analyze strong EQ precursor anomalies.

2.3. Geomagnetic and Solar Conditions Data

The space weather condition from August to October 2018 is shown in Figure 2, including the 10.7 cm solar radio flux index (F10.7), the Dst, and the Kp index. The red vertical lines represent the times of four strong EQs.

The Dst and Kp indexes indicate that intense solar activities and geomagnetic perturbations occurred from 26–27 August, 10–11 September, 22 September, and 7–10 October. They provide us with opportunities to compare the differential performances of ionospheric disturbances caused by intense magnetic storms and strong EQs.

Furthermore, some researchers proposed that an increase in solar activity could also trigger EQ occurrence [22], and this could be the case, for example, for the first EQ of 29 August 2018, but the study of a possible influence of solar activity on the seismicity is beyond the scope of this study.

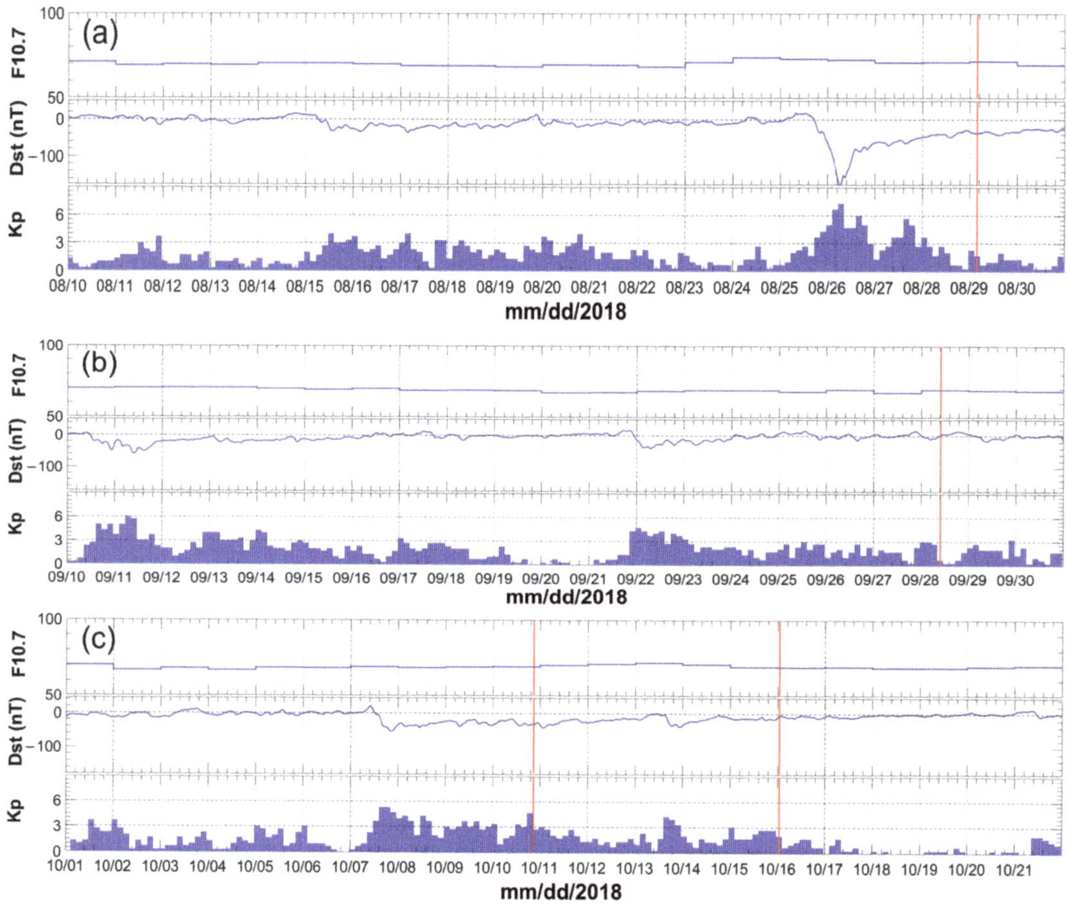

Figure 2. The space weather index from August to October 2018. From top to bottom: (**a**) the space weather index for August 2018, (**b**) the space weather index for September 2018, and (**c**) the space weather index for October 2018. The red vertical lines represent the times of four strong EQs.

2.4. Method

It is widely known that the ionospheric environment is easily influenced by a geomagnetic storm that mainly originates from solar activity. Thus, the relatively weak seismic precursors could be submerged by such powerful space weather disturbances [18,21]. Therefore, it is necessary to check the space weather conditions before looking for eventual seismic information. The satellite data recorded under the geomagnetic indexes Kp exceeded 3 or the Dst index lower than −30 nT were carefully examined to minimize the anomalous interference likely excited by strong geomagnetic activity. Moreover, the observation results affected by the satellite control experiments were also excluded, according to the satellite platform's data quality flags and information.

Based on previous studies [23], the orbits from the CSES-01 flying over the epicenters ±10° in longitude and latitude were chosen about 30 days before the EQs occurrence in these cases, as presented in Figure 3. Among these data, only those from ascending orbits (the satellite passed through the study areas around LT 2:00 a.m.) were used in this research, considering that the ionospheric environment of descending orbits (in the daytime) was strongly affected by the variations of the solar radiation [18].

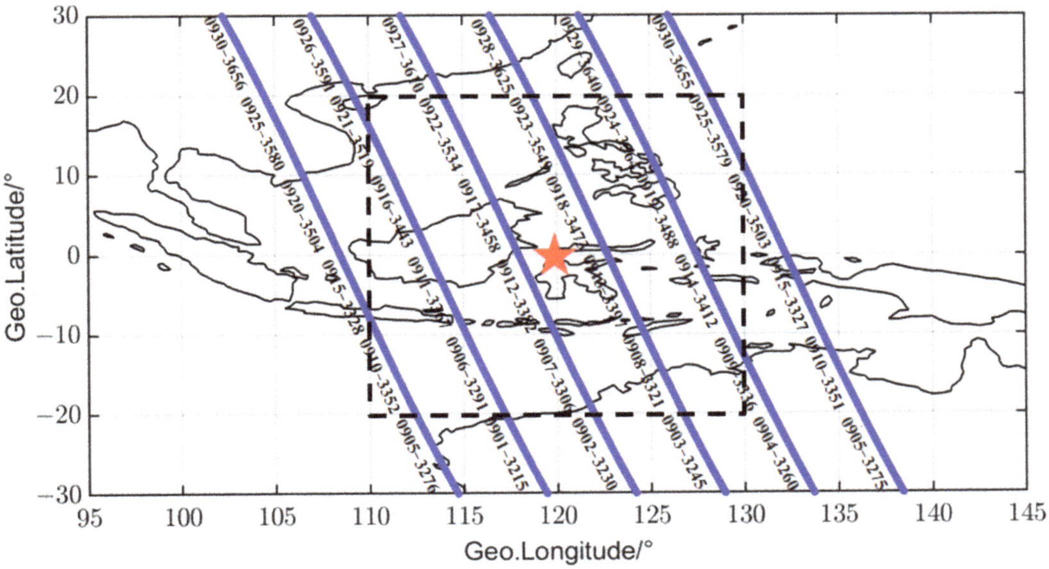

Figure 3. The range of orbital regions selected for data analysis is shown in the black dashed-bordered rectangle, taking the Ms 7.4 Indonesia EQ as an example. The red star represents the epicenter, and the blue lines are the ascending orbits of the CSES-01 during the period. Format of annotation information: mmdd-orbit number.

According to long-term data processing and previous analysis results, plasma values vary significantly in different regions, while the disturbance caused by the EQ is relatively small [20]. In addition, considering there is only one satellite, its observation ability is limited in spatial and time resolutions, considering the distance between adjacent tracks in one day is about 2600 km. Therefore, it is not recommended to analyze the seismo-ionospheric disturbances in large areas, which may mix with other disturbance phenomena induced by other factors. In this research, we have monitored the critical areas for a long time, checked the relative abnormal changes of multiple plasma parameters through the revisiting orbit quartile analysis method, and then extracted the seismic anomalies.

Furthermore, considering that there are numerical deviations in the absolute values of No^+ and V_z and their relative variations were reliable, we did not use their absolute values directly in the analysis in Section 3, but normalized them, i.e., firstly, we calculated the maximum values of No^+ and V_z in the region of the latitude range shown in the figures, and then divided the observations by the maximum values in order to obtain their relative change states. In this way, we will focus on their relative variations during the seismic anomaly analysis process rather than their absolute values. Moreover, we compared the simultaneous variations of multiple parameters, such as No^+, V_z, Ne, Te, and electric field waveform, to ensure the accuracy of seismic anomaly information extraction.

3. Results

3.1. Single-Orbit Analysis

The physical data of PAP and LAP around the epicenters within ±10° in longitude, a period of about 30 days before the main shocks, were selected to be arranged as a time sequence to check the apparent disturbance variations that may be induced by seismic activity. To keep the original abnormal information, the data used in the time series analysis was not smoothed or interpolated.

As shown in Figure 1 and Table 1, the epicenter locations of the Loyalty Islands (Ms 7.1 EQ) and the southeast of the Loyalty Islands (Ms 6.9 EQ) are very close. The occurrence time interval is very short—only about 50 days. Therefore, we think that the EQ preparation conditions are similar, allowing us to compare their common characteristics of pre-EQ ionospheric anomalies. In addition, the latitude distance between the Indonesia Ms 7.4 EQ and the Papua New Guinea Ms 7.1 EQ is only 5.5°, while their occurrence time interval is within 15 days, so we also have ideal conditions to carry out a comparison of common features.

It is worth noting in Figure 4 that the profile of orbit 2939_1 on 14 August 2018, 15 days before the Loyalty Island Ms 7.1 EQ, is different from other curves, with significant synchronous variation in the area ~200–550 km northwest of the epicenter. The values of *Ne* and No^+ increased (*Te* decreased) evidently, and V_z, which is vertical ion drift velocity, and positive (negative) values indicate downward (upward), increased strongly along the path of ground-to-space simultaneously, exhibiting the consistency of the plasma multi-parameter anomaly characteristics. Based on those reported by previous studies, before the EQ, the *Ne* and *Te* in the ionosphere displayed a negative correlation change, and the relative variations of *Ne* and *Ni* at a height of ~500 km were similar, being consistent with our observations. With the increase of *Ne* and *Ni*, the coulombic and elastic collisions of electrons and ions increase, and thermal energy is converted into kinetic energy to compensate for the lost energy, resulting in the decrease of *Te*. Among them, the relative variation of No^+ and V_z is more prominent, and the spatial range of the perturbation is ~350 km. After leaving the EQ abnormal area, these parameters returned to their original state.

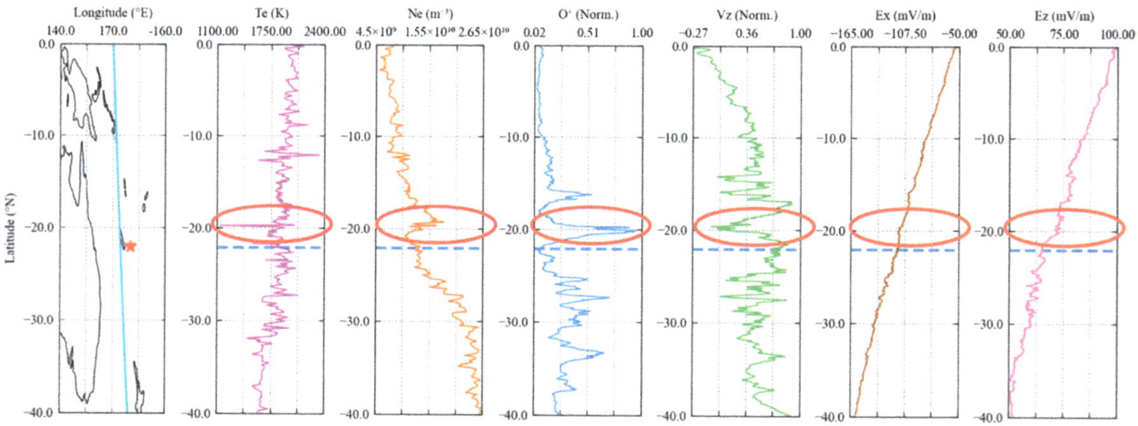

Figure 4. The synchronous abnormal perturbations of *Te*, *Ne*, No^+, V_z, *Ex*, and *Ez* of orbit 2939_1 occurred on 14 August 2018, 15 days before the Ms 7.1 Loyalty Islands EQ influential area. The red star represents the epicenter position, the blue oblique line in the left diagram is the orbit track, the red circles indicate the regions where the perturbation occur for each parameter, and the dotted lines represent the latitude location of the epicenter.

Figure 5 presents the abnormal perturbation phenomenon recorded by orbit 3775_1 on 8 October 2018, 8 days before the southeast of Loyalty Islands Ms 6.9 EQ, which is very similar to the one shown in Figure 4. In the area just ~100 km northwest of the epicenter, the *Te*, *Ne*, No^+, and V_z data show evident synchronous variations. The regional extension of the perturbation was also very similar, covering about 350 km. The difference from Figure 4 is that the northward shift distance of the disturbed area is shortened by ~100 km compared with the epicentral latitude, and the disturbance variations of the *Te*, *Ne*, No^+, and V_z are more prominent. Among them, the intensity of No^+ in orbit 3775_1 is about five times higher than the one in orbit 2939_1. After investigation, we identified that the phenomenon could be due to the moderate magnetic storm that happened on 8 October, which led to the overall significant increase of No^+, while the order of magnitude of electron parameters did not change significantly. Despite this, the variation is spatially coincident with the incoming EQs, so a possible interaction with the seismo-induced phenomenon could not be excluded, even if it is more plausible than the external perturbation source.

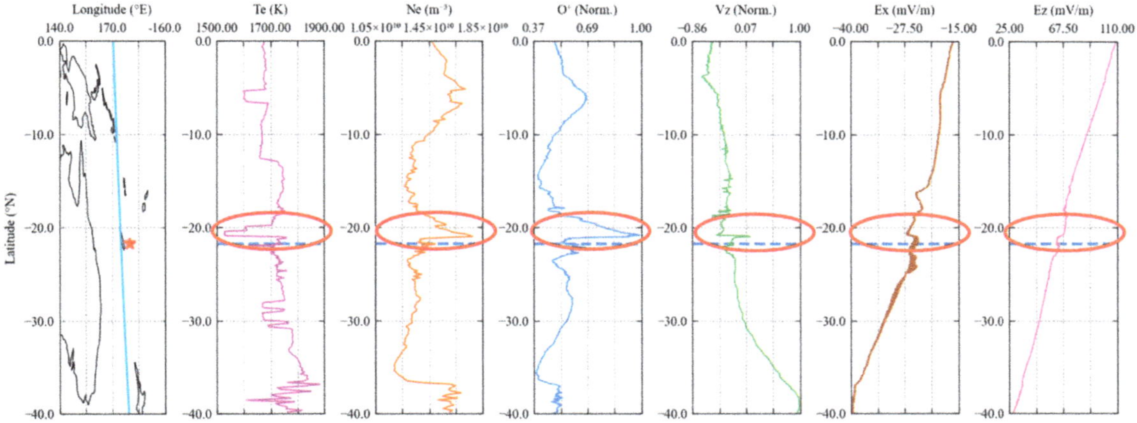

Figure 5. Same format as Figure 4, but shows that the synchronous abnormal perturbations recorded by orbit 3775_1 on 8 October 2018, 8 days before the Ms 6.9 Southeast of Loyalty Islands EQ influential area.

In this study, we analyzed the electric field data recorded by the Electric Field Detector (EFD) onboard the CSES-01 to determine the possible driving sources of plasma perturbations. The electric field data use the geographic coordinate system (+X indicates the direction that the center of the earth points to the prime meridian along the equatorial plane, and +Z indicates the direction that the center of the earth points to the North Pole) [24]. As presented in Figures 4 and 5, there were disturbances in the vertical direction of the Ultra Low Frequency band (ULF; the range is 0.5–16 Hz, with a resolution of 1 μV/m) waveform of the EQs' anomalous orbits, which were completely synchronized with the plasma perturbations at the same positions.

Figures 6 and 7 show the simultaneous perturbation of multiple plasma parameters recorded in orbits 3458_1 on 17 September 2018, 11 days before the Ms 7.4 Indonesia EQ, and orbit 3806_1 on 10 October 2018, 1 day before the Ms 7.1 Papua New Guinea EQ, which have similar characteristics as Figures 4 and 5. Certain synchronous variations in *Te*, *Ne*, No^+, V_z, *Ex*, and *Ez* data above the area of ~100 km southeast of the epicenter are discernible. On other days, no clear perturbations are shown. Among them, the relative variation characteristics of No^+ and V_z are the most significant plasma, compared to the relative smoothy feature of *Te* and *Ne* values in this area. A noticeable difference between Figures 4 and 5 is that the locations of the disturbances in these two EQs occurred south of the epicenters. At the same time, the two Loyalty Islands EQs appeared in the north of the epicenters. The different geomagnetic latitudes of these EQs, as shown in Figure 1, could explain this apparent discrepancy, even if they were all in the same geomagnetic southern hemisphere.

3.2. Multi-Orbit Analysis

The special orbital design of CSES-01 allows it to revisit identical locations every five days at the same LT [21]. The analysis in Section 3.1 illustrates that the pre-EQ characteristics of No^+ and V_z are more significant than the electron parameters. This phenomenon is probably due to the 1 s sampling rate of PAP, which is higher than the 3 s sampling rate of LAP, resulting in a higher detection sensitivity for PAP [14]. Therefore, in this section, we further analyzed the performance of No^+ and V_z based on the revisiting orbit data (multi-orbit data) by using a quartile analysis method to test and extract the possible seismic anomalies.

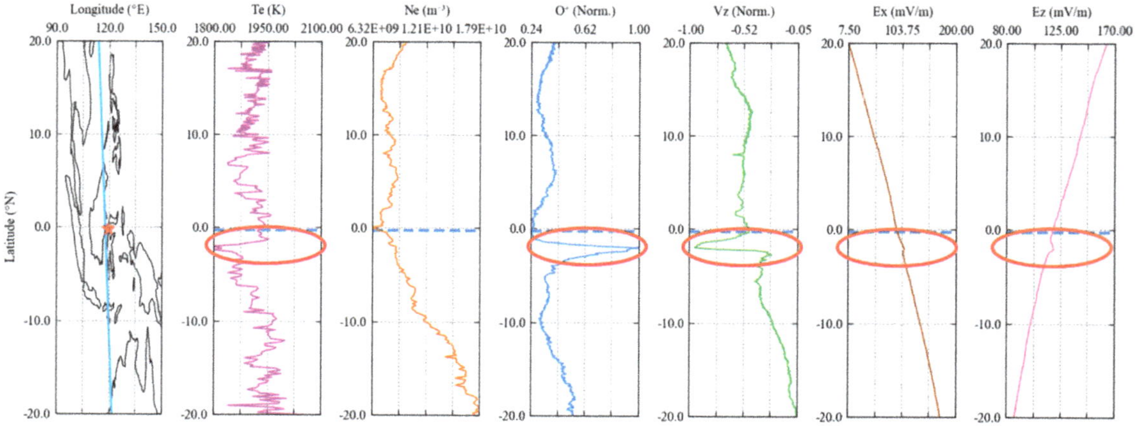

Figure 6. Same format as Figure 4, but the synchronous abnormal perturbations of *Te*, No^+, V_z, *Ex*, and *Ez* recorded by orbit 3458_1 on 17 September 2018, 11 days before the Ms 7.4 Indonesia EQ influential area.

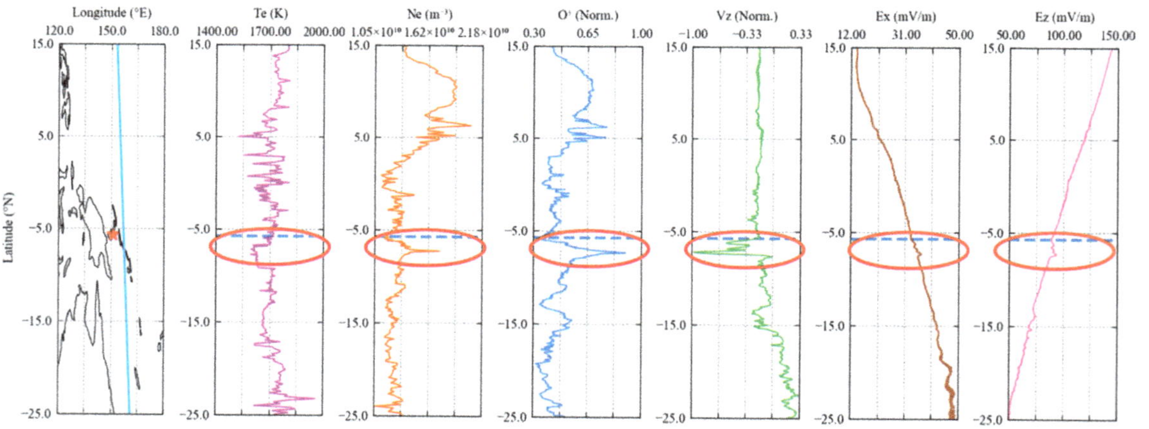

Figure 7. Same format as Figure 4, but for the synchronous abnormal perturbations of *Te*, *Ne*, No^+, V_z, *Ex*, and *Ez* of orbit 3806_1 on 10 October 2018, 1 day before the Ms 7.1 Papua New Guinea EQ influential area.

Considering the PAP's contamination, we selected revisiting orbits in the short term, about 25–30 days around the EQs, to ensure the dataset's reliability and reduce some factors related to seasonal changes, longitude position differences, etc. Their relative changes can somewhat reflect the precursor characteristics of EQ events that occurred in close locations in this short time.

Taking Figure 8 as an example, five revisiting orbits of orbit 2939_1 (shown in Figure 4) on 14 August were chosen to build a background trend, which are 2787_1 (ten days before, 4 August), 2863_1 (five days before, 9 August), 3015_1 (five days after, 19 August), 3243_1 (twenty-five days after, 3 September), and 3319_1 (thirty days after, 8 September). In the meantime, on 24 and 29 August, the observations were greatly disturbed by the strong geomagnetic storm around 26 August (as shown in Figure 2), so they were not selected for analysis of seismo-ionospheric anomalies. We used the inter-quartile ranges (IQR) method for comparing the No^+ and V_z data of orbit 2939_1 with their backgrounds and thus identified the anomalous data. The median and IQR values are statistical parameters, which are equal to the second quartile and the difference between the third (Q3) and first quartiles (Q1) of these revisiting orbits, respectively. The pre-defined allowed ranges were calculated by adding Q3 and Q1 and subtracting multiple IQR values to extract the anomaly for No^+ and V_z. Normally, it is best to choose the multiple as 1.5 or 2.0, but no more than 2.0. Otherwise, some samples would not be classified as anomalous, although they are [25].

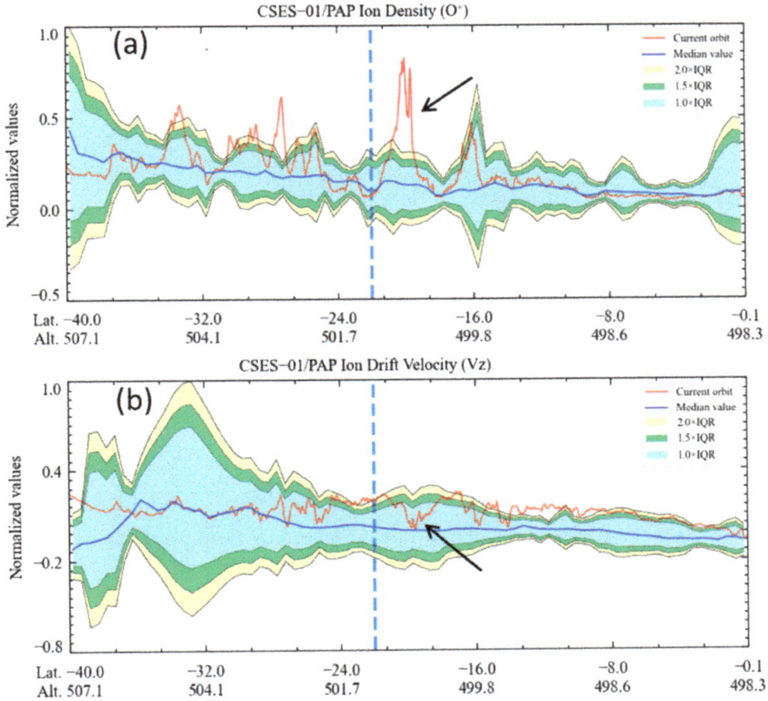

Figure 8. The variation of the (**a**) No^+ and (**b**) V_z detected by orbit 2939_1 with its revisiting orbits from 4 August to 8 September, which are impacted by the Ms 7.1 Loyalty Islands EQ, instead of the geomagnetic storm.

Figures 8–11 separately show the No^+ and V_z results for anomalous orbits 2939_1, 3775_1, 3458_1, and 3806_1, represented by red curves. The median values of the background trends are plotted by blue lines, the upper/lower bounds by different color blocks, and the blue dotted lines denote the locations of the epicentral latitudes. The observations of the current orbit that significantly exceeded the given thresholds were judged as anomalous parts, denoted by arrow indications.

It is noted that the absolute values of No^+ in orbit 3775_1 and orbit 3806_1 (shown in Figures 10 and 11) are far beyond the upper boundaries, and the latitude ranges affected are up to about 40°, while in orbit 2939_1 and orbit 3458_1 (shown in Figures 8 and 9), the No^+ data did not exceed the upper boundary in such a large range except for the candidate seismo-ionospheric anomaly area. Meanwhile, the V_z values in these orbits are stable, rarely exceeding the threshold ranges of 2.0 times IQR, except for the sudden disturbances in the EQ abnormal areas.

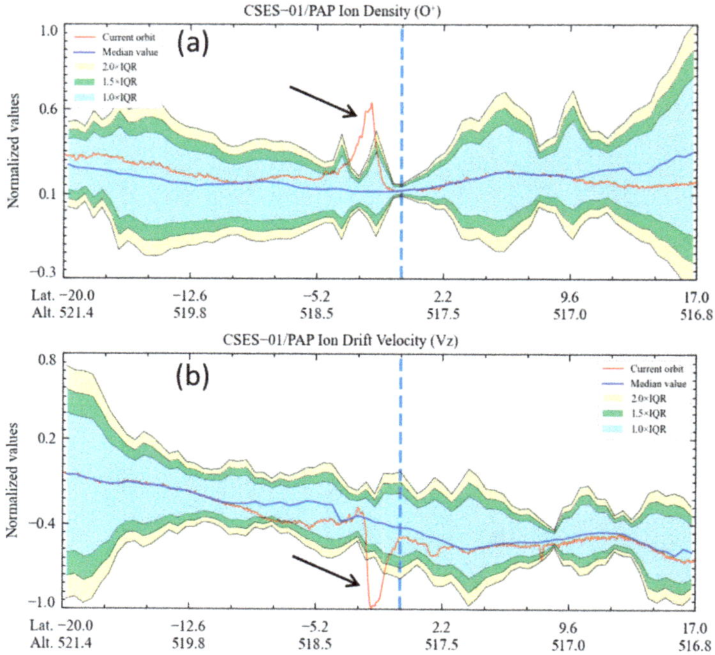

Figure 9. Same format as Figure 8, but for the variation of the (**a**) No^{+} and (**b**) V_z detected by orbit 3458_1 with its revisiting orbits from 2 September to 27 September, which are impacted by the Ms 7.4 Indonesia EQ, instead of the geomagnetic storm.

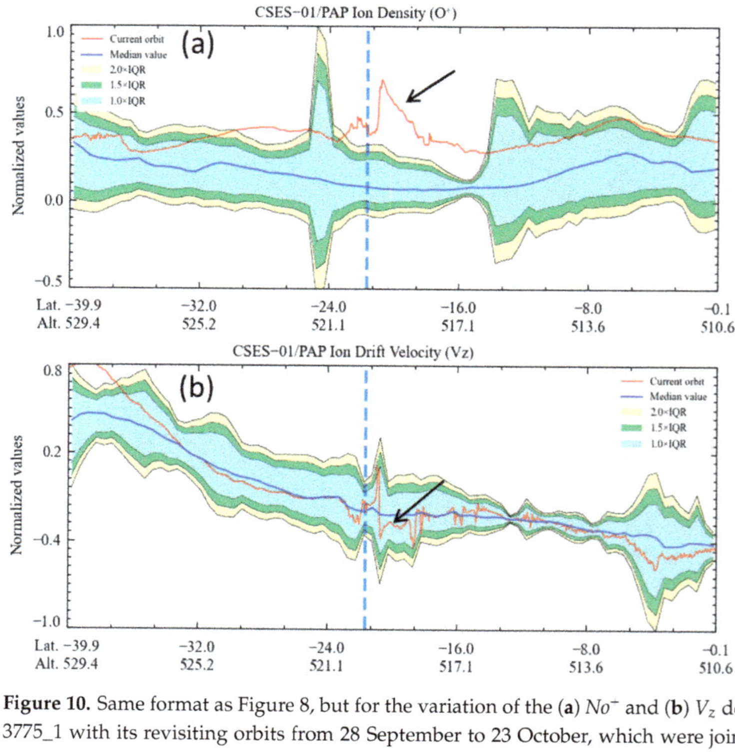

Figure 10. Same format as Figure 8, but for the variation of the (**a**) No^{+} and (**b**) V_z detected by orbit 3775_1 with its revisiting orbits from 28 September to 23 October, which were jointly affected by strong geomagnetic storms and the Ms 6.9 Southeast of Loyalty Islands EQ.

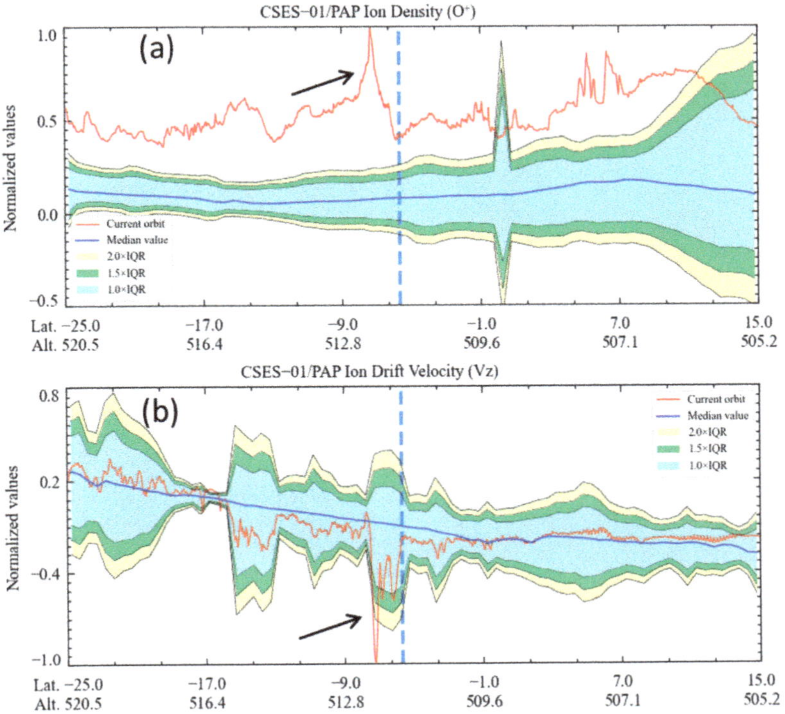

Figure 11. Same format as Figure 8, but for the variation of the (**a**) No^+ and (**b**) V_z detected by the orbit 3806_1 with its revisiting orbits from 15 September to 10 October, which are jointly affected by a strong geomagnetic storm and the Ms 7.1 Papua New Guinea EQ.

4. Discussion

4.1. Spatial Distributions of EQ Precursor

According to the seismo-ionospheric coupling mechanism, the ionospheric disturbances caused by EQs should not only show time series characteristics but also particular spatial distribution features [26–28]. Compared with the anomaly information in the longitude direction reported by the previous research, this study pays more attention to the spatial anomaly distribution characteristics in the latitude direction. The reason is that the distance in longitude between adjacent satellite orbits around the EQ epicenter is about 24° (~2600 km) each day, which is insufficient to monitor longitudinal spatial variations. Although the longitude distance between adjacent tracks in 5 days is reduced to about 4.7° (~500 km), the time span is too large to detect whole-EQ precursory information. On the other hand, the CSES-01's flight speed is 7.6 km per second, so it only takes tens of seconds to observe hundreds of kilometers in the latitude direction, which can comprehensively investigate the possible pre-EQ anomalies in the flighting orbit area.

As shown in Figures 4 and 5, the significant perturbations of *Te*, *Ne*, No^+, and V_z before Ms 7.1 EQ in Loyalty islands and Ms 6.9 EQ in the southeast of Loyalty islands appeared synchronously in the area 100–200 km northwest of the epicenters, and the spatial extension of these perturbations is about 350 km. According to Liu and Wan's statistical analysis of the spatial-temporal distribution characteristics of the ionospheric anomalous disturbances before Ms $\geq$ 6.0 EQs in China [28], the main seismo-ionospheric perturbations do not occur directly above the epicenter locations but shift towards the magnetic equator, which is consistent with these detection results.

Moreover, as a known fixed-position high-power artificial source, the impact of the terrestrial very low-frequency (VLF) transmitters on the ionospheric environment could be an important reference for studying the seismo-ionospheric disturbance effect.

Liu et al. [14] have presented the observation results when CSES-01 passes above the NAA (44.65°N, −67.28°E, transmission frequency 24 kHz, transmission power 885 kW) and NWC (−21.82°N, 114.17°E, transmission frequency 19.8 kHz, transmission power 1000 kW) stations [29,30], which are currently the most powerful VLF sources in the northern and southern hemispheres. Under the influence of the ground signal, the plasma parameters (e.g., *Ne*, *Te*, No^+, and V_z) of the ionosphere F layer appear to exhibit significantly synchronous heating disturbances not above the stations but in the corresponding area approximately 205 km south of the NAA station and 220 km north of the NWC station [14], which is very similar to the observed phenomena in this pre-EQ study in Figures 4 and 5.

Notably, the abnormal perturbations before Ms 7.4 EQ in Indonesia and Ms 7.1 EQ in Papua New Guinea, as shown in Figures 6 and 7, are located in the area of ~100 km southeast of the epicenters, which are opposite to those in Figures 4 and 5. A reason for the difference could be that the epicenter latitudes of Indonesia EQ and Papua New Guinea EQ are lower and are located south of the magnetic equatorial region.

As is well known, the fountain effect, i.e., the ionosphere plasma above the magnetic equator, drifts along the magnetic lines to form a double-peak structure, which is the so-called "Equatorial Ionospheric Anomaly (EIA)" [31,32]. Under the effect of the fountain effect, the ionospheric plasma anomalies, possibly excited by the strong EQs in the south of the magnetic equator, could be further transported to the area south of the epicenters. According to Hanson and Abdu et al. [33,34], the fountain effect caused by the eastward electric field generally ends at 17–19 LT. Still, the bimodal structure of EIA will continue until even early morning, and then the plasma disturbance phenomena can be recorded by CSES-01 flying over the areas.

4.2. Mechanism of the Seismo-Ionospheric Coupling

At present, the research on the mechanism of seismo-ionospheric coupling is mostly in the qualitative analysis stage, and some numerical simulation theories are still constantly developing. There are three main transmission channels that describe the seismo-ionospheric coupling model, such as the acoustic channel, the geochemical channel, and the electromagnetic channel [27,35–37].

Despite some "up-down" evidence regarding solar activity and geomagnetic storms before some strong earthquakes, it is observed that the triggering mechanism can be a result of seismic/tectonic processes at depth, i.e., a "bottom-up" process, as defined by Freund et al. [38–40]. Before strong earthquakes, the igneous and high-grade metamorphic rocks are subjected to defect-positive holes in the oxygen anion sublattice. These charge carriers are highly mobile, able to flow out of stressed rocks into surrounding unstressed rocks. They form electric currents, which emit electromagnetic radiation, sometimes in pulses, sometimes sustained, and may cause ionospheric perturbations and earthquake lights.

Using numerical simulations, Kuo et al. [27,41] demonstrated that observations of nighttime plasma bubbles within the affected region can be used as precursors for EQ prediction. Pressurized rocks can stimulate charged hole carriers and form currents along the stress gradient direction, causing an increase in atmospheric conductivity and jointly triggering the upward electric current over the seismo-active region, ultimately causing ionospheric plasma density enhancement.

Ruzhin et al. [42], based on observations of total electron content (TEC) perturbations over the seismic region, reported an enhancement of the electric field detected in the ionsphere over the epicenter a few days prior to the main shock, and the plasma density can either increase or decrease (or both) in the perturbed region. The extra electric field E that is generated in the ionosphere over the seismic preparation zone is considered to be the main source of these ionospheric anomalies. The plasma transfer caused by the extra electric field in the direction of E × B drift may produce regions of TEC enhancement (plasma aggregation) and weakening (plasma outflow).

Based on the previous research results and synchronous perturbations of the ULF electric field waveform data and the plasma multi-parameter data presented in Section 3.1, we suggest that large-scale crustal deformation preceded the strong EQ-induced electric current changes in the subsurface, thus giving rise to the upward low-frequency electromagnetic radiation over the epicenter aera, and the electrodynamic effect of the electric field on the plasma induced the plasma perturbation in the ionosphere. The energy propagated along the direction of the magnetic field line in the ionosphere, making the location of the plasma disturbance center shift in a latitude direction [43,44], which can be detected by CSES-01. The possible plasma perturbation process induced by the seismo-ionospheric coupling is shown in Figure 12.

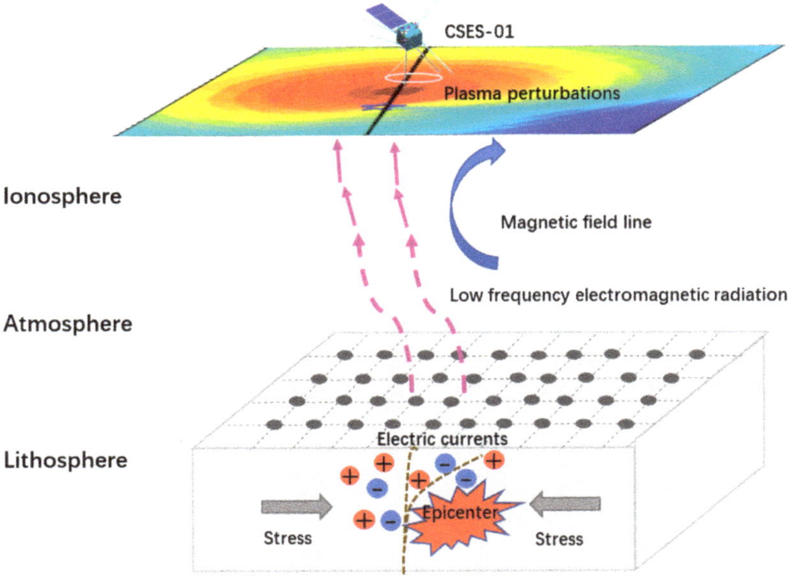

Figure 12. The possible schematic diagram of the plasma perturbation process induced by the seismo-ionospheric coupling.

4.3. Difference with Magnetic Storm Anomaly

The geomagnetic and solar data shown in Figure 2 indicate that several geomagnetic storms occurred from August to October 2018, resulting in a complex ionospheric environment. Based on the data of four seismic candidate anomalous orbits during this period and their revisiting orbits, we compared the different performances of No^+ and V_z under the influence of intense geomagnetic storms and strong EQs.

As shown in Figure 2, the Dst index on 8 and 10 October in 2018 is lower than -30 nT, and the Kp index is greater than 3, which indicates that a moderate geomagnetic storm occurred on these two days. Furthermore, from Figures 10a and 11a, we can find that due to the influence of a strong geomagnetic storm, the values of No^+ in orbit 3775_1 on 8 October and orbit 3806_1 on 10 October in 2018 are far beyond the upper boundaries, about 200–300% higher than the median values, and the affected latitude ranges are as large as more than 40°. Meanwhile, in the local areas of seismic anomaly disturbances, the values of No^+ increased significantly based on the influence of the geomagnetic storm, which seems to reflect the combined effects of strong EQs and the intense geomagnetic storm on the ionospheric plasma environment.

On the other hand, orbit 2939_1 on 29 August in 2018 and orbit 3458_1 on 28 September in 2018, as presented in Figures 8a and 9a, were not affected by the geomagnetic storm, but they were preceded by a few days of strong EQs. The No^+ data did not exceed the upper boundary in such a large range but only greatly enhanced in the local EQ anomaly areas that extended for hundreds of kilometers, which increased about 100–200% to the median values. According to previous studies, the seismic anomaly is a regional and local anomaly that is unlikely to cause a global disturbance [21,28]. These observation results are consistent with this view.

Based on the data sets obtained from the MICS spectrometer onboard the CRRES satellite, Fu et al. [45] made a statistical analysis of 12 cases of intense geomagnetic storms in 1991. The results proved that oxygen ions (O^+) in the ionosphere are the main component of the ring current particles during the main phases of intense geomagnetic storms. The increase and decrease of O^+ in the ionosphere are the primary factors leading to the rapid decline and growth of the Dst index during the main and recovery phases of the geomagnetic storm. Although the observation height of the CRRES is higher than the CSES-01, the role of the O^+ in the ionospheric environment could be an important reference for studying the seismo-ionospheric disturbance effect.

Ouyang and Liu et al. [12,14] reported significant disturbances of No^+ before Ms7.9 Tonga Islands EQ in 2006 detected by DEMETER satellite and Ms 7.3 Venezuela EQ in 2018 recorded by CSES-01, indicating that No^+ is a promising parameter for analyzing seismo-ionospheric perturbation. Therefore, monitoring the relative variation of No^+ is important for analyzing the ionospheric anomaly.

It is interesting to find that, as represented in Figures 10b and 11b, the trends of V_z in orbit 3775_1 on 8 October and orbit 3806_1 on 10 October in 2018, during the geomagnetic storm, are pretty different from those of No^+. V_z is almost within the upper and lower boundaries, and there is no particular change compared with the background trend except for the sudden reverse perturbation over the seismic area, which seems consequently to be related to the incoming strong EQ. The vertical ion drift velocity seems not to be affected by the geomagnetic storm that occurred in the past two days but is susceptible to seismic activities. This character of V_z is very precious and important for the research of strong EQ precursors, which can directly indicate the vertical motion characteristic of plasma and is not easily influenced by the geomagnetic storm.

5. Conclusions

In this work, we analyzed the characteristics of ionospheric plasma perturbations before four strong EQs with magnitudes ranging from Ms 6.9 to Ms 7.4 occurred in southeast Asia and northern Oceania (in the south of the circum-Pacific seismic belt) from August to October 2018, which were observed by PAP and LAP onboard CSES-01 and were supported by the ULF waveform results of EFD. The different variations of No^+ and V_z under intense geomagnetic storms and strong EQ were compared. Our findings support the possibility of applying in-situ plasma observation of LEO satellites to examine ionospheric pre-EQ signatures, which could be useful to understand seismo-ionospheric precursors better. The main results of this study are summarized as follows:

(1) The No^+, Ne, Te, and V_z values show significant synchronous disturbances at the same position near the epicenters from 1 to 15 days before the series of strong EQs occurred in southeast Asia and northern Oceania within 50 days. Regarding time and space characteristics, these plasma perturbations are possibly explained as seismo-ionospheric precursors.

(2) The No^+ and V_z seemed like promising parameters for analyzing ionospheric disturbances excited by strong seismic activity. Local ionospheric perturbations caused by strong EQs are often accompanied by a change in V_z, which is not easily influenced by the geomagnetic storm. Meanwhile, the variation of No^+ could reflect the combined effects induced by strong EQ and intense geomagnetic storms on the ionospheric plasma environment.

(3) Based on the long-term data analysis results of CSES-01, the absolute value differences of plasma in different locations are large. It may be more effective to extract precursory seismic anomalies by long-term monitoring of the critical areas and checking the relative abnormal changes of the revisiting orbits.

Author Contributions: D.L., Z.Z. and X.S. contributed to the conception and methodology. D.L. analyzed the data and wrote the original draft. H.H., Q.W. and D.Y. wrote some programs. C.L., R.Y. and Y.G. checked and organized the database. Z.Z. reviewed the expression. All authors have read and agreed to the published version of the manuscript.

Funding: This research is supported by the National Natural Science Foundation of China (grant number 41874174), the Natural Science Foundation of Beijing (grant number 8184091), and the ISSI-BJ project.

Data Availability Statement: The CSES-01's plasma observation data used in this paper can be downloaded from the CSES data service website: https://www.leos.ac.cn (accessed on 1 September 2022). The earthquake data were downloaded at: http://news.ceic.ac.cn (accessed on 10 February 2023) and https://earthquake.usgs.gov (accessed on 8 May 2023), and the geomagnetic index data were downloaded at: http://wdc.kugi.kyoto-u.ac.jp (accessed on 5 March 2023).

Acknowledgments: This work utilizes the data from the CSES mission. We acknowledge the experts who helped improve the quality of this paper.

Conflicts of Interest: The authors declare no conflict of interest.

References

1. Pulinets, S.A.; Boyarchuk, K.A.; Hegai, V.V.; Kim, V.P.; Lomonosov, A.M. Quasielectrostatic model of atmosphere-thermosphere-ionosphere coupling. *Adv. Space Res.* **2000**, *26*, 1209–1218. [CrossRef]
2. Yokoyama, T.; Su, S.Y.; Fukao, S. Plasma blobs and irregularities concurrently observed by ROCSAT-1 and Equatorial At-mosphere Radar. *J. Geophys. Res. Space Phys.* **2007**, *112*, 515–526. [CrossRef]
3. Liu, J.Y.; Chen, C.H.; Lin, C.H.; Tsai, H.F.; Kamogawa, M. Ionospheric disturbances triggered by the 11 March 2011 M9.0 Tohoku earthquake. *J. Geophys. Res. Space Phys.* **2011**, *116*, A06319. [CrossRef]
4. Pulinets, S.; Ouzounov, D. Lithosphere–Atmosphere–Ionosphere Coupling (LAIC) model—An unified concept for earthquake precursors validation. *J. Asian Earth Sci.* **2011**, *41*, 371–382. [CrossRef]
5. Macmillan, S.; Olsen, N. Observatory data and the Swarm mission. *Earth Planets Space* **2013**, *65*, 1355–1362. [CrossRef]
6. Maruyama, T.; Shinagawa, H. Infrasonic sounds excited by seismic waves of the 2011 Tohoku-oki earthquake as visualized in ionograms. *J. Geophys. Res. Space Phys.* **2014**, *119*, 4094–4108. [CrossRef]
7. Parrot, M.; Benoist, D.; Berthelier, J.; Błęcki, J.J.; Chapuis, Y.; Colin, F.; Elie, F.; Fergeau, P.; Lagoutte, D.; Lefeuvre, F.; et al. The magnetic field experiment IMSC and its data processing onboard DEMETER: Scientific objectives, description and first results. *Plan. Space Sci.* **2006**, *54*, 441–455. [CrossRef]
8. Cussac, T.; Clair, M.A.; Ultré-Guerard, P.; Buisson, F.; Rey, N. The DEMETER microsatellite and ground segment. *Plan. Space Sci.* **2006**, *54*, 413–427. [CrossRef]
9. Berthelier, J.; Godefroy, M.; Leblanc, F.; Seran, E.; Peschard, D.; Gilbert, P.; Artru, J. IAP, the thermal plasma analyzer on DEMETER. *Plan. Space Sci.* **2006**, *54*, 487–501. [CrossRef]
10. De Santis, A.; Marchetti, D.; Pavón-Carrasco, F.J.; Cianchini, G.; Perrone, L.; Abbattista, C.; Alfonsi, L.; Amoruso, L.; Campuzano, S.A.; Carbone, M.; et al. Precursory worldwide signatures of earthquake occurrences on Swarm satellite data. *Sci. Rep.* **2019**, *9*, 20287. [CrossRef]
11. Liu, J.Y.; Chen, Y.I.; Pulinets, S.A.; Tsai, Y.B.; Chuo, Y.J. Seismo-ionospheric signatures prior to M $\geq$ 6.0 Taiwan earthquakes. *Geophys. Res. Lett.* **2000**, *27*, 3113–3116. [CrossRef]
12. Ouyang, X.; Zhang, X.; Shen, X.; Huang, J.; Liu, J.; Zeren, Z.; Zhao, S. Disturbance of O+ Density Before Major Earthquake Detected by DEMETER Satellite. *Chin. J. Space Sci.* **2011**, *31*, 607–617. [CrossRef]
13. Piersanti, M.; Materassi, M.; Battiston, R.; Carbone, V.; Cicone, A.; D'angelo, G.; Diego, P.; Ubertini, P. Magnetospheric–Ionospheric–Lithospheric Coupling Model. 1: Observations during the 5 August 2018 Bayan Earthquake. *Remote Sens.* **2020**, *12*, 3299. [CrossRef]
14. Liu, D.; Zeren, Z.; Shen, X.; Zhao, S.; Yan, R.; Wang, X.; Liu, C.; Guan, Y.; Zhu, X.; Miao, Y.; et al. Typical ionospheric disturbances revealed by the plasma analyzer package onboard the China Seismo-Electromagnetic Satellite. *Adv. Space Res.* **2021**, *68*, 3796–3805. [CrossRef]
15. Li, M.; Shen, X.; Parrot, M.; Zhang, X.; Zhang, Y.; Yu, C.; Yan, R.; Liu, D.; Lu, H.; Guo, F.; et al. Primary Joint Statistical Seismic Influence on Ionospheric Parameters Recorded by the CSES and DEMETER Satellites. *J. Geophys. Res. Space Phys.* **2020**, *125*, e2020JA028116. [CrossRef]
16. Li, Z.; Yang, B.; Huang, J.; Yin, H.; Yang, X.; Liu, H.; Zhang, F.; Lu, H. Analysis of pre-earthquake space electric field disturbance observed by CSES. *Atmosphere* **2022**, *13*, 934. [CrossRef]
17. Song, R.; Hattori, K.; Zhang, X.; Sanaka, S. Seismic-ionospheric effects prior to four earthquakes in Indonesia detected by the China seismo-electromagnetic satellite. *J. Atmos. Terr. Phys.* **2020**, *205*, 105291. [CrossRef]
18. Zeren, Z.; Yan, R.; Lin, J.; Wang, Q.; Yang, Y.; Lv, F.; Huang, J.; Cui, J.; Liu, Q.; Zhao, S.; et al. The Possible Seismo-Ionospheric Perturbations Recorded by the China-Seismo-Electromagnetic Satellite. *Remote Sens.* **2022**, *14*, 905. [CrossRef]

19. Liu, J.; Qiao, X.; Zhang, X.; Wang, Z.; Zhou, C.; Zhang, Y. Using a Spatial Analysis Method to Study the Seismo-Ionospheric Disturbances of Electron Density Observed by China Seismo-Electromagnetic Satellite. *Front. Earth Sci.* **2022**, *10*, 811658. [CrossRef]
20. Oyama, K.-I.; Devi, M.; Ryu, K.; Chen, C.H.; Liu, J.Y.; Liu, H.; Bankov, L.; Kodama, T. Modifications of the ionosphere prior to large earthquakes: Report from the Ionosphere Precursor Study Group. *Geosci. Lett.* **2016**, *3*, 6. [CrossRef]
21. Yang, Y.; Zhima, Z.; Shen, X.; Chu, W.; Huang, J.; Wang, Q.; Yan, R.; Xu, S.; Lu, H.; Liu, D. The First Intense Geomagnetic Storm Event Recorded by the China Seismo-Electromagnetic Satellite. *Space Weather.* **2020**, *18*, e2019SW002243. [CrossRef]
22. Marchitelli, V.; Harabaglia, P.; Troise, C.; De Natale, G. On the correlation between solar activity and large earthquakes worldwide. *Sci. Rep.* **2020**, *10*, 11495. [CrossRef] [PubMed]
23. Dobrovolsky, I.P.; Zubkov, S.I.; Miachkin, V.I. Estimation of the size of earthquake preparation zones. *Pure Appl. Geophys.* **1979**, *117*, 1025–1044. [CrossRef]
24. Huang, J.; Lei, J.; Li, S.; Zeren, Z.; Li, C.; Zhu, X.; Yu, W. The Electric Field Detector (EFD) onboard the ZH-1 satellite and first observational results. *Earth Planet. Phys.* **2018**, *2*, 469–478. [CrossRef]
25. Zhang, Y.; Wang, T.; Chen, W.; Zhu, K.; Marchetti, D.; Cheng, Y.; Fan, M.; Wang, S.; Wen, J.; Zhang, D.; et al. Are There One or More Geophysical Coupling Mechanisms before Earthquakes? The Case Study of Lushan (China) 2013. *Remote Sens.* **2023**, *15*, 1521. [CrossRef]
26. Takeuchi, A.; Lau, B.; Freund, F. Current and surface potential induced by stress-activated positive holes in igneous rocks. *Phys. Chem. Earth* **2006**, *31*, 240–247. [CrossRef]
27. Kuo, C.L.; Huba, J.D.; Joyce, G.; Lee, L.C. Ionosphere plasma bubbles and density variations induced by pre-earthquake rock currents and associated surface charges. *J. Geophys. Res. Space Phys.* **2011**, *116*, A10317. [CrossRef]
28. Liu, J.; Wan, W. Spatial-temporal distribution of the ionospheric perturbations prior to MS $\geq$ 6.0 earthquakes in China main land. *Chin. J. Geophys.* **2014**, *57*, 2181–2189. [CrossRef]
29. Bell, T.; Graf, K.; Inan, U.; Piddyachiy, D.; Parrot, M. DEMETER observations of ionospheric heating by powerful VLF transmitters. *Geophys. Res. Lett.* **2011**, *38*, L11103. [CrossRef]
30. Tao, D.; Liu, W.; Ma, Y. Plasma perturbations in the coexisting environment of VLF transmitter emission, lightning strokes and seismic activity. *Sci. China Technol. Sci.* **2018**, *61*, 678–686. [CrossRef]
31. Appleton, E.V. Two Anomalies in the Ionosphere. *Nature* **1946**, *157*, 691. [CrossRef]
32. Whalen, J.A. Mapping a bubble at dip equator and anomaly with oblique ionospheric soundings of range spread F. *J. Geophys. Res.* **1996**, *101*, 5185–5193. [CrossRef]
33. Hanson, W.B.; Moffett, R.J. Ionization transport effects in the equatorial F region. *J. Geophys. Res.* **1966**, *71*, 5559–5572. [CrossRef]
34. Abdu, M.A. Major phenomena of the equatorial ionosphere-thermosphere system under disturbed conditions. *J. Atmos. Terres. Phys.* **1997**, *59*, 1505–1519. [CrossRef]
35. Wu, L.; Qi, Y.; Mao, W.; Lu, J.; Ding, Y.; Peng, B.; Xie, B. Scrutinizing and rooting the multiple anomalies of Nepal earthquake sequence in 2015 with the deviation–time–space criterion and homologous lithosphere–coversphere–atmosphere–ionosphere coupling physics. *Nat. Hazards Earth Syst Sci.* **2023**, *23*, 231–249. [CrossRef]
36. Guo, Y.; Zhang, X.; Liu, J.; Yang, M.; Yang, X.; Du, X.; Lü, J.; Xiao, J. Seismo-Ionospheric Effects Prior to Two Earthquakes in Taiwan Detected by the China Seismo-Electromagnetic Satellite. *Atmosphere* **2022**, *13*, 1523. [CrossRef]
37. Akhoondzadeh, M.; Parrot, M.; Saradjian, M.R. Electron and ion density variations before strong earthquakes (M > 6.0) using DEMETER and GPS data. *Nat. Hazards Earth Syst. Sci.* **2010**, *10*, 7–18. [CrossRef]
38. Freund, F.T.; Kulahci, I.G.; Cyr, G.; Ling, J.; Winnick, M.; Tregloan-Reed, J.; Freund, M.M. Air ionization at rock surfaces and pre-earthquake signals. *J. Atmos. Solar-Terr. Phys.* **2009**, *71*, 1824–1834. [CrossRef]
39. Freund, F.T. Earthquake forewarning—A multidisciplinary challenge from the ground up to space. *Acta Geophys.* **2013**, *61*, 775–807. [CrossRef]
40. Freund, F.T.; Daneshvar, M.M.; Ebrahimi, M. Atmospheric Storm Anomalies Prior to Major Earthquakes in the Japan Region. *Sustainability* **2022**, *14*, 10241. [CrossRef]
41. Kuo, C.L.; Lee, L.C.; Huba, J.D. An improved coupling model for the lithosphere-atmosphere-ionosphere system. *J. Geophys. Res. Space Phys.* **2014**, *119*, 3189–3205. [CrossRef]
42. Ruzhin, Y.Y.; Sorokin, V.M.; Yashchenko, A.K. Physical mechanism of ionospheric total electron content perturbations over a seismoactive region. *Geomagn. Aeron.* **2014**, *54*, 337–346. [CrossRef]
43. Wait, J.R. Mode conversion and refraction effects in the Earth-ionosphere waveguide for VLF radio waves. *J. Geophys. Res. Atmos.* **1968**, *73*, 3537–3548. [CrossRef]
44. Zhao, S.; Liao, L.; Zhang, X. Trans-ionospheric VLF wave power absorption of terrestrial VLF signal. *Chin. J. Geophys.* **2017**, *60*, 3004–3014. [CrossRef]
45. Fu, S.; Pu, Z.; Zong, Q.; Xiao, C.; Xie, L.; Wilken, B. Ion composition variations in the ring current during intense magnetic storms and their relationship with evolution of storms. *Chin. J. Geophys.* **2001**, *44*, 1–11. [CrossRef]

remote sensing

MDPI

Article

Are There One or More Geophysical Coupling Mechanisms before Earthquakes? The Case Study of Lushan (China) 2013

Yiqun Zhang, Ting Wang, Wenqi Chen, Kaiguang Zhu, Dedalo Marchetti *, Yuqi Cheng, Mengxuan Fan, Siyu Wang, Jiami Wen, Donghua Zhang and Hanshuo Zhang

College of Instrumentation and Electrical Engineering, Jilin University, Changchun 130061, China
* Correspondence: dedalomarchetti@jlu.edu.cn; Tel.: +86-189-3686-6083

Abstract: Several possible lithosphere–atmosphere–ionosphere coupling mechanisms before earthquake occurrence are presented in the literature. They are described by several models with different interaction channels (e.g., electromagnetic, mechanics, chemical, thermal), sometimes in conflict with each other. In this paper, we search for anomalies six months before the Lushan (China) 2013 earthquake in the three geo-layers looking for a possible view of the couplings and testing if one or another is more reliable to describe the observations. The Lushan earthquake occurred in China's Sichuan province on 20 April 2013, with a magnitude of Mw = 6.7. Despite the moderate magnitude of the event, it caused concern because its source was localized on the southwest side of the same fault that produced the catastrophic Wenchuan event in 2008. This paper applies a geophysical multi-layer approach to search for possible pre-earthquake anomalies in the lithosphere, atmosphere, and ionosphere. In detail, six main increases in the accumulated seismic stress were depicted. Anomalous geomagnetic pulsations were recorded in the Chengdu observatory, sometimes following the increased stress. Atmosphere status and composition were found to be anomalous in several periods before the earthquake, and, spatially, the anomalies seem to appear firstly far from the upcoming earthquakes and later approaching the Longmenshan fault where the Lushan earthquakes nucleated. The Formosat-3 data identified interesting anomalies in the altitude or electron content of the ionospheric F2 peak in correspondence with seismic and atmospheric anomalies 130 days before the earthquake. In addition, the total electron content showed high anomalous values from 12 to 6 days before the earthquake. We compared the anomalies and tried to explain their correspondences in different geo-layers by the lithosphere–atmosphere–ionosphere coupling models. In particular, we identified three possible couplings with different mechanisms: a first, about 130 days before the earthquake, with a fast (order of one day) propagation delay; a second, about 40 days before the earthquake occurrence, with a propagation delay of few days and a third from 2.5 weeks until one week before the event. Such evidence suggests that the geo-layers could interact with different channels (pure electromagnetic or a chain of physical-chemical processes) with specific propagation delays. Such results support the understanding of the preparation for medium and large earthquakes globally, which is necessary (although not sufficient) knowledge in order to mitigate their impact on human life.

Keywords: 2013 Lushan earthquake; multi-layer study; pre-earthquake anomalies; lithosphere; atmosphere; ionosphere; LAIC

Citation: Zhang, Y.; Wang, T.; Chen, W.; Zhu, K.; Marchetti, D.; Cheng, Y.; Fan, M.; Wang, S.; Wen, J.; Zhang, D.; et al. Are There One or More Geophysical Coupling Mechanisms before Earthquakes? The Case Study of Lushan (China) 2013. *Remote Sens.* **2023**, *15*, 1521. https://doi.org/10.3390/rs15061521

Academic Editor: Salvatore Stramondo

Received: 7 February 2023
Revised: 4 March 2023
Accepted: 8 March 2023
Published: 10 March 2023

1. Introduction

The Lushan earthquake occurred in China on 20 April 2013 at 00:02:47 UT (8:02:47 Chinese time). The hypocenter was localized at 30.308° N, 102.888° E, and a 14 km depth. The largest city close to the epicenter was Chengdu (120 km East-North-East), the capital of Sichuan province with about 20 million inhabitants. Xie et al. [1] determined the source parameters with several techniques estimating a moment magnitude Mw of about 6.66. From a tectonic point of view, the earthquake occurred due to a thrust fault on the Longmenshan fault, which was where the catastrophic Mw = 7.9 Wenchuan 2008 earthquake occurred five years before [2].

The possibility that there are precursors to earthquakes is a debated scientific question. Concerning the recent past in China, it is paradoxical that the successful prediction of the M7.3 Haicheng earthquake that occurred on 4 February 1975 was followed in the next year by a wholly unexpected and tragic M7.5 Tangshan earthquake on 27 July 1976 [3]. These predictions have been based mainly on observing strange animal behavior, which has also been observed before more recent earthquakes, for example, the Mw = 6.1 earthquake in Colfiorito, Central Italy, in 1997 [4]. If the animals, in some cases, feel some alteration of the environment that humans cannot perceive, the instrumentations developed in the last decades, especially remote sensing satellites, must be able to catch these possible alterations before an earthquake. So, the right research path for a future prediction of earthquakes is firstly to understand the possible environmental changes that typically occur before their occurrence. This is the topic of this paper.

Several pieces of evidence have been provided in the last few decades showing possible candidates for earthquake precursors. Preliminary studies focused on the investigation of one parameter before one earthquake, such as Frase-Smith et al. [5], who reported Ultra Low Frequency (ULF) electromagnetic emissions before the M7.1 Loma-Prieta (US) 1989 earthquake. Other studies, such as Piscini et al. [6], applied a multi-parametric approach to study the atmosphere in the specific case of Skin Temperature, Total Column Water Vapour, and Total Column Ozone before the Italian Seismic sequence of 2016–2017. The advantage of a multi-parametric approach is a better understanding of the physical phenomenon, in this case, in the atmosphere. Still, to have a complete picture of the Earth, a multi-layer approach applied to the lithosphere, atmosphere, and ionosphere is required. Such an approach can better understand the possible interactions between the geo-layers in the preparation phase of an earthquake. Interesting claiming of chains of the possible lithosphere–atmosphere–ionosphere couplings before the Mw = 9.1 great Tohoku (Japan) 2011, Mw = 7.8 Gorkha (Nepal) 2015, Mw = 6.0 Amatrice (Italy) 2016, Mw = 7.5 Indonesia 2018, Mw = 7.1 Ridgecrest (USA), and Mw = 7.2 Kermadec (New Zealand) 2019 earthquakes were reported, respectively, by Chen et al. [7], Han et al. [8], Ouzounov et al. [9], Fan et al. [10], Marchetti et al. [11,12], and De Santis et al. [13,14]. A similar approach with further interesting findings was applied before the La Palma 2021 and Hunga Tonga-Hunga-Ha'Apai 2022 volcano eruptions by Marchetti et al. [15,16] and D'Arcangelo et al. [17]. This multi-layer and multi-parametric approach is the one we followed in the present study.

Some single parameters used in such studies were statistically investigated in specific areas or worldwide. In particular, Hattori et al. [18] propose a 10-years statistical investigation of ULF electromagnetic signals from ground geomagnetic observatories in Japan. Genzano et al. [19] instead made a statistical assessment of Thermal InfraRed (TIR) anomalies detected from satellites, including Japan. The worldwide ionospheric perturbations detected by DEMETER, Swarm, and China Seismo Electromagnetic Satellite were systematically correlated with earthquakes of magnitude 4.8+, 5.5+, or 6.0+ by Yan et al. [20], Ouyang et al. [21], De Santis et al. [22,23], He et al. [24], and Marchetti et al. [16,25]. These studies demonstrated statistically that ionospheric perturbations preceded some earthquakes by days, weeks, or even months. They also identified that the anticipation time increases with the magnitude of the upcoming events and that the location under the sea or on land can influence the frequency of the electromagnetic signal [22,25].

The existence of pre-earthquake processes is also supported by different Lithosphere–Atmosphere–Ionosphere Coupling (LAIC) models [26]. Unfortunately, the different models contradict each other, describing the coupling with different physical and chemical mechanisms. Essentially, we can outline three coupling mechanisms:

1. Pure electromagnetic channel. In this case, the electromagnetic pre-earthquake disturbances propagate using the Maxwell equations of electromagnetism. These models include the ULF electromagnetic waves induced by the separation of charges at the

fault level as predicted by Molchanov and Hayakawa [27,28] or an induced current in the ionosphere for variation of resistivity in the ground as indicated by Enomoto [29].
2. Chain of mechanical, chemical, thermal, and electromagnetic phenomena. In this mechanism, one process induces another. For example, an increase in mechanical stress can release some radon gas at the Earth's surface that increases the temperature, drops the humidity, ionizes the air, creates aerosols and clouds, and perturbs electric circuits as proposed by Pulinets and Ouzounov [30], Pulinets and Khachikyan [31], and Pulinets et al. [32,33].
3. Acoustic gravity waves. Hayakawa et al. [34] proposed that thermal heating of the ground for the accumulation of the stress on the fault could induce some air pressure oscillations that generate acoustic gravity waves potentially reaching the ionosphere.

It is necessary to note that the above classification is a simplification even of the cited theories. Still, it could help to discriminate in observing what is more likely due to one coupling mechanism rather than another Finally, it is important to note that other explanations are also possible; for example, Freund et al. [35,36] proposed that increased stress can break peroxy-links in the rock, releasing free positive charges (positive-holes or simply p-holes) as the electron can quickly recombine in other ways. These p-holes can potentially explain all the pre-earthquake processes, even the increase in the thermal infrared (IR) radiation detected in several cases before earthquakes. Still, in this theory, such emission would be due to IR photons emitted during p-holes recombination just under the Earth's surface. If this assumption is valid, the IR emission will not correspond to an Earth's surface temperature increase. To confirm or exclude such a hypothesis, high-resolution spectra from satellites would be required to discriminate if the IR radiation comes from a "black body" or line emissions, as proposed by Freund et al. [36].

In this paper, we investigate the preparation phase of the Lushan 2013 earthquake through a multi-parametric and multi-layer study. In particular, data from the lithosphere, atmosphere, and ionosphere were analyzed to search for possible geophysical couplings between the geo-layers in the preparation phase of the seismic event.

2. Materials and Methods

This section presents the source and processing of the lithospheric, atmospheric (from the climatological archive), geomagnetic ground observatories, and ionospheric datasets.

2.1. Lithosphere

To study the variations in the lithosphere, we acquired an earthquake catalog. We analyzed cumulative Benioff strain and the released energy as the function of the distance, i.e., the Es parameter as defined by Hattori et al. [18] and Han et al. [37]. The earthquake catalog was retrieved from China National Earthquake Data Center (http://data.earthquake.cn, last accessed on 7 January 2023) and covered the period from 22 October 2012 to 20 April 2013 (i.e., a period of six months before the Lushan earthquake). We selected the seismic events that occurred within the Dobrovolsky radius D ($D = 10^{0.43M}$ km, where M is the earthquake magnitude) [38], and their geographical localization is shown in Figure 1.

We first used the ZMap toolbox provided by ETH (Zurich, Switzerland) [39] to calculate the completeness magnitude Mc and found that Mc was equal to 0.6 within the Dobrovolsky radius, as shown in Figure 2. The completeness magnitude Mc is the minimum magnitude of a seismic event that can be surely detected by a particular seismic network in a specific space-time scale [40]. Only the seismic events with magnitude M ≥ Mc were considered in the following lithospheric analyses.

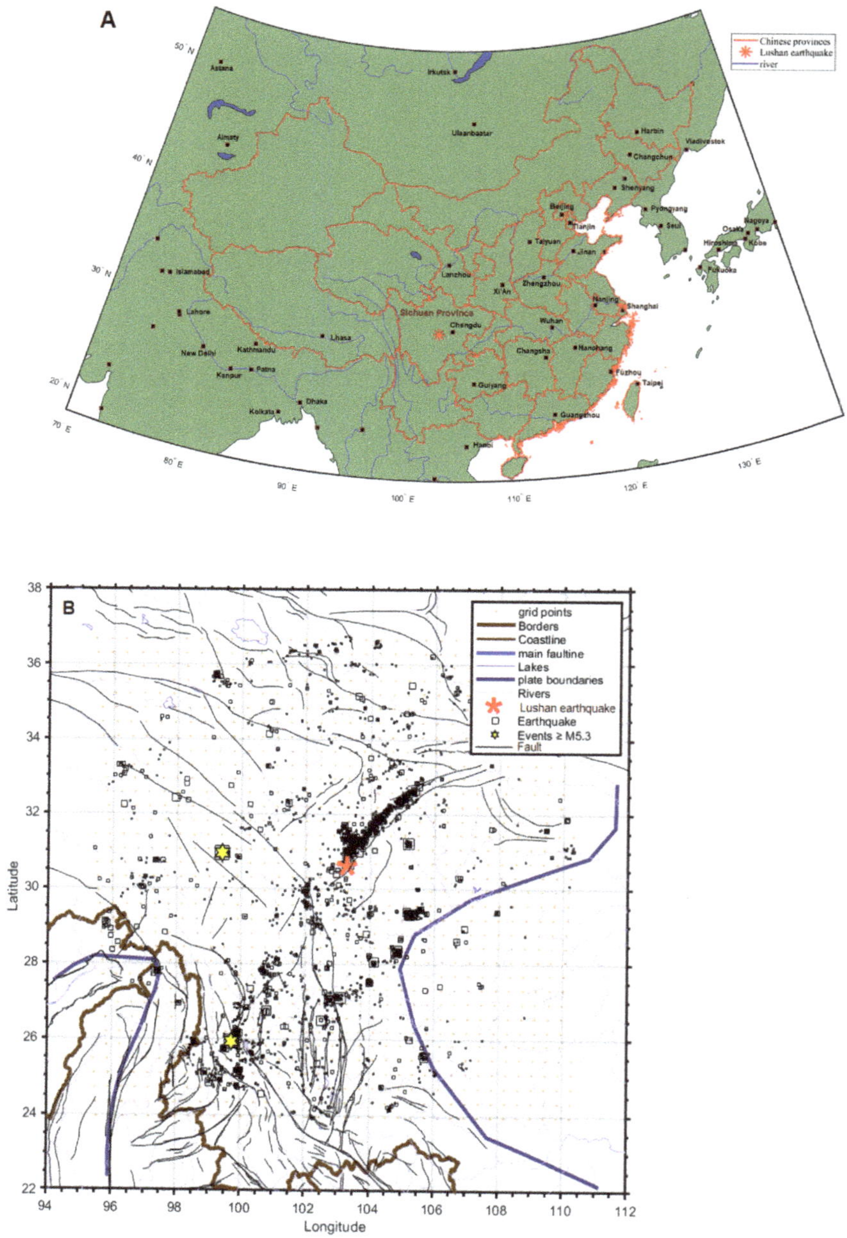

Figure 1. Geographical location of the Lushan earthquake within China (**A**) and map of seismic events (**B**) within the Dobrovolsky radius from 22 October 2012 to 20 April 2013 (events before the Lushan earthquake). Yellow stars represent earthquakes with a magnitude M $\geq$ 5.3. The Lushan 2013 earthquake is marked by a red asterisk.

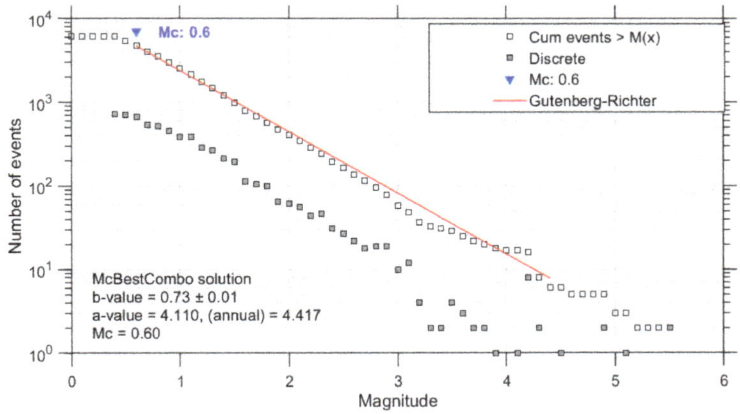

Figure 2. Gutenberg–Richter distributions of the magnitude of the earthquakes within the Dobrovolsky radius.

We calculated the cumulative Benioff strain "S" and "Es" parameter to analyze the lithospheric activity before the Lushan earthquake. The Benioff cumulative strain [41] can be estimated as follows:

$$S(t) = \sum_{i=1}^{N(t)} \sqrt{10^{(1.5 \cdot M_i + 4.8)}}$$

where "*Mi*" is the magnitude of the "i-th" seismic event and "*N*(*t*)" is the total number of seismic events that occurred until the time "*t*". The "Es" parameter is the daily sum of the local earthquake energy (E_s') defined by the following equation:

$$E_s(t) = \sum_{i=1}^{N(t)} E_s'(t,i)$$

$$E_s'(t,i) = \sqrt{\frac{10^{(1.5 \cdot M_i + 4.8)}}{r_i^2}}$$

where "r_i" and Es' are the epicentral distance in km and the local energy in J/km^2 of the "i-th" seismic event, respectively.

2.2. Geomagnetic Ground Data Processing

We investigated ground geomagnetic data from an observatory station relatively close to the Lushan earthquake, Chengdu FGM01 (CDP), and compared it with the far observatory station, Beijing Ming Tombs (BMT), which was more than 1500 km away from CDP and the epicenter. In particular, CDP was inside the Lushan Dobrovolsky area and near the epicenter, 653.3 m above sea level, while BMT (116.2°E and 40.3°N, 183 m above sea level) was outside the Dobrovolsky area. BMT was taken as a reference, as it was unlikely to be influenced by the Lushan earthquake.

Despite previous studies reporting seismo-induced electromagnetic phenomena from geomagnetic ground stations, as Ultra Low Frequency (ULF) electromagnetic waves recorded from the Z-Vertical component of the geomagnetic field from ground observatories (e.g., [37]), we think that, in this case, the geographical location of the CDP observatory could be not optimal to study this component. We note that Han et al. [37] and Zhuang et al. [42] analyzed statistically the geomagnetic data selecting the closest earthquake to the observatory (in the function of their magnitude). In this paper, we explored one earthquake, the Lushan 2013 seismic event, and the closest geomagnetic observatory was about 106.6 km away. Considering the hypocentral depth of the earthquake was about 14 km, a seismo-induced disturbance coming from the source zone of this earthquake would impact more the horizontal components of the geomagnetic field at the CDP observatory than the vertical one. Consequently, we provide a full analysis of the horizontal and vertical components at the CDP observatory compared with the recording at BMT observatory taking into account the geometry constraints from the earthquake and the geographical position of the geomagnetic observatories.

To avoid inevitable gaps in the data, the original data were cleaned of invalid samples (labeled −99,999) and interpolated by a Piecewise Cubic Hermite Interpolating Polynomial [43]. In addition, the data from CDP obtained from the China Earthquake Networks Center, National Earthquake Data Center (http://data.earthquake.cn) consists of relative values of the total horizontal (ΔH), declination (ΔD), and vertical (ΔZ) components with a sampling rate of 1 Hz. Conversely, the data from BMT (INTERMAGNET) contains absolute values of the north (X), east (Y), and vertical (Z) components with 1 min sampling. Thus, mean values of the CDP data were calculated within a window of 1 min and unified (added baseline and rotated in the other reference frame) by data format conversion:

$$
\begin{gathered}
\mathrm{H(t)} = \Delta\mathrm{H(t)} - \Delta\mathrm{H(0)} + \mathrm{H(0)} \\
\mathrm{D(t)} = \Delta\mathrm{D(t)}/60 - \Delta\mathrm{D(0)}/60 + \mathrm{D(0)} \\
\mathrm{Z(t)} = \Delta\mathrm{Z(t)} - \Delta\mathrm{Z(0)} + \mathrm{Z(0)} \\
\mathrm{X(t)} = \mathrm{H(t)} * \cos[D(t)] \\
\mathrm{Y(t)} = \mathrm{H(t)} * \sin[D(t)]
\end{gathered}
$$

where $\Delta\mathrm{H(t)}$, $\Delta\mathrm{D(t)}$, $\Delta\mathrm{Z(t)}$ are relative values of the total horizontal, declination (ΔD), and vertical (ΔZ) components, respectively, detected by CDP. "ΔH(0)", "ΔD(0)", "ΔZ(0)" are the original data in the research period. "H(0)", "D(0)", "Z(0)" are magnetic field components estimated for CDP location on the first research day (22 October 2012) by the International Geomagnetic Reference Field (IGRF [44]) model (acquired from NCEI Geomagnetic Calculators, https://www.noaa.gov/, last accessed on 30 November 2022), $\mathrm{H(0)} = 34018.8\mathrm{nT}$, $\mathrm{D(0)} = -1.8661°$, $\mathrm{Z(0)} = 37519.9\mathrm{nT}$.

We filtered the geomagnetic data by five frequency filters, which are defined as A: a high-pass filter of 1 h; B: a band-pass filter between 1 h and 6 h; C: a band-pass filter between 6 h and 1 day; D: a band-pass filter between 1 day and 1 week; and E: a low-pass filter of 1 week analogous to [15]. We compared the CDP and BMT observatories for X, Y, and Z geomagnetic field components based on five frequency bands divided by these filters. We considered that, if CDP and BMT were in accordance, it should be a response to a global perturbation. At the same time, we suggest a local effect for differences between the two observatories.

2.3. *Atmosphere: Climatological Archive Data Analysis*

In this paper, six parameters were selected to monitor the status of the atmosphere before the Lushan earthquake. They are surface Air temperature, sulfur dioxide, aerosol optical thickness (AOT), carbon monoxide, surface total latent energy flux, and surface specific humidity. Several authors have reported surface total latent energy flux as a possible earthquake precursor [45–47]. The relative (or specific) humidity at the surface level is predicted to decrease by the LAIC model of Pulinets and Ouzounov [30], so we also included it to complete our framework of atmospheric analyses.

The data were retrieved from the Modern-Era Retrospective Analysis for Research and Applications, Version 2 (MERRA-2) climatological model provided by NASA [48]. The dataset was conceived to investigate possible variations in the Earth's climate. Still, it constitutes a precious resource of data that has extended from 1980 to the present, updated monthly based on several atmospheric parameters interpolated on a regular grid in space and time. Its spatial resolution is 0.625° in longitude, 0.50° in latitude, and 1 h in time. Among the other purposes for which MERRA-2 can be used, it has already been used in the study of the pre-earthquake processes in several cases (the Mw = 6.0 & 6.5 Italy 2016 [11], Mw = 7.8 Ecuador 2016 [45], Mw = 7.3 Iran-Iraq border 2017 [46], Mw = 7.5 Indonesia

2018 [12], Mw = 7.2 Kermadec Islands 2019 [14], and Mw = 7.1 Japan 2021 [47] earthquakes) by the CAPRI and MEANS algorithms defined by Piscini et al. [6,49]. The main difference between the two algorithms is that CAPRI considers possible global warming, performs a linear fit over the data, and removes the obtained slope, correcting the time series. Here, checking the original data for atmospheric parameters in Sichuan, as reported in Figure 3, it can be seen that SO_2 and CO substantially increased between 2000 and 2010, and CO became low after 2020. The decrease in CO after 2020 could be due to ecological policies implemented by China to reduce pollution (for example, promoting electric vehicles instead of combustion-engined vehicles) or an effect of a reduction in production in factories due to COVID pandemic prevention rules. It is out of the scope of this manuscript to determine the cause of such trends, but we want to remove them from the original data as they look to be related to air pollution and not to possible pre-earthquake phenomena. At the same time, the trends are very different from a linear trend, so a polynomial of sufficient degree could fit better such trends.

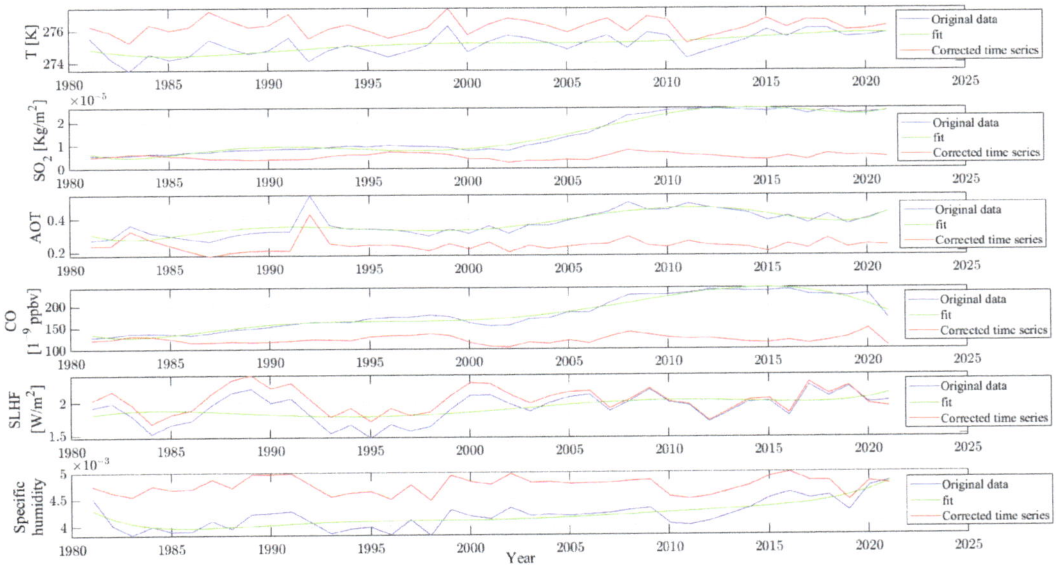

Figure 3. Atmospheric time series (blue line) of the yearly mean value (during the six analyzed months) for Temperature (T), SO_2, aerosol (AOT), CO, surface latent energy flux (SLHF), and specific humidity from 1981 to 2021. The year of the earthquake (2013) is excluded from these graphs. A polynomial fit of degree 6 was performed and plotted as a green line, and the original data without the fit, i.e., the corrected time series, are plotted as red lines.

In Figure 4, it is possible to see the residual sum of the square of a polynomial with a degree from 1 (linear) to 11. Initially, only odd degrees were tested, and then, the even degrees were added around the minimum. Degree 6 is a minimum of the residuals for almost all the parameters and was selected as the best fit for all the atmospheric analyses presented in this paper. The time series are again represented after subtracting the 6-degree polynomial in Figure 3 with red lines. The value of the first year was added to the full-time series to have a real absolute value. We noted that the absolute value of these curves doesn't influence the classification of a particular day as anomalous or not. In fact, the anomaly criterium is based on the standard deviation, and, thus, is independent of the mean absolute value of the whole time series.

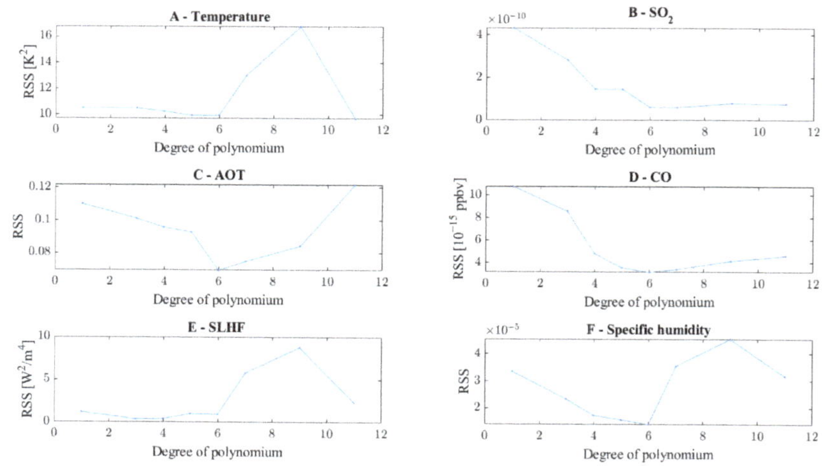

Figure 4. Residual Sum of Square (RSS) of the polynomial fits using different degrees from 1 to 11 applied to atmospheric time series of Temperature (**A**), SO_2 (**B**), aerosol (**C**), CO (**D**) surface latent energy flux (**E**), and specific humidity (**F**).

After this data pre-processing, the standard MEANS algorithm [49] was applied, the process of which can be briefly summarized as:

1. Calculation of preliminary historical mean and exclusion of years that contain some outliers, i.e., one or more values greater than the mean plus 10 standard deviations of all the data.
2. Calculation of historical time series (from 1980 to 2021), in terms of mean and standard deviations of the values of a 3-degree side square centered on the epicenter excluding the year of the earthquakes (2013).
3. Overplot of the year of the earthquake (2013) above the mean plus or minus 1.0, 1.5, and 2.0 standard deviations of each specific day of the historical time series.
4. Defining the values outside the 2-standard deviation thresholds as anomalous and, for such days, extracting the map and spatial information of locations of high values of the investigated quantity.

The calculation of the mean and standard deviation day-by-day assures that the result is independent of any seasonal trends.

2.4. Ionosphere

At the time of the occurrence of the Lushan earthquake, the Low Earth Orbit (LEO) satellites, usually used for pre-earthquake study, were unavailable. In fact, the DEMETER and CHAMP satellites were decommissioned some years before (in 2010), Swarm was launched only at the end of 2013, and the China Seismo Electromagnetic satellite (CSES) in 2018. For this reason, we relied on the Total Electron Content (TEC) global maps obtained from ground stations and vertical electron density profiles measured by the FormoSat-3 satellite to investigate the ionosphere.

2.4.1. TEC

TEC data was retrieved from the Global Ionospheric Map (GIM). This dataset is available worldwide with a spatial resolution of 5° longitude, 2.5° latitude, and 2 h in time at precise UT. The data are available systematically from 2000, and we retrieved up to 2022 to construct a background. Only the six months before the earthquake for each year were selected from 22 October (of the year before) until 20 April. The data processing here applied was initially prepared to analyze another natural hazard: the "La Palma" volcano eruption of 19 September 2021 by Marchetti et al. [15]. In that paper, the solar cycle modulation of TEC was removed by normalization using a sinusoidal curve. Even if the accordance of the sine curve with the TEC data was found to be good (with a correlation coefficient greater than 0.9), the sine curve could not follow the variations in the maxima and minima of the solar cycle that are different cycle-by-cycle. This paper introduces the daily Sunspot number to follow the solar cycle trend better. The use of Sunspot number improved the remotion of the TEC modulation induced by solar activity, surely not due to eventual pre-earthquake processes. Three different TEC dependencies as a function of Sunspot number were tested: linear, 2-order polynomial (Figure S5 in Supplementary Materials), and exponential (Figure 5B). Considering the minimum residual balanced with the degrees of freedom of the fits, i.e., the maximum adjusted coefficient of determination, the exponential fit was selected.

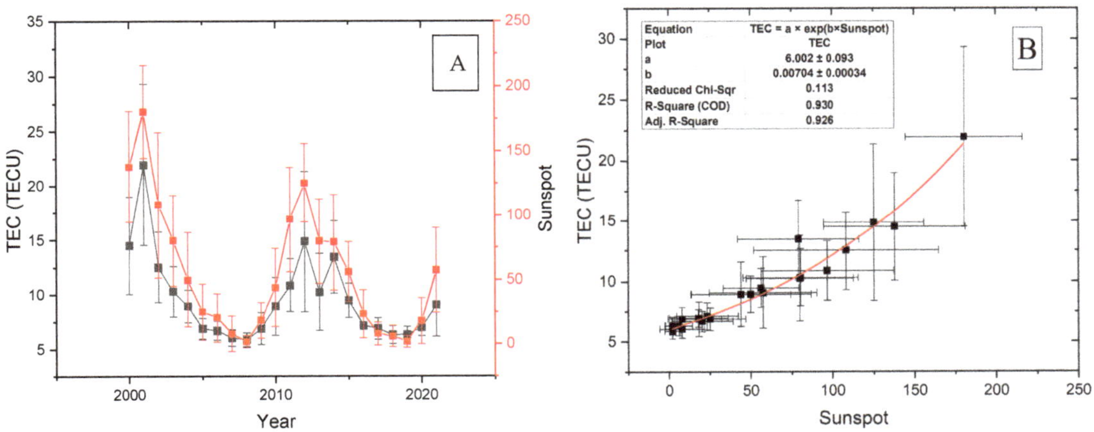

Figure 5. (**A**) TEC and Sunspot number time trend in the investigated period from 2000 to 2021. Each year is the average value from 22 October until 20 April of the following year (e.g., 2000 = 22 October 2000/20 April 2001, … , 2021 = 22 October 2021/20 April 2022). (**B**) TEC versus Sunspot number. Each error bar represents one standard deviation of the original data distribution in the six months used for the average calculus. An exponential fit is superposed as a red curve, and their coefficients and statistical goodness parameters are reported in the embedded table.

The following steps can summarise the procedure of the algorithm:

1. Select a couple of TEC maps: one preceding, and the other following, the selected local time to be analyzed.
2. Spatially interpolate the value of TEC at the epicenter latitude and longitude for each map by a cubic surface.
3. Temporal linear interpolate the TEC at the selected local time with the previous two values from step 2.
4. Check the geomagnetic conditions by Dst and ap indexes rejecting the samples recorded during disturbed conditions (|Dst| > 20 nT or ap > 10 nT).
5. Repeat this process for each day/year to be analyzed, contributing to the background or the time series to be analyzed if inside the period of six months preceding the Lushan earthquake.
6. For each year, the average TEC and Sun Spot Number index (also selected in quiet geomagnetic time, i.e., the same days used for TEC) is calculated and reported in Figure 5A.
7. An exponential fit of the TEC versus the Sunspot number is performed and shown in Figure 5B.
8. The TEC is finally normalized by dividing each value for the value estimated for that year by the fit calculated in Step 7.

9. For each day, the background mean and standard deviation of the normalized TEC are calculated, excluding the six months before the Lushan earthquake
10. A time series graph is produced, and the values before the Lushan earthquake that overpassed the background mean plus two standard deviations are defined as anomalous.

2.4.2. Vertical Electron Density profiles from FormoSat-3

In addition to TEC data, we retrieved the available electron density ionospheric profiles measured by the FormoSat-3/COSMIC-1 (Formosa Satellite 3/Constellation Observing System for Meteorology, Ionosphere, and Climate) satellite during the same six months before the Lushan 2013 earthquake. FormoSat-3/COSMIC-1 was launched on 14 April 2006 and worked until 6 May 2020. Previous authors have used FormoSat satellites to search for possible pre-earthquake ionospheric disturbances [50,51].

The vertical electron density profiles were retrieved within the Dobrovoslky radius and a comparison annular surrounding region. Figure 6 shows two examples of electron density profiles from FormoSat, the first (Figure 6A) shows only the typical electron density peak F2. The second (Figure 6B) presents a second peak, i.e., a small inversion of the vertical profile at around 100 km altitude, known as the sporadic E layer, formed under particular circumstances, including pre-seismic activity [52,53]. In the electron density profile of Figure 6B, there are some negative electron density values under the F2 peak. Such values are obviously physically impossible as the electron density cannot be physically negative as the minimum possible value is zero. However, as the quantity is not directly measured but indirectly estimated by inversion of GNSS Radio Occultations data, the result of the calculation could provide some negative data. In addition, visible from the figure, the shape of the Ne profile is still reliable, and probably it is just underestimated. Such a weakness in the inversion technique for Formosat is known in the literature, and some improved methods were proposed, for example, by Pedatella et al. [54]. They provided an upgraded electron density profile starting from 2015 data, so we still rely on this paper's original electron density profiles. Furthermore, the analysis proposed in this paper compares the measurements inside the Dobrovolsky area with a surrounding area observed by the same satellite, so an underestimation of the values would not affect the results of this paper as the data and background are based on the same dataset.

The process of elaboration of electron density profiles consists of the following steps:

1. Smoothing of the original Ne vertical profile with a moving average calculated sample-by-sample in a window of 11 samples. This operation assures that any spike in the original vertical profiles can be removed.
2. Rejection of the profile below 50 km altitude as the F2 peak cannot be under such an altitude. However, the satellite often measured very high electron density values close to the ground, which were probably some bias in the measurements or data inversion.
3. Extraction of the F2 maximum electron density value (NmF2), its altitude (hmF2), and its geographical position. It is important to note that the vertical profile is inclined, and any point corresponds to a different latitude and longitude, and we obtained the F2 value.
4. Using the International Reference Ionosphere Model IRI-2016 [55], the information was reported to the epicentral coordinates by the proportion of the model computed at the measurement point and the epicenter.

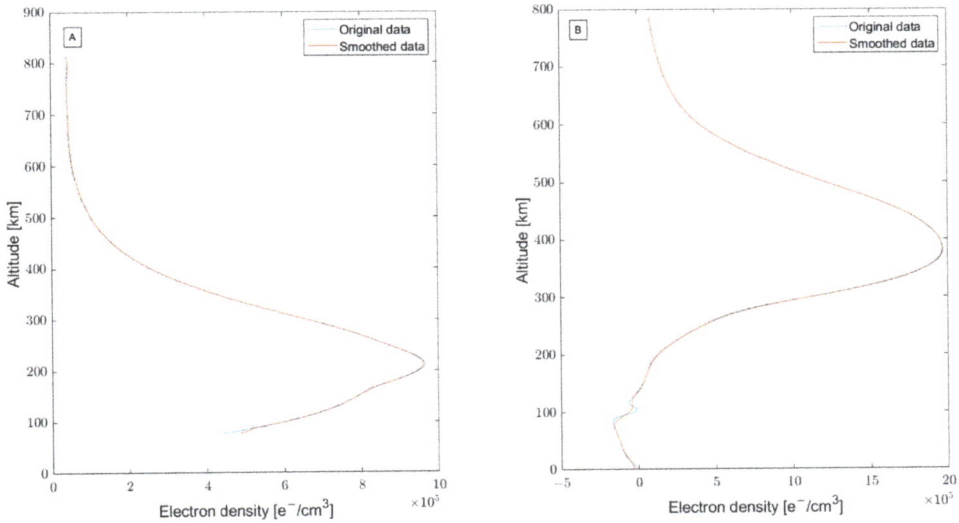

Figure 6. Examples of Formosat3 electron density profiles. Original data and smoothed data after processing is presented. (**A**) represents a standard electron density profiles while (**B**) shows a particular electron density profile with negative values and inversion of vertical profile at 100 km (ionospheric sporadic E-layer).

3. Results

3.1. Lithosphere

The earthquake catalog data from six months before the Lushan earthquake were systematically analyzed using the method described in the previous section. According to the cumulative result of Benioff strain "S" shown in Figure 7, there are five accelerations in the trend, especially ~92 and ~48 days before the earthquake, and particularly notable is the last growth three days before the earthquake. The accelerated increases could be connected to activities of the fault system before the Lushan earthquake. Calculating the Es parameter, we obtained the graph shown in Figure 8.

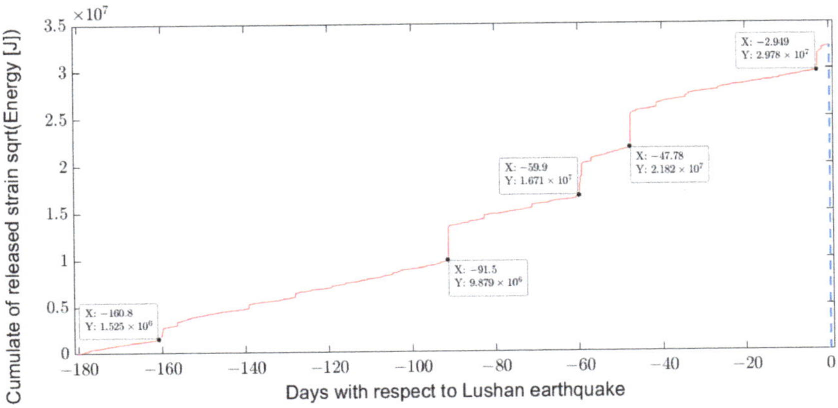

Figure 7. Cumulative Benioff strain S calculated from the earthquake catalog six months before the Lushan earthquake. Only the events equal to or greater than the completeness magnitude of 0.6 were considered in the graph.

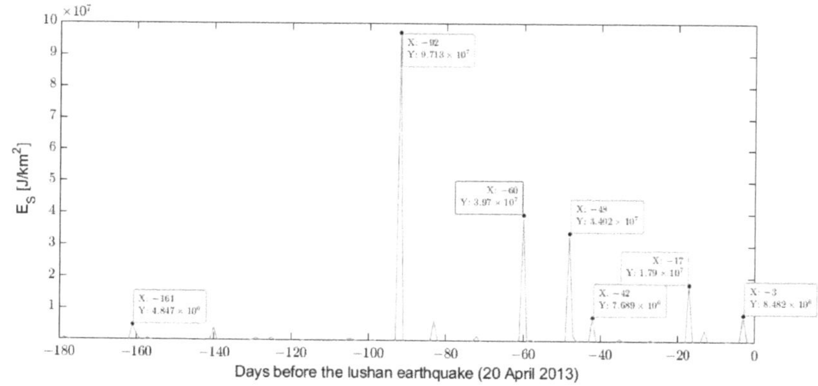

Figure 8. E_S parameter calculated from the earthquake catalog six months before the Lushan earthquake. Only the events equal to or greater than the completeness magnitude of 0.6 were considered in the graph.

3.2. Geomagnetic Data from Earth's Surface

Figure 9, Figure 10, and Figure S1 in Supplementary Materials show the North (X), East (Y), and vertical (Z) components of geomagnetic data measured by CDP and BMT ground observatories, respectively. The data were filtered into five frequency bands. These figures demonstrate that the two observatories were generally consistent with each other, especially in terms of the Z component. At the same time, we indicate some differences between CDP and BMT with red circles, which may be produced by the preparation phase of the Lushan earthquake. We applied correlation analysis numerically based on Spearman rank correlation coefficient estimation to validate the uncorrelation between two datasets of CDP and BMT. The window used to compute correlation coefficient r is shifted by 20% of its length for every step. The lengths were selected as twice the higher cut-off period or longer, 2 h, 12 h, 2 days, 14 days, and 35 days, respectively, for the five filters. Then we defined quiet geomagnetic conditions as $|Dst| \leq 20$ nT and $ap \leq 10$ nT. Only windows with more than 10 samples in quiet geomagnetic conditions were considered in this research. Moreover, because of invalid samples without information, labeled −99,999 in the data, we rejected windows with more than 10% invalid samples. Figure S2 in Supplementary Materials shows the behavior of the correlation coefficient r, whose values range from −1 to +1. When the r value is close to 0 inside the window, it demonstrates that the datasets from CDP and BMT presented different trends; when r is a positive value (close to 1), it implies proportional trends in the two datasets, while 1 means 100% concordance; when r is a negative value (close to −1), it designates negative correlation. Furthermore, uncorrelated data were selected as candidate pre-earthquake anomalous signals by the window with $|r| \leq 0.3$. In fact, as we judge those differences between the two observatories will appeal to local phenomena, hopefully, these are possible pre-earthquake anomalies.

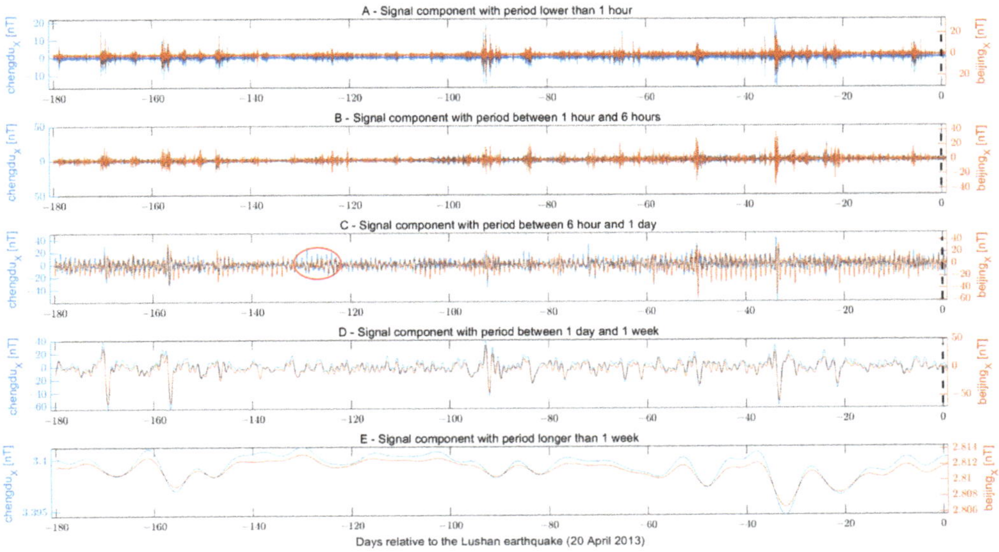

Figure 9. X (North) component of geomagnetic data from CDP (Chengdu, blue lines) and BMT (Beijing, orange lines) ground observatories. The data were filtered into five frequency bands. The red oval highlights more signals from the Chengdu observatory with respect to the Beijing observatory.

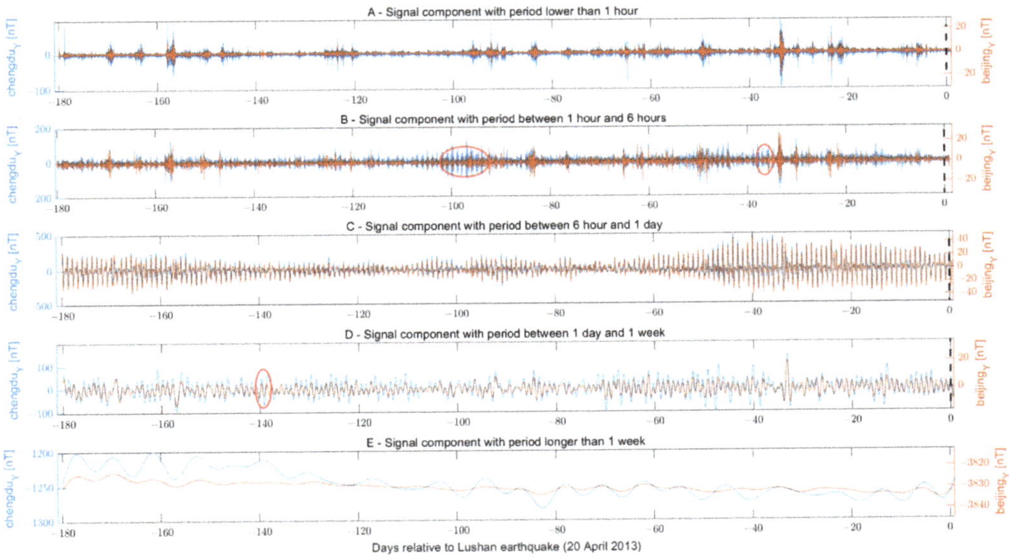

Figure 10. Y (East) component of geomagnetic data from CDP (Chengdu, blue lines) and BMT (Beijing, orange lines) ground observatories. The data were filtered into five frequency bands. The red ovals highlight more signals from the Chengdu observatory with respect to the Beijing observatory.

3.3. Atmospheric Temperature and Composition

The atmospheric composition was investigated by analyzing surface Air temperature, sulfur dioxide mixing ratio, Aerosol Optical Thickness (AOT), carbon monoxide surface concentration, surface total latent energy flux, and surface specific humidity.

The temperature time series is shown in Figure 11, and three anomalies (i.e., values above the two-standard deviation of the historical mean) are depicted by red ovals. Two of these have persistence as defined by Piscini et al. [6], i.e., their occurrence lasts at least two consecutive days. The temperature is very high on the day before the earthquake but inside the two standard deviation thresholds for just 0.1 K.

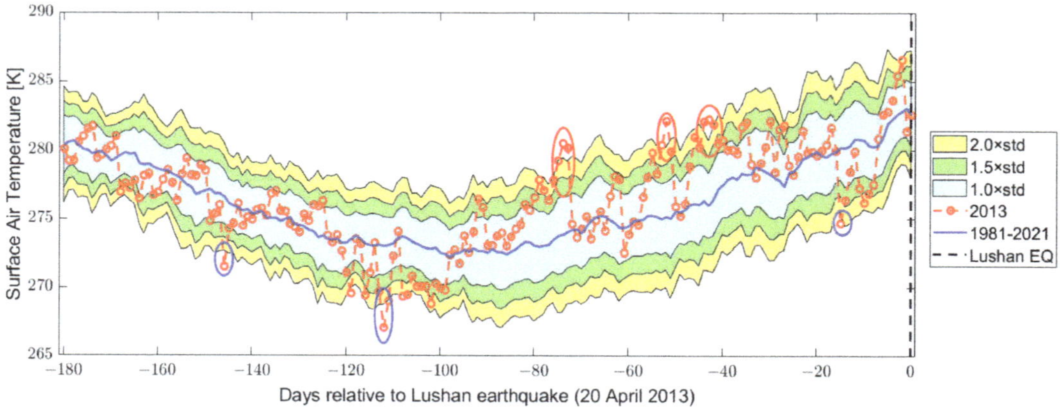

Figure 11. Surface Air temperature from the six months before the Mw 6.7 Lushan 2013 earthquake.

The time series in the six months before the Lushan earthquake of carbon monoxide surface concentration is shown in Figure 12. Two consecutive anomalous days are depicted and highlighted by a red circle on 10 and 11 December 2012 (i.e., −130 and −129 days with respect to the earthquake). Several days show negative anomalies underlined by blue circles. Still, no LAIC models predict a decrease in CO in the atmosphere as a precursor of earthquakes; consequently, we don't consider these events further.

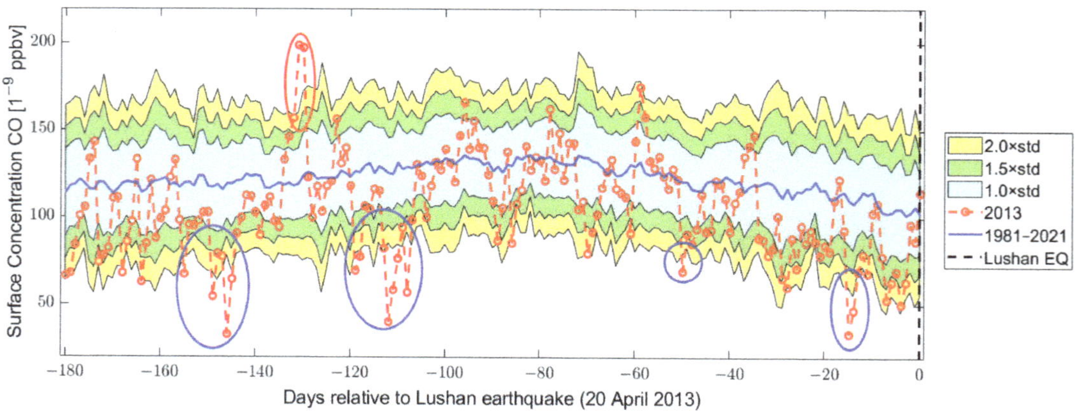

Figure 12. Carbon monoxide levels from the six months before the Mw 6.7 Lushan 2013 earthquake.

In Figure 13, we show the time series of aerosol levels in the six months before the Lushan 2013 earthquake. The anomalous values are greater than the previous time series and are highlighted by red circles. In particular, some anomalies were persistent for three days on 4–6 November 2012 (about −165 days with respect to the earthquake), for two days plus one close day on 8, 9, and 11 December 2012 (about −130 days with respect to the earthquake), for three days plus one close day on 9, 11–13 March 2013 (about −40 days with respect to the earthquake) and two days on 8–9 April 2013, so just 10 days before the earthquake. The aerosol level on 10 April 2013 is a little inside the 2 standard deviations, so it cannot be considered an anomalous day, even though its value is very much higher than the normal value and following two anomalous days. The AOD abnormal value is about 1.5 on the earthquake day. This result is consistent with the study of Liu et al. [56].

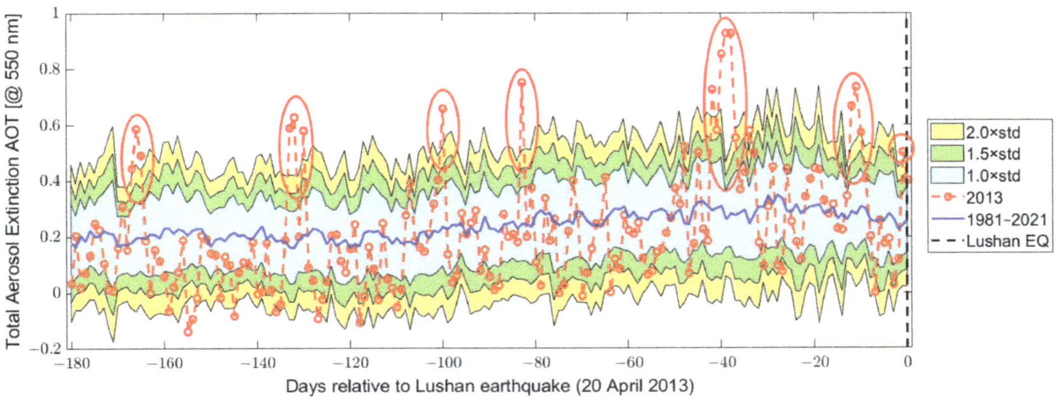

Figure 13. Aerosol levels from the six months before the Mw 6.7 Lushan 2013 earthquake.

The time series for the six months before the Lushan 2013 earthquake of sulfur dioxide levels is shown in Figure 14, and red circles depict the positive anomalous values. In particular, one value (7 February 2013, 71 days before the earthquake) is exceptionally high but doesn't persist for more than one day. Other days present persistent anomalies, such as 1–2 November 2012 (−169 days with respect to the earthquake), 9–10 December 2012 (−130 days with respect to the earthquake), and 19 and 20 February 2013 (−59 days with respect to the earthquake).

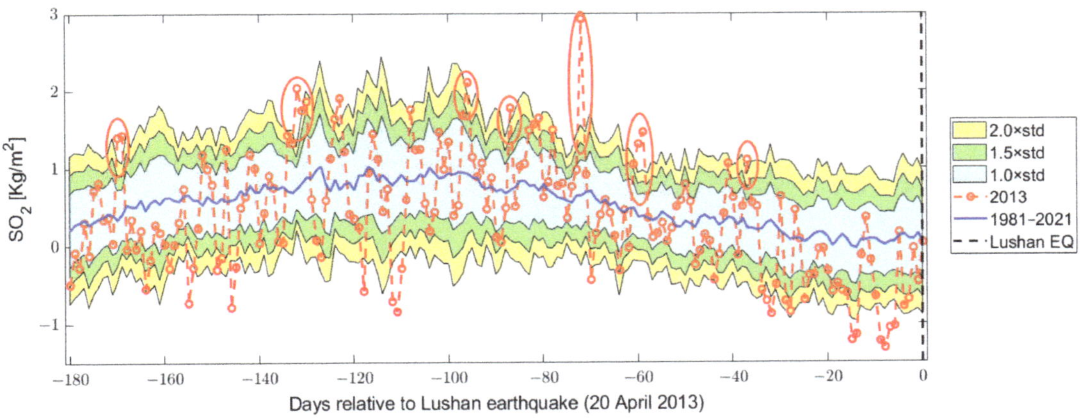

Figure 14. SO_2 levels from the six months before the Mw 6.7 Lushan 2013 earthquake.

In Figure 15, a map for each atmospheric parameter on an anomalous day is reported. In particular, CO levels showed two anomalous days: 10 and 11 December, and both are reported. It is possible to see that on the first day, the highest pixels were located on the main plate boundary and closer to the future epicenter. In contrast, only the anomaly closest to the future epicenter persisted on the second day, suggesting a possible relation with the upcoming seismic event.

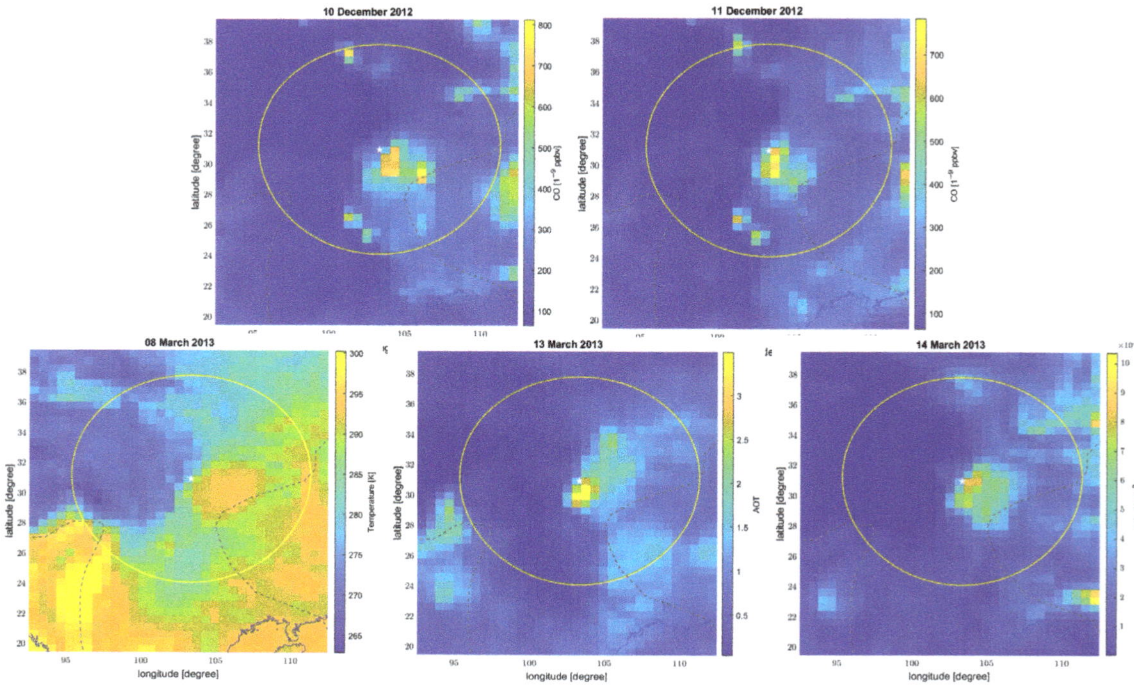

Figure 15. Maps of CO levels on 10 and 11 December 2012, Temperature on 8 March 2013, aerosol levels on 13 March 2013, and SO2 levels on 14 March 2013. The yellow circle represents the Dobrovosky radius of the Mw 6.7 Lushan 2013 earthquake. The white star indicates the epicenter of Mw 6.7 Lushan 2013 earthquake.

For temperature, aerosol, and SO_2, three closer and anomalous days (8, 13, and 14 March 2013) were selected, respectively.

In addition to the previous parameters, in Figures 16 and 17, we analyzed the time series of the surface total latent energy flux and the surface specific humidity in the six months prior to the Lushan earthquake. The total latent heat flux is more commonly known as surface latent heat flux, but in the label, we used the parameter's name used in the MERRA-2 archive. It is possible to note that total latent energy flux has multiple anomalous days: 29 October, 15 and 24 November, 22, 28, and 29 December 2012, 2 January, 7 February (the most anomalous), 28 February, 9, 25 and 30 March and 4 April 2013. They are, respectively, −173, −156, −147, −119, −113, −112, −108, −72, −51, −42, −26, −21, and −16 days with respect to the event. It is striking that, among all the anomalous days (see Figure S4 in the Supplementary Materials), the closer to the time of the upcoming earthquake, i.e., the 4 April 2013 anomaly (16 days prior to the event), the higher the values of total latent energy flux (see Figure 18) aligned with the Longmenshan fault where the seismic source was located.

For the specific humidity, we extracted the negative anomalies (i.e., the values lower than the mean minus two standard deviations), which are four values: 3 and 16 November, 23 December 2012 and 5 April 2013 (−168, −155, −118, −15 days with respect to the earthquake day). We further note that 17 November 2012 is lower than 1.5 standard deviations as well as 6 April 2013, so both of these anomalies could be considered as partially persistent for two consecutive days.

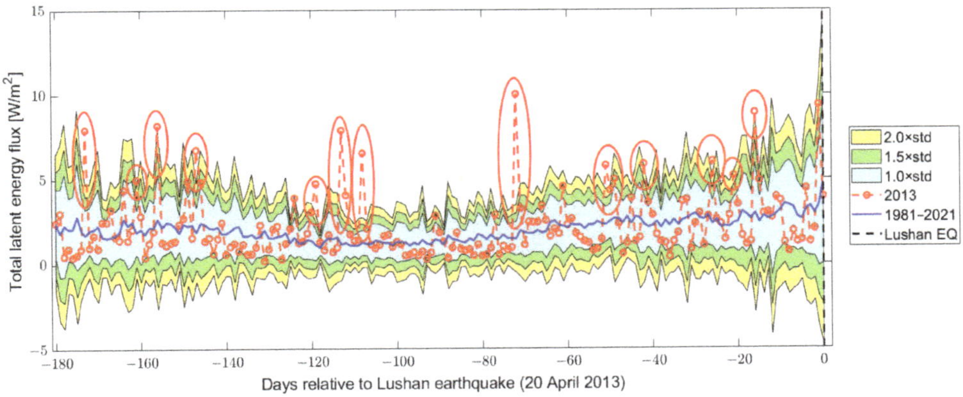

Figure 16. Total latent energy flux from the six months before the Mw 6.7 Lushan 2013 earthquake.

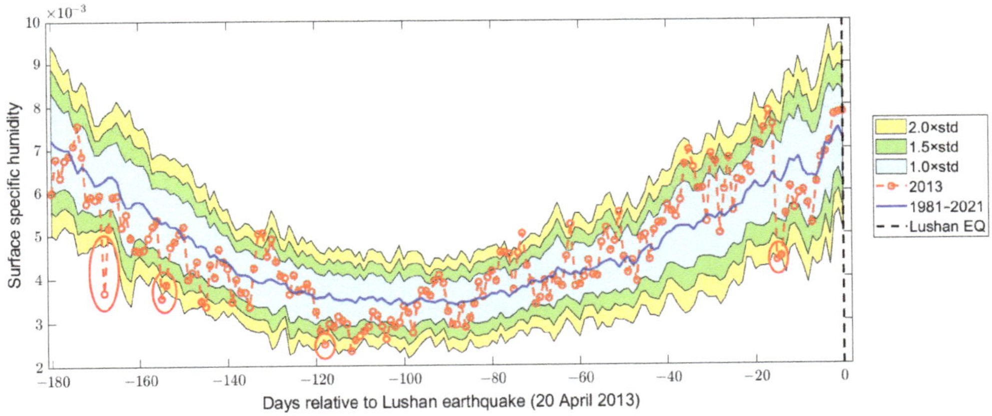

Figure 17. Surface specific humidity from the six months before the Mw 6.7 Lushan 2013 earthquake.

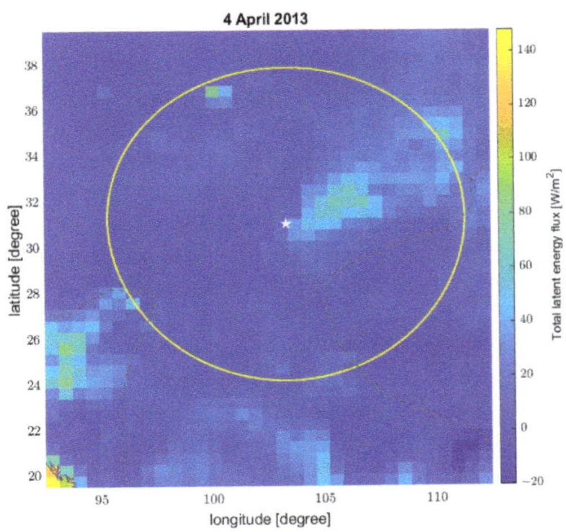

Figure 18. Map of the total latent energy flux on 4 April 2013. The white star indicates the epicenter of Mw 6.7 Lushan 2013 earthquake.

3.4. Ionosphere

The ionosphere was investigated by Total Electron Content (TEC) data and electron density profiles measured by FormoSat-3 (also known as COSMIC-1). Figure 19 shows the time series of normalized TEC interpolated above the Lushan earthquake at 2:00 local time. This time was chosen to directly compare the results with the present China Seismo Electromagnetic Satellite (CSES), even though it was not yet launched at the time of the Lushan earthquake occurrence. Despite this, it is interesting to repeat the technique for natural hazards occurring during CSES fly time like the La Palma 19 September 2021 volcanic eruption [15].

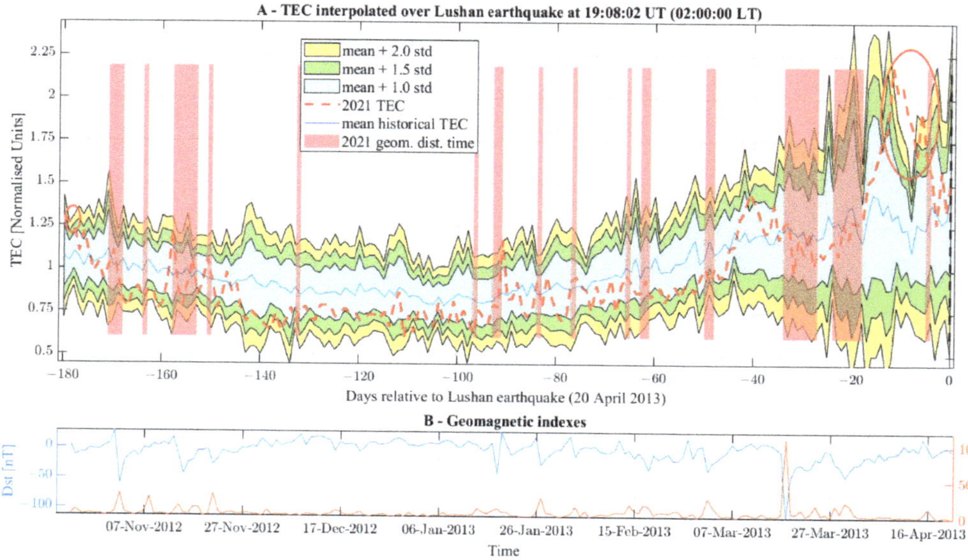

Figure 19. TEC time series from the six months before the Mw 6.7 Lushan 2013 earthquake. The original TEC data were retrieved by GIM-TEC and interpolated at 2 LT above the epicenter (**A**). The disturbed geomagnetic time ($|Dst| > 20$ nT and/or ap > 10 nT) is highlighted by light red boxes. The geomagnetic indices Dst and ap are shown in the lower panel (**B**).

The time series depicted one small anomaly (slightly above the threshold) on 23 October 2012 and a more interesting group of anomalies from 12 to 6 days before the Lushan earthquake. In particular, five consecutive anomalous days were recorded from 8 to 12 April 2013 and 14 April 2014, i.e., a little more than one week in advance of the seismic event, which agrees with the study by Dai et al. [57]. It is worth noting that the TEC value on 13 April 2013 is high, but it did not overpass the threshold of 2 standard deviations, so from the objective point of view, it cannot be classified as an anomalous day. On these days, geomagnetic conditions were extremely quiet, as visible from the lower panel (-1 nT $\leq$ Dst ≤ 7 nT and 2nT $\leq$ ap $\leq$ 3nT). This group of three days with a high TEC value (including two anomalous days) followed the increased aerosol levels in the atmosphere detected by the previous analysis.

The electron density profiles of Formosat-3 were investigated as described in Section 2.4, extracting the altitude of the electron density peak (hmF2) and its electron density (NeF2) for all the profiles acquired in the circle investigation area centered on the earthquake and with a radius r_{max} defined by the following equation:

$$r_{max} = \sqrt{2 \cdot (r_{Dob})^2 + (r_{Dob} + g)^2}$$

where r_{Dob} is the Dobrovolsky radius [38] and g is a gap set equal to 300 km. The scope is to have two areas separated by this gap: the inner is the Dobrovolsky area to search for possible pre-earthquake effects, and the outer is an annulus surrounding the epicenter of an equivalent area of two times the Dobrovoslky area used to define a background value. Six months before the Lushan 2013 earthquake, 2465 points were selected in the above-selected area. The histogram of the distribution of hmF2 is reported in Figure 20. The samples below 200 km are not only unrealistic but also not expected, and from the investigation of the original profiles, they are clearly due to unrealistic electron density profiles. Consequently, we reject all the points below 200 km, reducing the dataset to 2352 samples.

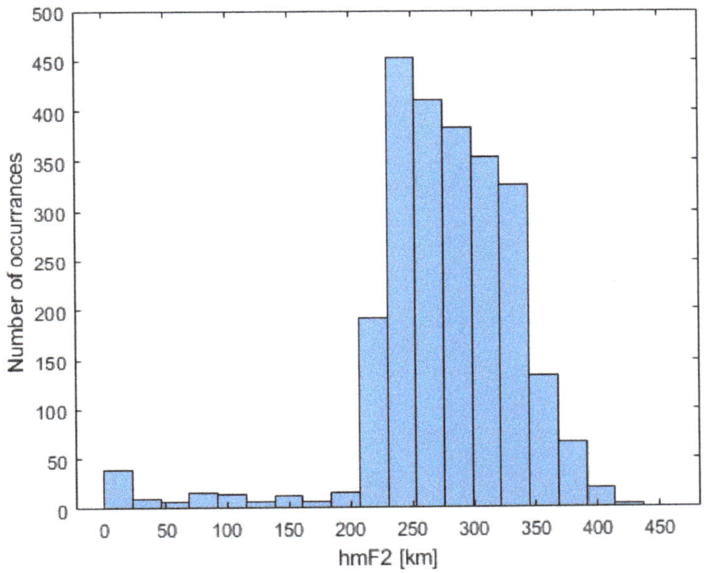

Figure 20. Distribution of the hmF2 extracted from Formosat3 ionospheric profiles.

For each sample, the local time LT was calculated from the UT time and the longitude "lon" of acquisition:

$$LT = UT + \frac{lon}{15}$$

Factor 15 is the ratio between the degree and hours, considering that half circumference is 180° and 12 h. For each day, the samples acquired in the Dobrovoslky area with a maximum difference of 1 h in local time with the samples in the background area are selected. Finally, such samples are compared with the closer sample in the background area, obtaining a total of 281 relative estimations of the variations of F2 reported in Figure 21 and defined as:

$$\Delta hmF2 = \frac{hmF2_{Dob} - hmF2_{bg}}{hmF2_{bg}},\ \Delta NeF2 = \frac{NeF2_{Dob} - NeF2_{bg}}{NeF2_{bg}}$$

As can be seen, the variations in the altitude of the F2 peak follow a more Gaussian distribution than the electron density variation, which is biased toward positive values. Some values are clearly anomalous and are highlighted by red circles in Figure 21. The precise values are provided in Table 1, and the electron density vertical profiles of the most representative are shown in Figure 22. In particular, the geomagnetic indexes Dst, ap, and AE recorded during each anomalous value are also reported in Table 1. For all the anomalies, it can be excluded that they could be induced by the geomagnetic activity, except for the one recorded on 22 February 2013, as it could be induced by penetrating electric currents from the high poles where high activity was measured (AE = 634 nT). Contrariwise, the anomalies observed on 31 October, 8,10,12 December 2012, and 5 January 2013 cannot be explained by geomagnetic activity and could be, in principle, induced by the preparation phase of the earthquake. The high value of NeF2 recorded in the Dobrovolsky on 20 December 2012 with respect to the one in the background is unlikely to be due to geomagnetic activity even if at pole auroral electrojet was a little high with AE = 208 nT.

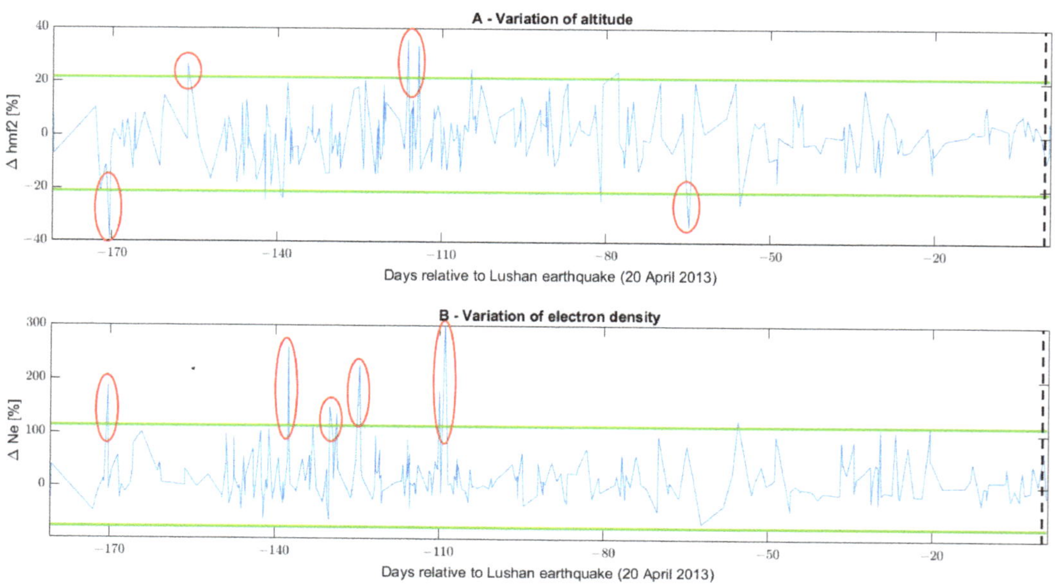

Figure 21. Variation in the altitude (**A**) and electron density value (**B**) of the ionospheric peak F2 in the six months preceding the Mw = 6.7 Lushan 2013 earthquake. The estimation of the two parameters comes from the ionospheric profiles measured by the Formosat3 satellite, and the variations are relative to the background area. Two green lines show the mean plus or minus two standard deviations of the same values.

Table 1. Anomalous values of the F2 ionospheric peak as extracted from the time series shown in Figure 21. In the first rows, the anomalous altitude days of the F2 peak are presented, and in the following rows, the anomalous days for an increase in electron density in the F2 peak are listed.

Date and Time UT	Days with Respect to the EQ	hmF2 in the Dobrovolsky [km]	hmF2 in the Background [km]	Δ hmF2	Local Time	Geomagnetic Indexes
31 October 2012 13:08:24	−170.5	202. 5	333.1	−39.2%	20:00	Dst = 12 nT, ap = 4 nT, AE = 44 nT
14 November 2012 14:47:09	−156.4	372.8	294.6	26.6%	21:39	Dst = − 64nT, ap = 7 nT, AE = 83 nT
24 December 2012 18:59:09	−116.2	348.1	256.4	35.7%	1:51	Dst = 2 nT, ap = 2 nT, AE = 84 nT
26 December 2012 18:14:25	−114.2	334.4	250.5	33.5%	1:06	Dst = 7 nT, ap = 0 nT, AE = 17 nT
14 February 2012 4:56:41	−64.8	258.0	391.6	-34.1%	11:48	Dst = −21 nT, ap = 18 nT, AE = 338 nT

Table 1. *Cont.*

Date	Days with Respect to the EQ	NeF2 in the Dobrovolsky [e^-/cm^3]	NeF2 in the Background [e^-/cm^3]	Δ NeF2	Local Time	Geomagnetic Indexes
31 October 2012 13:08:24	−170.5	1.48×10^6	5.18×10^5	187%	20:00	Dst = 12 nT, ap = 4 nT, AE = 44 nT
3 December 2012 11:13:45	−137.5	9.13×10^5	2.54×10^5	260%	18:05	Dst = 8 nT, ap = 4 nT, AE = 35 nT
11 December 2012 0:08:34	−130.0	3.22×10^5	1.29×10^5	150%	7:00	Dst = 10 nT, ap = 5 nT, AE = 12 nT
16 December 2012 8:28:51	−124.6	2.35×10^6	7.22×10^5	226%	15:20	Dst = −7 nT, ap = 6 nT, AE = 239 nT
31 December 2012 23:49:33	−109.0	3.06×10^5	7.57×10^4	304%	6:41	Dst = 8 nT, ap = 2 nT, AE = 26 nT

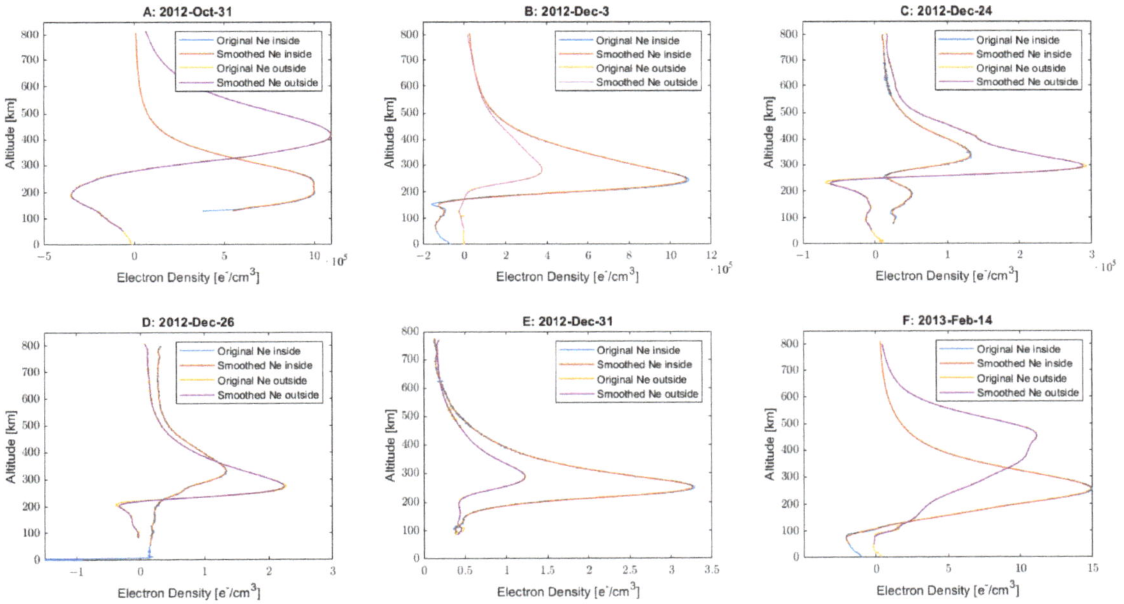

Figure 22. Electron density profiles inside and outside the Dobrovoslky area for the most anomalous days depicted in the time series, i.e., 31 October 2012 (**A**), 3 December 2012 (**B**), 24 December 2013 (**C**), 26 December 2013 (**D**), 31 December 2012 (**E**), and 14 February 2013 (**F**).

4. Discussion

4.1. Three Possible Lithosphere–Atmosphere–Ionosphere Couplings before the Earthquake

Comparing the results in the analyzed three geo-layers, it is possible to depict a common anomalous time about 130 days before the earthquake. In fact, the lithosphere presents a very slight increase in stress, while the aerosol, sulfur dioxide, and carbon monoxide levels, and surface total latent energy flux are increased in the atmosphere; at around the same time, the Chengdu geomagnetic station recorded more magnetic pulses between 6 and 24 h in the X component and the electron density of the F2 ionospheric peak was higher. Furthermore, the atmospheric aerosol level became anomalous three days before the ionospheric anomaly suggesting a bottom-up coupling mechanism.

To investigate better the possible coupling 130 days before the earthquake, Figure 23 represents the difference in surface air temperature on 11 December 2012 with respect to the typical values of the region. In addition, Figure 23 shows higher temperature values in the mountain region, northwest of the upcoming epicenter. Higher temperature values at the high-altitude area are explainable by Freund et al. [36], which predicts that positive holes migrate towards geographical high peak regions and emit infrared radiation due to charges recombination at or just below the Earth's surface. Even though this would not be a real increase in temperature, 'it's impossible to distinguish from present satellite remote sensing observations without spectra of the radiation. Despite this possible explanation, 'it is important to note that no pixels reached an increase in temperature greater than 2 standard deviations, so according to our definition of "anomaly", there was no temperature anomaly on 11 December 2012 inside the Dobrovolsky area.

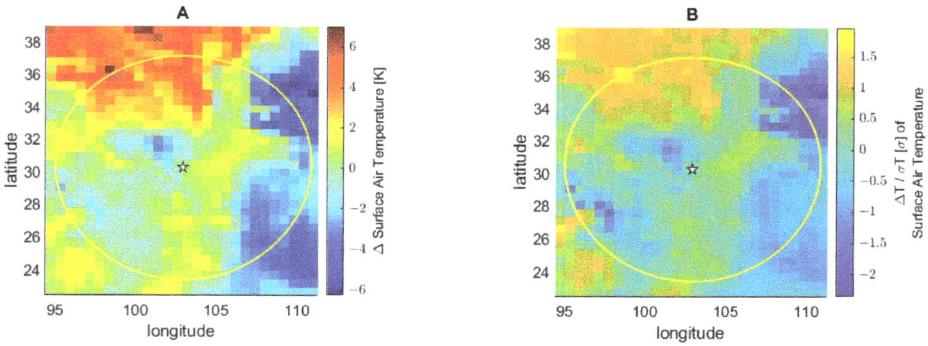

Figure 23. Maps of differences in surface air temperature, ΔT (**A**) and ΔT/σT (**B**) on 11 December 2012. The yellow circle represents the Dobrovosky radius of the Mw 6.7 Lushan 2013 earthquake. The white star indicates the epicenter of Mw 6.7 Lushan 2013 earthquake.

The almost synchronous increase in these quantities requires a fast-coupling mechanism. Due to the absence of surface air temperature anomalies, we would exclude the possibility of an acoustic gravity wave to describe this possible lithosphere–atmosphere–ionosphere coupling as proposed in other cases [34,58]. Similar synchronous anomalies in the atmosphere and ionosphere in coincidence with some foreshocks of the Mw = 7.5 Indonesia 2018 were reported by Marchetti et al. [12], even though the anticipation time was different (slightly less than two months before the Indonesia 2018 earthquake).

A second possible coupling between the lithosphere and atmosphere was identified about forty days before the earthquake. A strong increase in the Benioff cumulative stress curve and Es parameter happened 47~48 days before the seismic event. This was followed by an increase in temperature and surface total latent energy flux after a large emission of aerosol, SO_2, and, finally, geomagnetic anomalies about 40 days before the Lushan earthquake. We would underline that this possible coupling has a longer propagation delay than the previous case; in fact, it requires about a week, while the first eventual coupling involved all the layers on the same day, so this is about 7/8 times slower.

Comparing the results in the last few weeks before the earthquake, it is very interesting to note that seismicity recorded a peak on about 3 April 2013, followed, on 4 April (see Figure 17), by a high surface total energy flux, the day after by a decrease in humidity, then the atmosphere showed a high increase in aerosol from 8 to 10 April 2013, and, finally, the ionosphere TEC was high from 8 to 14 April 2013. As discussed in the Section 3, 10 April 2013 is high but not anomalous for aerosol and 13 April for TEC. Nevertheless, we could explain the detected increase in aerosol as a potential sign of air ionization that, in about two days, propagates in the ionosphere, inducing an increase in total electron content as predicted by the Pulinets and Ouzounov [30] model. Their model suggests that radon release induces air ionization if this is the case. Ye et al. [59] investigated radon concentration in water at six stations around the Longmenshan fault from 2003 to 2014, including the Wenchuan 2008 and Lushan 2013 earthquakes. The authors claimed that long-term variations in radon concentrations are associated with seismic trends, while short-term variations produced contradictory results. However, we notice that the Ganzi station showed high radon concentrations from about 1 March 2013 until the Lushan 2013 earthquake, followed by a gradual decrease immediately afterward [59]. A small peak was identified around 1 April 2013, just one week before the increase in aerosol and TEC that we identified, possibly explaining it. Unfortunately, we cannot draw a final conclusion due to the unconfirmed radon results from the other stations.

4.2. Space-Time Distribution of the Anomalies in Lithosphere Atmosphere and Ionosphere

In order to better show the identified anomalies or events in the lithosphere, atmosphere, and ionosphere, we provide a four-dimensional map (longitude, latitude, depth, and time) and a three-dimensional map (longitude, time, and latitude) in Figures 24 and 25. We divided the depth of the anomalies in the ionosphere by 10 to avoid large vertical distances in the figure. In both Figures 24 and 25, a triangle represents anomalies in the atmosphere; a square indicates anomalies in the ionosphere and geomagnetic ground data; dots depict anomalies in the lithosphere; and the Lushan earthquake is marked with a diamond. It could be seen that there are a large number of anomalies in three geo-layers clustered along the Longmenshan fault, which is the seismogenic fault zone of the Lushan earthquake. While this is expected for earthquakes in the lithosphere, it is remarkable for atmospheric and ionospheric anomalies. From Figure 24, it is also possible to depict a suggested space-time order of anomalies and interactions between the three geo-layers that we marked with ellipses and arrows and numbered from 1 to 4. In addition, we circled small areas where anomalies appeared in all three geo-layers in Figure 25.

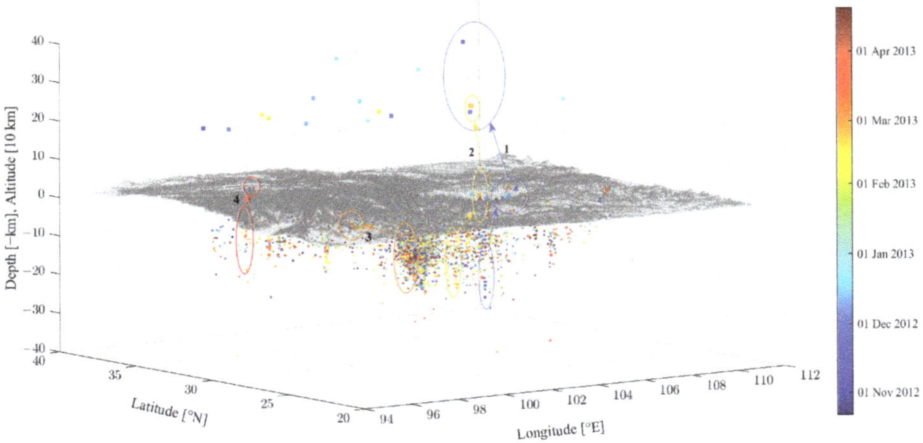

Figure 24. Four-dimensional map of the lithosphere, atmosphere, and ionosphere anomalies or events. The triangle represents anomalies in the atmosphere, the square represents anomalies in the ionosphere and geomagnetic ground data, and the dots represent anomalies in the lithosphere. The number from 1 to 4 with corresponding ellipses represents possible geophysical interactions in time-order of occurrence.

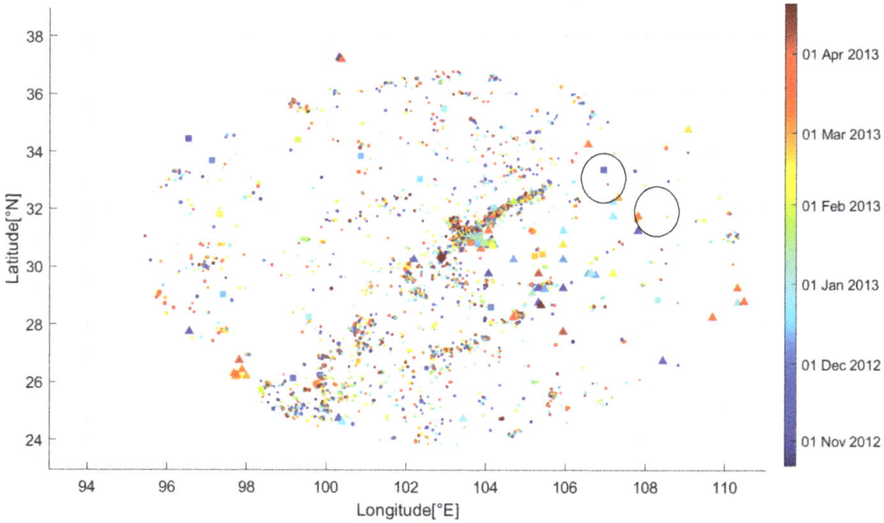

Figure 25. Three-dimensional map of anomalies or events in the lithosphere, atmosphere, and ionosphere. The diamond represents the Lushan earthquake.

4.3. Possible Explanation from the LAIC Models for the Identified Anomalies

The anomalies identified about 130 days before the Lushan earthquake could be explained with a degassing induced by fluid migrations as proposed for other earthquakes by Ventura and Di Giovambattista [60] together with a release of p-holes [36] at the Earth's surface. The release of p-holes can explain the anomalous pulsations recorded by the geomagnetic ground stations. Their accumulation could have induced the formation of plasma bubbles up to the ionosphere, as numerically simulated by Kuo et al. [61]. Such a simulation opened a scientific debate about its validity [62,63], but also other authors supported the theoretical existence of pre-earthquake ionospheric plasma irregularities [26]. At the same time, the degassing could explain the anomalous concentrations of SO_2 and CO in the atmosphere, which could have facilitated the formation of aerosol [30] and ionized the air increasing the free charges in the atmosphere and contributing to the eventual p-holes. Figure 26A presents a graphic sketch of these mechanisms.

A similar explanation could explain the possible coupling, identified about 40 days before the Lushan earthquake, probably without p-holes, due to the lack of evidence for ionospheric anomalies. The magnetic anomalies detected at the end of this possible coupling could be produced by electrical alteration of the atmosphere for a slower air ionization with respect to the previous coupling. We notice a crucial difference between the possible coupling 130 days and the one 40 days before the Lushan earthquake: the lithospheric source of the first seems to be a parallel fault located South-East of the first fault. Instead, the second group of anomalies seems to have been induced by the Longmenshan fault, which is responsible for the upcoming earthquake. This also shows how the anomalies could approach the upcoming event of interested fault as the event is closer in time, as proposed by Wu et al. [64].

The last group of anomalies starting about 2.5 weeks before the earthquakes could be well explained with a release of radon and following implications in the atmosphere and ionosphere proposed by the Pulinets and Ouzounov model [30] and represented in Figure 26B. The lithospheric source of these anomalies seems to be located Southwest on the same Longmenshan fault system; even though it was spatially not so close, it was the one responsible for the upcoming Lushan earthquake. The fact that the place of the upcoming earthquake was placed in the gap between the previous events is also well in agreement with other seismological simulations and investigations for other tectonic contexts [65,66].

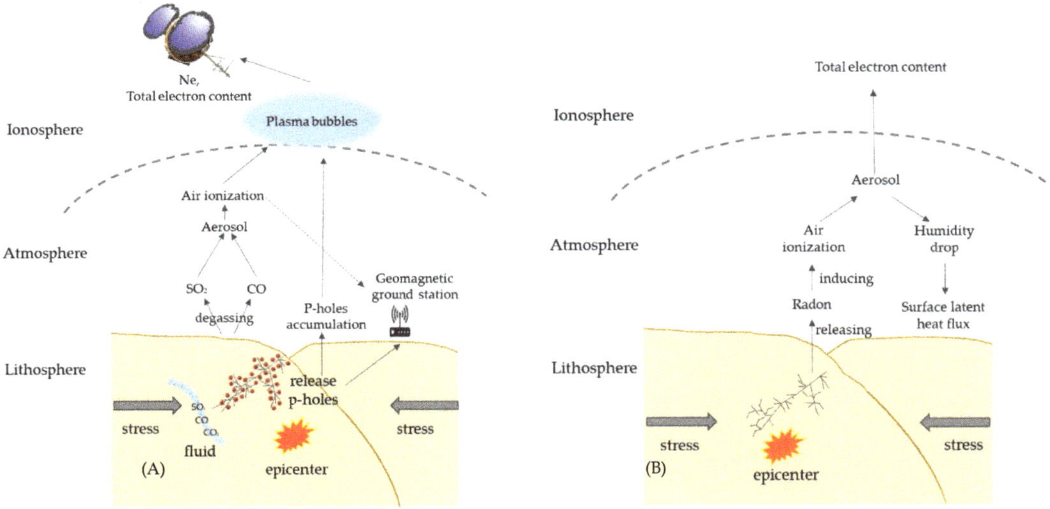

Figure 26. Graphic sketch of mechanisms. (**A**) Electromagnetic coupling driven by p-holes with also possible degassing process; (**B**) Chain of lithosphere atmosphere and ionosphere coupling (chemical-physical mechanisms) driven by radon release and air ionization.

5. Conclusions

In this paper, anomalies in the lithosphere, atmosphere, and ionosphere in the six months before the Mw6.7 Lushan (China) 2013 earthquake were extracted. In particular, increases in seismic stress were depicted 160, 92, 60, 48, 17, and 3 days before the Lushan earthquake. The geomagnetic ground station at Chengdu displayed anomalous signals not seen at the Beijing station, 140, 130, 100, and 35 days before the main shock. Surface air temperature, CO, Aerosol, and SO_2 levels, total latent energy flux, and specific humidity displayed several anomalies in the investigated period. The most interesting feature is their spatial organization, which converged toward the Longmenshan fault where the Lushan earthquake nucleated. The Formosat-3 satellite detected anomalies in the F2 ionospheric peak 170, 140, 130, and 110 days before the earthquake. TEC increased above the two standard deviations of its historical value from 12 to 6 days prior to the seismic event. The identified anomalies were further discussed, and possible couplings between the three geo-layers are presented in the view of the several lithosphere–atmosphere–ionosphere coupling models available in the literature.

We identify three possible bottom-up couplings with different mechanisms between three geo-layers. The first was about 130 days before the earthquake, with a fast (order of one day) propagation delay. The lithosphere presented a very slight increase in the stress, while the aerosol, sulfur dioxide, and carbon monoxide levels, and surface total latent energy flux were increased in the atmosphere; at the similar time, the Chengdu geomagnetic station recorded more magnetic pulses, and the electron density of the F2 ionospheric peak was higher.

The second happened about 40 days before the earthquake between the lithosphere and the atmosphere. A propagation delay of a few days suggests a slower bottom-up interaction between the two layers. A substantial stress increase occurred 47~48 days before the seismic event, followed by a rise in temperature and surface total latent energy flux after a large emission of aerosol, SO_2, and, finally, geomagnetic anomalies about 40 days before the Lushan earthquake.

The third started 2.5 weeks before, and lasted until one week before, the event. Consequently, this eventual coupling was also characterized by a longer interaction time from the bottom layers to the upper ones. In addition, all three geo-layers showed good bottom-up delay anomalies. Seismicity recorded a peak on about 3 April 2013; then the atmosphere depicted a high surface total energy flux on 4 April, followed the day after by a decrease in humidity, a high increase in aerosol from 8 to 10 April 2013, and, finally, the ionosphere TEC was high from 8 to 14 April 2013.

We can conclude that lithosphere–atmosphere–ionosphere couplings before Lushan (China) 2013 earthquake seemed to occur, but with different propagation mechanisms, locations (starting far from the involved fault and approaching it in the last few months), and delays. Further studies are necessary to confirm such possible mechanisms for other earthquakes in China and the rest of the world, considering the different tectonic and continental (as Lushan) or subduction contexts.

Our evidence suggests that the geo-layers could interact with different channels with specific propagation delays. Despite this, we cannot exclude alternative explanations of the identified anomalies, and future statistical studies applied to the lithosphere, atmosphere, and ionosphere are required to validate these results further.

We think that the "state of the art" is still insufficient to predict future earthquakes, but these studies help reduce the knowledge gap we need to implement in future prediction systems. The following step could be implementing a real-time geophysical platform monitoring system such as the one proposed in the conclusions provided in [67].

Supplementary Materials: The following are available online at https://www.mdpi.com/article/10.3390/rs15061521/s1. Figure S1: Geomagnetic ground observatory Z—vertical component; Figure S2: Correlation coefficients of Geomagnetic ground observatories; Figure S3: Cumulative graphs for geomagnetic ground observatories; Figure S4: Maps of the total latent energy flux on all the anomalous days, Figure S5: Linear and 2-order polynomial fits of TEC as a function of Sunspot numbers. Animation (gif file) of the 4 dimensional distributions of Figure 24 of earthquakes and identified anomalies (3 space and color for time).

Author Contributions: Conceptualization, methodology, software, writing—original draft preparation, D.M. and Y.Z.; validation, K.Z. and T.W.; formal analysis, investigation, data curation, Y.Z. and D.M.; visualization, Y.Z., D.M. and W.C.; resources, supervision, project administration, funding acquisition, K.Z. and D.M.; writing—review and editing Y.Z., K.Z., W.C., D.M., Y.C., M.F., T.W., S.W., J.W., D.Z. and H.Z. All authors have read and agreed to the published version of the manuscript.

Funding: This research was funded by the National Natural Science Foundation of China, grant number 41974084; the China Postdoctoral Science Foundation, grant number 2021M691190; the International Cooperation Project of the Department of Science and Technology of Jilin Province, grant number 20200801036GH.

Data Availability Statement: The authors would like to acknowledge the data support from the China Earthquake Networks Center, National Earthquake Data Center (http://data.earthquake.cn, last accessed on 20 September 2022, where we obtained the earthquake catalog and geomagnetic ground data of CPD). We thank International Real-time Magnetic Observatory Network for the open access to geomagnetic ground data of Beijing Ming Tombs (BMT) (intermagnet.github.io, last accessed on 12 September 2022). MERRA-2 data can be downloaded from https://disc.gsfc.nasa.gov/datasets?project=MERRA-2 (last accessed on 23 July 2022) with Earth Observation NASA free credential. We acknowledge the Taiwan Analysis Center for COSMIC for the free availability of FormoSat3 ionospheric profiles (ionPrf) at the web portal https://tacc.cwb.gov.tw/v2/en/fs3_download.html (last accessed on 7 February 2023). We acknowledge the International GNSS Service (IGS) for the work to provide GIM-MAP and maintain the archive where they can be freely downloaded: https://cddis.nasa.gov/archive/gnss/products/ionex/(last accessed on 7 February 2023).

Acknowledgments: We would acknowledge Alessandro Piscini, Rita Di Giovambattista, Guido Ventura, Gianfranco Cianchini, Saioa Arquero Campuzano, Dario Sabbagh, Loredana Perrone, Maurizio Soldani, and Angelo De Santis for their contributions in the preparation, writing, and optimization of some codes re-used in this work and constructive scientific discussions on previous work whose concepts are here applied again. D.M. also acknowledges the International Space Science Institute (ISSI at Bern, Switzerland) for supporting International Team 553 led by Essam Ghamry and Zeren Zhima.

Conflicts of Interest: The authors declare no conflict of interest. The funders had no role in the design of the study; in the collection, analyses, or interpretation of data; in the writing of the manuscript, or in the decision to publish the results.

References

1. Xie, Z.; Jin, B.; Zheng, Y.; Ge, C.; Xiong, X.; Xiong, C.; Hsu, H. Source Parameters Inversion of the 2013 Lushan Earthquake by Combining Teleseismic Waveforms and Local Seismograms. *Sci. China Earth Sci.* **2013**, *56*, 1177–1186. [CrossRef]
2. Zhao, G.; Unsworth, M.J.; Zhan, Y.; Wang, L.; Chen, X.; Jones, A.G.; Tang, J.; Xiao, Q.; Wang, J.; Cai, J.; et al. Crustal Structure and Rheology of the Longmenshan and Wenchuan Mw 7.9 Earthquake Epicentral Area from Magnetotelluric Data. *Geology* **2012**, *40*, 1139–1142. [CrossRef]
3. Geller, R.J. Earthquake Prediction: A Critical Review. *Geophys. J. Int.* **1997**, *131*, 425–450. [CrossRef]
4. Fidani, C.; Freund, F.; Grant, R. Cows Come Down from the Mountains before the (Mw = 6.1) Earthquake Colfiorito in September 1997; A Single Case Study. *Animals* **2014**, *4*, 292–312. [CrossRef] [PubMed]
5. Fraser-Smith, A.C.; Bernardi, A.; McGill, P.R.; Ladd, M.E.; Helliwell, R.A.; Villard, O.G. Low-Frequency Magnetic Field Measurements near the Epicenter of the M_S 7.1 Loma Prieta Earthquake. *Geophys. Res. Lett.* **1990**, *17*, 1465–1468. [CrossRef]
6. Piscini, A.; De Santis, A.; Marchetti, D.; Cianchini, G. A Multi-Parametric Climatological Approach to Study the 2016 Amatrice–Norcia (Central Italy) Earthquake Preparatory Phase. *Pure Appl. Geophys.* **2017**, *174*, 3673–3688. [CrossRef]
7. Chen, H.; Han, P.; Hattori, K. Recent Advances and Challenges in the Seismo-Electromagnetic Study: A Brief Review. *Remote Sens.* **2022**, *14*, 5893. [CrossRef]
8. Han, P.; Hattori, K.; Huang, Q.; Hirooka, S.; Yoshino, C. Spatiotemporal Characteristics of the Geomagnetic Diurnal Variation Anomalies Prior to the 2011 Tohoku Earthquake (Mw 9.0) and the Possible Coupling of Multiple Pre-Earthquake Phenomena. *J. Asian Earth Sci.* **2016**, *129*, 13–21. [CrossRef]
9. Ouzounov, D.; Pulinets, S.; Davidenko, D.; Rozhnoi, A.; Solovieva, M.; Fedun, V.; Dwivedi, B.N.; Rybin, A.; Kafatos, M.; Taylor, P. Transient Effects in Atmosphere and Ionosphere Preceding the 2015 M7.8 and M7.3 Gorkha–Nepal Earthquakes. *Front. Earth Sci.* **2021**, *9*, 757358. [CrossRef]
10. Fan, M.; Zhu, K.; De Santis, A.; Marchetti, D.; Cianchini, G.; Piscini, A.; He, X.; Wen, J.; Wang, T.; Zhang, Y.; et al. Analysis of Swarm Satellite Magnetic Field Data for the 2015 Mw 7.8 Nepal Earthquake Based on Nonnegative Tensor Decomposition. *IEEE Trans. Geosci. Remote Sens.* **2022**, *60*, 2006119. [CrossRef]
11. Marchetti, D.; De Santis, A.; D'Arcangelo, S.; Poggio, F.; Piscini, A.; Campuzano, S.A.; De Carvalho, W.V.J.O. Pre-Earthquake Chain Processes Detected from Ground to Satellite Altitude in Preparation of the 2016–2017 Seismic Sequence in Central Italy. *Remote Sens. Environ.* **2019**, *229*, 93–99. [CrossRef]
12. Marchetti, D.; De Santis, A.; Shen, X.; Campuzano, S.A.; Perrone, L.; Piscini, A.; Di Giovambattista, R.; Jin, S.; Ippolito, A.; Cianchini, G.; et al. Possible Lithosphere-Atmosphere-Ionosphere Coupling Effects Prior to the 2018 Mw = 7.5 Indonesia Earthquake from Seismic, Atmospheric and Ionospheric Data. *J. Asian Earth Sci.* **2020**, *188*, 104097. [CrossRef]
13. De Santis, A.; Cianchini, G.; Marchetti, D.; Piscini, A.; Sabbagh, D.; Perrone, L.; Campuzano, S.A.; Inan, S. A Multiparametric Approach to Study the Preparation Phase of the 2019 M7.1 Ridgecrest (California, United States) Earthquake. *Front. Earth Sci.* **2020**, *8*, 540398. [CrossRef]
14. De Santis, A.; Perrone, L.; Calcara, M.; Campuzano, S.A.; Cianchini, G.; D'Arcangelo, S.; Di Mauro, D.; Marchetti, D.; Nardi, A.; Orlando, M.; et al. A Comprehensive Multiparametric and Multilayer Approach to Study the Preparation Phase of Large Earthquakes from Ground to Space: The Case Study of the June 15 2019, M7.2 Kermadec Islands (New Zealand) Earthquake. *Remote Sens. Environ.* **2022**, *283*, 113325. [CrossRef]
15. Marchetti, D.; Zhu, K.; Zhang, H.; Zhima, Z.; Yan, R.; Shen, X.; Chen, W.; Cheng, Y.; He, X.; Wang, T.; et al. Clues of Lithosphere, Atmosphere and Ionosphere Variations Possibly Related to the Preparation of La Palma 19 September 2021 Volcano Eruption. *Remote Sens.* **2022**, *14*, 5001. [CrossRef]
16. Marchetti, D.; Zhu, K.; Yan, R.; ZeRen, Z.; Shen, X.; Chen, W.; Cheng, Y.; Fan, M.; Wang, T.; Wen, J.; et al. Ionospheric Effects of Natural Hazards in Geophysics: From Single Examples to Statistical Studies Applied to M5.5+ Earthquakes. In Proceedings of the 4th International Electronic Conference on Geosciences, Online, 1–15 December 2022; MDPI: Basel, Switzerland, 2022.
17. D'Arcangelo, S.; Bonforte, A.; De Santis, A.; Maugeri, S.R.; Perrone, L.; Soldani, M.; Arena, G.; Brogi, F.; Calcara, M.; Campuzano, S.A.; et al. A Multi-Parametric and Multi-Layer Study to Investigate the Largest 2022 Hunga Tonga–Hunga Ha'apai Eruptions. *Remote Sens.* **2022**, *14*, 3649. [CrossRef]
18. Hattori, K.; Han, P.; Yoshino, C.; Febriani, F.; Yamaguchi, H.; Chen, C.-H. Investigation of ULF Seismo-Magnetic Phenomena in Kanto, Japan During 2000–2010: Case Studies and Statistical Studies. *Surv. Geophys.* **2013**, *34*, 293–316. [CrossRef]
19. Genzano, N.; Filizzola, C.; Hattori, K.; Pergola, N.; Tramutoli, V. Statistical Correlation Analysis Between Thermal Infrared Anomalies Observed From MTSATs and Large Earthquakes Occurred in Japan (2005–2015). *J. Geophys. Res. Solid Earth* **2021**, *126*, e2020JB020108. [CrossRef]
20. Yan, R.; Parrot, M.; Pinçon, J.-L. Statistical Study on Variations of the Ionospheric Ion Density Observed by DEMETER and Related to Seismic Activities: Ionospheric Density and Seismic Activity. *J. Geophys. Res. Space Phys.* **2017**, *122*, 12421–12429. [CrossRef]
21. Ouyang, X.Y.; Parrot, M.; Bortnik, J. ULF Wave Activity Observed in the Nighttime Ionosphere Above and Some Hours Before Strong Earthquakes. *J. Geophys. Res. Space Phys.* **2020**, *125*, e2020JA028396. [CrossRef]

22. De Santis, A.; Marchetti, D.; Pavón-Carrasco, F.J.; Cianchini, G.; Perrone, L.; Abbattista, C.; Alfonsi, L.; Amoruso, L.; Campuzano, S.A.; Carbone, M.; et al. Precursory Worldwide Signatures of Earthquake Occurrences on Swarm Satellite Data. *Sci. Rep.* **2019**, *9*, 20287. [CrossRef] [PubMed]
23. De Santis, A.; Marchetti, D.; Perrone, L.; Campuzano, S.A.; Cianchini, G.; Cesaroni, C.; Di Mauro, D.; Orlando, M.; Piscini, A.; Sabbagh, D.; et al. Statistical Correlation Analysis of Strong Earthquakes and Ionospheric Electron Density Anomalies as Observed by CSES-01. *Il Nuovo Cim. C* **2021**, *44*, 1–4. [CrossRef]
24. He, Y.; Zhao, X.; Yang, D.; Wu, Y.; Li, Q. A Study to Investigate the Relationship between Ionospheric Disturbance and Seismic Activity Based on Swarm Satellite Data. *Phys. Earth Planet. Inter.* **2022**, *323*, 106826. [CrossRef]
25. Marchetti, D.; De Santis, A.; Campuzano, S.A.; Zhu, K.; Soldani, M.; D'Arcangelo, S.; Orlando, M.; Wang, T.; Cianchini, G.; Di Mauro, D.; et al. Worldwide Statistical Correlation of Eight Years of Swarm Satellite Data with M5.5+ Earthquakes: New Hints about the Preseismic Phenomena from Space. *Remote Sens.* **2022**, *14*, 2649. [CrossRef]
26. Liperovsky, V.A.; Pokhotelov, O.A.; Meister, C.-V.; Liperovskaya, E.V. Physical Models of Coupling in the Lithosphere-Atmosphere-Ionosphere System before Earthquakes. *Geomagn. Aeron.* **2008**, *48*, 795–806. [CrossRef]
27. Molchanov, O.A.; Hayakawa, M. Generation of ULF Electromagnetic Emissions by Microfracturing. *Geophys. Res. Lett.* **1995**, *22*, 3091–3094. [CrossRef]
28. Molchanov, O.A.; Hayakawa, M. On the Generation Mechanism of ULF Seismogenic Electromagnetic Emissions. *Phys. Earth Planet. Inter.* **1998**, *105*, 201–210. [CrossRef]
29. Enomoto, Y. Coupled Interaction of Earthquake Nucleation with Deep Earth Gases: A Possible Mechanism for Seismo-Electromagnetic Phenomena. *Geophys. J. Int.* **2012**, *191*, 1210–1214. [CrossRef]
30. Pulinets, S.; Ouzounov, D. Lithosphere–Atmosphere–Ionosphere Coupling (LAIC) Model–An Unified Concept for Earthquake Precursors Validation. *J. Asian Earth Sci.* **2011**, *41*, 371–382. [CrossRef]
31. Pulinets, S.; Khachikyan, G. The Global Electric Circuit and Global Seismicity. *Geosciences* **2021**, *11*, 491. [CrossRef]
32. Pulinets, S.; Ouzounov, D.; Karelin, A.; Boyarchuk, K. *Earthquake Precursors in the Atmosphere and Ionosphere: New Concepts*; Springer: Dordrecht, The Netherlands, 2022; ISBN 978-94-024-2170-5.
33. Pulinets, S.; Ouzounov, D.; Karelin, A.; Davidenko, D. Lithosphere-Atmosphere-Ionosphere-Magnetosphere Coupling-A Concept for Pre-Earthquake Signals Generation. In *Pre-Earthquake Processes: A Multidisciplinary Approach to Earthquake Prediction Studies*; Geophysical Monograph Series; Ouzounov, D., Pulinets, S., Hattori, K., Taylor, P., Eds.; John Wiley & Sons, Inc.: Hoboken, NJ, USA, 2018; pp. 77–98. ISBN 978-1-119-15694-9.
34. Hayakawa, M.; Kasahara, Y.; Nakamura, T.; Hobara, Y.; Rozhnoi, A.; Solovieva, M.; Molchanov, O.; Korepanov, V. Atmospheric Gravity Waves as a Possible Candidate for Seismo-Ionospheric Perturbations. *J. Atmospheric Electr.* **2011**, *31*, 129–140. [CrossRef]
35. Freund, F. Pre-Earthquake Signals: Underlying Physical Processes. *J. Asian Earth Sci.* **2011**, *41*, 383–400. [CrossRef]
36. Freund, F.; Ouillon, G.; Scoville, J.; Sornette, D. Earthquake Precursors in the Light of Peroxy Defects Theory: Critical Review of Systematic Observations. *Eur. Phys. J. Spec. Top.* **2021**, *230*, 7–46. [CrossRef]
37. Han, P.; Hattori, K.; Hirokawa, M.; Zhuang, J.; Chen, C.-H.; Febriani, F.; Yamaguchi, H.; Yoshino, C.; Liu, J.-Y.; Yoshida, S. Statistical Analysis of ULF Seismomagnetic Phenomena at Kakioka, Japan, during 2001–2010: ULF seismo-magnetic phenomena at Kakioka. *J. Geophys. Res. Space Phys.* **2014**, *119*, 4998–5011. [CrossRef]
38. Dobrovolsky, I.P.; Zubkov, S.I.; Miachkin, V.I. Estimation of the Size of Earthquake Preparation Zones. *Pure Appl. Geophys.* **1979**, *117*, 1025–1044. [CrossRef]
39. Wiemer, S. A Software Package to Analyze Seismicity: ZMAP. *Seismol. Res. Lett.* **2001**, *72*, 373–382. [CrossRef]
40. Wiemer, S. Minimum Magnitude of Completeness in Earthquake Catalogs: Examples from Alaska, the Western United States, and Japan. *Bull. Seismol. Soc. Am.* **2000**, *90*, 859–869. [CrossRef]
41. Mignan, A. The Stress Accumulation Model: Accelerating Moment Release and Seismic Hazard. In *Advances in Geophysics*; Elsevier: Amsterdam, The Netherlands, 2008; Volume 49, pp. 67–201. ISBN 978-0-12-374231-5.
42. Zhuang, J.; Vere-Jones, D.; Guan, H.; Ogata, Y.; Ma, L. Preliminary Analysis of Observations on the Ultra-Low Frequency Electric Field in the Beijing Region. *Pure Appl. Geophys.* **2005**, *162*, 1367–1396. [CrossRef]
43. Fritsch, F.N.; Carlson, R.E. Monotone Piecewise Cubic Interpolation. *SIAM J. Numer. Anal.* **1980**, *17*, 238–246. [CrossRef]
44. Alken, P.; Thébault, E.; Beggan, C.D.; Amit, H.; Aubert, J.; Baerenzung, J.; Bondar, T.N.; Brown, W.J.; Califf, S.; Chambodut, A.; et al. International Geomagnetic Reference Field: The Thirteenth Generation. *Earth Planets Space* **2021**, *73*, 49. [CrossRef]
45. Akhoondzadeh, M.; De Santis, A.; Marchetti, D.; Piscini, A.; Cianchini, G. Multi Precursors Analysis Associated with the Powerful Ecuador (MW= 7.8) Earthquake of 16 April 2016 Using Swarm Satellites Data in Conjunction with Other Multi-Platform Satellite and Ground Data. *Adv. Space Res.* **2018**, *61*, 248–263. [CrossRef]
46. Akhoondzadeh, M.; De Santis, A.; Marchetti, D.; Piscini, A.; Jin, S. Anomalous Seismo-LAI Variations Potentially Associated with the 2017 Mw = 7.3 Sarpol-e Zahab (Iran) Earthquake from Swarm Satellites, GPS-TEC and Climatological Data. *Adv. Space Res.* **2019**, *64*, 143–158. [CrossRef]
47. Akhoondzadeh, M.; De Santis, A.; Marchetti, D.; Wang, T. Developing a Deep Learning-Based Detector of Magnetic, Ne, Te and TEC Anomalies from Swarm Satellites: The Case of Mw 7.1 2021 Japan Earthquake. *Remote Sens.* **2022**, *14*, 1582. [CrossRef]
48. Gelaro, R.; McCarty, W.; Suárez, M.J.; Todling, R.; Molod, A.; Takacs, L.; Randles, C.A.; Darmenov, A.; Bosilovich, M.G.; Reichle, R.; et al. The Modern-Era Retrospective Analysis for Research and Applications, Version 2 (MERRA-2). *J. Clim.* **2017**, *30*, 5419–5454. [CrossRef] [PubMed]

49. Piscini, A.; Marchetti, D.; De Santis, A. Multi-Parametric Climatological Analysis Associated with Global Significant Volcanic Eruptions During 2002–2017. *Pure Appl. Geophys.* **2019**, *176*, 3629–3647. [CrossRef]
50. Shi, K.; Guo, J.; Zhang, Y.; Li, W.; Kong, Q.; Yu, T. Multi-Dimension and Multi-Channel Seismic-Ionospheric Coupling: Case Study of Mw 8.8 Concepcion Quake on 27 February 2010. *Remote Sens.* **2021**, *13*, 2724. [CrossRef]
51. Liu, J.-Y.; Chao, C.-K. An Observing System Simulation Experiment for FORMOSAT-5/AIP Detecting Seismo-Ionospheric Precursors. *Terr. Atmos. Ocean. Sci.* **2017**, *28*, 117–127. [CrossRef]
52. Perrone, L.; Korsunova, L.P.; Mikhailov, A.V. Ionospheric Precursors for Crustal Earthquakes in Italy. *Ann. Geophys.* **2010**, *28*, 941–950. [CrossRef]
53. Bogdanov, V.; Gavrilov, V.; Pulinets, S.; Ouzounov, D. Responses to the Preparation of Strong Kamchatka Earth-Quakes in the Lithosphere–Atmosphere–Ionosphere System, Based on New Data from Integrated Ground and Iono-Spheric Monitoring. *E3S Web Conf.* **2020**, *196*, 03005. [CrossRef]
54. Pedatella, N.M.; Yue, X.; Schreiner, W.S. An Improved Inversion for FORMOSAT-3/COSMIC Ionosphere Electron Density Profiles. *J. Geophys. Res. Space Phys.* **2015**, *120*, 8942–8953. [CrossRef]
55. Bilitza, D.; Altadill, D.; Truhlik, V.; Shubin, V.; Galkin, I.; Reinisch, B.; Huang, X. International Reference Ionosphere 2016: From Ionospheric Climate to Real-time Weather Predictions. *Space Weather.* **2017**, *15*, 418–429. [CrossRef]
56. Liu, Q.; Shen, X.; Zhang, J.; Li, M. Exploring the Abnormal Fluctuations of Atmospheric Aerosols before the 2008 Wenchuan and 2013 Lushan Earthquakes. *Adv. Space Res.* **2019**, *63*, 3768–3776. [CrossRef]
57. Dai, X.; Liu, J.; Zhang, H. Application of AR Model in the Analysis of Preearthquake Ionospheric Anomalies. *Math. Probl. Eng.* **2015**, *2015*, 157184. [CrossRef]
58. Sasmal, S.; Chowdhury, S.; Kundu, S.; Politis, D.Z.; Potirakis, S.M.; Balasis, G.; Hayakawa, M.; Chakrabarti, S.K. Pre-Seismic Irregularities during the 2020 Samos (Greece) Earthquake (M = 6.9) as Investigated from Multi-Parameter Approach by Ground and Space-Based Techniques. *Atmosphere* **2021**, *12*, 1059. [CrossRef]
59. Ye, Q.; Singh, R.P.; He, A.; Ji, S.; Liu, C. Characteristic Behavior of Water Radon Associated with Wenchuan and Lushan Earthquakes along Longmenshan Fault. *Radiat. Meas.* **2015**, *76*, 44–53. [CrossRef]
60. Ventura, G.; Di Giovambattista, R. Fluid Pressure, Stress Field and Propagation Style of Coalescing Thrusts from the Analysis of the 20 May 2012 M_L 5.9 Emilia Earthquake (Northern Apennines, Italy): Propagation Style of Coalescing Thrusts. *Terra Nova* **2013**, *25*, 72–78. [CrossRef]
61. Kuo, C.L.; Lee, L.C.; Huba, J.D. An Improved Coupling Model for the Lithosphere-Atmosphere-Ionosphere System. *J. Geophys. Res. Space Phys.* **2014**, *119*, 3189–3205. [CrossRef]
62. Prokhorov, B.E.; Zolotov, O.V. Comment on "An Improved Coupling Model for the Lithosphere-Atmosphere-Ionosphere System" by Kuo et al. [2014]: COMMENTS ON "AN IMPROVED COUPLING...". *J. Geophys. Res. Space Phys.* **2017**, *122*, 4865–4868. [CrossRef]
63. Kuo, C.-L.; Lee, L.-C. Reply to Comment by B. E. Prokhorov and O. V. Zolotov on "An Improved Coupling Model for the Lithosphere-Atmosphere-Ionosphere System": Reply to Comment. *J. Geophys. Res. Space Phys.* **2017**, *122*, 4869–4874. [CrossRef]
64. Wu, L.; Qi, Y.; Mao, W.; Lu, J.; Ding, Y.; Peng, B.; Xie, B. Scrutinizing and Rooting the Multiple Anomalies of Nepal Earthquake Sequence in 2015 with the Deviation–Time–Space Criterion and Homologous Lithosphere–Coversphere–Atmosphere–Ionosphere Coupling Physics. *Nat. Hazards Earth Syst. Sci.* **2023**, *23*, 231–249. [CrossRef]
65. Mogi, K. Sequential Occurrences of Recent Great Earthquakes. *J. Phys. Earth* **1968**, *16*, 30–36. [CrossRef]
66. Console, R.; Carluccio, R.; Papadimitriou, E.; Karakostas, V. Synthetic Earthquake Catalogs Simulating Seismic Activity in the Corinth Gulf, Greece, Fault System: Corinth Earthquakes Simulations. *J. Geophys. Res. Solid Earth* **2015**, *120*, 326–343. [CrossRef]
67. Marchetti, D.; Zhu, K.; Marchetti, L.; Zhang, Y.; Chen, W.; Cheng, Y.; Fan, M.; Wang, S.; Wang, T.; Wen, J.; et al. Quick Report on the ML = 3.3 on 1 January 2023 Guidonia (Rome, Italy) Earthquake: Evidence of a Seismic Acceleration. *Remote Sens.* **2023**, *15*, 942. [CrossRef]

remote sensing

MDPI

Technical Note

Developing a Fuzzy Inference System Based on Multi-Sensor Data to Predict Powerful Earthquake Parameters

Mehdi Akhoondzadeh [1] and Dedalo Marchetti [2,*]

1 Photogrammetry and Remote Sensing Department, School of Surveying and Geospatial Engineering, College of Engineering, University of Tehran, North Amirabad Ave., Tehran 1417614411, Iran; makhonz@ut.ac.ir
2 College of Instrumentation and Electrical Engineering, Jilin University, Changchun 130061, China
* Correspondence: dedalomarchetti@jlu.edu.cn

Abstract: Predicting the parameters of upcoming earthquakes has always been one of the most challenging topics in studies related to earthquake precursors. Increasing the number of sensors and satellites and consequently incrementing the number of observable possible earthquake precursors in different layers of the lithosphere, atmosphere, and ionosphere of the Earth has opened the possibility of using data fusion methods to estimate and predict earthquake parameters with low uncertainty. In this study, a Mamdani fuzzy inference system (FIS) was proposed and implemented in five case studies. In particular, the magnitude of Ecuador (16 April 2016), Iran (12 November 2017), Papua New Guinea (14 May 2019), Japan (13 February 2021), and Haiti (14 August 2021) earthquakes were estimated by FIS. The results showed that in most cases, the highest number of anomalies was usually observed in the period of about one month before the earthquake and the predicted magnitude of the earthquake in these periods was slightly different from the actual magnitude value. Therefore, based on the results of this study, it could be concluded that if a significant number of anomalies are observed in the time series of different precursors, it is likely that an earthquake of the magnitude predicted by the proposed FIS system within the Dobrovolsky area of the studied location will happen during the next month.

Keywords: earthquake precursor; fuzzy inference system (FIS); earthquake magnitude; swarm satellite data

Citation: Akhoondzadeh, M.; Marchetti, D. Developing a Fuzzy Inference System Based on Multi-Sensor Data to Predict Powerful Earthquake Parameters. *Remote Sens.* **2022**, *14*, 3203. https://doi.org/10.3390/rs14133203

Academic Editor: José Fernández

Received: 13 June 2022
Accepted: 1 July 2022
Published: 4 July 2022

1. Introduction

Due to the loss of life and heavy material damage of powerful earthquakes, many efforts have been made to predict the principal parameters of earthquakes, i.e., at least the date, magnitude, and location. However, this goal seems still far, and nowadays, it is still not possible to make a prediction of an earthquake. To be able to better understand the preparation phase of an earthquake, several studies have been conducted for decades. For example, in interpreting several possible precursors, a possible general model called "Dilatancy" was proposed by Scholz et al. [1]. For this purpose, any abnormal change in the physical and chemical observables in different layers of the Earth (lithosphere, atmosphere, and ionosphere) in the absence of other causes can probably be considered as one of the possible signs of an impending earthquake, and hence, it can be classified as an earthquake precursor candidate [2–5]. Therefore, for several decades, different parameters have been proposed to be altered before the occurrence of an earthquake: for example, Wyss [6] identified seismic, ground deformation, and magnetic alteration by investigating a very long time series before the M7.1 Sitka 1972 earthquake; Fraser-Smith et al. [7] claimed that extremely low frequency (ELF)/very low frequency (VLF) anomalies were detected from ground observatories some weeks before the M7.1 Loma Prieta earthquake occurred in 1989; Molchanov et al. [8] reported disturbances in VLF signal transmissions before the M7.2 Kobe 1995 earthquake. Wu and Tikhonov [9] searched for a systematic occurrence of atmospheric jet streams in China and the northwestern part of the Pacific seismic belt before the M6+ earthquakes from 2006 to 2021. Shou and Fang [10] formulated a model proposing that earthquakes can release thermal vapor (geoeruption), inducing the formation of clouds. Such "earthquake clouds" can be distinguished from other geological or weather phenomena as they are supposed to suddenly appear, being vapor-based, having a fixed source, and displaying a high-temperature and high-pressure state. The location of geoeruption would predict the epicentre, the amount of vapor, the magnitude, and the time after a complete geoeruption empirically could predict the time of the earthquake within days of precision [10]. With the launch of various satellites with different sensors such as optical and radar as well as accelerometers, magnetometers, electric field detectors, particle tracers, calorimeters, etc., the number of readable precursors for earthquakes has increased. The large availability of big data archives has permitted researchers to identify a chain of processes in the lithosphere, the atmosphere, and the ionosphere before the occurrence of several earthquakes in the world, such as M7.8 Nepal 2015 [11], the M6.0 and 6.5 Italy 2016 seismic sequence [12], the M7.5 Indonesia 2018 earthquake [13], and the M7.1 Ridgecrest, California 2019 earthquake [14]. Statistical proof of the existence of ionospheric precursors analysing DEMETER and Swarm satellites has also prompted some worldwide investigations of M5.5+ (or M4.8+ in the case of DEMETER) shallow earthquakes [15–19]. Creating cloud computing systems such as Google Earth Engine (GEE) has greatly helped to analyse various precursors without the necessity of downloading their corresponding raw data [20]. It can be reasonable that as the number of anomalies observed in different precursors increases, the degree of uncertainty in predicting the parameters of upcoming earthquakes decreases. An investigation of 12 earthquakes from magnitude 6.1 to 8.3 that occurred between 2014 and 2016 identified a linear relationship between ionospheric anomalies identified by Swarm and the magnitude of the incoming earthquake [21]. Fortunately, the number of classical and intelligent anomaly detection methods in the time series of various precursors has improved considerably [22,23]. Therefore, it is necessary to use methods to fuse the results of these predictors to estimate earthquake parameters. In this study, a fuzzy inference system (FIS) was proposed to estimate the magnitude of the impending powerful earthquakes.

Case Studies

In particular, we investigated here five strong earthquakes to evaluate the proposed fuzzy system for earthquake magnitude estimation. Table 1 summarises the real characteristics of the five analysed earthquakes and Figure 1 shows their location and their focal mechanism. A powerful earthquake of magnitude Mw = 7.8 occurred on 16 April 2016 at 23:58:36 UTC (LT = UTC + $\lambda_{epicenter}/15$ = 18:38:55) on the coast of Ecuador, approximately 27 km south-southeast of Muisne in the province of Esmeraldas at a depth of 20.6 km [24]. We also analysed as another case study an M_w = 7.3 strong earthquake that took place at 18:18:17 UTC (LT = 21:22:07) on 12 November 2017 along the border region between Iran and Iraq close to the town of Sarpol-e Zahab (34.911°N, 45.959°E, 19.00 km depth) [25]. The third analysed earthquake had a magnitude M_w = 7.6 and happened at 12:58:25 UTC (LT = 23:08:48) on 14 May 2019, 46 km South-Southeast of Namatanai in Papua New Guinea (4.051°S, 152.597°E) at a shallow estimated depth of about 10 km [26]. The fourth strong earthquake of M_w = 7.1 occurred at 14:07:50 UTC (LT = 23:34:55) on 13 February 2021 near the east coast of Honshu, Japan (37.727°N, 141.775°E, 44 km depth) as the result of thrust faulting near the subduction zone interface plate boundary between the Pacific and North America plates [23]. We analysed another strong earthquake of M_w = 7.2 that happened at 12:29:08 UTC (LT = 7:35:12) on 14 August 2021 (18.434°N, 73.482°W, 10 km depth).

Table 1. The characteristics of the investigated earthquakes (reported by United States Geological Survey, http://earthquake.usgs.gov/ (accessed on 5 June 2022)).

Region	Date	Time (UTC)	Geographic Latitude, Longitude	Magnitude (M_W)	Focal Depth (km)
Ecuador	16 April 2016	23:58:36	0.382°N, 79.922°W	7.8	20.6
Iran	12 November 2017	18:18:17	34.911°N, 45.959°E	7.3	19
Papua New Guinea	14 May 2019	12:58:25	4.051°S, 152.597°E	7.6	10
Japan	13 February 2021	14:07:50	37.73°N, 141.77°E	7.1	44
Haiti	14 August 2021	12:29:08	18.434°N, 73.482°W	7.2	10

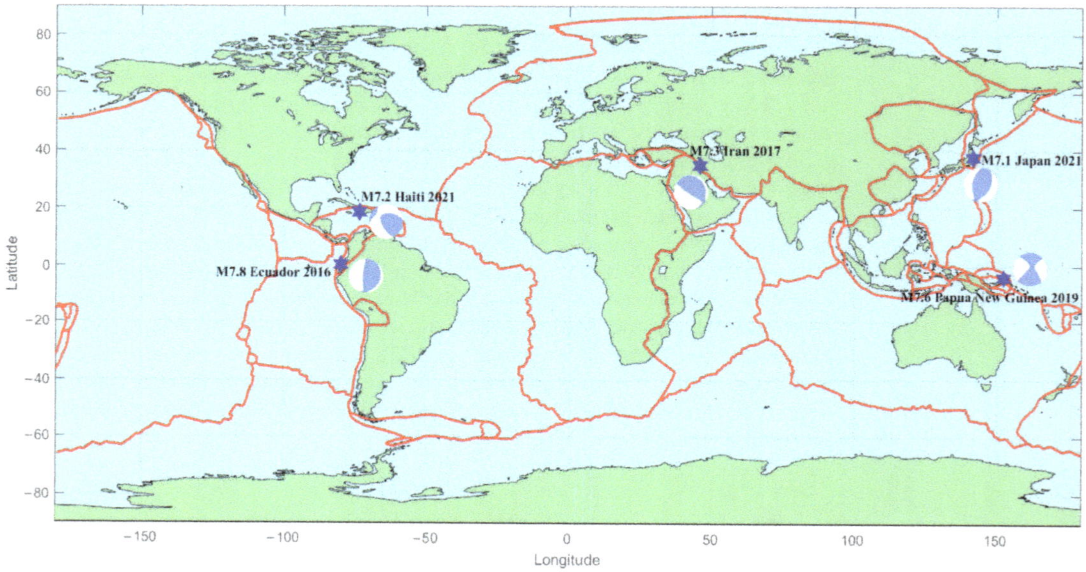

Figure 1. Map of the investigated earthquakes with overplot of main plate boundaries (red lines). For each seismic event, the "beachball" with the estimated focal mechanism solution from USGS is reported.

2. Methodology

Fuzzy logic is a set of mathematical principles for knowledge representation based on degrees of membership. Unlike two-valued Boolean logic, fuzzy logic is multi-valued. It deals with degrees of membership and degrees of truth [27]. Fuzzy logic uses a continuum of logical values between 0 (completely false) and 1 (completely true). In 1965, Lotfi Zadeh published his famous paper "Fuzzy sets" [28]. Zadeh extended the work on possibility theory into a formal system of mathematical logic and introduced a new concept for applying natural language terms. This new logic for representing and manipulating fuzzy terms was called fuzzy logic [27].

A fuzzy inference system (FIS) is a way of mapping an input space to an output space using fuzzy logic. An FIS tries to formalise the reasoning process of the human language by means of fuzzy logic (that is, by building fuzzy IF-THEN rules). FIS methods are used to solve decision problems, i.e., to make a decision and act accordingly. The FIS is based on converting a set of numerical variables (i.e., crisp) to a set of fuzzy variables. This is obtained using a series of logical rules and linguistic variables. These rules are organised in the "IF-THEN" format and constitute the FIS's core. Before applying these rules, the input and output variables must be fuzzified, and the conversion is conducted using membership functions. Membership functions are mathematical functions that illustrate an element's membership in a fuzzy set [29].

The structure of an FIS is as follows (see also Figure 2):

- Fuzzification module: transforms the system inputs, which are crisp numbers, into fuzzy sets. This is done by applying a fuzzification function.
- Knowledge base: stores IF-THEN rules provided by experts.
- Inference engine: simulates the human reasoning process by making fuzzy inferences on the inputs and IF-THEN rules.
- Defuzzification module: transforms the fuzzy set obtained by the inference engine into a crisp value.

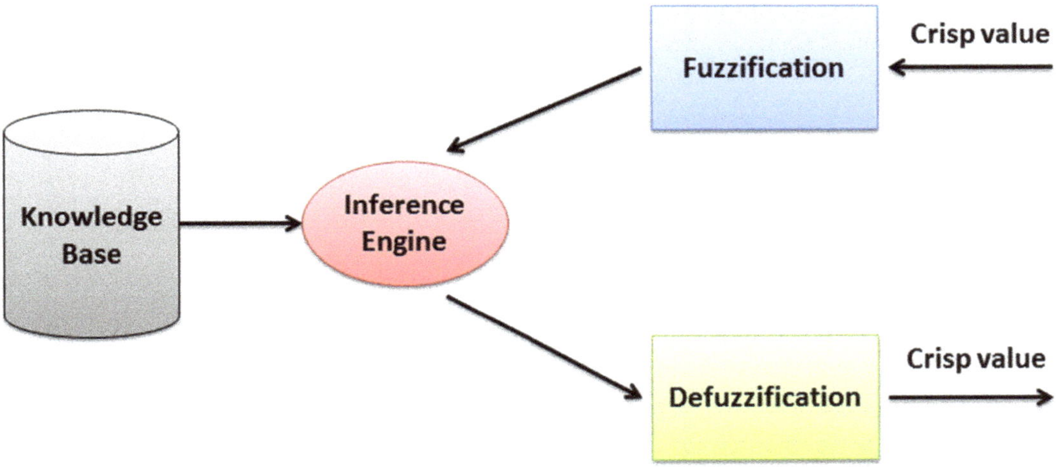

Figure 2. Structure of a fuzzy inference system.

Fuzzy inference methods are classified into direct methods and indirect methods. Direct methods, such as Mamdani's approach, are the most commonly used. In 1974, British E.H. Mamdani applied fuzzy logic and fuzzy reasoning to real life for the first time [30].

The realisation process of the Mamdani FIS flow chart in this study is shown in Figure 3. The choice of the membership function for input variables depends on user knowledge about the system behaviour. The Gaussian membership function "gaussmf" is applied to map the daily value of every precursor into [0 ... 1] numerical range.

It should be noted that for each precursor time series, the median and the inter-quartile range of data are calculated to construct their upper and lower bound in order to separate potentially seismic anomalies from the background of natural variations. The upper and lower bounds of the mentioned range can be calculated using the following equations:

$$x_{high} = m + k \times iqr \quad (1)$$

$$x_{low} = m - k \times iqr \quad (2)$$

$$Dx = (\left(\frac{x - m}{iqr}\right) - 1) * 100 \quad (3)$$

where x, x_{high}, x_{low}, m, iqr, and Dx are the parameter value, upper bound, lower bound, median value, inter-quartile range, and differential of x, respectively. It should be noted that the Dx value is considered the input value in FIS. Therefore, the input (precursor) ranges from 0 to 100. It is clear that the number of input variables is equal to the number of analysed precursors for each case study. The range of input variables for each precursor is divided into four parts corresponding to semantic subsets: weak, moderate, intense, and very intense (Figure 4). The Gaussian membership function is also applied for the earthquake magnitude output value. The predicted earthquake magnitude ranges from 0 to 9. This range is divided into four orders of magnitude, which are semantic sets composed of the following four subsets: weak, moderate, strong, and powerful (Figure 5). This study used the Mamdani reasoning method, established rules, and then applied the "min" implication method. All rules were applied to input and output values, and then the results of implications (conclusions) were aggregated using the "max" method. The defuzzification method used in this study was largest of maximum (LOM). After defuzzification, the final output was proposed as the predicted earthquake magnitude.

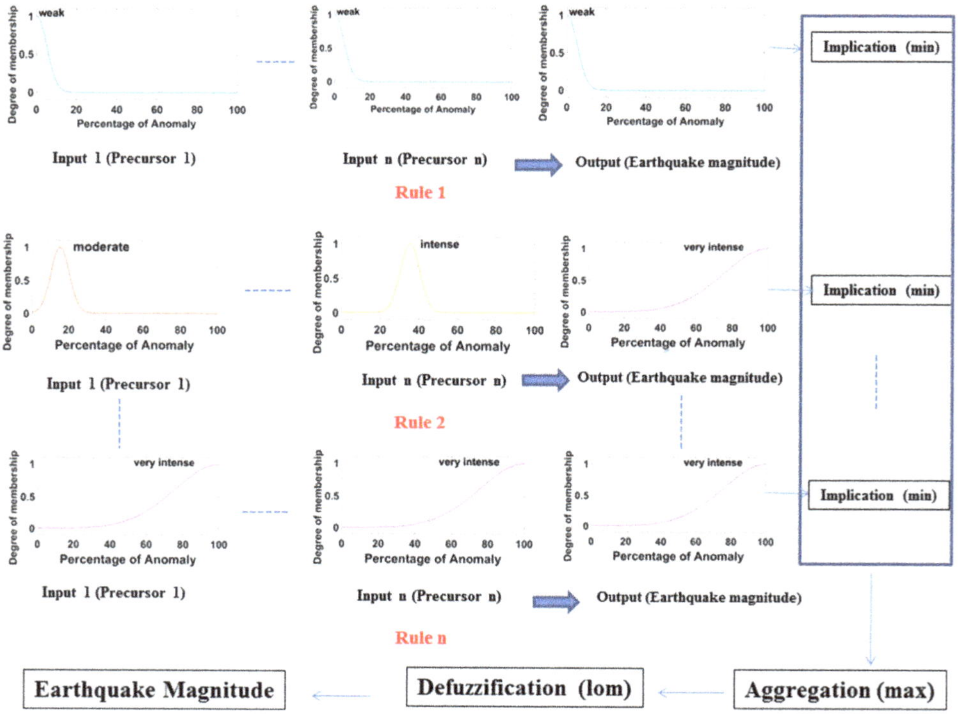

Figure 3. Structure of a Mamdani FIS model for estimation of earthquake magnitude.

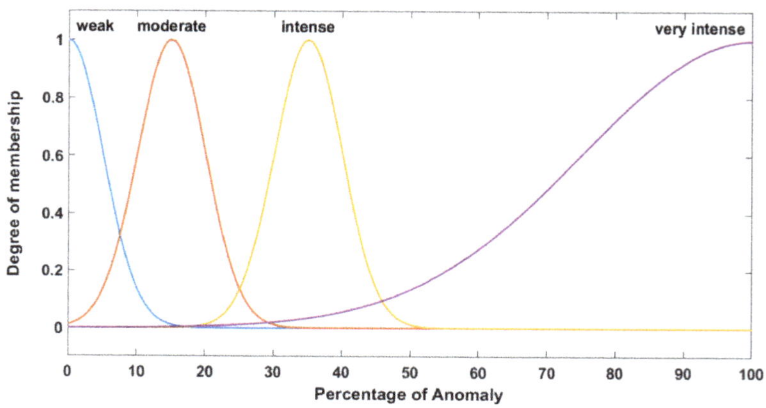

Figure 4. Membership functions of the input data.

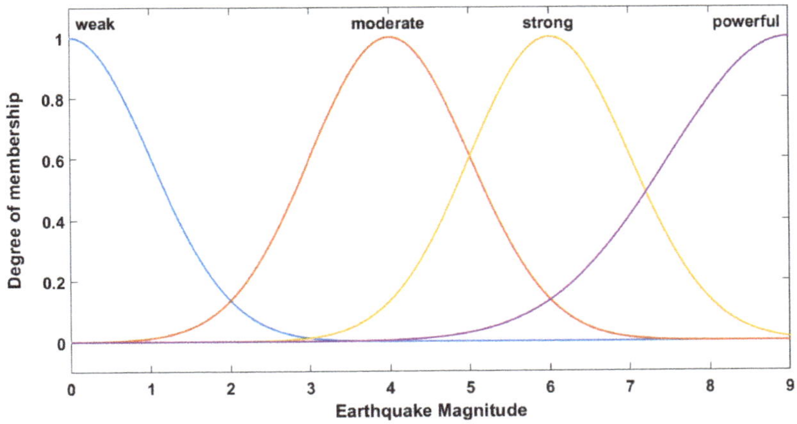

Figure 5. Membership functions of the output data.

3. Observations

Table 2 shows the analysed precursors concerning each case study. It is seen that 50, 47, 61, 58, and 51 precursors were investigated for Ecuador, Iran, Papua New Guinea, Japan and Haiti earthquakes, respectively. The details of the investigated data and implemented methods are found in [23–26]. As mentioned before in the methodology section, the *Dx* values for all detected anomalies during the quiet solar and geomagnetic activities were obtained and applied as input data for the proposed fuzzy inference system.

Table 2. The analysed precursors (D: day, N: night, Ne: electron density, Te: electron temperature, STEC: slant TEC, VTEC: vertical TEC, LST: Land Surface Temperature, AOD: Aerosol Optical Depth, AOT: Aerosol Optical Thickness).

Precursor			Ecuador [24]	Iran [25]	Papua New Guinea [26]	Japan [23]	Haiti
Layer	**Satellite**	**Parameter**					
Ionosphere (plasma)	Swarm A	Ne (D&N)	√	√	√	√	√
		Te (D&N)	√	√	√	√	√
		STEC (D&N)	-	-	√	√	-
		VTEC (D&N)	-	-	√	√	-
	Swarm B	Ne (D&N)	√	√	√	√	√
		Te (D&N)	√	√	√	√	√
		STEC (D&N)	-	-	√	√	-
		VTEC (D&N)	-	-	√	√	-
	Swarm C	Ne (D&N)	√	√	√	√	√
		Te (D&N)	√	√	√	√	√
		STEC (D&N)	-	-	√	√	-
		VTEC (D&N)	-	-	√	√	-
	Swarm A–C	Ne (D&N)	√	√	√	√	√
	Swarm A–C	Te (D&N)	√	√	√	√	√
	CSES	Ne (D&N)	-	-	√	-	√
		Te (D&N)	-	-	-	-	√

Table 2. *Cont.*

Precursor			Ecuador [24]	Iran [25]	Papua New Guinea [26]	Japan [23]	Haiti
Layer	**Satellite**	**Parameter**					
	GPS	TEC	√	√	√	-	√
Ionosphere (magnetic)	Swarm A	MS (D&N)	√	√	√	√	√
		MVx (D&N)	√	√	√	√	√
		Mvy (D&N)	√	√	√	√	√
		MVz (D&N)	√	√	√	√	√
	Swarm B	MS (D&N)	√	√	√	√	√
		MVx (D&N)	√	√	√	√	√
		Mvy (D&N)	√	√	√	√	√
		MVz (D&N)	√	√	√	√	√
	Swarm C	MS (D&N)	√	√	√	√	√
		MVx (D&N)	√	√	√	√	√
		Mvy (D&N)	√	√	√	√	√
		MVz (D&N)	√	√	√	√	√
Atmosphere	MODIS	AOD	√	-	-	-	-
	Climatological data	AOT	√	√	√	√	√
		CO	√	√	√	√	√
		SO_2	√	√	√	√	√
		DMS	√	-	√	√	√
		CH4	-	√	-	-	-
Atmosphere (surface)	MODIS	LST (D&N)	√	-	-	-	
Lithosphere	Seismic data	Eqs magnitude	√	√	√	√	√
		Eqs number	√	√	√	√	√
Number of Precursors (D&N)			**50**	**47**	**61**	**58**	**51**

In order to investigate the three geo-layers, we included data from earthquake catalogues or lithosphere and from Modern-Era Retrospective Analysis for Research and Applications, Version 2 (MERRA-2 [31]) and methane data for the Iran earthquake retrieved from the AIRS instrument onboar the NASA Earth observation system satellite Aqua [32] and from low Earth orbit (LEO) satellites such as Swarm [33] or CSES [34] missions to investigate the ionosphere. A sketch of the techniques of the investigation in this paper of earthquake catalogues and atmospheric data is given in the following.

3.1. Earthquake Catalogues

In order to obtain higher quality results, regional catalogues are preferred over global ones as they are able to provide a more complete catalogue, i.e., to have a lower magnitude of completeness. Despite this, for Papua New Guinea and Haiti case studies, we did not find local catalogues, so we used the United States Geological Survey (USGS) global earthquake catalogue. We note that Haiti is in a region well-covered by the seismic network of USGS, and Papua New Guinea provides a sufficient number of events due to its intrinsically high seismicity rate. For Ecuador, we retrieved the earthquake catalogue from "Instituto Geofísico de la Escuela Politécnica Nacional" (download from https://www.igepn.edu.ec/ (accessed on 30 April 2022). For the Iran earthquake catalogue, we used the one provided by Mousavi-Bafrouei and Mahani [35], which includes the period investigated in this paper. Finally, we retrieved the Japan Meteorological Agency (JMA) Unified Hypocenter Catalog for the Japan earthquake. For all the catalogues, we extracted the events within 1000 km distance from the investigated case study, and we checked the completeness of magnitude (Mc) by the software Z-Map [36]. We found that the best completeness was achieved by the Japanese catalogue (Mc = 1.1), and the worst was in the case of Papua New Guinea (Mc = 4.3) as reported in Supplementary Materials Figure S1 with the Gutenberg–Richter [37] distribution for each case study. After selecting only the earthquakes with magnitude $M \geq Mc$, we computed two time series for each case study: one with the daily number of seismic events and a second one with the maximum recorded daily magnitude. Such time series were then truncated on the day before the mainshock and passed to the FIS algorithm. For the figures, we reported the observed magnitudes also during and after the mainshock.

3.2. Atmospheric Data Preparation

To investigate eventual atmospheric anomalies, we selected four parameters from the MERRA-2 climatological archive: aerosol optical thickness (AOT), SO_2, CO, and dimethyl sulfide (DMS). The last parameter had negligible concentration above land areas, so for the Iran earthquake, we substituted it with CH4 measurements. All the parameters were analysed by the algorithm MERRA-2 analysis to search for seismic precursors (MEANS) fully described in [38]. Essentially, we retrieved the time series of the specific parameter in the year affected by the earthquake and a historical time series from the mean values estimated in the other available years (generally from 1980 to 2021). The obtained time series are included in the Supplementary Materials in Figures S2–S21, but the specific discussion is considered out of the scope of this paper. In fact, with such a time series, we provided four additional inputs from the atmosphere for the FIS algorithm as the daily deviation of the specific parameter, i.e., the time series in the year of the earthquake minus the historical one.

3.3. Results Applying FIS to the Five Investigated Earthquakes

3.3.1. Ecuador 2016

Figure 6a illustrates the variation in the number of anomalies (green curve), (b) the variation in the predicted earthquake magnitude (red curve), and (c) the variation in the maximum daily observed earthquake magnitude (blue curve) for the Ecuador earthquake (16 April 2016) from 1 November 2015 to 30 April 2016. In all panels, the *x*-axis represents the day relative to the earthquake day which is indicated as a vertical dotted line. Figure 6b indicates that the FIS predicted a magnitude M_w = 7.02 earthquake 38 days before the event (the black dotted ellipsoid in Figure 6b). It should be noted that the maximum number of anomalies (N = 10) was observed on this date (Figure 6a). The reported earthquake magnitude was M_w = 7.8 (Figure 6c).

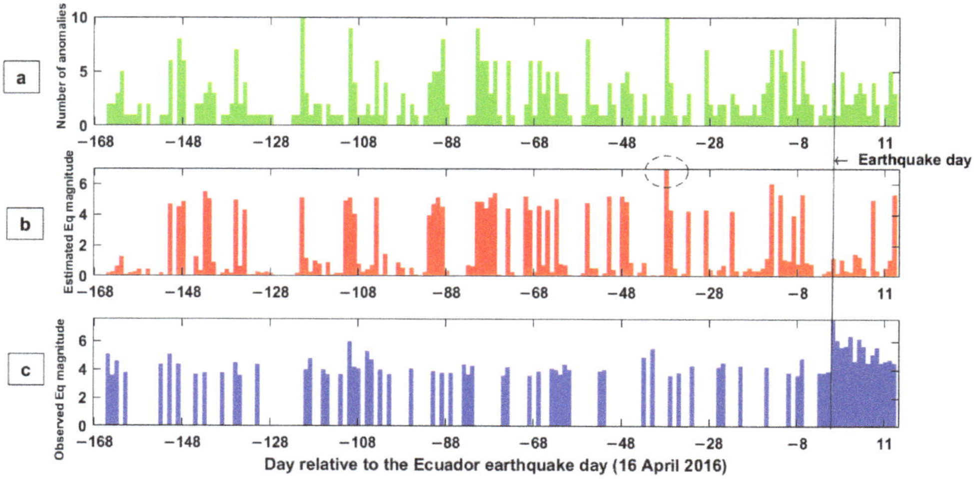

Figure 6. (**a**) Variation in the number of anomalies (green curve), (**b**) variation in the predicted earthquake magnitude (red curve), and (**c**) variation in the observed earthquake magnitude (blue curve) for the Ecuador earthquake (16 April 2016) from 1 November 2015 to 30 April 2016. In both panels, the *x*-axis represents the day relative to the earthquake day, which is indicated as a vertical dotted line.

3.3.2. Iran 2017

Figure 7a shows the daily variation in the number of detected anomalies (green curve) among 47 precursors and (b) the variation in the predicted earthquake magnitude (red curve) for the Iran earthquake (12 November 2017) from 1 August to 30 November 2017. The maximum number of detected anomalies (N = 13) was observed 19 days before the earthquake (Figure 7a). FIS estimated an earthquake with magnitude Mw = 7.02 (the black dotted ellipsoid in Figure 7b). The registered magnitude of the actual earthquake was Mw = 7.3.

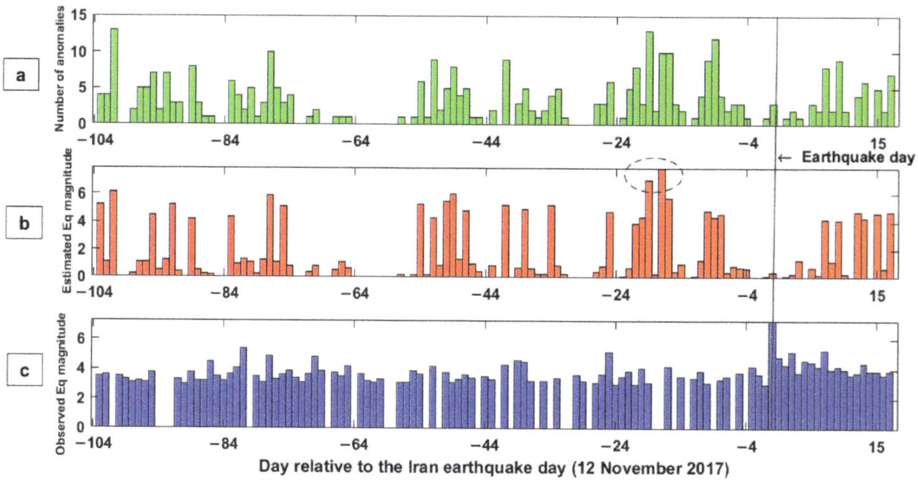

Figure 7. (**a**) Variation in the number of anomalies (green curve), (**b**) variation in the predicted earthquake magnitude (red curve), and (**c**) variation in the observed earthquake magnitude (blue curve) for the Iran earthquake (12 November 2017) from 1 August to 30 November 2017. In both panels, the *x*-axis represents the day relative to the earthquake day, indicated as a vertical dotted line.

3.3.3. Papua New Guinea 2019

Figure 8a indicates the daily variation in the number of anomalies (green curve) and (b) the variation in the predicted earthquake magnitude (red curve) for the Papua New Guinea earthquake (14 May 2019) from 1 January to 30 June 2019. Figure 8a shows that the maximum number of anomalies (N = 21) was observed 32 days prior to the earthquake. FIS predicted an impending earthquake with a magnitude of 9 32, 33, and 35 days before the mainshock.

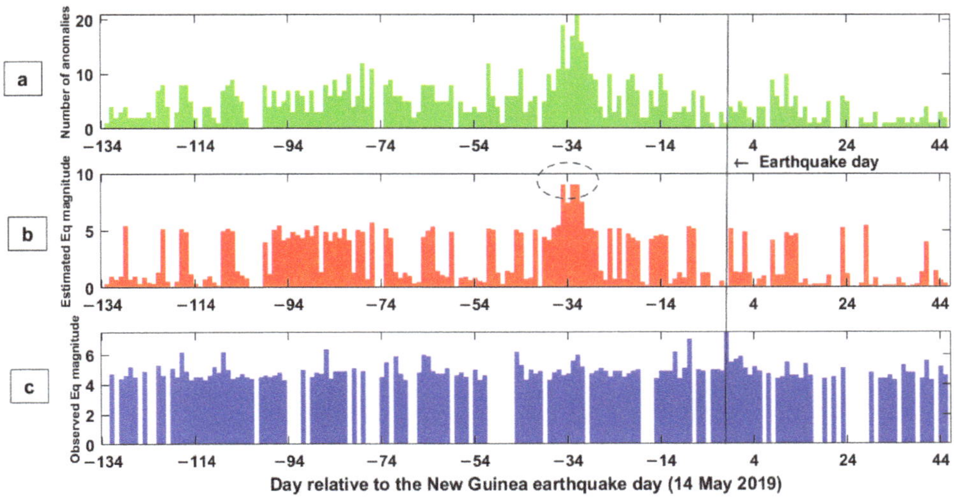

Figure 8. (**a**) Variation in the number of anomalies (green curve), (**b**) variation in the predicted earthquake magnitude (red curve), and (**c**) variation in the observed earthquake magnitude (blue curve) for the Papua New Guinea earthquake (14 May 2019) from 1 January to 30 June 2019. In both panels, the x-axis represents the day relative to the earthquake day, indicated as a vertical dotted line.

3.3.4. Japan 2021

Figure 9a shows the variations in the number of anomalies (green curve) and (b) the variation in the predicted earthquake magnitude (red curve) for the Japan earthquake (13 February 2021) from 1 September 2020 to 18 February 2021. FIS predicted the magnitude of the impending earthquake with the values of Mw = 7.30, 6.85, 7.51, 7.47, and 7.51 on 6, 8, 19, 31, and 32 days preceding the earthquake, respectively (marked by the black dotted ellipsoid in Figure 9b). A noticeable number of anomalies (N = 16) was observed 31 and 32 days before the earthquake. The maximum number of detected anomalies (N = 20) was observed 119 days before the earthquake when FIS predicted an earthquake with magnitude Mw = 7.508. The reported magnitude of the mentioned earthquake was Mw = 7.1. It should be noted that FIS predicted an impending earthquake with a magnitude of Mw = 7.75 3 days after the earthquake and 32 days before the next powerful earthquake (Mw = 7.0) that happened close to the first epicentre at 38.452°N, 141.648°E, 43.0 km depth, 09:09:43 UTC on 20 March 2021 (indicated by the black arrow (4) in Figure 9b).

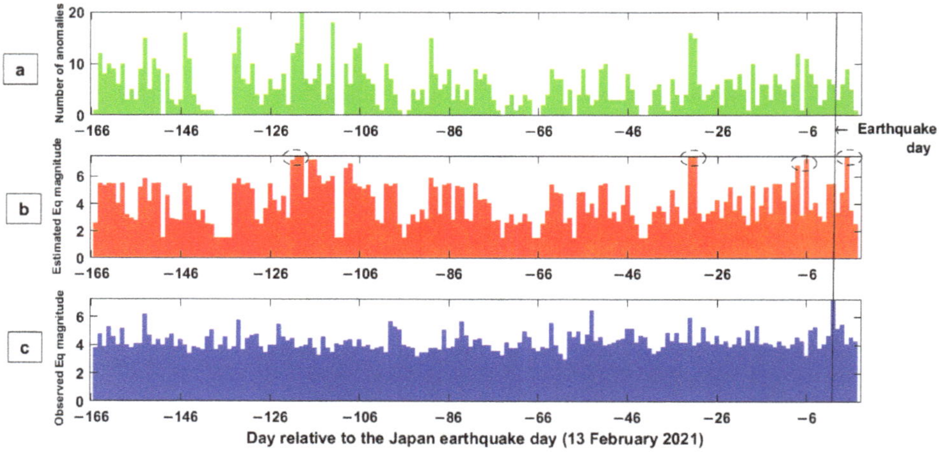

Figure 9. (**a**) Variation in the number of anomalies (green curve), (**b**) variation in the predicted earthquake magnitude (red curve), and (**c**) variation in the observed earthquake magnitude (blue curve) for the Japan earthquake (13 February 2021) from 1 September 2020 to 18 February 2021. In both panels, the x-axis represents the day relative to the earthquake day, indicated as a vertical dotted line.

3.3.5. Haiti 2021

Figure 10a illustrates the variation in the number of anomalies (green curve), (b) the variation in the predicted earthquake magnitude (red curve), and (b) the variation in the observed earthquake magnitude (blue curve) for the Haiti earthquake (14 August 2021) from 1 June to 31 August 2021. Figure 10b shows that the FIS predicted magnitude Mw = 7.24, 6.44, and 7.60 earthquakes 38, 40, and 45 days before the event, respectively (marked by a black dotted ellipsoid in Figure 9b). It should be noted that the maximum number of anomalies (N = 19) was observed 40 and 45 days before the main shock (Figure 10a). Actually, the reported earthquake magnitude was Mw = 7.2 (Figure 10c).

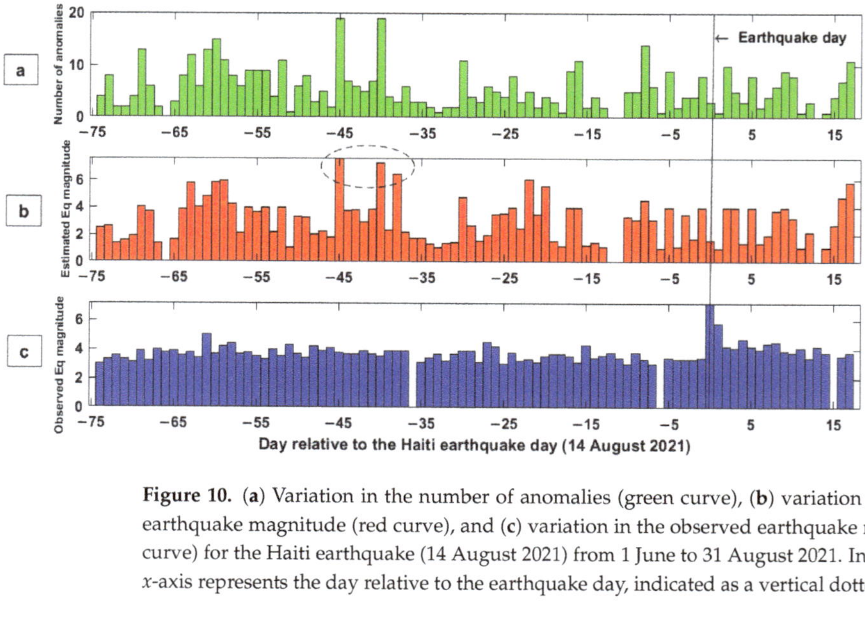

Figure 10. (**a**) Variation in the number of anomalies (green curve), (**b**) variation in the predicted earthquake magnitude (red curve), and (**c**) variation in the observed earthquake magnitude (blue curve) for the Haiti earthquake (14 August 2021) from 1 June to 31 August 2021. In both panels, the x-axis represents the day relative to the earthquake day, indicated as a vertical dotted line.

3.4. Validation of FIS

In order to evaluate the efficiency of the FIS system, a confutation analysis was performed. Among the analysed case studies, the Iran earthquake was selected, and the time series of 41 different precursors in the same time period set one year before the earthquake was analysed. Figure 11a shows the variations in the number of anomalies (green curve) and Figure 11b the variations in the predicted earthquake magnitude (red curve) from 1 August to 30 November 2016. It is worth noting that the maximum predicted earthquake magnitude by FIS was M_W = 3.42. Figure 11a shows no noticeable anomalies close to the selected day (12 November 2016), i.e., one year before the Iran 2017 mainshock.

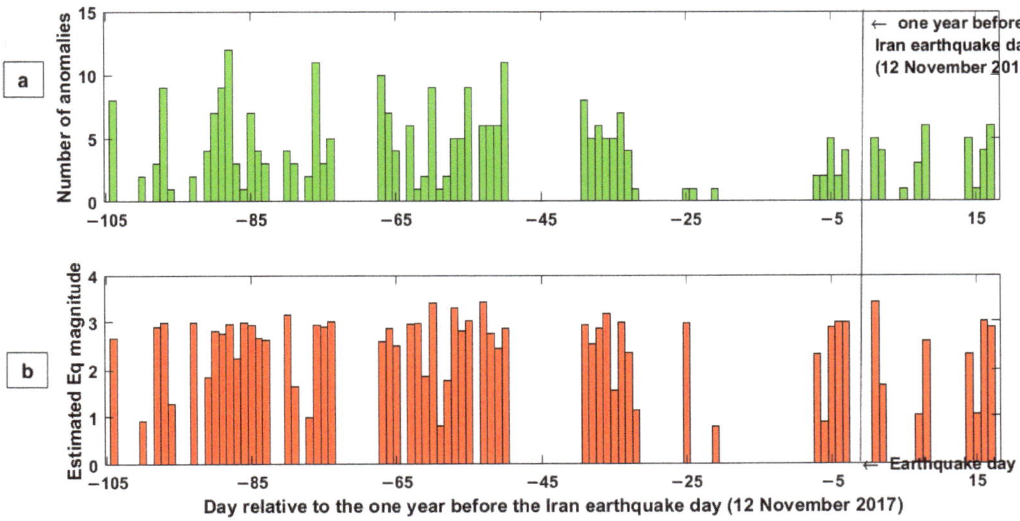

Figure 11. Validation analysis that was conducted in the same region as Figure 7 but from 1 August to 30 November 2016 for the same location. (**a**) the variations in the number of anomalies (green curve) and (**b**) the variations in the predicted earthquake magnitude (red curve).

In order to finally validate and refine the obtained result, a plot of the estimated earthquake magnitude versus the observed one is reported in Figure 12. Here, all five case studies are shown with a black dot, and in addition, the validation case was taken into account. For this last case, we considered the maximum observed magnitude as the one of the event that occurred on 26 October 2016 at 39.485°N, 54.509°E, 26 km depth and with a magnitude of 5.4. We performed a robust linear fit of the six experimental points shown as the red line. The M7.8 Ecuador 2016 case study was considered an outlier by the linear fit and thus excluded by the computation. The goodness of the fit was evaluated by the adjusted coefficient of determination that even with the low number of degrees of freedom (dof = 4) confirmed the reliability of the fit as R^2 was slightly greater than 0.9.

Finally, we tried to use the fit to apply a correction to the estimated magnitude obtained from the FIS algorithm, and we report in Table 3 the calculated values. The correction was applied by inverting the linear fit equation, i.e.:

$$\text{corrected magnitude} = \frac{\text{estimated magnitude} + 9.37}{2.37} \tag{4}$$

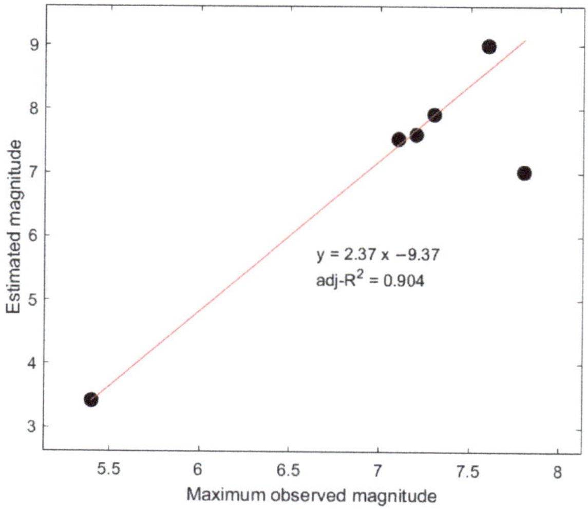

Figure 12. Comparison of the estimated magnitude by FIS algorithm and the real observed magnitude. The validation case study was also included. A linear fit with its output coefficients and adjusted coefficient of determination is also provided.

Table 3. Estimated magnitudes and comparison with the real one after correction.

Case Study	Estimated Magnitude by FIS	Corrected Magnitude	Real Magnitude	Error in the Estimation of the Magnitude after Correction
Ecuador 2016	7.02	6.92	7.8	0.88
Iran 2017	7.92	7.30	7.3	0.0046
Papua New Guinea 2019	9.00	7.75	7.6	0.15
Japan 2021	7.53	7.13	7.1	0.032
Haiti 2021	7.60	7.16	7.2	0.038

It is outstanding to note that after the correction, all the predicted magnitudes were equal to the real one within one decimal digit, except for the case of Ecuador, where the estimated magnitude remained too low with a magnitude that was estimated to be 0.9 less than the real one.

4. Conclusions

Creating an earthquake prediction system requires estimating the event location, future time, and magnitude with low uncertainty. Fortunately, with the increase in the number of geosphere observables and various anomaly detection methods, significant progress has been made in studies related to the possible earthquake precursors. Using data fusion algorithms leads to a more accurate estimation of earthquake parameters. In the first step, the deviation values (Dx) of different possible precursors' time series from the allowable limits were calculated for each studied area for several parameters of the lithosphere, atmosphere, and ionosphere. In most case studies, the highest number of anomalies in different precursors occurred about one month before the investigated M > 7 earthquakes. Therefore, it can be mentioned that by observing a significant number of anomalies (N > 5) in a study area, an earthquake could have a higher probability of occurrence in the next 1 to 40 days. To estimate the magnitude of the earthquake, a data fusion system based on the Mamdani fuzzy inference system was proposed. The Dx values were considered as input data to the FIS. The results showed that the predicted magnitude values differed slightly from the recorded ones. It is worth noting that the two most recent earthquakes that occurred in 2021 in Japan and Haiti had a good estimation of the magnitude, probably due to the improved satellite coverage by joint analysis of Swarm and CSES missions. The Dobrovolsky equation ($R = 10^{0.43 \cdot M}$, where R is the radius in kilometres of the earthquake preparation zone and M is the earthquake magnitude [39]) based on the estimated magnitude of the FIS could also be used to predict the affected area of the impending earthquake.

It is important to note that in this work, we tried to estimate only the magnitude of the impending earthquake, but due to differences in physical and tectonic settings of each event, the pattern of eventual precursors can be different. For example, the focal mechanism is expected to play a role in the preparation phase of each earthquake as several works have proposed and searched for preliminary pieces of evidence [18,40]. The present work investigated a limited number of earthquakes with different focal mechanisms. Future research needs to clarify and better explore such aspects, including the geological, topographic, and overall tectonic settings to study the pre-earthquake anomaly patterns. We expected that the estimated values of the incoming earthquake magnitude could be improved. To conclude, the FIS method with the applied correction seems a very promising tool, and future studies can confirm or improve the methodology, expand the prediction to other parameters of the earthquakes, and extend it to smaller magnitude events following the approach successfully applied to worldwide M5.5+ by De Santis et al. [16] and Marchetti et al. [18].

Supplementary Materials: The following supporting information can be downloaded at: https://www.mdpi.com/article/10.3390/rs14133203/s1, Figure S1: Gutenberg-Richter distributions of the investigated earthqaukes title. Figures S2 to S21: Atmospheric time series of the analysed parameters before the five earthquake case studies.

Author Contributions: Conceptualisation, methodology, software, data curation, and visualisation, M.A. and D.M.; writing—original draft preparation, M.A.; writing—reviewing and editing, M.A. and D.M.; project administration and funding acquisition, D.M. All authors have read and agreed to the published version of the manuscript.

Funding: This research was funded by Chinese Postdoctoral Science Foundation, grant number 2021M691190 for the project "Multiparameter pre-earthquake anomaly study of lithosphere-atmosphere-ionosphere"; National Natural Science Foundation of China, grant number 41974084 for project "Research on anomaly extraction technology of seismic electromagnetic satellite data based on blind source separation". APC was funded by Chinese Postdoctoral Science Foundation, grant 2021M691190.

Data Availability Statement: Swarm satellite data are freely available by ESA https/ftp server at https://swarm-diss.eo.esa.int/ (accessed on 21 March 2022). The Ecuador earthquake catalogue is freely available from https://www.igepn.edu.ec/ (accessed on 30 April 2022) provided by Instituto Geofísico de la Escuela Politécnica Nacional, Quito, Ecuador. The Global USGS earthquake catalogue is freely available from https://earthquake.usgs.gov/earthquakes (accessed on 5 June 2022). The Japanese earthquake catalogue freely available upon registration is produced by the Japan Meteorological Agency in cooperation with the Ministry of Education, Culture, Sports, Science and Technology. The catalogue is based on seismic data provided by the National Research Institute for Earth Science and Disaster Resilience, the Japan Meteorological Agency, Hokkaido University, Hirosaki University, Tohoku University, the University of Tokyo, Nagoya University, Kyoto University, Kochi University, Kyushu University, Kagoshima University, the National Institute of Advanced Industrial Science and Technology, the Geographical Survey Institute, Tokyo Metropolis, Shizuoka Prefecture, Hot Springs Research Institute of Kanagawa Prefecture, Yokohama City, and Japan Agency for Marine-Earth Science and Technology.

Acknowledgments: The authors would like to acknowledge the European Space Agency (ESA) for the Swarm data and the NASA Jet Propulsion Laboratory for the solar and geomagnetic indices. One of the results presented in this paper relies on the data collected at "Kakioka", so we thank the Japan Meteorological Agency (JMA) for supporting its operation and INTERMAGNET for promoting high standards of magnetic observatory practise (www.intermagnet.org (accessed on 5 June 2022)). We acknowledge Angelo De Santis for the suggestions that he provided to improve this work.

Conflicts of Interest: The authors declare no conflict of interest. The funders had no role in the design of the study; in the collection, analyses, or interpretation of data; in the writing of the manuscript, or in the decision to publish the results.

References

1. Scholz, C.H.; Sykes, L.R.; Aggarwal, Y.P. Earthquake Prediction: A Physical Basis. *Science* **1973**, *181*, 803–810. [CrossRef] [PubMed]
2. Pulinets, S.; Ouzounov, D. *The Possibility of Earthquake Forecasting: Learning from Nature*; IOP Publishing: Bristol, UK, 2018.
3. Pulinets, S.; Boyarchuk, K.A. *Ionospheric Precursors of Earthquakes*; Springer: Berlin/Heidelberg, Germany, 2004.
4. Parrot, M. Use of satellites to detect seismo-electromagnetic effects, main phenomenological features of ionospheric precursors of strong earthquakes. *Adv. Space Res.* **1995**, *15*, 1337–1347. [CrossRef]
5. Wu, H.C.; Tikhonov, I.N.; Césped, A.R. Multi-parametric analysis of earthquake precursors. *Russ. J. Earth Sci.* **2015**, *15*, ES3002. [CrossRef]
6. Wyss, M. A search for precursors to the Sitka, 1972, earthquake: Sea level, magnetic field, and *P*-residuals. *Pure Appl. Geophys.* **1975**, *113*, 297–309. [CrossRef]
7. Fraser-Smith, A.C.; Bernardi, A.; McGill, P.R.; Ladd, M.E.; Helliwell, R.A.; Villard, O.G., Jr. Low-frequency magnetic field measurements near the epicenter of the Ms7.1 Loma Prieta Earthquake. *Geophys. Res. Lett.* **1990**, *17*, 1465–1468. [CrossRef]
8. Molchanov, O.A.; Hayakawa, M.; Oudoh, T.; Kawai, H. Precursory effects in the subionospheric VLF signals for the Kobe earthquake. *Phys. Earth Planet. Inter.* **1998**, *105*, 239–248. [CrossRef]
9. Wu, H.-C.; Tikhonov, I.N. Jet streams anomalies as possible short-term precursors of earthquakes with M > 6.0. *Res. Geophys.* **2014**, *4*, 4939. [CrossRef]
10. Shou, Z.; Fang, Y. *Earthquake Vapor Model and Precise Prediction*, 1st ed.; Shou, W., Ed.; Earthquake Prediction Center: Menlo Park, CA, USA, 2016.
11. Ouzounov, D.; Pulinets, S.; Davidenko, D.; Rozhnoi, A.; Solovieva, M.; Fedun, V.; Dwivedi, B.N.; Rybin, A.; Kafatos, M.; Taylor, P. Transient Effects in Atmosphere and Ionosphere Preceding the 2015 M7.8 and M7.3 Gorkha–Nepal Earthquakes. *Front. Earth Sci.* **2021**, *9*, 757358. [CrossRef]
12. Marchetti, D.; De Santis, A.; D'Arcangelo, S.; Poggio, F.; Piscini, A.; Campuzano, S.A.; De Carvalho, W.V.J.O. Pre-earthquake chain processes detected from ground to satellite altitude in preparation of the 2016–2017 seismic sequence in Central Italy. *Remote Sens. Environ.* **2019**, *229*, 93–99. [CrossRef]
13. Marchetti, D.; De Santis, A.; Shen, X.; Campuzano, S.A.; Perrone, L.; Piscini, A.; Di Giovambattista, R.; Jin, S.; Ippolito, A.; Cianchini, G.; et al. Possible Lithosphere-Atmosphere-Ionosphere Coupling effects prior to the 2018 Mw = 7.5 Indonesia earthquake from seismic, atmospheric and ionospheric data. *J. Asian Earth Sci.* **2020**, *188*, 104097. [CrossRef]
14. De Santis, A.; Cianchini, G.; Marchetti, D.; Piscini, A.; Sabbagh, D.; Perrone, L.; Campuzano, S.A.; Inan, S. A Multiparametric Approach to Study the Preparation Phase of the 2019 M7.1 Ridgecrest (California, United States) Earthquake. *Front. Earth Sci.* **2020**, *8*, 540398. [CrossRef]
15. Yan, R.; Parrot, M.; Pinçon, J.-L. Statistical Study on Variations of the Ionospheric Ion Density Observed by DEMETER and Related to Seismic Activities. *J. Geophys. Res. Space Phys.* **2017**, *122*, 12421–12429. [CrossRef]
16. De Santis, A.; Marchetti, D.; Pavón-Carrasco, F.J.; Cianchini, G.; Perrone, L.; Abbattista, C.; Alfonsi, L.; Amoruso, L.; Campuzano, S.A.; Carbone, M.; et al. Precursory worldwide signatures of earthquake occurrences on Swarm satellite data. *Sci. Rep.* **2019**, *9*, 20287. [CrossRef] [PubMed]
17. Ouyang, X.Y.; Parrot, M.; Bortnik, J. ULF Wave Activity Observed in the Nighttime Ionosphere above and Some Hours before Strong Earthquakes. *J. Geophys. Res. Space Phys.* **2020**, *125*, e2020JA028396. [CrossRef]
18. Marchetti, D.; De Santis, A.; Campuzano, S.A.; Zhu, K.; Soldani, M.; D'Arcangelo, S.; Orlando, M.; Wang, T.; Cianchini, G.; Di Mauro, D.; et al. Worldwide Statistical Correlation of Eight Years of Swarm Satellite Data with M5.5+ Earthquakes: New Hints about the Preseismic Phenomena from Space. *Remote Sens.* **2022**, *14*, 2649. [CrossRef]
19. Xiong, P.; Marchetti, D.; De Santis, A.; Zhang, X.; Shen, X. SafeNet: SwArm for Earthquake Perturbations Identification Using Deep Learning Networks. *Remote Sens.* **2021**, *13*, 5033. [CrossRef]
20. Akhoondzadeh, M. Advances in Seismo-LAI anomalies detection within Google Earth Engine (GEE) cloud platform. *Adv. Space Res.* **2022**, *69*, 4351–4357. [CrossRef]
21. De Santis, A.; Marchetti, D.; Spogli, L.; Cianchini, G.; Pavón-Carrasco, F.J.; Franceschi, G.D.; Di Giovambattista, R.; Perrone, L.; Qamili, E.; Cesaroni, C.; et al. Magnetic Field and Electron Density Data Analysis from Swarm Satellites Searching for Ionospheric Effects by Great Earthquakes: 12 Case Studies from 2014 to 2016. *Atmosphere* **2019**, *10*, 371. [CrossRef]
22. Akhoondzadeh, M. A MLP neural network as an investigator of TEC time series to detect seismo-ionospheric anomalies. *Adv. Space Res.* **2013**, *51*, 2048–2057. [CrossRef]

23. Akhoondzadeh, M.; De Santis, A.; Marchetti, D.; Wang, T. Developing a Deep Learning-Based Detector of Magnetic, Ne, Te and TEC Anomalies from Swarm Satellites: The Case of Mw 7.1 2021 Japan Earthquake. *Remote Sens.* **2022**, *14*, 1582. [CrossRef]
24. Akhoondzadeh, M.; De Santis, A.; Marchetti, D.; Piscini, A.; Cianchini, G. Multi precursors analysis associated with the powerful Ecuador (M_W = 7.8) earthquake of 16 April 2016 using Swarm satellites data in conjunction with other multi-platform satellite and ground data. *Adv. Space Res.* **2018**, *61*, 248–263. [CrossRef]
25. Akhoondzadeh, M.; De Santis, A.; Marchetti, D.; Piscini, A.; Jin, S. Anomalous seismo-LAI variations potentially associated with the 2017 Mw = 7.3 Sarpol-e Zahab (Iran) earthquake from Swarm satellites, GPS-TEC and climatological data. *Adv. Space Res.* **2019**, *64*, 143–158. [CrossRef]
26. Akhoondzadeh, M.; De Santis, A.; Marchetti, D.; Shen, X. Swarm-TEC satellite measurements as a potential earthquake precursor together with other Swarm and CSES data: The case of Mw 7.6 2019 Papua New Guinea seismic event. *Front. Earth Sci.* **2022**, *10*, 820189. [CrossRef]
27. Zadeh, L.A. Probability measures of fuzzy events. *J. Math. Anal. Appl.* **1968**, 23, 421–427. [CrossRef]
28. Zadeh, L.A. Fuzzy sets. *Inf. Control* **1965**, *8*, 338–353. [CrossRef]
29. Rostami, A.H.; Akhoondzadeh, M.; Amani, M. A fuzzy-based flood warning system using 19-year remote sensing time series data in the Google Earth Engine cloud platform. *Adv. Space Res.* **2021**. *corrected proof.* [CrossRef]
30. Mamdani, E.H. Application of fuzzy algorithms for control of simple dynamic plant. *Proc. Inst. Electr. Eng.* **1974**, *121*, 1585–1588. [CrossRef]
31. Gelaro, R.; McCarty, W.; Suárez, M.J.; Todling, R.; Molod, A.; Takacs, L.; Randles, C.A.; Darmenov, A.; Bosilovich, M.G.; Reichle, R.; et al. The Modern-Era Retrospective Analysis for Research and Applications, Version 2 (MERRA-2). *J. Clim.* **2017**, *30*, 5419–5454. [CrossRef]
32. Fetzer, E.; McMillin, L.M.; Tobin, D.; Aumann, H.H.; Gunson, M.R.; McMillan, W.W.; Hagan, D.E.; Hofstadter, M.D.; Yoe, J.; Whiteman, D.N.; et al. AIRS/AMSU/HSB validation. *IEEE Trans. Geosci. Remote Sens.* **2003**, *41*, 418–431. [CrossRef]
33. Friis-Christensen, E.; Lühr, H.; Hulot, G. *Swarm*: A constellation to study the Earth's magnetic field. *Earth Planets Space* **2006**, *58*, 351–358. [CrossRef]
34. Shen, X.; Zhang, X.; Yuan, S.; Wang, L.; Cao, J.; Huang, J.; Zhu, X.; Picozza, P.; Dai, J. The state-of-the-art of the China Seismo-Electromagnetic Satellite mission. *Sci. China Technol. Sci.* **2018**, *61*, 634–642. [CrossRef]
35. Mousavi-Bafrouei, S.H.; Mahani, A.B. A comprehensive earthquake catalogue for the Iranian Plateau (400 B.C. to December 31, 2018). *J. Seismol.* **2020**, *24*, 709–724. [CrossRef]
36. Wiemer, S. A software package to analyze seismicity: ZMAP. *Seismol. Res. Lett.* **2001**, *72*, 373–382. [CrossRef]
37. Gutenberg, B.; Richter, C.F. *Seismicity of the Earth and Associated Phenomena*, 2nd ed.; Princeton University Press: Princeton, NJ, USA, 1954.
38. Piscini, A.; Marchetti, D.; De Santis, A. Multi-parametric climatological analysis associated with global significant volcanic eruptions during 2002–2017. *Pure Appl. Geophys.* **2019**, *176*, 3629–3647. [CrossRef]
39. Dobrovolsky, I.R.; Zubkov, S.I.; Myachkin, V.I. Estimation of the size of earthquake preparation zones. *Pure Appl. Geophys.* **1979**, *117*, 1025–1044. [CrossRef]
40. Doglioni, C.; Barba, S.; Carminati, E.; Riguzzi, F. Fault on-off versus strain rate and earthquakes energy. *Geosci. Front.* **2015**, *6*, 265–276. [CrossRef]

remote sensing

MDPI

Article

Analysis of Ocean–Lithosphere–Atmosphere–Ionosphere Coupling Related to Two Strong Earthquakes Occurring in June–September 2022 on the Sea Coast of Philippines and Papua New Guinea

Xitong Xu [1,2], Lei Wang [2] and Shengbo Chen [1,*]

1 College of Geo-Exploration Science and Technology, Jilin University, Changchun 130026, China; xuxt21@mails.jlu.edu.cn
2 British Geological Survey, Keyworth, Nottingham NG12 5GG, UK; lei.wang@bgs.ac.uk
* Correspondence: chensb@jlu.edu.cn

Abstract: Scientific progress in the context of seismic precursors reveals a systematic mechanism, namely lithosphere–atmosphere–ionosphere coupling (LAIC), to elaborate the underlying physical processes related to earthquake preparation phases. In this study, a comprehensive analysis was conducted for two earthquakes that occurred on the sea coast through tidal force fluctuation to investigate ocean–lithosphere–atmosphere–ionosphere coupling (OLAIC), based on oceanic parameters (i.e., sea potential temperature and seawater salinity), air temperature and electron density profiles. The interrupted enhancement and diffusion process of thermal anomalies indicate that the intensity of seismic anomalies in the atmosphere is affected by the extent of land near the epicenter. By observing the evolution of the ocean interior, we found that the deep water was lifted and formed upwelling, which then diffused along the direction of plate boundaries with an "intensification-peak-weakening" trend under the action of the accelerated subduction of tectonic plates. Furthermore, the analysis shows that the seismic anomalies have two propagation paths: (i) along active faults, with the surface temperature rising as the initial performance, then the air pressure gradient being generated, and finally the ionosphere being disturbed; (ii) along plate boundaries, upwelling, which is the initial manifestation, leading to changes in the parameters of the upper ocean. The results presented in this study can contribute to understanding the intrinsic characteristics of OLAIC.

Keywords: earthquake; ocean–lithosphere–atmosphere–ionosphere coupling; thermal anomalies; upwelling; tidal force

Citation: Xu, X.; Wang, L.; Chen, S. Analysis of Ocean–Lithosphere–Atmosphere–Ionosphere Coupling Related to Two Strong Earthquakes Occurring in June–September 2022 on the Sea Coast of Philippines and Papua New Guinea. *Remote Sens.* **2023**, *15*, 4392. https://doi.org/10.3390/rs15184392

Academic Editor: Stephan Havemann

Received: 9 August 2023
Revised: 30 August 2023
Accepted: 5 September 2023
Published: 6 September 2023

1. Introduction

The issue of pronounced changes in different geospheres possibly correlated with seismic activity has been a consistent focal point for numerous researchers [1–7]. Many studies have furnished scientific evidence to elucidate diverse forms of pre-seismic and post-seismic comprehensive perturbations as manifestations of lithosphere–atmosphere–ionosphere coupling (LAIC) [8–11]. Until now, several theories have emerged to clarify the origin of seismic anomalies and the formation of coupling relationships [12–15], but this topic still requires further exploration. At present, few studies have correlated seismic anomalies in the ocean with LAIC. The evolution of seawater can facilitate a deeper comprehension of the transmission of subsurface anomalies and their association with higher-level anomalies. Therefore, it is expected that ocean–lithosphere–atmosphere–ionosphere coupling (OLAIC) will be developed for the enhancement of its theoretical basis.

A thermal anomalous signal is commonly regarded as the direct manifestation of earthquake precursor processes, and a variety of methods and data have found extensive utility in examining thermal anomalies linked to seismic events [3,16–19]. The elevated level of attention directed toward thermal anomalies is attributed to their association with tectonic movements and their potential to trigger other precursor phenomena [20]. To understand the temporal and spatial variation, as well as the transfer process of thermal anomalies, the tidal force fluctuation analysis (TFFA) was developed and then employed to identify atmospheric thermal anomalies linked to seismic activity [16,21–24]. This approach capitalizes on the distinctive traits of tidal force, that is, the indicator function in the seismic time domain [25]. Previously, TFFA was confined to detecting atmospheric anomalies, while the transmission of anomalies from the lithosphere to the ionosphere implies that tidal force fluctuations could potentially identify seismic anomalies across multiple layers and with a broader range of parameters [24]. Applying TFFA across the ocean, lithosphere, atmosphere and ionosphere could yield more significant observations.

Ionospheric perturbations have been highlighted as potential seismic precursors in numerous studies [2,8–10,26–28]. Researchers have focused on the different distributions of ionospheric anomaly, including temporal, horizontal and vertical spatial scales. Throughout earthquake preparation phases, the propagation of the resulting disturbance is a continuous process [10], and thus changes in the ionospheric vertical structure over time within a certain distance from the epicenter are indicative. Utilizing observations encompassing both land and sea areas proximate to the epicenter is essential. Applying the radio occultation technique, which can detect ionospheric structures around the world, the FORMOSAT-7/COSMIC-2 (F7/C2) mission provides sufficient soundings to compensate for the absence of terrestrial instruments extending over oceanic expanses.

Anomalies linked to earthquakes occurring on the sea coast have been previously documented to occasionally exhibit greater prominence [29,30] and have a temporal and spatial correlation with terrestrial and atmospheric anomalies [31,32]. The mobility of seawater and possible solid Earth–seafloor interactions related to earthquakes may lead to the wide diffusion and high intensity of seismic disturbances in the ocean. To verify the interactions, multi-parameter fluctuation on the sea surface needs to be explained by the evolution process of the ocean interior. Some studies have attributed seismic anomalies on the sea surface to the upwelling of cold water induced by tectonic activity [30,33,34]. However, the detection and analysis of seismic anomalies covering the large-scale horizontal and vertical space of ocean interior evolution are still vacant in the continuous time interval. In addition, some studies have undertaken the comparison of parameter fluctuations within the ocean near the occurrence of earthquakes with those observed in other geospheres. Comparative analysis reveals a discernible synchronicity between anomalies in sea surface temperature and land surface temperature [35]. Spatial-temporal analysis involving multiple parameters indicates a correlation between sea surface salinity concentration and atmospheric disturbances [36]. Evidence has also been provided by recent research regarding the potential link between seismicity-induced upwelling and surface atmospheric temperature [33]. Past studies have gradually established the embryonic framework of OLAIC. To further enhance the comprehensiveness of OLAIC, an investigation into the anomalous overall propagation process is imperative.

In the current work, we conducted a complementary analysis by integrating parameters of two earthquakes that occurred on the sea coast to understand the mechanisms of seismic anomalies in the ocean, lithosphere, atmosphere and ionosphere. In the pursuit of creating a comprehensive anomaly detection approach, the anomalies across multiple layers and their subsequent transmission processes were discerned using TFFA. The electron density (Ne) profiles covering land and sea from the F7/2C mission were used to report the vertical variation in the ionospheric structure resulting from earthquakes. The possible solid Earth–seafloor interactions related to earthquakes were analyzed using large-scale ocean interior evolution. This study aims to investigate OLAIC and to expose its manifestation and action mode on the spatiotemporal scale.

2. Materials and Methods

2.1. Seismic Data

As the largest island in the Philippine archipelago, Luzon is subjected to more pronounced fault activity owing to the relative movement between the Sunda plate and the Philippine Sea plate. In the southern sector of Luzon Island, the Philippine Sea Plate subducts westward beneath the Sunda Plate. In northern Luzon, particularly where the Mw = 7.0 earthquake occurred on 27 July 2022, the Sunda Plate subducts eastward beneath the Philippine Sea Plate. In contrast to the Mw = 7.0 earthquake, the plate motion in the vicinity of the Mw = 7.6 earthquake on 10 September 2022 was characterized by a greater degree of complexity. The earthquake was connected with the extensive converging movement of the Australia plate and the Pacific plate and with the interactions of other microplates, including the North Bismarck plate, the South Bismarck plate, the Solomon Sea plate and the Woodlark plate. The earthquakes studied here include the Mw = 7.0 and the Mw = 7.6 earthquakes, which arose from oblique reverse faulting and dipping fault mechanisms, respectively, and the location is shown in Figure 1.

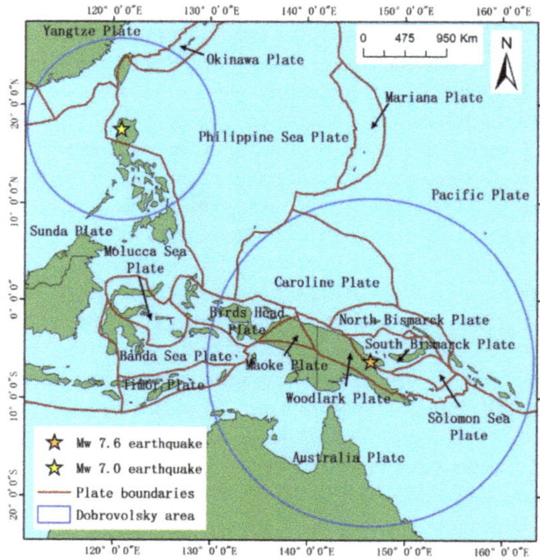

Figure 1. The geographical location of the two earthquakes that occurred on the sea coast. The red lines delineate plate boundaries. The yellow star and orange star represent epicenters of two earthquakes that occurred on the sea coast.

Additionally, the coupling anomalies in multi-layers pertaining to the two earthquakes that occurred on the sea coast were examined. The earthquake data are sourced from the United States Geological Survey (https://earthquake.usgs.gov/earthquakes, accessed on 7 August 2023), as shown in Table 1.

Table 1. The fundamental traits of studied earthquakes.

Date	Time (UTC)	Latitude	Longitude	The Altitude of the Epicenter (m)	Depth (km)	Mw
27 July 2022	00:43:27	17.5207°N	120.8181°E	433	34	7.0
10 September 2022	23:47:00	6.2949°S	146.5025°E	584	116	7.6

2.2. Oceanic Data

In this study, sea potential temperature (SPT) and seawater salinity (SWS) are used to reflect anomalous changes in the ocean around the occurrence time of earthquakes. The oceanic data were acquired from Copernicus Marine Environmental Monitoring Service (CNEMS), which provides global ocean analysis and forecasts. We used daily updated data from the Global Ocean Physical Analysis and Forecasting Product (https://data.marine.copernicus.eu/products, accessed on 29 August 2023). Data with a horizontal resolution of $1/12^\circ \times 1/12^\circ$ (approx. 8 km × 8 km) at 50 vertical levels (range from 0 to 5500 m with logarithmic space) assimilated temperature and salinity profiles. The mean error of SPT and SWS across 50 vertical levels, in comparison to on-site observations, scarcely surpassed 0.5 K and 0.01 PSU, respectively [37].

2.3. Air Temperature

The air temperature data were gathered from National Center for Environmental Prediction (NCEP) reanalysis data, which are collated by the National Center for Atmospheric Research (NCAR) and NCEP [38]. The reanalysis data with $1^\circ \times 1^\circ$ horizontal resolution and 6 h temporal resolution included the ground surface and 26 isobaric levels [39]. We selected the ground surface and five isobaric levels. Specifically, the isobaric levels at 950 hPa, 925 hPa, 900 hPa, 850 hPa and 800 hPa were used to detect anomalies for the Mw = 7.0 earthquake, and the isobaric levels at 925 hPa, 900 hPa, 850 hPa, 800 hPa and 750 hPa were used to detect anomalies for the Mw = 7.6 earthquake.

2.4. Electron Density

As a low-orbit cluster of six low-Earth-orbit (LEO) satellites, F7/C2 was launched on 25 June 2019 with an inclination of 24° and the ability to provide over 4000 ionospheric soundings per day at low latitudes. The six LEO satellites, namely Flight Models 1 through 6 (FM1-FM6), collaboratively furnish holistic understanding regarding the vertical electron density profile encompassing the ionosphere. Each LEO satellite in this constellation is equipped with advanced GPS radio occultation receivers. The Ne profiles are retrieved by utilizing variations in a radio signal transmitted by the GNSS satellite while passing through the ionosphere.

Due to the intrinsic properties of the ionosphere, the diurnal and nocturnal Ne profiles exhibit significant disparities. The photochemical process dominates the characteristics of the ionosphere at E and F1 layers in the daytime, and thus Ne profiles drop dramatically to vanishing at night because of the high loss rate [40]. The loss rate in the F2 layer (mainly higher than 200 km altitude) and above decreases with rising altitude, which reduces the weakening amplitude of Ne profiles at night [41]. Typically, Ne profiles in the daytime are significantly enhanced overall compared to at night [40,42]. Therefore, in this study, we separated diurnal and nocturnal Ne. The time range from 15:00 UT to 23:00 UT was used to select nocturnal Ne profiles for the Mw = 7.0 earthquake, and the time range from 13:00 UT to 21:00 UT was used to select nocturnal Ne profiles for the Mw = 7.6 earthquake. Figure 2 shows the radio occultation soundings around two earthquakes on 22 August 2022 and 6 October 2022.

2.5. Tidal Force and Anomaly Detection

In our pursuit of understanding the transmission chain of seismic disturbances, we examined the amalgamation of processes through which anomalies traverse from the ocean to the ground surface, propagate through the atmosphere and subsequently influence the ionosphere. The ocean, atmosphere and ionosphere are subject to various disturbances, which produce a large number of subtle and messy fluctuations in each variable. Hence, there is a need to formulate a consistent method that can extract the abnormal evolution process with temporal and spatial rules. Tidal force plays an important role in this process, pushing the rocks to a state of criticality [43]. Researchers have established the basis for detecting thermal anomalies through tidal force variation [16,21–24]. Tidal Force

Fluctuant Analysis (TFFA) is an approach used to identify thermal anomalies in the atmosphere. The method for computing tidal force potential has been detailed and described in references [16,21]. The initial TFFA is constrained by the requirement to establish the correlation between the timing of earthquakes and the tidal force potential. Given this problem, Xu et al. modified the TFFA method [24], which is the basis of anomaly detection through tidal force variation in this study.

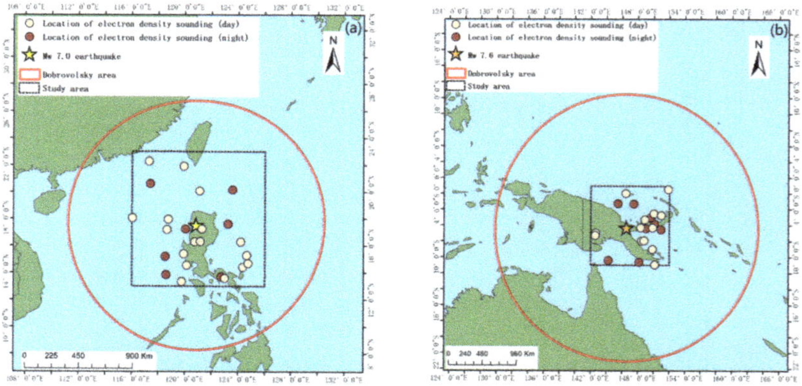

Figure 2. Location of the electron density profiles on (**a**) 22 August 2022 and (**b**) 6 October 2022 within the study area of earthquakes. The light brown dots represent the mean longitude and latitude of radio occultation soundings in the daytime, and the dark brown dots represent the mean longitude and latitude of radio occultation soundings in the nighttime.

We first applied the TFFA to detect atmospheric thermal anomalies. After calculating tidal force potential, the peak was taken as the first day of the next period, and the day before the next peak was taken as the last day of the period. The mean value of the whole period was set as the background at each pixel. The difference between the air temperature and the corresponding background value was used for the next analysis. The specific criteria [24] for detecting thermal anomalies related to earthquakes are as follows:

- Thermal anomalies emerge from the bottom. As the altitude increases, thermal anomalies tend to diminish.
- The distribution of thermal anomalies at each level coincides with the fault pattern.
- Thermal anomalies spread on the time scale, and therefore thermal anomalies should last two consecutive days at least.

Similarly, we performed background value removal on SPT and SWS on the basis of TFFA. Within a tidal force period, the mean values of SPT and SWS for each pixel were established as background values. The corresponding difference between the original value and the background was obtained. From this foundation, daily variations in SPT anomalies and SWS anomalies were generated for the analysis of the upwelling formation.

Then, we conducted anomaly identification for Ne based on tidal force periods. The radio occultation soundings within ±5° latitude and longitude away from the epicenter were extracted, with a spatial range consistent with the study of air temperature, SPT and SWS. Note that such a study area was located within the Dobrovolsky radius [44] of both earthquakes. As described in Section 2.4, the daily data were divided into two groups, daytime and nighttime, due to significant differences in Ne profiles. The mean values of Ne profiles, separated by daytime and nighttime were calculated for each day, denoted as Ne_{day} and Ne_{night}, and Ne_{dif} was obtained using $Ne_{day} - Ne_{night}$. Based on the same tidal force periods as above, the mean profiles of each period were calculated as the background. The difference between Ne_{dif} and background was defined as ΔNe. The confidence bounds were computed through the mean value and interquartile range of ΔNe, as shown in Equation (1).

$$\begin{cases} UB = \mu + 2 \times IQR \\ LB = \mu - 2 \times IQR \end{cases} \quad (1)$$

where UB represents the upper bound, LB represents the lower bound, μ is the mean value of ΔNe and IQR is the interquartile range of ΔNe.

3. Results

We examined various anomalies across multiple layers during six distinct periods of each earthquake, uncovering a spatiotemporal coupling phenomenon aligned with the propagation patterns of seismic anomalies. The investigated time spans for the two earthquakes that occurred on the sea coast encompassed 88 days and 89 days, as illustrated in Figure 3.

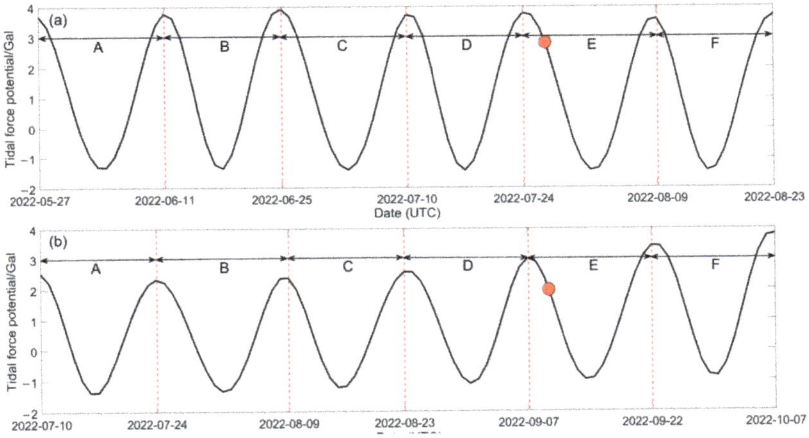

Figure 3. The fluctuation of tidal force potential for (**a**) Mw = 7.0 earthquake and (**b**) Mw = 7.6 earthquake. The red points represent seismic dates.

3.1. Seismic Thermal Anomalies

The temperature fluctuations at different isobaric levels throughout the analyzed time frames are depicted in the Supplementary Materials (Figures S1 and S2). Figures A1 and A2 demonstrate the identified thermal anomalies related to the two earthquakes. We identified a total of seven seismic thermal anomalies spanning from 3 days to 6 days for the Mw = 7.0 earthquake, and eight seismic thermal anomalies ranging from 2 days to 19 days for the Mw = 7.6 earthquake, as listed in Table 2. It can be seen that the seismic thermal anomalies adhered to the sequential pattern of "initial rise-intensification-attenuation", which correlates with the thermal infrared radiation process associated with rock fracturing under stress loading [45]. Other disorganized anomalies might arise from intricate meteorological activity and can be readily differentiated from seismic thermal anomalies based on their distinct patterns of variation and distribution [24]. The anomalies associated with the two earthquakes exhibited substantial variations in duration. The earthquakes (Mw = 7.0) with a smaller land area near the epicenter produced thermal anomalies that were temporally more dispersed compared to the earthquake (Mw = 7.6) with a larger land area near the epicenter. Similarly, the earthquake with an Mw = 7.6 magnitude generated broader and more intense anomalies on a spatial scale compared to the earthquake with an Mw = 7.0 magnitude. Additionally, the spatial evolution process demonstrated an intriguing tendency, wherein seismic thermal anomalies propagated from land towards the surrounding oceans.

Table 2. The duration of thermal anomalies associated with two earthquakes.

	Mw = 7.0 Earthquake	**Mw = 7.6 Earthquake**
Duration of seismic thermal anomalies	28 May–29 May 2022 11 June–12 June 2022 25 June–26 June 2022 3 July–5 July 2022 9 July–11 July 2022 22 July–27 July 2022 12 August–15 August 2022	11 July–13 July 2022 20 July–23 July 2022 1 August–6 August 2022 9 August–10 August 2022 13 August–16 August 2022 21 August–22 August 2022 2 September–20 September 2022 29 September–4 October 2022

3.2. *Sea Potential Temperature (SPT) and Sea Water Salinity (SWS)*

3.2.1. Identification of Abnormal Upwelling

In order to offer robust evidence concerning oceanic anomalies associated with earthquakes and their interplay with atmospheric seismic thermal anomalies, we delved into the vertical evolution process of SPT and SWS subsequent to background removal via the TFFA method. In this study, we focused on the generation of upwelling in the ocean near the earthquake breeding zones. Therefore, the negative anomalies of SPT and the positive anomalies of SWS were identified for subsequent analysis. The comprehensive evolution process of two parameters for the Mw = 7.0 earthquake and Mw = 7.6 earthquake is shown in the Supporting Information (Figures S3–S6). Throughout the studied periods, we observed three patterns of SPT variation: (1) no anomalies emerging or weak anomalies emerging with a short spatial span; (2) anomalies emerging with variable strength and discontinuous distribution; (3) intense anomalies emerging with a trend of "intensification-peak-weakening" or "peak-weakening" in time, wide spatial span and continuous spatial distribution. The corresponding time ranges of the three anomalous variation patterns are shown in Table 3. It is evident that the third pattern exhibits a consistent trend of variation and pronounced spatial continuity, whereas the anomalous distribution of the first and second patterns appears obscure and irregular. Hence, further analysis was carried out on the third pattern, and the evolution process of SPT and SWS over the third pattern duration is shown in Figure A3 (Mw = 7.0 earthquake) and Figure A4 (Mw = 7.6 earthquake). The SWS anomalies maintain a high degree of concurrence with SPT anomalies for most of the third pattern. In particular, SWS anomalies manifest near or above the SPT anomalies, exhibiting a tendency of "intensification-peak-weakening" or "peak-weakening". The simultaneous shifts in SPT decrease and SWS increase suggest that external driving forces are causing an uplift of deep water. As depicted in Figure A3, in the vicinity of the northern region (22°N, 116°S), a pronounced and enduring enhancement of SWS is spatially concentrated, which contrasts with the behavior of other anomalies. This is related to seasonal variation in summer; that is, the enhanced penetration of Kuroshio water via the Luzon Strait [46–48].

Table 3. The occurrence time of three anomalous variation patterns of sea potential temperature.

	Mw = 7.0 Earthquake	**Mw = 7.6 Earthquake**
First pattern	31 May–10 June 2022 16 June–21 June 2022 29 June–4 July 2022 13 July–20 July 2022 29 July–5 August 2022 13 August–22 August 2022	28 July–2 August 2022 11 August–16 August 2022 22 August–3 September 2022 24 September–4 October 2022
Second pattern	22 June–24 June 2022 21 July–28 July 2022 6 August–8 August 2022	15 July–19 July 2022 24 July–27 July 2022 17 August–21 August 2022 15 September–23 September 2022 6 October 2022

Table 3. *Cont.*

	Mw = 7.0 Earthquake	Mw = 7.6 Earthquake
Third pattern	27 May–30 May 2022 11 June–15 June 2022 25 June–28 June 2022 5 July–12 July 2022 9 August–12 August 2022	10 July–14 July 2022 20 July–23 July 2022 3 August–10 August 2022 4 September–14 September 2022 4 October–5 October 2022

3.2.2. Confutation Analysis

The circulation in the ocean varies at seasonal time scales. Investigating the variation in SPT and SWS across different years within the same spatial range can contribute to confirming the distinctiveness of the upwelling generated before and after the two earthquakes. We conducted the TFFA on SPT and SWS for the vicinity of the two earthquakes during comparable time periods in 2021. Note that due to the intrinsic periodic nature of tidal force, the selection of an entirely congruent time range was unattainable. The time span from 6 June 2021 to 18 August 2021, encompassing five tidal force cycles, and the period from 20 July 2021 to 2 August 2021, spanning five tidal force cycles, were, respectively, employed as temporal windows for the confutation analysis pertaining to the two seismic events. Here, we present the key findings indicated by the confutation analysis (Figures S7–S10): (1) For the Mw = 7.0 earthquake, the SPT anomalies in 2021 were generally dispersed throughout most of the time period. An anomaly caused by force intruding from above was observed from July 16 to 18. Anomalies aligned with the aforementioned "the third pattern" occurred from July 20 to 23 and August 1 to 6, yet no corresponding increase in the salinity of overlying seawater at the same spatial locations was observed. This suggests that these anomalies are not induced by intense upwelling currents. Additionally, the enhanced SWS in the northern region was consistent with the results of 2022, further indicating its association with the enhanced penetration of Kuroshio water via the Luzon Strait. (2) For the Mw = 7.6 earthquake, the SPT anomalies were both dispersed and of relatively weak intensity for the majority of the time. Occasional SPT anomalies exhibited clustering tendencies, yet they dissipated swiftly within two to three days. Also, there was no apparent spatiotemporal synchrony observed with salinity anomalies. This provides evidence that the seasonal influence of oceanic currents consistently leads to anomalies during comparable timeframes across different years. In contrast, the intense and prolonged upwelling associated with earthquakes is distinct and distinguishable.

3.3. *Vertical Electron Density Variation*

In order to examine the vertical perturbations in the ionosphere associated with earthquakes, anomaly detection on Ne profiles was performed based on TFFA. Figure 4 illustrates the numerical distance between daily ΔNe and boundaries within the examined timeframes for both the Mw = 7.0 earthquake and the Mw = 7.6 earthquake. Subsequently, we isolated the profiles that intersect the boundaries, as shown in Figure 5. In general, ΔNe anomalies show an obvious aggregation phenomenon on the time scale. In particular, four short-term consecutive anomalies emerged in the case of the Mw = 7.0 earthquake (i.e., 10 June and 11 June, 24 June and 26 June, 11 July and 12 July, 19 July and 21 July). Also, two short-term consecutive anomalies and one long-term consecutive anomaly emerged in the case of the Mw = 7.6 earthquake (i.e., 1 August and 5 August, from 2 September to 11 September, 3 October and 5 October). It is interesting to notice that continuous anomalies aligned with one or both of the following variations: (1) profiles maintaining a similar shape and the amplitude of fluctuation increasing; (2) the peak moving to a higher altitude. This suggests that the ionospheric disturbance is enduring and propagates from lower to upper layers, with an evolutionary trajectory consistent with seismic thermal anomalies and upwelling driven by external forces. Additionally, conducting a comparative analysis of the temporal distribution is essential to demonstrate that the three types of anomalies do not occur in isolation.

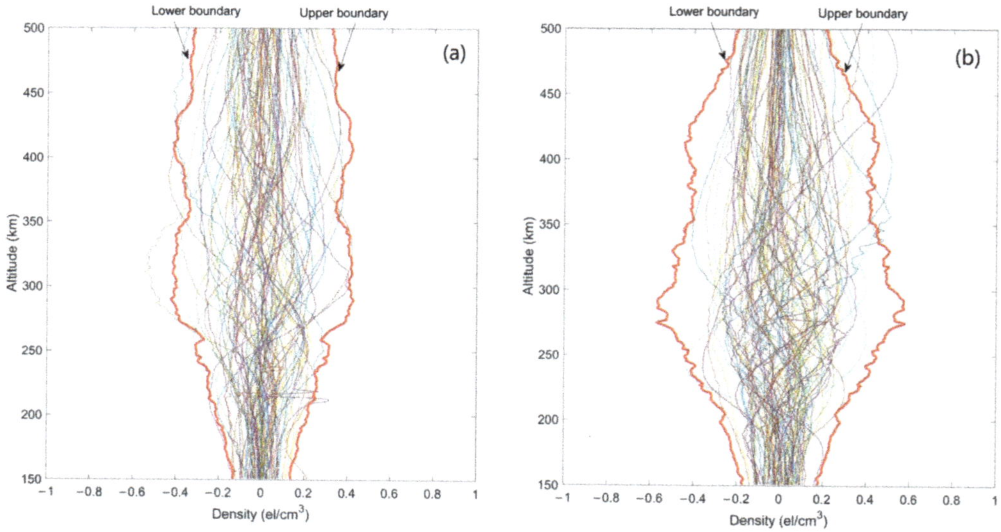

Figure 4. The electron density profiles processed by tidal force fluctuation analysis during studied periods of (**a**) Mw = 7.0 earthquake and (**b**) Mw = 7.6 earthquake. The red lines represent lower and upper boundaries.

Figure 6 shows the temporal distribution of anomalies across the ocean, lithosphere, atmosphere and ionosphere. Anomalies associated with the Mw = 7.6 earthquake exhibited stronger temporal continuity compared to those linked with the Mw = 7.0 earthquake. This contrast is particularly evident in the context of atmospheric thermal anomalies and ionospheric perturbations. The three types of anomalies exhibited notable temporal synchronization, with the majority of anomalies in the ocean and ionosphere occurring within the timeframe of thermal anomalies. Meanwhile, anomalies in the ocean tended to initiate around the same time as thermal anomalies. This observation implies a more profound intrinsic interrelation among the multi-layer disturbances induced by earthquakes.

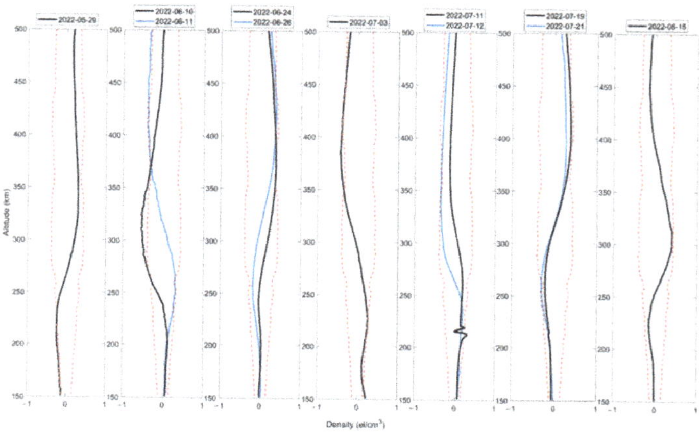

(**a**)

Figure 5. *Cont.*

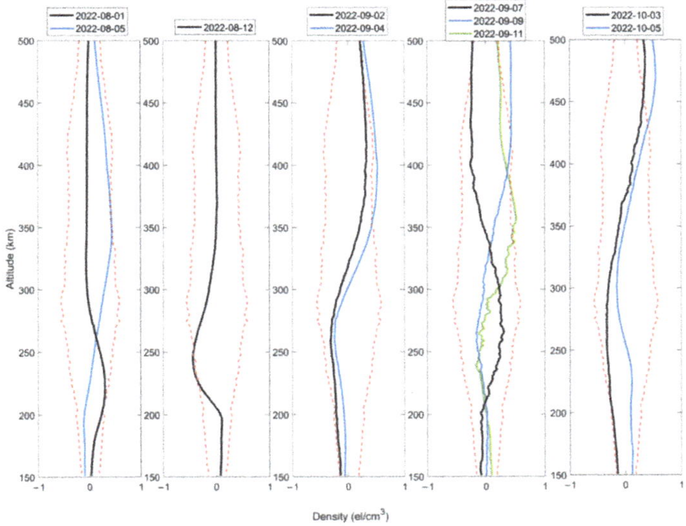

(b)

Figure 5. The isolated anomalous electron density profiles of (**a**) Mw = 7.0 earthquake and (**b**) Mw = 7.6 earthquake. The red lines represent the lower and upper boundaries.

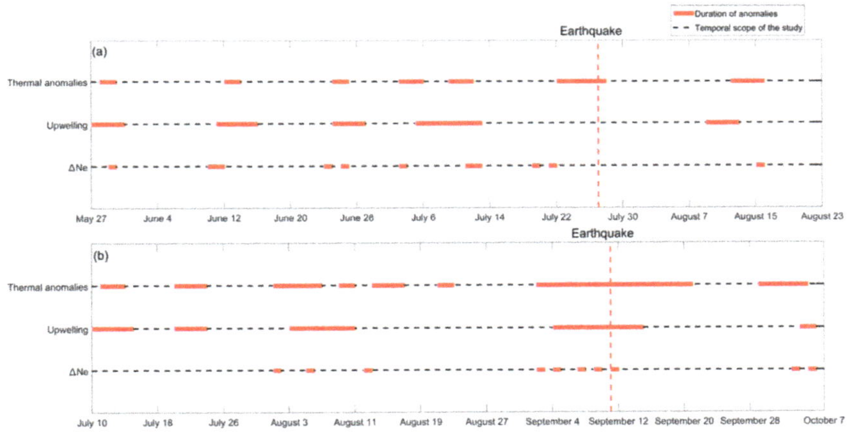

Figure 6. The time distribution diagram showing thermal anomalies, upwelling and ΔNe for (**a**) Mw = 7.0 earthquake and (**b**) Mw = 7.6 earthquake.

4. Discussion

As an external factor, tidal force exerts an inducing impact on an active fault with tectonic stress in a critical state [21]. Both earthquakes studied, which occurred along coastal areas, took place three days after the tidal force peaked (Figure 1). This observation implies that tidal force might have played a substantial role in triggering the occurrence of these two major earthquakes. The seismic thermal anomalies observed recurred several times through the TFFA method. This is consistent with the reduplicative process of rock fracture and energy release [49–51]. However, the conspicuous temporal discontinuity of the thermal anomaly is a noteworthy aspect, with this discontinuity being more pronounced in the case of the Mw = 7.0 earthquake. Examining the specifics reveals that the ongoing enhancement and diffusion process is periodically disrupted by external factors, giving rise to short-term anomalies. For instance, during the Mw = 7.0 earthquake, the fourth thermal anomaly (from 3 July 2022 to 5 July 2022) and the fifth anomaly (from 9 July 2022 to 11 July 2022) exemplified this pattern. This interruption can be attributed to the environmental context of the studied earthquakes, particularly the relatively lower land-to-sea ratio.

The trend of thermal anomaly diffusion can provide additional support for further analysis. The seismic thermal anomalies associated with both earthquakes emanate from active faults situated on the land and in proximity, subsequently extending to the surrounding sea. This concurs with findings from alternative methodologies employed to detect seismic thermal anomalies [52–54]. The energy accumulation in seismogenic zones is dominated by squeezed rock under tectonic stress in active faults [55]. The process is intricately connected to the dynamic interplay of geological forces and convective thermal processes within the Earth's subsurface. In this scenario, convective thermal fluids transform into thermal water vapor, gaseous emissions and heat flux. These elements ascend along geological structures and eventually reach the surface, subsequently leading to an increase in ground surface temperature, which signifies the presence of this process [12,56,57], as shown in Figure 7. The origin of atmospheric seismic thermal anomalies lies in the increased surface temperature over land. Consequently, the discernibility of thermal anomalies can be hindered by various overlapping environmental influences. The smaller land area renders anomalies less intense and more susceptible to interference.

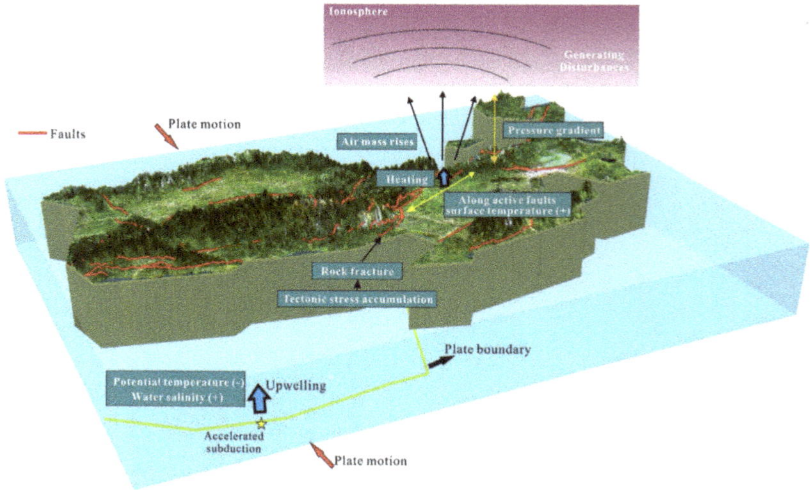

Figure 7. The schematic of ocean–lithosphere–atmosphere–ionosphere coupling (OLAIC).

In contrast to anomalies observed on the surface and in the atmosphere, the seismic anomalies within the ocean display analogous evolutionary patterns, yet distinctive manifestations. We observed SPT and SWS anomalies in the two earthquakes, most of which occurred within the duration of seismic thermal anomalies. The synchronous process of the decreasing SPT and increasing SWS indicated that the deep water was uplifted by external driving forces, resulting in the formation of upwelling. These occurrences were accompanied by unique time frames and spatial spans of observed upwelling. For the Mw = 7.0 earthquake, frequent upwelling was concentrated around the epicenter (northwestern Luzon island) between June and August, with higher intensity observed in the northern regions. The coastal upwelling along the northern shores of Luzon Island was predominantly influenced by a cold eddy known as the Luzon cold eddy [58]. This cyclonic eddy was detected using multiple data, including long-term observational data in the South China Sea [59], temperature and salinity data [60,61] and altimeter data [62]. Analysis reveals that this eddy followed a distinct "development-intensification-decay" pattern from October to the subsequent May [58]. In the case of the Mw = 7.6 earthquake, upwelling was reduplicative and continuous in the Bismarck Sea and the Solomon Sea from July to September. During July and August, the temperature and salinity patterns in the Bismarck Sea and Solomon Sea were predominantly influenced by horizontal advection rather than upwelling. In contrast, in January and February, coastal upwelling driven by northwest monsoon winds significantly impacted temperature and salinity dynamics [63]. The confutation analysis yielded evidence of consistent seasonal anomalies induced by oceanic currents within comparable timeframes across different years, while in contrast, distinct and prolonged upwelling events linked to earthquakes were discernible. Hence, the upwelling events before and after earthquakes exhibited distinct features and were attributed to seismic driving mechanisms. Furthermore, the upwelling was distributed in a north–south direction near the epicenter of the Mw = 7.0 earthquake, while its diffusion in the east–west direction was not apparent. Conversely, the upwelling observed near the epicenter of the Mw = 7.6 earthquake demonstrated a tendency to propagate in an east–west direction. This spatial distribution corresponded to the adjacent tectonic plate boundaries near the epicenters (Figure 1). Therefore, the spatial distribution, temporal progression and evolution pattern of SPT and SWS anomalies suggest that the earthquake preparation process impacts the ocean dynamics (Figure 7). This effect is manifested by the energy imparted to the ocean floor during the acceleration of tectonic plates into subduction, subsequently leading to the uplifting of colder water.

The soundings provided by F7/C2 had a relatively dense distribution and covered land and ocean, thereby unveiling intriguing vertical variations in seismic perturbations in the ionosphere. The anomalies, emerging from a few days to approximately 60 days before earthquakes, coincided with ionospheric perturbations related to earthquakes detected by alternative data and methodologies [64–66]. Most Ne anomalies occur in the duration of seismic thermal anomalies, consistent with our past research [24]. We noticed that continuous Ne anomalies exhibited an increasing amplitude or an upward shift in the peak. This vertical structure change reflects the same bottom-up disturbances as the seismic thermal anomalies. Additionally, similar to the interruption observed in the analysis of seismic thermal anomalies discussed earlier and its correlation with earthquakes occurring in regions with smaller land area, analogous patterns were also discernible in ionospheric seismic anomalies. In comparison to the Mw = 7.0 earthquake, the Mw = 7.6 earthquake was characterized by a higher ratio of land to sea, resulting in a more concentrated occurrence of Ne anomalies. These observations substantiate the significant correlation between ionospheric disturbances induced by earthquakes and seismic thermal anomalies. The heating of the lower isobaric level from the ground surface produces an air pressure gradient [23]. The pressure gradient propagates toward the upper atmosphere and disturbs the ionosphere [67–69].

In light of the above analysis, the clear anomalous propagation pattern that enhances the understanding of OLAIC has been interconnected and presented, as illustrated in Figure 7. Furthermore, seismic anomalies across the ocean, lithosphere, atmosphere and ionosphere were effectively identified through tidal force fluctuation, substantiating their coupled relationship. In previous studies, tidal force fluctuation has only been used to detect seismic anomalies in the atmosphere and on the surface [16,21–24]. Through amalgamating multi-source data and employing refined parameter criteria, tidal force fluctuation has the potential to emerge as a viable approach for delineating ocean–lithosphere–atmosphere–ionosphere coupling (OLAIC).

5. Conclusions

In the current investigation, we explored coupling anomalies in the ocean, atmosphere and ionosphere through tidal force fluctuation based on oceanic parameters (i.e., sea potential temperature and sea water salinity), air temperature and electron density profiles. The findings reveal that the intensity of seismic anomalies in the atmosphere is influenced by the land-to-sea ratio near the epicenter. Seismic thermal anomalies initiate along the active faults and adhere to the pattern of change characterized by 'initial rise-intensification-attenuation', attributed to the thermal radiation resulting from rock fracture. Similarly, seismic upwelling, driven by the uplift of cold water, propagates along plate boundaries with a discernible 'intensification-peak-weakening' trend under the action of the accelerated subduction of tectonic plates. The evolution process and temporal discontinuity of seismic thermal anomalies and electron density (Ne) anomalies suggest a potential link between ionospheric perturbations and atmospheric pressure gradients. The occurrence of Ne anomalies and seismic upwelling is effectively encompassed within the duration of seismic thermal anomalies, and there is a synchronous evolution tendency among multi-layers. This collective behavior offers insights into ocean–lithosphere–atmosphere–ionosphere coupling (OLAIC) and the intricate transmission of anomalies associated with earthquakes that occur on the sea coast. During earthquake preparation phases, the acceleration of plate subduction leads to two main processes: (1) The energy released by rock fracture under enhanced tectonic stress in active faults acts on convective thermal fluids. Subsequently, the fluid breaks down, rises and heats the land surface, causing atmospheric thermal anomalies, creating the air pressure gradient that disturbs the ionosphere. (2) The energy deposited into the ocean floor along the plate boundaries uplifts cold water to drive upwelling, which promotes a decline in potential temperature and an increase in the salinity in the upper ocean. Furthermore, the tidal force fluctuation method, which is excellent in the identification of multi-layer anomalies, is worthy of further study.

Supplementary Materials: The following supporting information can be downloaded at: https://www.mdpi.com/article/10.3390/rs15184392/s1, Figure S1: Air temperature change in (a) tidal force period A, (b) tidal force period B, (c) tidal force period C, (d) tidal force period D, (e) tidal force period E and (f) tidal force period F for Mw = 7.0 earthquake. The epicenter and faults are shown by black point and red lines, respectively; Figure S2: Air temperature change in (a) tidal force period A, (b) tidal force period B, (c) tidal force period C, (d) tidal force period D, (e) tidal force period E and (f) tidal force period F for Mw = 7.6 earthquake. The epicenter and faults are shown by black point and red lines, respectively; Figure S3: The variation of sea potential temperature anomalies during the studies periods for Mw = 7.0 earthquake; Figure S4: The variation of sea potential temperature anomalies during the studies periods for Mw = 7.6 earthquake; Figure S5: The variation of sea water salinity anomalies during the studies periods for Mw = 7.0 earthquake; Figure S6: The variation of sea water salinity anomalies during the studies periods for Mw = 7.6 earthquake; Figure S7: The variation of sea potential temperature anomalies from 6 June 2021 to 18 August 2021 within 13–23°N and 116–126°E; Figure S8: The variation of sea potential temperature anomalies from 20 July 2021 to 2 October 2021 within 11–1°S and 142–152°E; Figure S9: The variation of sea water salinity anomalies from 6 June 2021 to 18 August 2021 within 13–23°N and 116–126°E; Figure S10: The variation of sea water salinity anomalies from 20 July 2021 to 2 October 2021 within 11–1°S and 142–152°E.

Author Contributions: Conceptualization, X.X.; methodology, X.X.; software, X.X.; validation, X.X., L.W.; formal analysis, X.X.; investigation, X.X.; resources, X.X.; data curation, X.X.; writing—original draft preparation, X.X.; writing—review and editing, S.C., L.W.; visualization, X.X.; supervision, S.C.; project administration, S.C. All authors have read and agreed to the published version of the manuscript.

Funding: This research was funded by China Scholarship Council (CSC).

Data Availability Statement: The earthquake data are available at USGS (https://earthquake.usgs.gov/earthquakes, accessed on 7 August 2023). The air temperature data are available at NCAR (https://rda.ucar.edu/datasets/ds083.2/, accessed on 7 August 2023). The sea potential temperature data and seawater salinity data are available at CNEMS (https://data.marine.copernicus.eu/products, accessed on 29 August 2023). The FORMOSAT-7/COSMIC-2 soundings are available at TACC (tacc.cwb.gov.tw/v2/trops_download.html, accessed on 7 August 2023).

Acknowledgments: The authors express gratitude to NCEP for supplying air temperature data and to the CNEMS community for sharing oceanic data. The authors also extend their appreciation to TACC for sharing electron density data and to USGS for providing earthquake data.

Conflicts of Interest: The authors declare no conflict of interest.

Appendix A

The evolution processes of seismic thermal anomalies of the two earthquakes are shown in Figures A1 and A2.

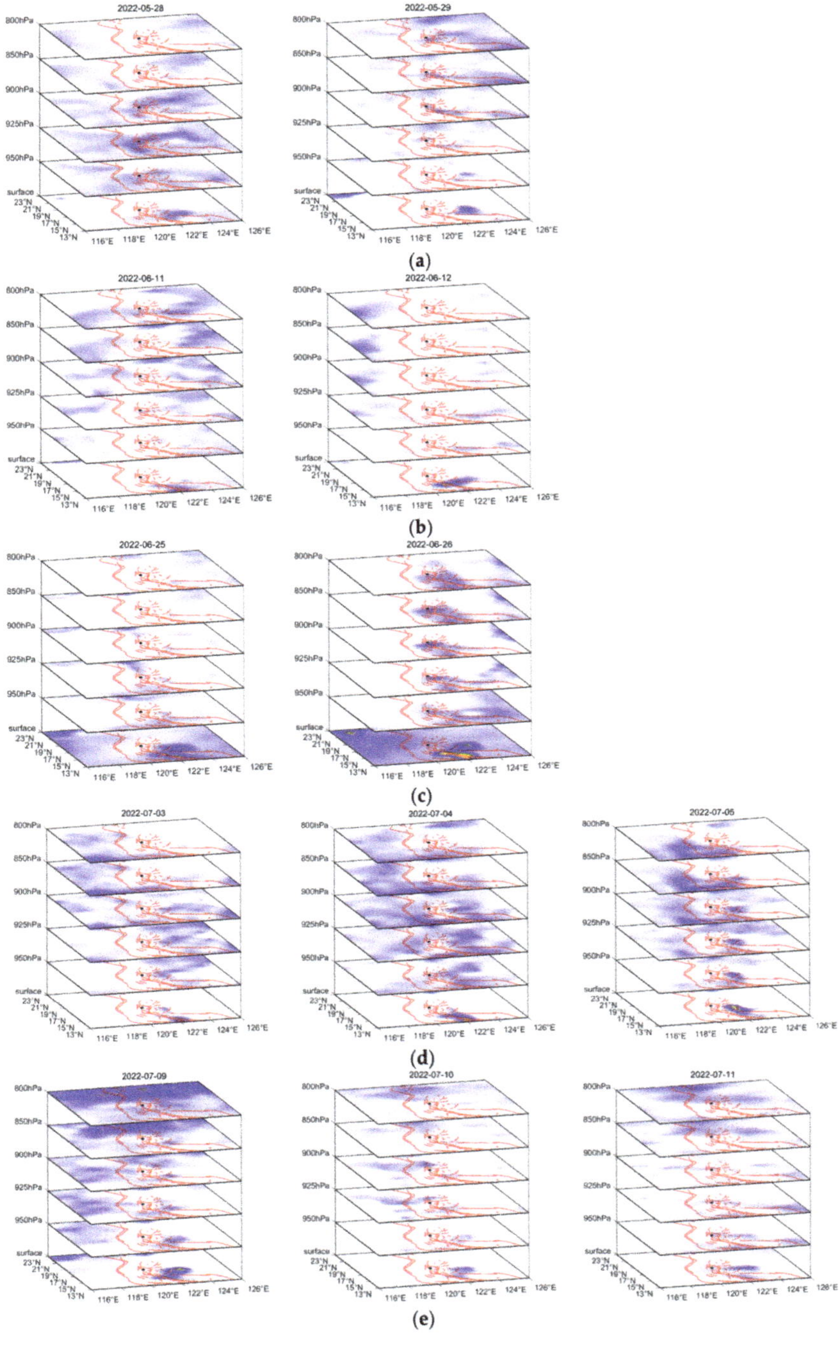

Figure A1. *Cont.*

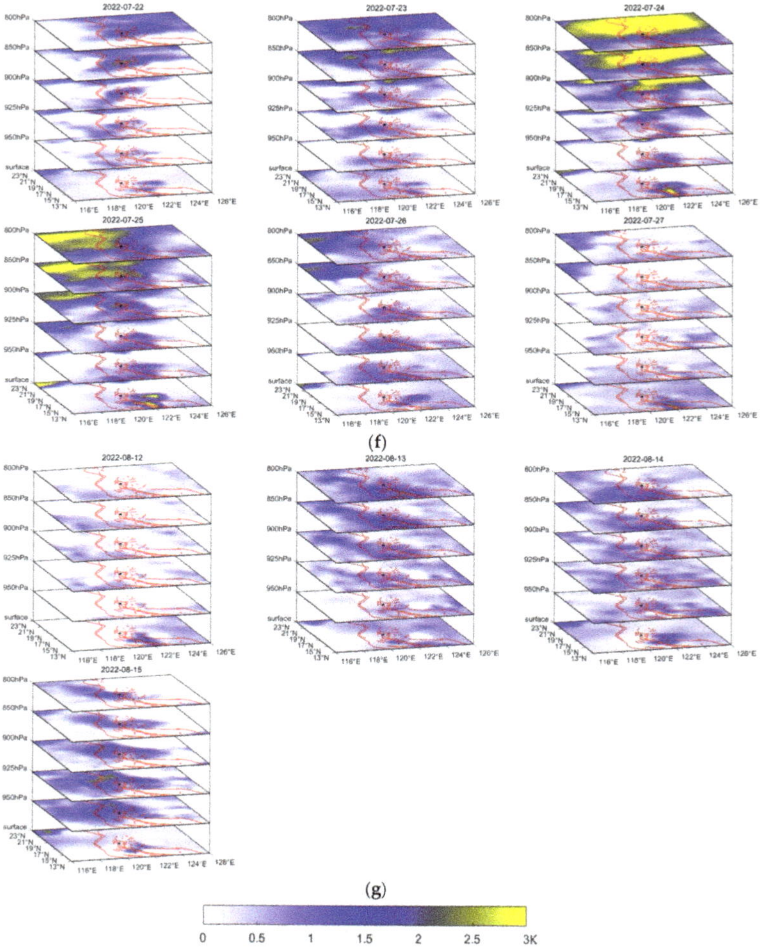

Figure A1. Seismic thermal anomalies that occurred (**a**) from 28 May to 29 May, (**b**) from 11 June to 12 June, (**c**) from 25 June to 26 June, (**d**) from 3 July to 5 July, (**e**) from 9 July to 11 July, (**f**) from 22 July to 27 July and (**g**) from 12 August to 15 August.

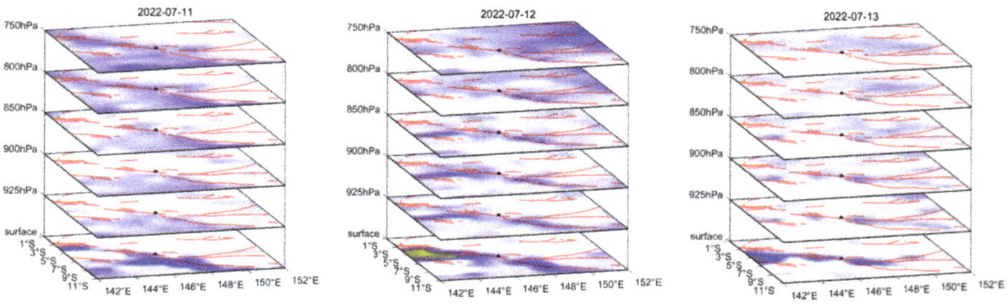

(**a**)

Figure A2. *Cont.*

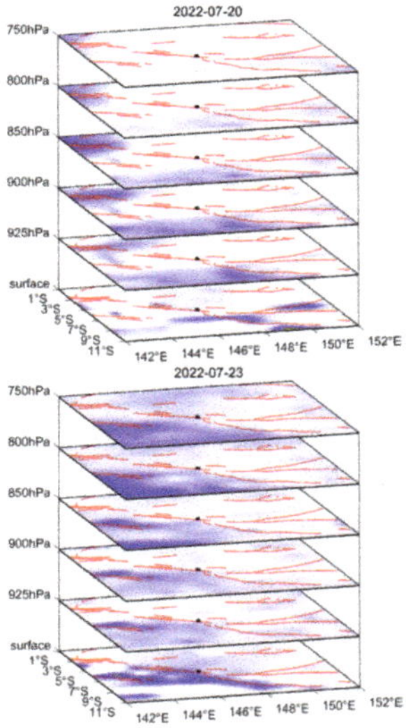

(b)

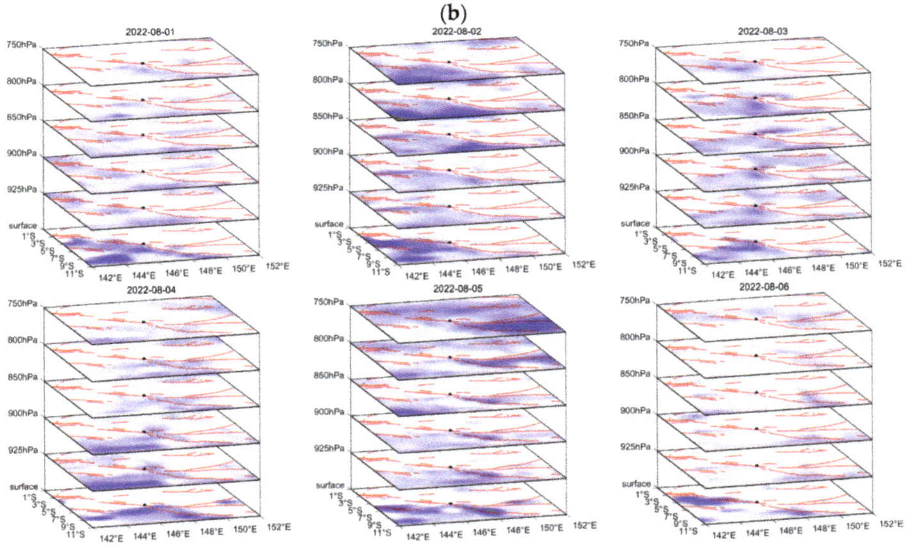

(c)

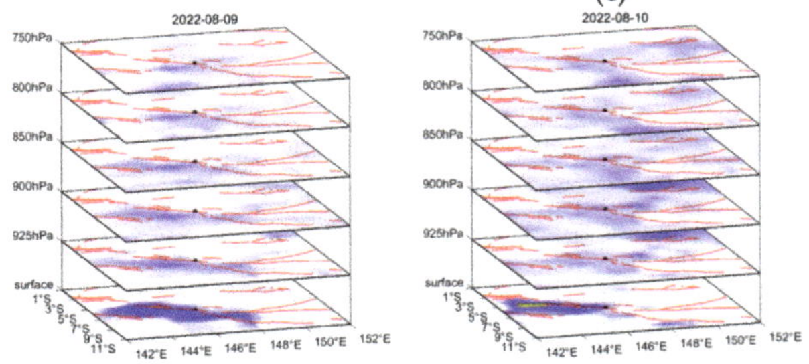

(d)

Figure A2. *Cont.*

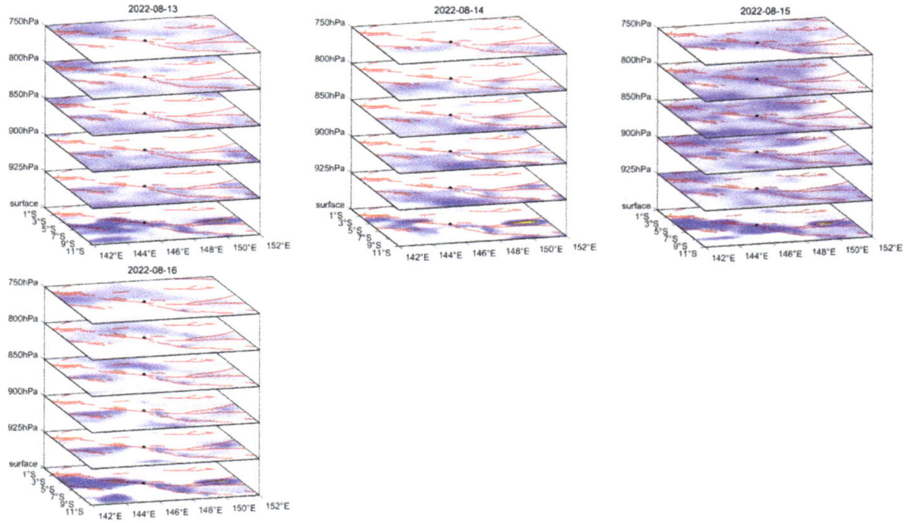

(e)

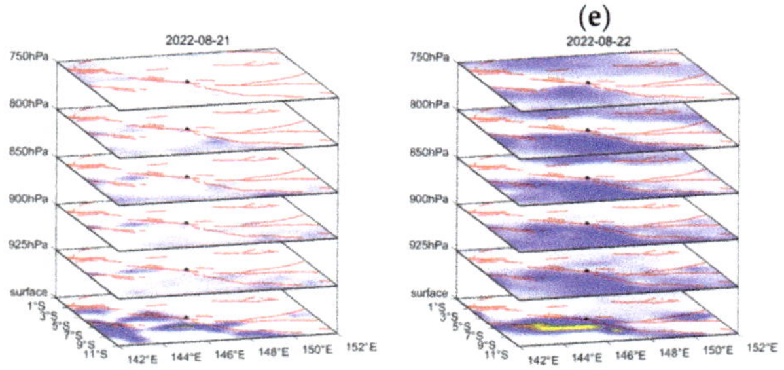

(f)

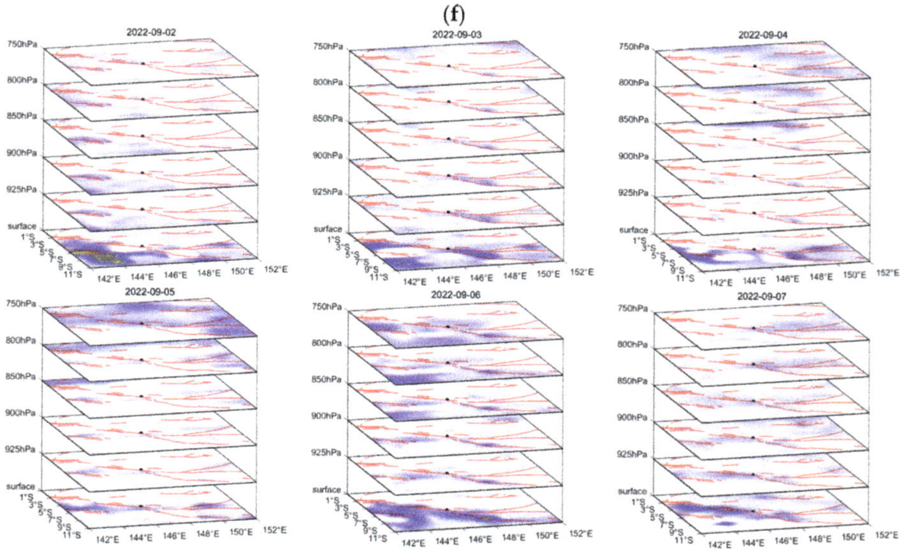

Figure A2. *Cont.*

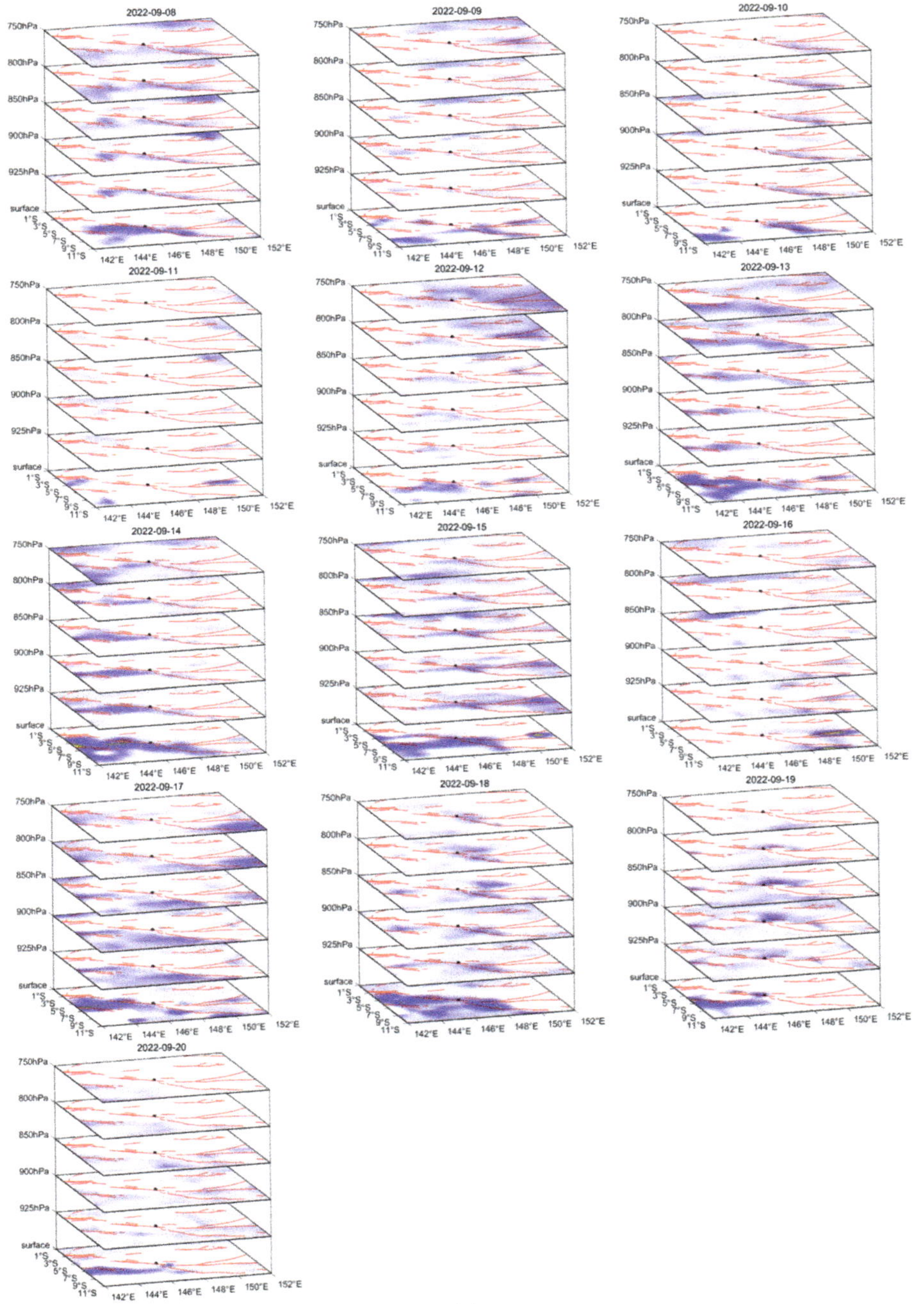

(g)

Figure A2. *Cont.*

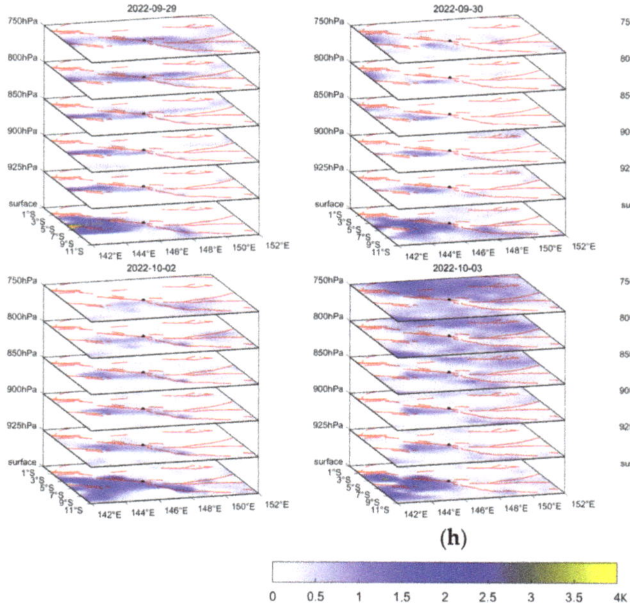

Figure A2. Seismic anomalies that occurred (**a**) from 11 July to 13 July, (**b**) from 20 July to 23 July, (**c**) from 1 August to 6 August, (**d**) from 9 August to 10 August, (**e**) from 13 August to 16 August, (**f**) 21 August to 22 August, (**g**) 2 September to 20 September and (**h**) 29 September to 4 October.

Appendix B

The evolution processes of SPT anomalies and SWS anomalies over the third pattern duration of the two earthquakes are shown in Figures A3 and A4.

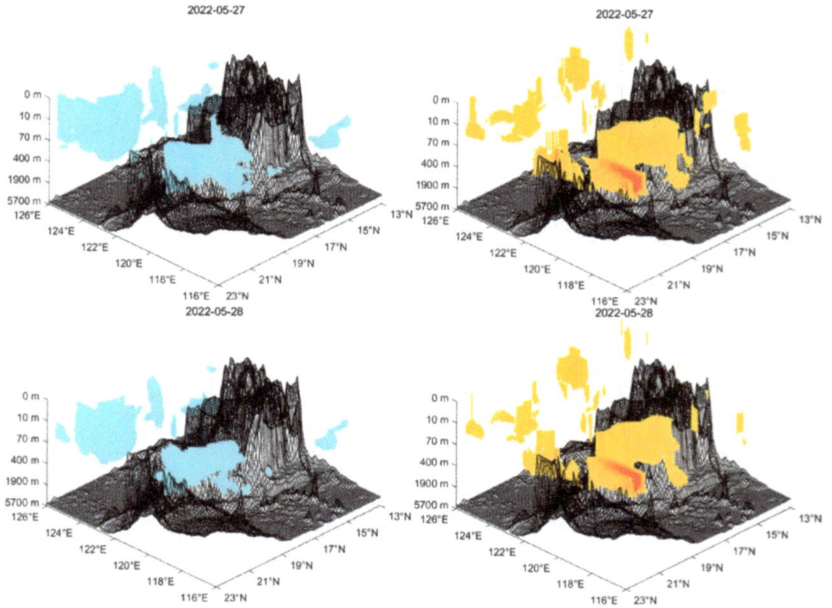

Figure A3. *Cont.*

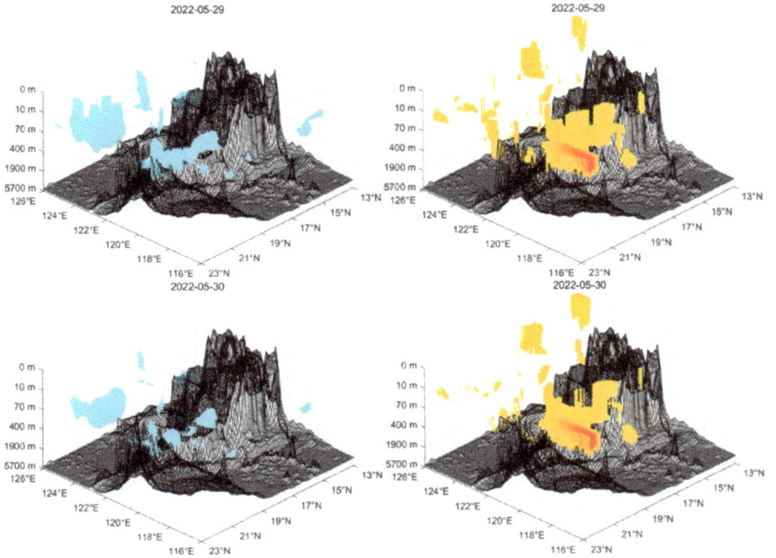

(a)

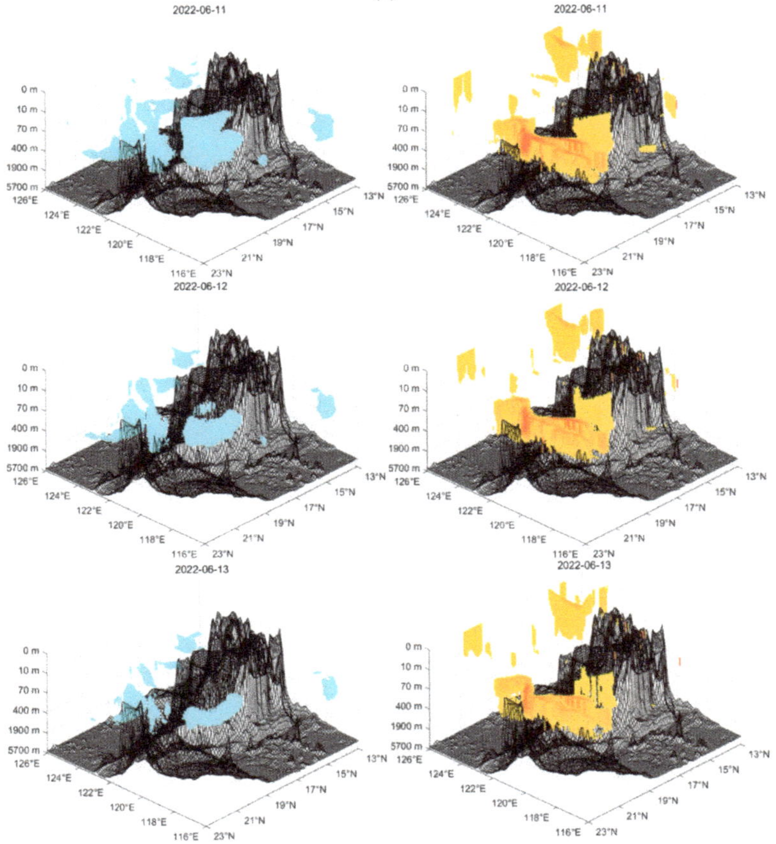

Figure A3. *Cont.*

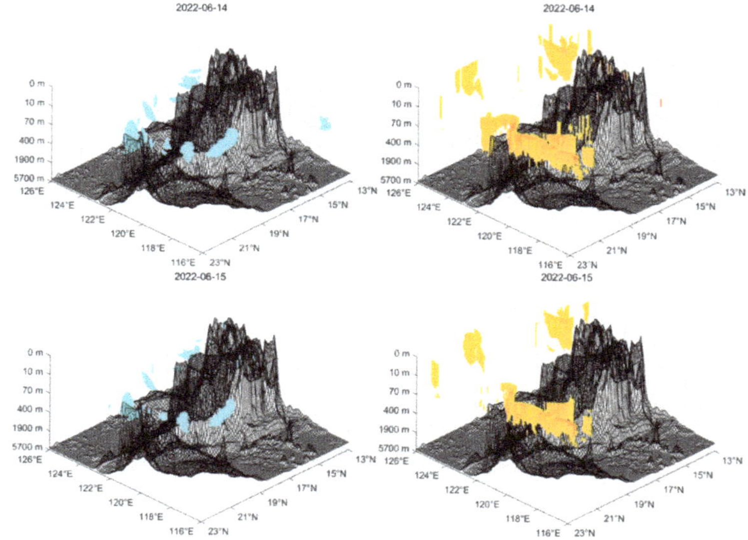

(**b**)

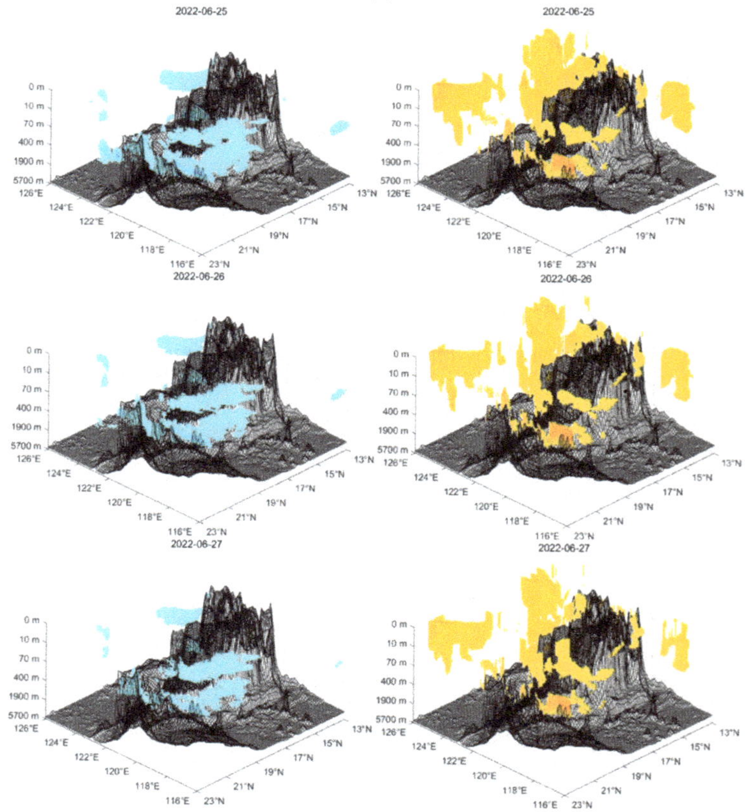

Figure A3. *Cont.*

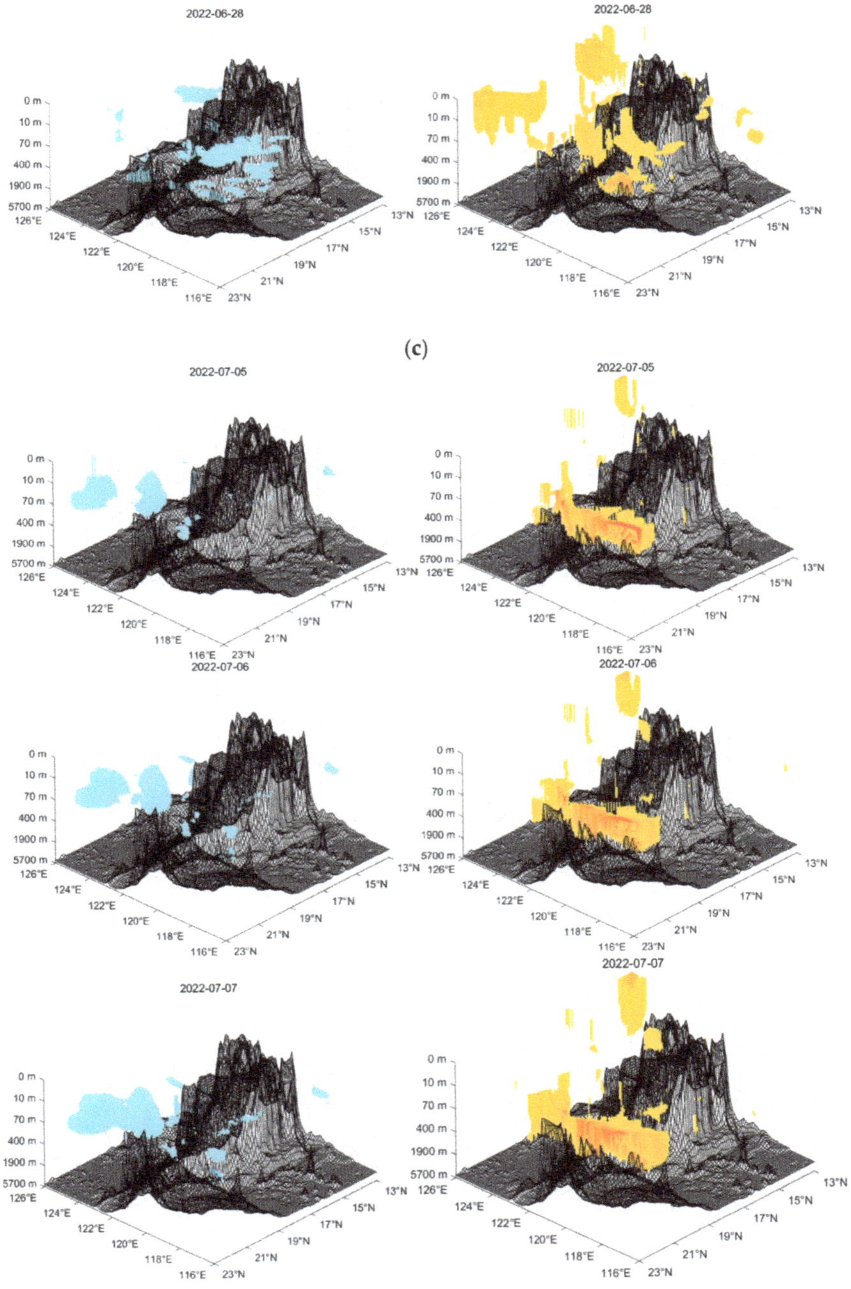

Figure A3. *Cont.*

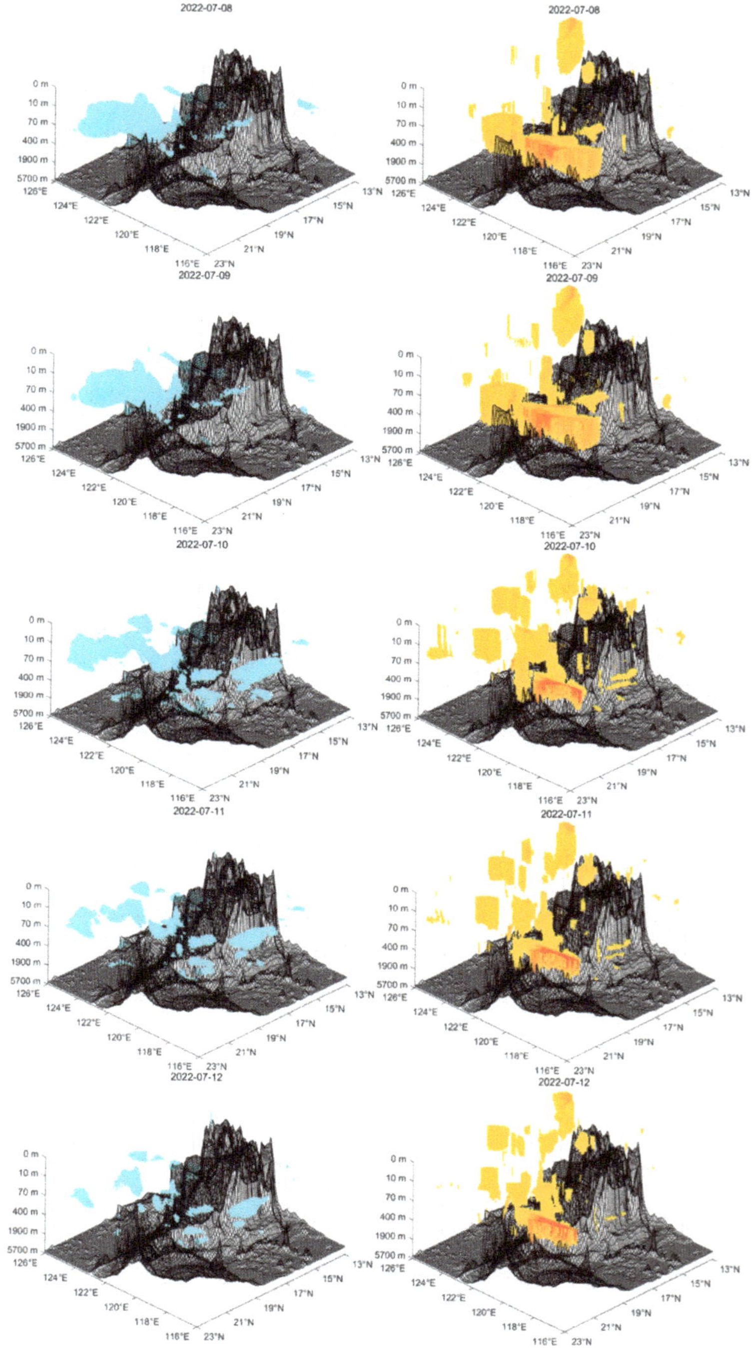

(**d**)

Figure A3. *Cont.*

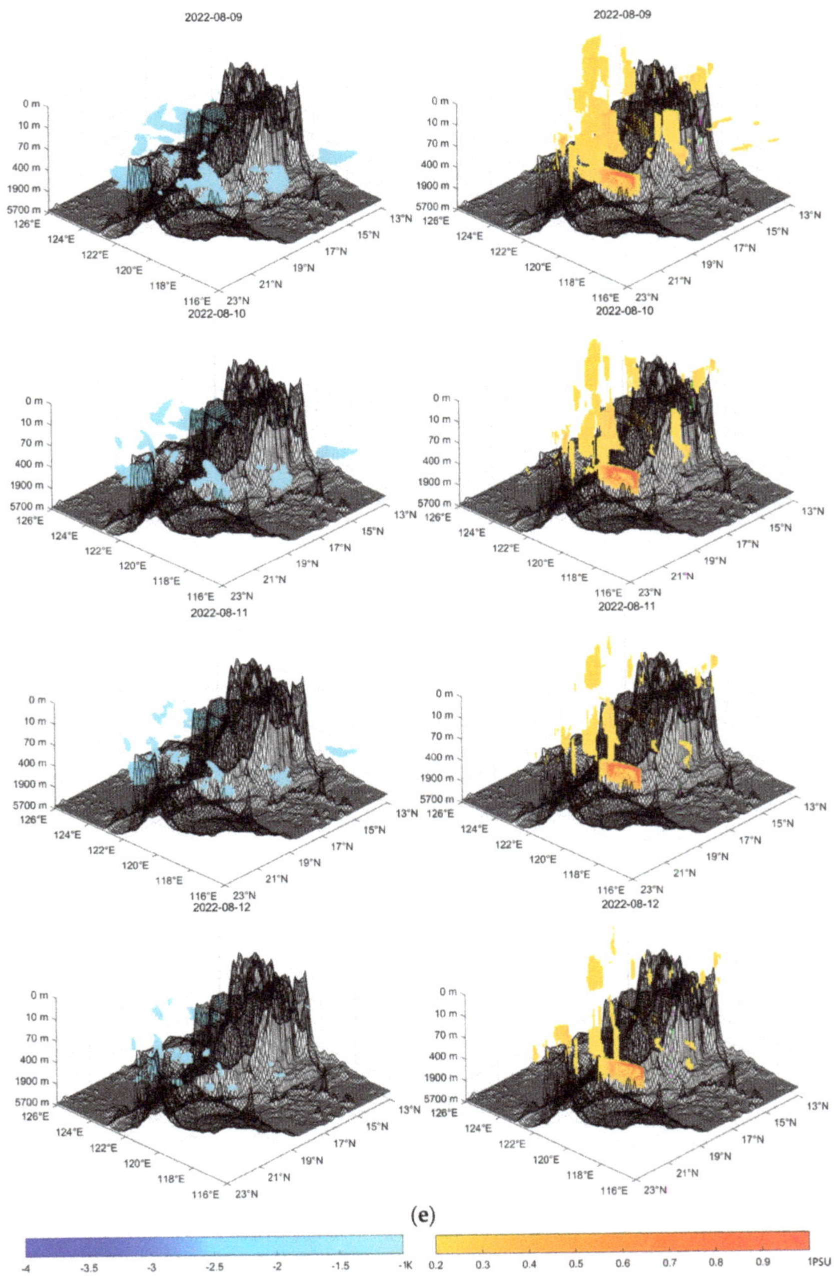

Figure A3. The third anomalous variation pattern of sea potential temperature and corresponding anomalous variation in seawater salinity (**a**) from 27 May to 30 May, (**b**) from 11 June to 15 June, (**c**) from 25 June to 28 June, (**d**) from 5 July to 12 July and (**e**) from 9 August to 12 August.

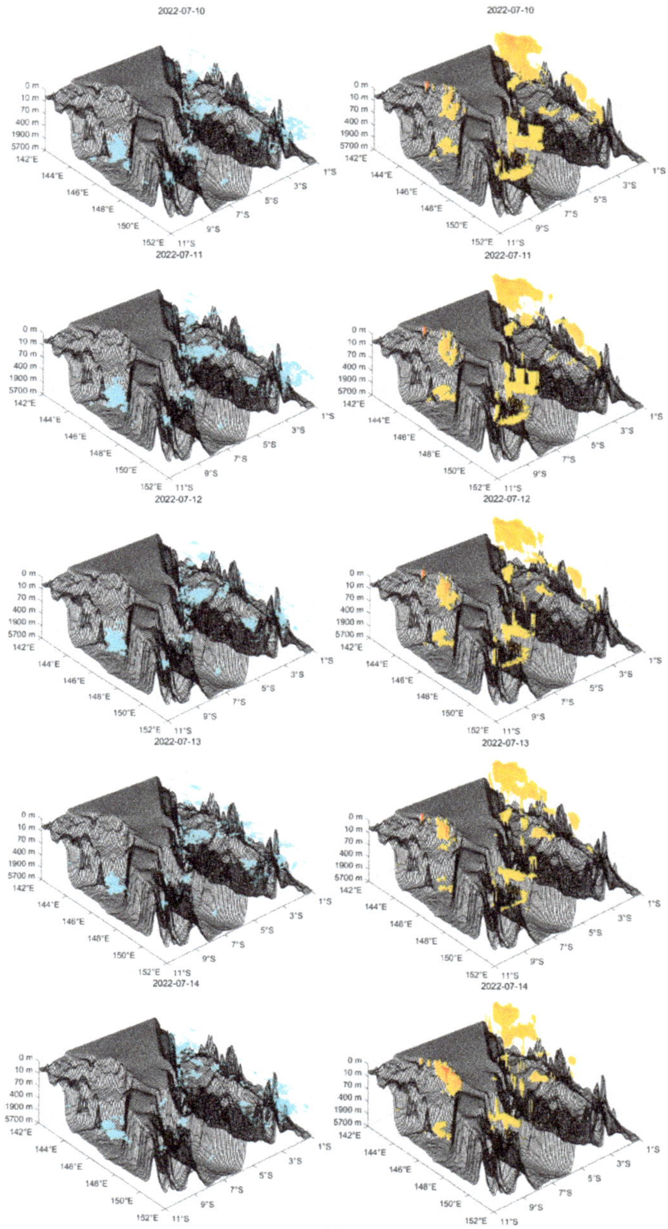

(a)

Figure A4. *Cont.*

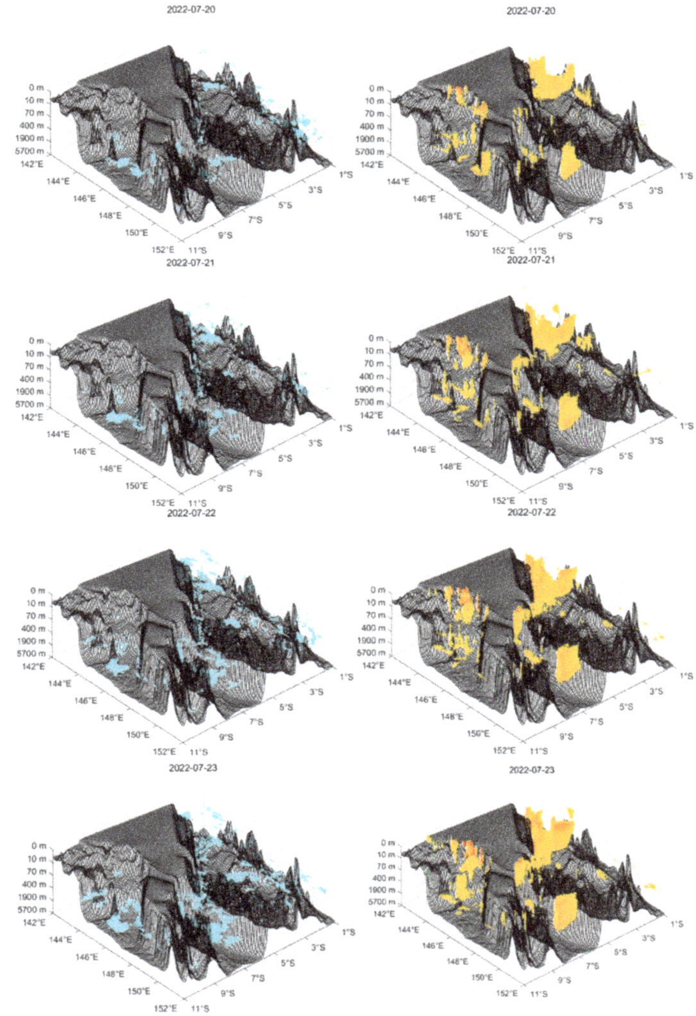

(**b**)

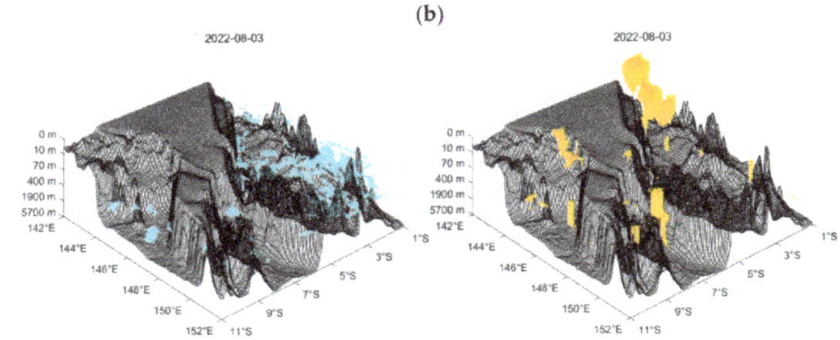

Figure A4. *Cont.*

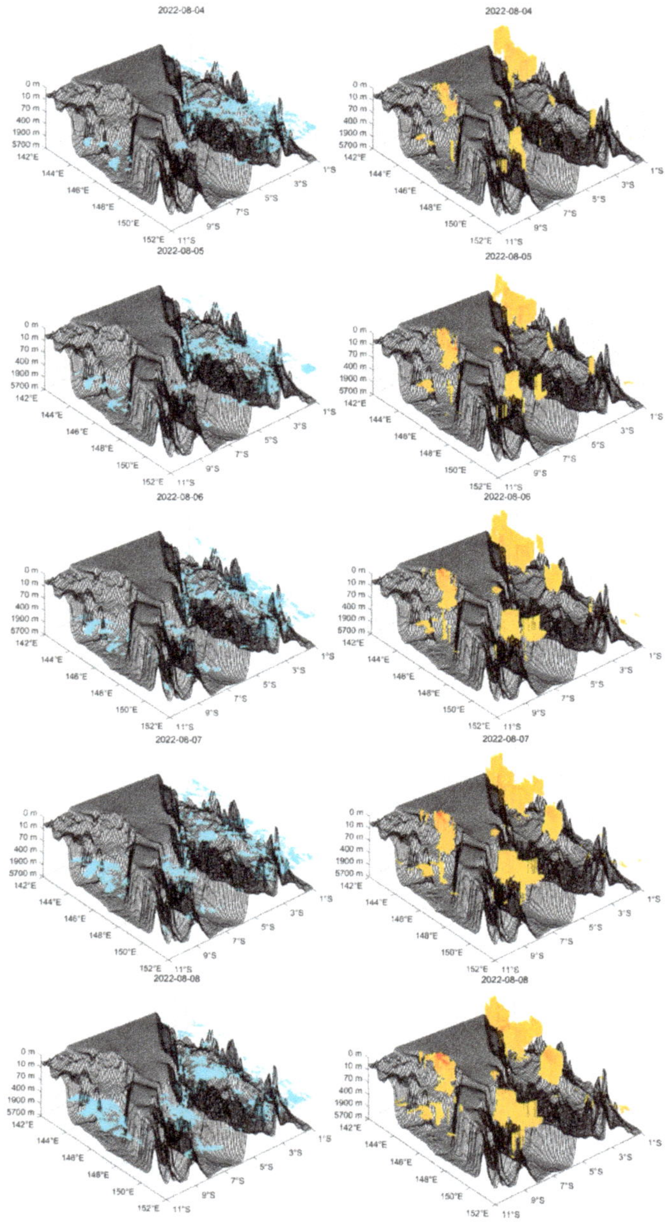

Figure A4. *Cont.*

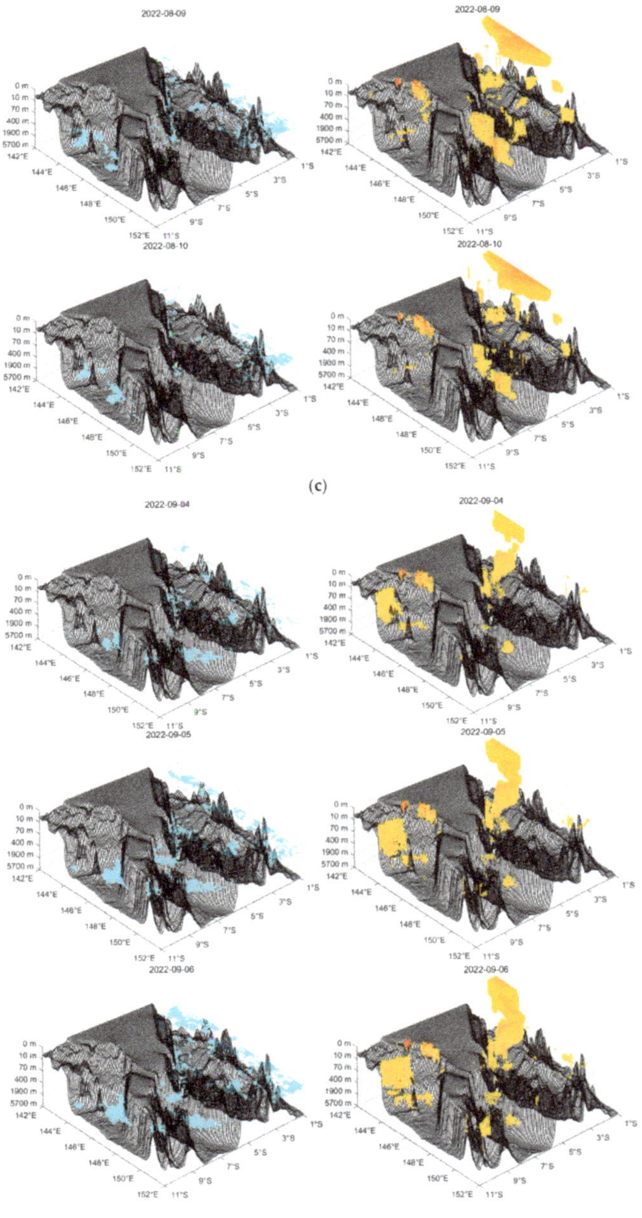

Figure A4. *Cont.*

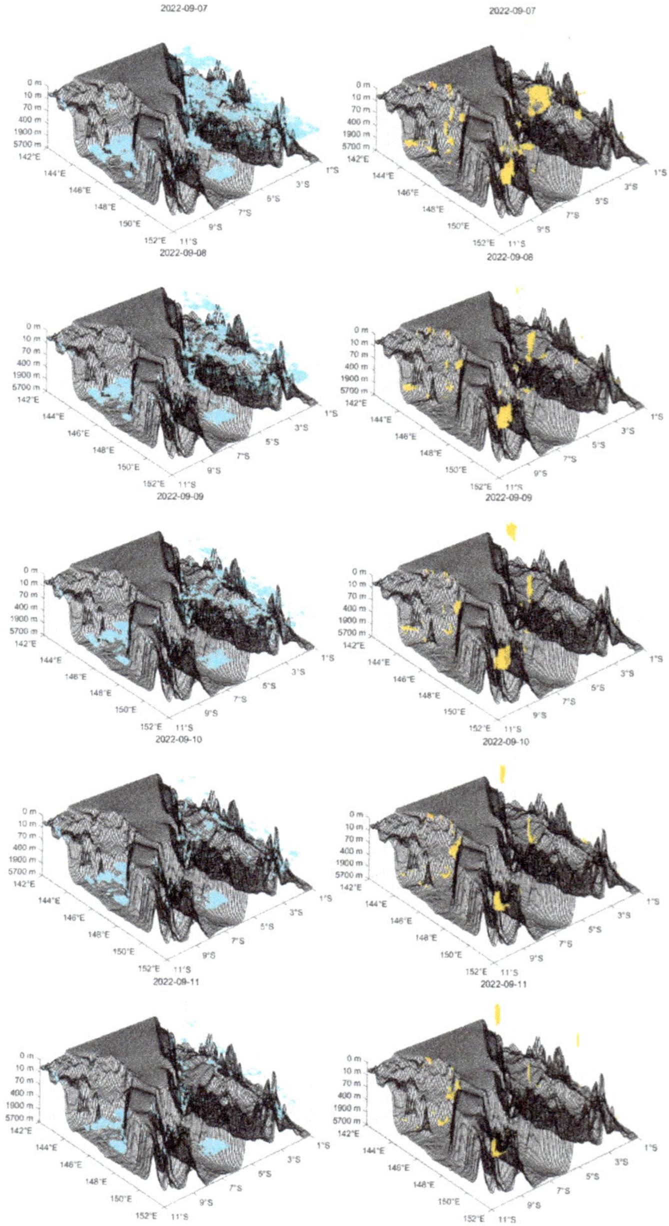

Figure A4. *Cont.*

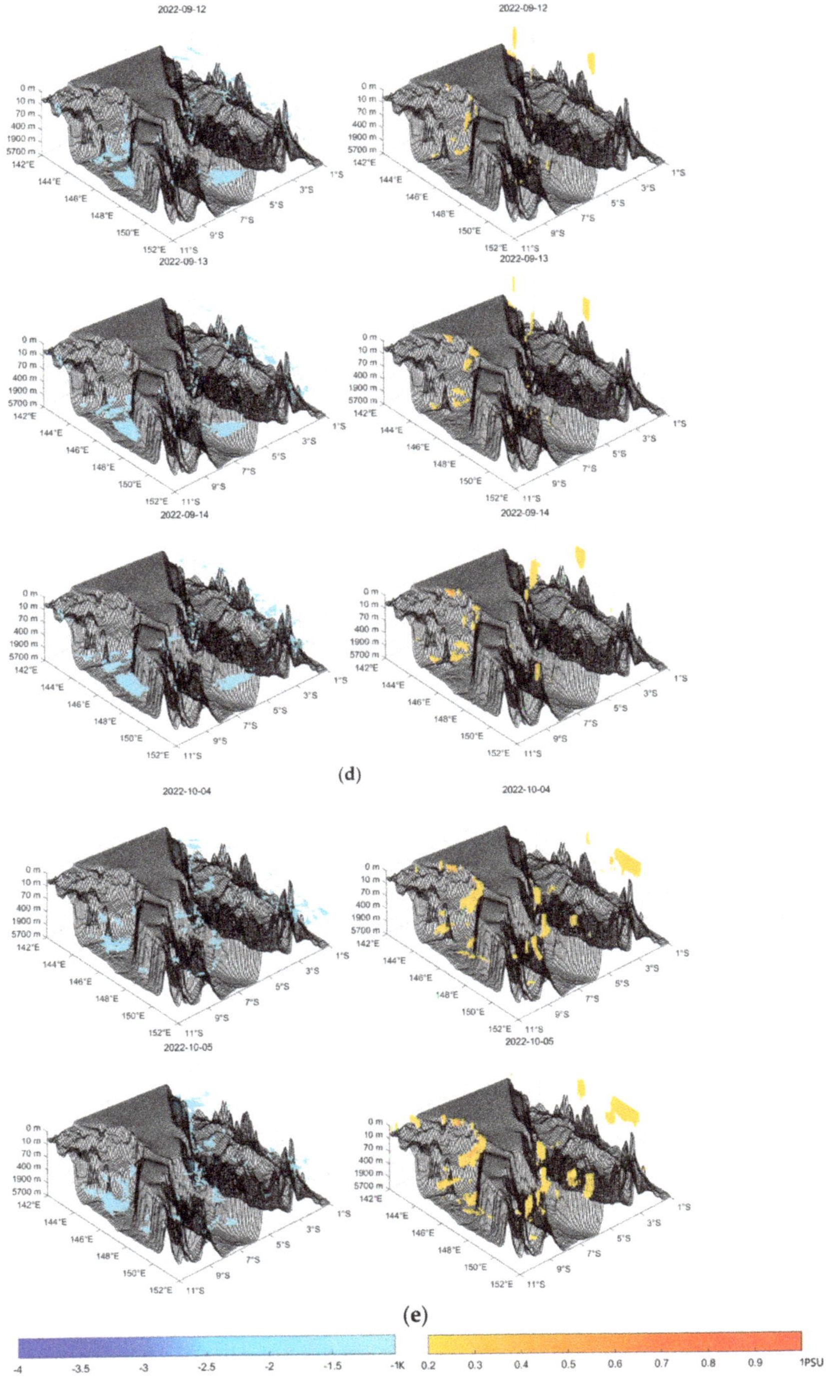

Figure A4. The third anomalous variation pattern of sea potential temperature and corresponding anomalous variations in seawater salinity (**a**) from 10 July to 14 July, (**b**) 20 July to 23 July, (**c**) from 3 August to 10 August, (**d**) from 4 September to 14 September and (**e**) from 4 October to 5 October.

References

1. Davidenko, D.; Pulinets, S. Deterministic variability of the ionosphere on the eve of strong ($M \geq 6$) earthquakes in the regions of Greece and Italy according to long-term measurements data. *Geomagn. Aeron.* **2019**, *59*, 493–508. [CrossRef]
2. Shah, M.; Tariq, M.A.; Ahmad, J.; Naqvi, N.A.; Jin, S. Seismo ionospheric anomalies before the 2007 M7. 7 Chile earthquake from GPS TEC and DEMETER. *J. Geodyn.* **2019**, *127*, 42–51. [CrossRef]
3. Choudhury, S.; Dasgupta, S.; Saraf, A.K.; Panda, S. Remote sensing observations of pre-earthquake thermal anomalies in Iran. *Int. J. Remote Sens.* **2006**, *27*, 4381–4396. [CrossRef]
4. Cui, Y.; Ouzounov, D.; Hatzopoulos, N.; Sun, K.; Zou, Z.; Du, J. Satellite observation of CH4 and CO anomalies associated with the Wenchuan MS 8.0 and Lushan MS 7.0 earthquakes in China. *Chem. Geol.* **2017**, *469*, 185–191. [CrossRef]
5. Chakraborty, S.; Sasmal, S.; Basak, T.; Chakrabarti, S.K. Comparative study of charged particle precipitation from Van Allen radiation belts as observed by NOAA satellites during a land earthquake and an ocean earthquake. *Adv. Space Res.* **2019**, *64*, 719–732. [CrossRef]
6. Singh, R.P.; Dey, S.; Bhoi, S.; Sun, D.; Cervone, G.; Kafatos, M. Anomalous increase of chlorophyll concentrations associated with earthquakes. *Adv. Space Res.* **2006**, *37*, 671–680. [CrossRef]
7. Genzano, N.; Filizzola, C.; Hattori, K.; Pergola, N.; Tramutoli, V. Statistical correlation analysis between thermal infrared anomalies observed from MTSATs and large earthquakes occurred in Japan (2005–2015). *J. Geophys. Res. Solid Earth* **2021**, *126*, e2020JB020108. [CrossRef]
8. Adil, M.A.; Şentürk, E.; Pulinets, S.A.; Amory-Mazaudier, C. A lithosphere–atmosphere–ionosphere coupling phenomenon observed before M 7.7 Jamaica Earthquake. *Pure Appl. Geophys.* **2021**, *178*, 3869–3886. [CrossRef]
9. Mehdi, S.; Shah, M.; Naqvi, N.A. Lithosphere atmosphere ionosphere coupling associated with the 2019 M w 7.1 California earthquake using GNSS and multiple satellites. *Environ. Monit. Assess.* **2021**, *193*, 501. [CrossRef]
10. Shah, M.; Aibar, A.C.; Tariq, M.A.; Ahmed, J.; Ahmed, A. Possible ionosphere and atmosphere precursory analysis related to Mw> 6.0 earthquakes in Japan. *Remote Sens. Environ.* **2020**, *239*, 111620. [CrossRef]
11. Jing, F.; Singh, R.P.; Shen, X. Land–atmosphere–meteorological coupling associated with the 2015 Gorkha (M 7.8) and Dolakha (M 7.3) Nepal earthquakes. *Geomat. Nat. Hazards Risk* **2019**, *10*, 1267–1284. [CrossRef]
12. Pulinets, S.; Ouzounov, D. Lithosphere–Atmosphere–Ionosphere Coupling (LAIC) model–An unified concept for earthquake precursors validation. *J. Asian Earth Sci.* **2011**, *41*, 371–382. [CrossRef]
13. Freund, F. Pre-earthquake signals: Underlying physical processes. *J. Asian Earth Sci.* **2011**, *41*, 383–400. [CrossRef]
14. Kuo, C.; Huba, J.; Joyce, G.; Lee, L. Ionosphere plasma bubbles and density variations induced by pre-earthquake rock currents and associated surface charges. *J. Geophys. Res. Space Phys.* **2011**, *116*, A10317. [CrossRef]
15. Kuo, C.; Lee, L.; Huba, J. An improved coupling model for the lithosphere-atmosphere-ionosphere system. *J. Geophys. Res. Space Phys.* **2014**, *119*, 3189–3205. [CrossRef]
16. Weiyu, M.; Xuedong, Z.; Liu, J.; Yao, Q.; Zhou, B.; Yue, C.; Kang, C.; Lu, X. Influences of multiple layers of air temperature differences on tidal forces and tectonic stress before, during and after the Jiujiang earthquake. *Remote Sens. Environ.* **2018**, *210*, 159–165. [CrossRef]
17. Tramutoli, V. Robust satellite techniques (RST) for natural and environmental hazards monitoring and mitigation: Theory and applications. In Proceedings of the 2007 International Workshop on the Analysis of Multi-Temporal Remote Sensing Images, Leuven, Belgium, 18–20 July 2007; pp. 1–6.
18. Ouzounov, D.; Pulinets, S.; Romanov, A.; Romanov, A.; Tsybulya, K.; Davidenko, D.; Kafatos, M.; Taylor, P. Atmosphere-ionosphere response to the M 9 Tohoku earthquake revealed by multi-instrument space-borne and ground observations: Preliminary results. *Earthq. Sci.* **2011**, *24*, 557–564. [CrossRef]
19. Xiong, P.; Shen, X. Outgoing longwave radiation anomalies analysis associated with different types of seismic activity. *Adv. Space Res.* **2017**, *59*, 1408–1415. [CrossRef]
20. Tramutoli, V.; Corrado, R.; Filizzola, C.; Genzano, N.; Lisi, M.; Pergola, N. From visual comparison to Robust Satellite Techniques: 30 years of thermal infrared satellite data analyses for the study of earthquake preparation phases. *Boll. Geofis. Teor. Appl.* **2015**, *56*, 167–202.
21. Zhang, X.; Kang, C.; Ma, W.; Ren, J.; Wang, Y. Study on thermal anomalies of earthquake process by using tidal-force and outgoing-longwave-radiation. *Therm. Sci.* **2018**, *22*, 767–776. [CrossRef]
22. Zhang, Y.; Meng, Q.; Wang, Z.; Lu, X.; Hu, D. Temperature variations in multiple air layers before the Mw 6.2 2014 Ludian earthquake, Yunnan, China. *Remote Sens.* **2021**, *13*, 884. [CrossRef]
23. Xu, X.; Chen, S.; Yu, Y.; Zhang, S. Atmospheric anomaly analysis related to Ms> 6.0 earthquakes in China during 2020–2021. *Remote Sens.* **2021**, *13*, 4052. [CrossRef]
24. Xu, X.; Chen, S.; Zhang, S.; Dai, R. Analysis of Potential Precursory Pattern at Earth Surface and the Above Atmosphere and Ionosphere Preceding Two Mw≥ 7 Earthquakes in Mexico in 2020–2021. *Earth Space Sci.* **2022**, *9*, e2022EA002267. [CrossRef]
25. Heaton, T.H. Tidal triggering of earthquakes. *Geophys. J. Int.* **1975**, *43*, 307–326. [CrossRef]
26. Su, Y.-C.; Liu, J.-Y.; Chen, S.-P.; Tsai, H.-F.; Chen, M.-Q. Temporal and spatial precursors in ionospheric total electron content of the 16 October 1999 Mw7. 1 Hector Mine earthquake. *J. Geophys. Res. Space Phys.* **2013**, *118*, 6511–6517. [CrossRef]
27. Yan, X.; Sun, Y.; Yu, T.; Liu, J.Y.; Qi, Y.; Xia, C.; Zuo, X.; Yang, N. Stratosphere perturbed by the 2011 Mw9. 0 Tohoku earthquake. *Geophys. Res. Lett.* **2018**, *45*, 10050–10056. [CrossRef]

28. Ahmed, J.; Shah, M.; Zafar, W.A.; Amin, M.A.; Iqbal, T. Seismoionospheric anomalies associated with earthquakes from the analysis of the ionosonde data. *J. Atmos. Sol. Terr. Phys.* **2018**, *179*, 450–458. [CrossRef]
29. Dey, S.; Singh, R. Surface latent heat flux as an earthquake precursor. *Nat. Hazards Earth Syst. Sci.* **2003**, *3*, 749–755. [CrossRef]
30. Ouzounov, D.; Freund, F. Mid-infrared emission prior to strong earthquakes analyzed by remote sensing data. *Adv. Space Res.* **2004**, *33*, 268–273. [CrossRef]
31. Okada, Y.; Mukai, S.; Singh, R. Changes in atmospheric aerosol parameters after Gujarat earthquake of January 26, 2001. *Adv. Space Res.* **2004**, *33*, 254–258. [CrossRef]
32. Akhoondzadeh, M. Ant Colony Optimization detects anomalous aerosol variations associated with the Chile earthquake of 27 February 2010. *Adv. Space Res.* **2015**, *55*, 1754–1763. [CrossRef]
33. Zhang, L.; Jiang, M.; Jing, F. Sea temperature variation associated with the 2021 Haiti M w 7.2 earthquake and possible mechanism. *Geomat. Nat. Hazards Risk* **2022**, *13*, 2840–2863. [CrossRef]
34. Alvan, H.V.; Azad, F.H.; Omar, H.B. Chlorophyll concentration and surface temperature changes associated with earthquakes. *Nat. Hazards* **2012**, *64*, 691–706. [CrossRef]
35. Mohamed, E.K.; Elrayess, M.; Omar, K. Evaluation of thermal anomaly preceding northern red sea earthquake, the 16th June 2020. *Arab. J. Sci. Eng.* **2022**, *47*, 7387–7406. [CrossRef]
36. Liu, Y.; Wu, L.; Qi, Y.; Ding, Y. Very-Short-Term Variations of Sea Surface and Atmospheric Parameters Before the Ms 6.2 Zhangbei (China) Earthquake in 1998. *Front. Environ. Sci.* **2022**, *10*, 906455. [CrossRef]
37. Lellouche, J.; Legalloudec, O.; Regnier, C.; Levier, B.; Greiner, E.; Drevillon, M. *Quality Information Document for Globa L Sea Physical Analysis and Forecasting Product, Global_Analysis_ForecaST_PHY_001_024*; EU Copernicus Marine Service European Union: Brussels, Belgium, 2019; Volume 81.
38. Kalnay, E.; Kanamitsu, M.; Kistler, R.; Collins, W.; Deaven, D.; Gandin, L.; Iredell, M.; Saha, S.; White, G.; Woollen, J. The NCEP/NCAR 40-year reanalysis project. *Bull. Am. Meteorol. Soc.* **1996**, *77*, 437–472. [CrossRef]
39. Kistler, R.; Kalnay, E.; Collins, W.; Saha, S.; White, G.; Woollen, J.; Chelliah, M.; Ebisuzaki, W.; Kanamitsu, M.; Kousky, V. The NCEP/NCAR 50-year reanalysis: Documentation and monthly-means CD-ROM. *Bull. Am. Meteorol. Soc* **2001**, *82*, 247–268. [CrossRef]
40. Sun, Y.-Y.; Chen, C.-H.; Lin, C.-Y. Detection of vertical changes in the ionospheric electron density structures by the radio occultation technique onboard the FORMOSAT-7/COSMIC2 mission over the eruption of the Tonga underwater volcano on 15 January 2022. *Remote Sens.* **2022**, *14*, 4266. [CrossRef]
41. Davies, K. *Ionospheric Radio*; IET: London, UK, 1990.
42. Wang, J.; Sun, Y.Y.; Yu, T.; Wang, Y.; Mao, T.; Yang, H.; Xia, C.; Yan, X.; Yang, N.; Huang, G. Convergence effects on the ionosphere during and after the annular solar eclipse on 21 June 2020. *J. Geophys. Res. Space Phys.* **2022**, *127*, e2022JA030471. [CrossRef]
43. Kilston, S.; Knopoff, L. Lunar–solar periodicities of large earthquakes in southern California. *Nature* **1983**, *304*, 21–25. [CrossRef]
44. Dobrovolsky, I.; Zubkov, S.; Miachkin, V. Estimation of the size of earthquake preparation zones. *Pure Appl. Geophys.* **1979**, *117*, 1025–1044. [CrossRef]
45. Wu, L.; Liu, S.; Wu, Y.; Wang, C. Precursors for rock fracturing and failure—Part II: IRR T-Curve abnormalities. *Int. J. Rock Mech. Min. Sci.* **2006**, *43*, 483–493. [CrossRef]
46. Qu, T.; Mitsudera, H.; Yamagata, T. Intrusion of the north Pacific waters into the South China Sea. *J. Geophys. Res. Ocean.* **2000**, *105*, 6415–6424. [CrossRef]
47. Dong, L.; Qi, J.; Yin, B.; Zhi, H.; Li, D.; Yang, S.; Wang, W.; Cai, H.; Xie, B. Reconstruction of Subsurface Salinity Structure in the South China Sea Using Satellite Observations: A LightGBM-Based Deep Forest Method. *Remote Sens.* **2022**, *14*, 3494. [CrossRef]
48. Yi, D.L.; Melnichenko, O.; Hacker, P.; Potemra, J. Remote sensing of sea surface salinity variability in the South China Sea. *J. Geophys. Res. Ocean.* **2020**, *125*, e2020JC016827. [CrossRef]
49. Tronin, A.A.; Hayakawa, M.; Molchanov, O.A. Thermal IR satellite data application for earthquake research in Japan and China. *J. Geodyn.* **2002**, *33*, 519–534. [CrossRef]
50. Panda, S.; Choudhury, S.; Saraf, A.K.; Das, J.D. MODIS land surface temperature data detects thermal anomaly preceding 8 October 2005 Kashmir earthquake. *Int. J. Remote Sens.* **2007**, *28*, 4587–4596. [CrossRef]
51. Fu, C.-C.; Lee, L.-C.; Ouzounov, D.; Jan, J.-C. Earth's outgoing longwave radiation variability prior to M$\geq$ 6.0 earthquakes in the Taiwan area during 2009–2019. *Front. Earth Sci.* **2020**, *8*, 364. [CrossRef]
52. Zoran, M. MODIS and NOAA-AVHRR l and surface temperature data detect a thermal anomaly preceding the 11 March 2011 Tohoku earthquake. *Int. J. Remote Sens.* **2012**, *33*, 6805–6817. [CrossRef]
53. Genzano, N.; Filizzola, C.; Paciello, R.; Pergola, N.; Tramutoli, V. Robust Satellite Techniques (RST) for monitoring earthquake prone areas by satellite TIR observations: The case of 1999 Chi-Chi earthquake (Taiwan). *J. Asian Earth Sci.* **2015**, *114*, 289–298. [CrossRef]
54. Wu, L.; Zheng, S.; De Santis, A.; Qin, K.; Di Mauro, R.; Liu, S.; Rainone, M.L. Geosphere coupling and hydrothermal anomalies before the 2009 M w 6.3 L'Aquila earthquake in Italy. *Nat. Hazards Earth Syst. Sci.* **2016**, *16*, 1859–1880. [CrossRef]
55. Ma, W.; Zhao, H.; Li, H. Temperature changing process of the Hokkaido (Japan) earthquake on 25 September 2003. *Nat. Hazards Earth Syst. Sci.* **2008**, *8*, 985–989. [CrossRef]
56. Piscini, A.; De Santis, A.; Marchetti, D.; Cianchini, G. A Multi-parametric climatological approach to study the 2016 Amatrice–Norcia (Central Italy) earthquake preparatory phase. *Pure Appl. Geophys.* **2017**, *174*, 3673–3688. [CrossRef]

57. Ouzounov, D.; Pulinets, S.; Hattori, K.; Taylor, P. *Pre-Earthquake Processes: A Multidisciplinary Approach to Earthquake Prediction Studies*; John Wiley & Sons: Hoboken, NJ, USA, 2018; Volume 234.
58. Hu, J.; Wang, X.H. Progress on upwelling studies in the China seas. *Rev. Geophys.* **2016**, *54*, 653–673. [CrossRef]
59. Liu, J.; Mao, K.; Yan, M.; Zhang, X.; Shi, Y. The general distribution characteristics of Luzon cold eddy. *Mar. Forecast.* **2006**, *23*, 39–44.
60. Qu, T. Upper-layer circulation in the South China Sea. *J. Phys. Oceanogr.* **2000**, *30*, 1450–1460. [CrossRef]
61. Udarbe-Walker, M.J.B.; Villanoy, C.L. Structure of potential upwelling areas in the Philippines. *Deep Sea Res. Part I Oceanogr. Res. Pap.* **2001**, *48*, 1499–1518. [CrossRef]
62. Hwang, C.; Chen, S.A. Circulations and eddies over the South China Sea derived from TOPEX/Poseidon altimetry. *J. Geophys. Res. Ocean.* **2000**, *105*, 23943–23965. [CrossRef]
63. Delcroix, T.; Radenac, M.H.; Cravatte, S.; Alory, G.; Gourdeau, L.; Léger, F.; Singh, A.; Varillon, D. Sea surface temperature and salinity seasonal changes in the western Solomon and Bismarck Seas. *J. Geophys. Res. Ocean.* **2014**, *119*, 2642–2657. [CrossRef]
64. Akhoondzadeh, M.; De Santis, A.; Marchetti, D.; Piscini, A.; Cianchini, G. Multi precursors analysis associated with the powerful Ecuador (MW = 7.8) earthquake of 16 April 2016 using Swarm satellites data in conjunction with other multi-platform satellite and ground data. *Adv. Space Res.* **2018**, *61*, 248–263. [CrossRef]
65. Marchetti, D.; Akhoondzadeh, M. Analysis of Swarm satellites data showing seismo-ionospheric anomalies around the time of the strong Mexico (Mw= 8.2) earthquake of 08 September 2017. *Adv. Space Res.* **2018**, *62*, 614–623. [CrossRef]
66. De Santis, A.; Marchetti, D.; Pavón-Carrasco, F.J.; Cianchini, G.; Perrone, L.; Abbattista, C.; Alfonsi, L.; Amoruso, L.; Campuzano, S.A.; Carbone, M. Precursory worldwide signatures of earthquake occurrences on Swarm satellite data. *Sci. Rep.* **2019**, *9*, 20287. [CrossRef] [PubMed]
67. Bondur, V.; Garagash, I.; Gokhberg, M.; Lapshin, V.; Nechaev, Y.V.; Steblov, G.; Shalimov, S. Geomechanical models and ionospheric variations related to strongest earthquakes and weak influence of atmospheric pressure gradients. *Dokl. Earth Sci.* **2007**, *414*, 666–669. [CrossRef]
68. Shinagawa, H.; Iyemori, T.; Saito, S.; Maruyama, T. A numerical simulation of ionospheric and atmospheric variations associated with the Sumatra earthquake on December 26, 2004. *Earth Planets Space* **2007**, *59*, 1015–1026. [CrossRef]
69. Astafyeva, E. Ionospheric detection of natural hazards. *Rev. Geophys.* **2019**, *57*, 1265–1288. [CrossRef]

MDPI AG
Grosspeteranlage 5
4052 Basel
Switzerland
Tel.: +41 61 683 77 34

Remote Sensing Editorial Office
E-mail: remotesensing@mdpi.com
www.mdpi.com/journal/remotesensing

www.ingramcontent.com/pod-product-compliance
Lightning Source LLC
Chambersburg PA
CBHW061619220926
51850CB00005B/547
* 9 7 8 3 7 2 5 8 2 1 0 3 7 *